Student Solutions Manual

Algebra: Introductory and Intermediate
An Applied Approach

FIFTH EDITION

Richard N. Aufmann
Palomar College

Joanne S. Lockwood
Nashua Community College

Prepared by

Pat Foard
South Plains College

Ellena Reda
Dutchess Community College

BROOKS/COLE
CENGAGE Learning™

Australia • Brazil • Japan • Korea • Mexico • Singapore • Spain • United Kingdom • United States

BROOKS/COLE
CENGAGE Learning™

ISBN-13: 978-0-538-49723-7
ISBN-10: 0-538-49723-8

Brooks/Cole
20 Davis Drive
Belmont, CA 94002-3098
USA

Cengage Learning is a leading provider of customized learning solutions with office locations around the globe, including Singapore, the United Kingdom, Australia, Mexico, Brazil, and Japan. Locate your local office at: **www.cengage.com/global**

Cengage Learning products are represented in Canada by Nelson Education, Ltd.

To learn more about Brooks/Cole, visit **www.cengage.com/brookscole**

Purchase any of our products at your local college store or at our preferred online store **www.CengageBrain.com**

Printed in the United States of America
1 2 3 4 5 6 7 14 13 12 11 10

Contents

Chapter 1: Real Numbers and Variable Expressions

Prep Test

1. 127.16

2. $3416 + 42{,}561 + 537 = 46{,}514$

3. $5004 - 487 = 4517$

4. $407 \times 28 = 11{,}396$

5. $11{,}684 \div 23 = 508$

6. 24

7. 4

8. $3 \cdot 7$

9. $\frac{4}{10} = \frac{2}{5}$

Section 1.1

Objective A Exercises

1. $8 > -6$

3. $-12 < 1$

5. $42 > 19$

7. $0 > -31$

9. $53 > -46$

11. False

13. True

15. False

17. True

19. False

21. $-23 < -8$

$-18 < 8$

$-8 = -8$

$0 > -8$

The elements -23 and -18 are less than -8.

23. $-33 < -10$

$-15 < -10$

$-1 > -10$

$37 > 10$

The elements -1 and 37 are greater than -10.

25. i

Objective B Exercises

27. -4

29. 9

31. 28

33. 14

35. -77

37. 0

39. 74

41. -82

43. -81

45. $|-83| > |58|$

47. $|43| < |-52|$

49. $|-68| > |-42|$

51. $|-45| < |-61|$

53. $-(-19) = 19$

$-(0) = 0$

$-(28) = -28$

55. $-|-45| = -45$

$-|0| = 0$

$-|17| = -17$

57. True

Objective C Exercises

59. $-3 + (-8) = -11$

61. $-8 + 3 = -5$

63. $-3 + (-80) = -83$

65. $-23 + (-23) = -46$

67. $16 + (-16) = 0$

69. $48 + (-53) = -5$

71. $-17+(-3)+29=-20+29=9$

73. $-3+(-8)+12=-11+12=1$

75. $16-8=16+(-8)=8$

77. $7-14=7+(-14)=-7$

79. $-7-2=-7+(-2)=-9$

81. $7-(-2)=7+2=9$

83. $-6-(-3)=-6+3=-3$

85. $6-(-12)=6+12=18$

87. $13+(-22)+4+(-5)=-9+4+(-5)$
$=-5+(-5)=-10$

89. $-16+(-17)+(-18)+10=-33+(-18)+10$
$=-51+10=-41$

91. $26+(-15)+(-11)+(-12)=11+(-11)+(-12)$
$=0+(-12)=-12$

93. $-14+(-15)+(-11)+40=-29+(-11)+40$
$=-40+40=0$

95. $-4-3-2=-4+(-3)+(-2)$
$=-7+(-2)=-9$

97. $12-(-7)-8=12+7+(-8)$
$=19+(-8)=11$

99. $-19-(-19)-18=-19+19+(-18)$
$=0+(-18)=-18$

101. $-17-(-8)-(-9)=-17+8+9$
$=-9+9=0$

103. $-30-(-65)-29-4=-30+65+(-29)+(-4)$
$=35+(-29)+(-4)$
$=6+(-4)=2$

105. $-16-47-63-12=-16+(-47)+(-63)+(-12)$
$=-63+(-63)+(-12)$
$=-126+(-12)=-138$

107. $-47-(-67)-13-15=-47+(67)+(-13)+(-15)$
$=20+(-13)+(-15)$
$=7+(-15)=-8$

109. $-19-17-(-36)-12=-19+(-17)+36+(-12)$
$=-36+36+(-12)$
$=0+(-12)=-12$

111. Negative

113. Positive

Objective D Exercises

115. $(14)3=42$

117. $-7\cdot 4=-28$

119. $(-12)(-5)=60$

121. $-11(23)=-253$

123. $(-17)(14)=-238$

125. $6(-19)=-114$

127. $12\div(-6)=-2$

129. $(-72)\div(-9)=8$

131. $-42\div 6=-7$

133. $(-144)\div 12=-12$

135. $48\div(-8)=-6$

137. $\frac{-49}{7}=-7$

139. $\frac{-44}{-4}=11$

141. $\frac{98}{-7}=-14$

143. $-\frac{-120}{8}=-(-15)=15$

145. $-\frac{-80}{-5}=-(16)=-16$

147. $0\div(-9)=0$

149. $\frac{-261}{9}=-29$

151. $9\div 0$ is undefined

153. $\frac{132}{-12}=-11$

155. $\frac{0}{0}$ is undefined

157. $7(5)(-3)=35(-3)=-105$

159. $9(-7)(-4)=-63(-4)=252$

161. $16(-3)(5)=-48(5)=-240$

163. $-4(-3)8=12(8)=96$

165. $-3(-8)(-9)=24(-9)=-216$

167. $(-9)7(5)=-63(5)=-315$

169. $7(-2)(5)(-6)=-14(5)(-6)=-70(-6)=420$

171. $(-14)9(-11)0=-126(-11)0=(1386)0=0$

173. Negative

Objective E Exercises

175. Strategy To find the difference between the record high temperature (117° F) and the record low temperature (–36° F), subtract the record low temperature from high temperature.

Solution $117-(-36)=117+36=153$

The difference between the record low temperature and the record low temperature is 153° F.

177. Strategy To find the temperature after the rise, add the rise in temperature (5°) to the beginning temperature (–19°).

Solution $-19+5=-14$

The temperature after a rise of 5° is –14° C.

179. Strategy To find the difference, subtract the depth of the Mariana Trench (–11,520 m) from the height of Mt. Everest (8850 m).

Solution

$8850-(-11{,}520)=8850+11{,}520=20{,}370$

The difference is 20,370 m.

181. a. Strategy To find the amount below par, subtract par from each score.

Solution

$67-72=-5$

$65-72=-7$

$66-72=-6$

b. Strategy To find the score for the first three days, add the three scores.

Solution $-5+(-7)+(-6)=-12+(-6)=-18$

Woods' score for the first three days was -18.

c. Strategy To find the score for the first four days, find the score for the fourth day and add the fourth day's score to the first three day's scores.

Solution

Score for the fourth day: $71-72=-1$

Score for the first four days:

$-18+(-1)=-19$

Wood's score for the first three days was -19.

183. Strategy To find the average daily temperature:

- Add the seven temperature readings.
- Divide by 7.

Solution

$-6+(-11)+1+5+(-3)+(-9)+(-5)$

$=-17+1+5+(-3)+(-9)+(-5)$

$=-16+5+(-3)+(-9)+(-5)$

$=-11+(-3)+(-9)+(-5)$

$=-14+(-9)+(-5)=-23+(-5)$

$=-28$

$-28\div 7=-4$

The average daily low temperature was -4° F.

185. False

187. Strategy To find the average, divide the total by 10.

Solution $-20 \div 10 = -2$

The average score was –2.

189. Strategy To find the score:

- Multiply the number of correct answers by 5.
- Multiply the number of incorrect answers by −5.
- Multiply the number of blank answers by –2.
- Add the results.

Solution

$20(5) = 100$

$5(-5) = -25$

$2(-2) = -4$

$100 + (-25) + (-4) = 75 + (-4) = 71$

The student's score was 71.

Applying the Concepts

191. Strategy To find the largest difference, subtract the smallest number form the largest number.

Solution $13 - (-9) = 13 + 9 = 22$

The largest difference is 22.

Strategy To find the smallest difference, subtract the two number that are the closest together.

Solution $-5 - (-7) = -5 + 7 = 2$

The smallest difference is 2.

193. **a.** True

b. True

Section 1.2

Objective A Exercises

1. $\frac{1}{8} = 1 \div 8 = 0.125$

3. $\frac{2}{9} = 2 \div 9 = 0.\overline{2}$

5. $\frac{1}{6} = 1 \div 6 = 0.1\overline{6}$

7. $\frac{9}{16} = 9 \div 16 = 0.5625$

9. $\frac{7}{12} = 7 \div 12 = 0.58\overline{3}$

11. $\frac{6}{25} = 6 \div 25 = 0.24$

13. $\frac{9}{40} = 9 \div 40 = 0.225$

15. $\frac{5}{11} = 5 \div 11 = 0.\overline{45}$

Objective B Exercises

17. 100% = 100(0.01) = 1, multiplying by 1 does not change the value of the number.

19. $40\% = 40\left(\frac{1}{100}\right) = \frac{40}{100} = \frac{2}{5}$

$40\% = 40(0.01) = 0.40$

21. $88\% = 88\left(\frac{1}{100}\right) = \frac{88}{100} = \frac{22}{25}$

$88\% = 88(0.01) = 0.88$

23. $160\% = 160\left(\frac{1}{100}\right) = \frac{160}{100} = \frac{8}{5}$

$160\% = 160(0.01) = 1.60$

25. $87\% = 87\left(\frac{1}{100}\right) = \frac{87}{100}$

$87\% = 87(0.01) = 0.87$

27. $450\% = 450\left(\frac{1}{100}\right) = \frac{450}{100} = \frac{9}{2}$

$450\% = 450(0.01) = 4.50$

29. $4\frac{2}{7}\% = 4\frac{2}{7}\left(\frac{1}{100}\right) = \frac{30}{7}\left(\frac{1}{100}\right) = \frac{3}{70}$

31. $37\frac{1}{2}\% = 37\frac{1}{2}\left(\frac{1}{100}\right) = \frac{75}{2}\left(\frac{1}{100}\right) = \frac{3}{8}$

33. $\frac{1}{4}\% = \frac{1}{4}\left(\frac{1}{100}\right) = \frac{1}{400}$

35. $6\frac{1}{4}\% = 6\frac{1}{4}\left(\frac{1}{100}\right) = \frac{25}{4}\left(\frac{1}{100}\right) = \frac{1}{16}$

37. $5\frac{3}{4}\% = 5\frac{3}{4}\left(\frac{1}{100}\right) = \frac{23}{4}\left(\frac{1}{100}\right) = \frac{23}{400}$

39. $9.1\% = 9.1(0.01) = 0.091$

41. $16.7\% = 16.7(0.01) = 0.167$

43. $0.9\% = 0.9(0.01) = 0.009$

45. $9.15\% = 9.15(0.01) = 0.0915$

47. $18.23\% = 18.23(0.01) = 0.1823$

49. $0.37 = 0.37(100\%) = 37\%$

51. $0.02 = 0.02(100\%) = 2\%$

53. $0.125 = 0.125(100\%) = 12.5\%$

55. $1.36 = 1.36(100\%) = 136\%$

57. $0.004 = 0.004(100\%) = 0.4\%$

59. $\frac{83}{100} = \frac{83}{100}(100\%) = \frac{8300}{100}\% = 83\%$

61. $\frac{3}{8} = \frac{3}{8}(100\%) = \frac{300}{8}\% = 37\frac{1}{2}\%$

63. $\frac{4}{9} = \frac{4}{9}(100\%) = \frac{400}{9}\% = 44\frac{4}{9}\%$

65. $\frac{9}{20} = \frac{9}{20}(100\%) = \frac{900}{20}\% = 45\%$

67. $2\frac{1}{2} = 2\frac{1}{2}(100\%) = \frac{5}{2}(100\%) = \frac{500}{2}\%$

$= 250\%$

69. Greater than 1%.

Objective C Exercises

71. $-\frac{6}{13} + \frac{17}{26} = \frac{-12}{26} + \frac{17}{36} = \frac{-12+17}{26} = \frac{5}{26}$

73. $\frac{5}{8} - \left(-\frac{3}{4}\right) = \frac{5}{8} + \frac{6}{8} = \frac{5+6}{8} = \frac{11}{8}$

75. $\frac{11}{12} - \frac{5}{6} = \frac{11}{12} - \frac{10}{12} = \frac{11-10}{12} = \frac{1}{12}$

77. $-\frac{5}{8} - \left(-\frac{11}{12}\right) = \frac{-15}{24} + \frac{22}{24} = \frac{-15+22}{24} = \frac{7}{24}$

79. $\frac{1}{2} - \frac{2}{3} + \frac{1}{6} = \frac{3}{6} - \frac{4}{6} + \frac{1}{6} = \frac{3-4+1}{6} = \frac{0}{6} = 0$

81. $\frac{1}{2} - \frac{3}{8} - \left(-\frac{1}{4}\right) = \frac{4}{8} - \frac{3}{8} + \frac{2}{8} = \frac{4-3+2}{8} = \frac{3}{8}$

83. $2.54 - 3.6 = -1.06$

85. $-16.92 - 6.925 = -23.845$

87. $6.9027 - 17.692 = -10.7893$

89. $-18.39 + 4.9 - 23.7 = -13.49 - 23.7 = -37.19$

91. $-3.07 - (-2.97) - 17.4 = -3.07 + 2.97 - 17.4$

$= -0.1 - 17.4 = -17.5$

93. $-3.09 - 4.6 - (-27.3) = -7.69 + 27.3 = 19.61$

95. $\frac{7}{8} + \frac{4}{5} \approx 2$

97. $-0.125 + 1.25 \approx 1$

Objective D Exercises

99. $\left(-\frac{3}{4}\right)\left(-\frac{8}{27}\right) = \frac{\overset{1}{\cancel{3}}\cdot\overset{1}{\cancel{2}}\cdot\overset{1}{\cancel{2}}\cdot 2}{\underset{1}{\cancel{2}}\cdot\underset{1}{\cancel{2}}\cdot\underset{1}{\cancel{3}}\cdot 3\cdot 3} = \frac{2}{9}$

101. $\left(\frac{5}{12}\right)\left(-\frac{8}{15}\right) = -\frac{\overset{1}{\cancel{5}}\cdot\overset{1}{\cancel{2}}\cdot\overset{1}{\cancel{2}}\cdot 2}{\underset{1}{\cancel{2}}\cdot\underset{1}{\cancel{2}}\cdot 3\cdot 3\cdot\underset{1}{\cancel{5}}} = -\frac{2}{9}$

103. $\frac{5}{12}\left(-\frac{8}{15}\right)\frac{1}{3} = -\frac{\overset{1}{\cancel{5}}\cdot\overset{1}{\cancel{2}}\cdot\overset{1}{\cancel{2}}\cdot 2}{\underset{1}{\cancel{2}}\cdot\underset{1}{\cancel{2}}\cdot 3\cdot 3\cdot\underset{1}{\cancel{5}}\cdot 3} = -\frac{2}{27}$

105. $\frac{3}{8} \div \frac{1}{4} = \frac{3}{8}\cdot\frac{4}{1} = \frac{3\cdot\overset{1}{\cancel{2}}\cdot\overset{1}{\cancel{2}}}{\underset{1}{\cancel{2}}\cdot\underset{1}{\cancel{2}}\cdot 2} = \frac{3}{2}$

107. $-\frac{5}{12} \div \frac{15}{32} = -\frac{5}{12}\cdot\frac{32}{15} = -\frac{\overset{1}{\cancel{5}}\cdot\overset{1}{\cancel{2}}\cdot\overset{1}{\cancel{2}}\cdot 2\cdot 2\cdot 2}{\underset{1}{\cancel{2}}\cdot\underset{1}{\cancel{2}}\cdot 3\cdot 3\cdot\underset{1}{\cancel{5}}} = -\frac{8}{9}$

109. $-\frac{4}{9} \div \left(-\frac{2}{3}\right) = -\frac{4}{9}\cdot\left(-\frac{3}{2}\right) = \frac{\overset{1}{\cancel{2}}\cdot 2\cdot\overset{1}{\cancel{3}}}{\underset{1}{\cancel{3}}\cdot 3\cdot\underset{1}{\cancel{2}}} = \frac{2}{3}$

111. $1.2(3.47) = 4.164$

113. $(-1.89)(-2.3) = 4.347$

115. $1.2(-0.5)(3.7) = (-0.6)(3.7) = -2.22$

117. $-1.27 \div (-1.7) \approx 0.75$

119. $0.0976 \div 0.042 \approx 2.32$

121. $-7.894 \div (-2.06) \approx 3.83$

123. a. Less than 1

b. Greater than 1

Objective E Exercises

125. $7^4 = 7 \cdot 7 \cdot 7 \cdot 7 = 2401$

127. $-4^3 = -(4 \cdot 4 \cdot 4) = -64$

129. $(-2)^3 = (-2)(-2)(-2) = -8$

131. $(-5)^3 = (-5)(-5)(-5) = -125$

133. $\left(-\frac{3}{4}\right)^3 = \left(-\frac{3}{4}\right)\left(-\frac{3}{4}\right)\left(-\frac{3}{4}\right) = -\frac{3 \cdot 3 \cdot 3}{4 \cdot 4 \cdot 4} = -\frac{27}{64}$

135. $(1.5)^3 = (1.5)(1.5)(1.5) = 3.375$

137. $\left(-\frac{1}{2}\right)^3 \cdot 8 = \left(-\frac{1}{2}\right)\left(-\frac{1}{2}\right)\left(-\frac{1}{2}\right) \cdot 2 \cdot 2 \cdot 2$

$$= -\frac{\overset{1}{\cancel{2}} \cdot \overset{1}{\cancel{2}} \cdot \overset{1}{\cancel{2}}}{\underset{1}{\cancel{2}} \cdot \underset{1}{\cancel{2}} \cdot \underset{1}{\cancel{2}}} = -1$$

139. $(-2) \cdot (-2)^2 = (-2)(-2)(-2) = -8$

141. $(-3)^3 \cdot 5^2 \cdot 10 = (-3)(-3)(-3) \cdot 5 \cdot 5 \cdot 10$

$= -27 \cdot 25 \cdot 10 = -675 \cdot 10 = -6750$

143. Negative

145. Positive

Objective F Exercises

147. $\sqrt{16} = 4$

149. $\sqrt{49} = 7$

151. $\sqrt{32} = \sqrt{16 \cdot 2} = \sqrt{16} \cdot \sqrt{2} = 4\sqrt{2}$

153. $\sqrt{8} = \sqrt{4 \cdot 2} = \sqrt{4} \cdot \sqrt{2} = 2\sqrt{2}$

155. $6\sqrt{18} = 6\sqrt{9 \cdot 2} = 6\sqrt{9} \cdot \sqrt{2}$

$= 6 \cdot 3\sqrt{2} = 18\sqrt{2}$

157. $5\sqrt{40} = 5\sqrt{4 \cdot 10} = 5\sqrt{4} \cdot \sqrt{10}$

$= 5 \cdot 2\sqrt{10} = 10\sqrt{10}$

159. $\sqrt{15} = \sqrt{3 \cdot 5} = \sqrt{15}$

161. $\sqrt{29}$

163. $-9\sqrt{72} = -9\sqrt{4 \cdot 9 \cdot 2} = -9\sqrt{4} \cdot \sqrt{9} \cdot \sqrt{2}$

$= -9 \cdot 2 \cdot 3\sqrt{2} = -54\sqrt{2}$

165. $\sqrt{45} = \sqrt{9 \cdot 5} = \sqrt{9} \cdot \sqrt{5} = 3\sqrt{5}$

167. $\sqrt{0} = 0$

169. $6\sqrt{128} = 6\sqrt{64 \cdot 2} = 6\sqrt{64} \cdot \sqrt{2}$

$= 6 \cdot 8\sqrt{2} = 48\sqrt{2}$

171. $\sqrt{240} \approx 15.492$

173. $\sqrt{288} \approx 16.971$

175. $\sqrt{256} = 16$

177. Between –11 and –10

179. Between 2 and 3

Objective G Exercises

181. Strategy To find the difference, subtract the low temperature (–48.9°) from the high temperature (6.67°).

Solution $6.67 - (-48.9) = 6.67 + 48.9 = 55.57$

The difference between the record high and record low temperature in Browing is 55.57° C.

183. Strategy To find the difference, subtract the low temperature (–41.8°) from the high temperature (104.5°).

Solution $104.5 - (-41.8) = 104.5 + 41.8 = 146.3$

The difference between the record high and record low temperature in Slovakia is 146.3° F.

185. Below

187. Strategy To find the difference, subtract the melting point (–218.4°) from the boiling point (–182.962°).

Solution

$-182.962-(-218.4)=-182.962+218.4=35.438$

The difference between the boiling point and melting point is 35.438° C.

189. a. Strategy To find the closing price for the previous day, subtract the change (–0.07) from price for September 4 (13.35).

Solution $13.35-(-0.07)=13.35+0.07=13.42$

The closing price for the previous day was $13.42.

b. Strategy To find the closing price for the previous day, subtract the change (–0.41) from price for September 4 (36.42).

Solution $36.42-(-0.41)=36.42+0.41=36.83$

The closing price for the previous day was $36.83.

Applying the Concepts

191. Answers will vary. For example:

a. 0.15

b. 1.05

c. 0.001

Section 1.3

Objective A Exercises

1. We need an Order of Operations Agreement to prevent there being more than one answer for a numerical expression.

3. $4-8\div 2=4-4=0$

5. $2(3-4)-(-3)^2=2(-1)-(-3)^2$
$=2(-1)-9$
$=-2-9$
$=-11$

7. $24-18\div 3+2=24-6+2=18+2=20$

9. $8-2(3)^2=8-2(9)$
$=8-18$
$=-10$

11. $12+16\div 4\cdot 2=12+4\cdot 2$
$=12+8$
$=20$

13. $27-18\div(-3^2)=27-18\div(-9)$
$=27+2$
$=29$

15. $16+15\div(-5)-2=16+(-3)-2$
$=13-2$
$=11$

17. $14-2^2-|4-7|=14-2^2-|-3|=14-2^2-3$
$=14-4-3=10-3=7$

19. $3-2[8-(3-2)]=3-2[8-(1)]$
$=3-2[7]$
$=3-14$
$=-11$

21. $6+\frac{16-4}{2^2+2}-2=6+\frac{12}{4+2}-2$
$=6+\frac{12}{6}-2$
$=6+2-2$
$=8-2$
$=6$

23. $18\div|9-2^3|+(-3)=18\div|9-8|+(-3)$
$=18\div 1+(-3)$
$=18+(-3)$
$=15$

25. $4[16-(7-1)]\div 10=4[16-6]\div 10$
$=4[10]\div 10$
$=40\div 10$
$=4$

27. $20\div(10-2^3)+(-5)=20\div(10-8)+(-5)$
$=20\div 2+(-5)=10+(-5)=5$

29. $4(-8)\div\left[2(7-3)^2\right]=4(-8)\div\left[2(4)^2\right]$
$=4(-8)\div[2(16)]=4(-8)\div 32$
$=-32\div 32=-1$

31. $$16-4\cdot\frac{3^3-7}{2^3+2}-(-2)^2=16-4\cdot\frac{27-7}{8+2}-(4)$$
$$=16-4\cdot\frac{20}{10}-4$$
$$=16-4\cdot 2-4$$
$$=16-8-4=8-4=4$$

33. $$0.3(1.7-4.8)+(1.2)^2=0.3(-3.1)+1.44$$
$$=-0.93+1.44=0.51$$

35. $$(1.65-1.05)^2\div 0.4+0.8=(0.6)^2\div 0.4+0.8$$
$$=0.36\div 0.4+0.8$$
$$=0.9+0.8=1.7$$

Applying the Concepts

37. Answers will vary. For example, $\frac{17}{24}$ and $\frac{33}{48}$.

39. Answers will vary. For example:

a. $\frac{1}{2}$

b. 1

c. 2

41. No, the Order of Operations Agreement was not followed in the given simplification of $6+2(4-9)$ because the addition was performed before the multiplication. The correct simplification is:
$$6+2(4-9)=6+2(-5)$$
$$=6+(-10)$$
$$=-4$$

Section 1.4

Objective A Exercises

1. $6b\div(-a)$
$$6(3)\div(-2)=18\div(-2)=-9$$

3. b^2-4ac
$$(3)^2-4(2)(-4)=9-4(2)(-4)$$
$$=9-(-32)=9+32$$
$$=41$$

5. b^2-c^2
$$3^2-(-4)^2=9-16=-7$$

7. a^2+b^2
$$2^2+3^2=4+9=13$$

9. $(b-a)^2+4c$
$$(3-2)^2+4(-4)=1^2+4(-4)$$
$$=1+(-16)=-15$$

11. $\frac{5ab}{6}-3cb$
$$\frac{5(2)(3)}{6}-3(-4)(3)=\frac{30}{6}-(-36)$$
$$=5-(-36)=41$$

13. $\frac{b+c}{d}$
$$\frac{4+(-1)}{3}=\frac{3}{3}=1$$

15. $\frac{2d+b}{-a}$
$$\frac{2(3)+4}{-(-2)}=\frac{6+4}{2}=\frac{10}{2}=5$$

17. $\frac{b-d}{c-a}$
$$\frac{4-3}{-1-(-2)}=\frac{1}{1}=1$$

19. $(b+d)^2-4a$
$$(4+3)^2-4(-2)=7^2-4(-2)$$
$$=49-(-8)=57$$

21. $(d-a)^2\div 5$
$$[3-(-2)]^2\div 5=5^2\div 5=25\div 5=5$$

23. $\frac{b-2a}{bc^2-d}$
$$\frac{4-2(-2)}{4(-1)^2-3}=\frac{4-(-4)}{4(1)-3}=\frac{8}{4-3}=\frac{8}{1}=8$$

25. $\frac{1}{3}d^2-\frac{3}{8}b^2$
$$\frac{1}{3}(3)^2-\frac{3}{8}(4)^2=\frac{1}{3}(9)-\frac{3}{8}(16)=3-6=-3$$

27. $\frac{-4bc}{2a-b}$
$$\frac{-4(4)(-1)}{2(-2)-4}=\frac{16}{-4-4}=\frac{16}{-8}=-2$$

29. $-\frac{2}{3}d-\frac{1}{5}(bd-ac)$

$$-\frac{2}{3}(3)-\frac{1}{5}[4(3)-(-2)(-1)]=-\frac{2}{3}(3)-\frac{1}{5}[12-2]$$
$$=-\frac{2}{3}(3)-\frac{1}{5}(10)$$
$$=-2-2=-4$$

31. Positive

33. Negative

Objective B Exercises

35. $6x+8x=14x$

37. $9a-4a=5a$

39. $4y-10y=-6y$

41. $7-3b=7-3b$

43. $-12a+17a=5a$

45. $5ab-7ab=-2ab$

47. $-12xy+17xy=5xy$

49. $-3ab+3ab=0$

51. $-\frac{1}{2}x-\frac{1}{3}x=-\frac{3}{6}x-\frac{2}{6}x=-\frac{5}{6}x$

53. $2.3x+4.2x=6.5x$

55. $x-0.55x=0.45x$

57. $5a-3a+5a=7a$

59. $-5x^2-12x^2+3x^2=-14x^2$

61. $\frac{3}{4}x-\frac{1}{3}x-\frac{7}{8}x=\frac{18}{24}x-\frac{8}{24}x-\frac{21}{24}x=-\frac{11}{24}x$

63. $7x-3y+10x=17x-3y$

65. $3a+(-7b)-5a+b=-2a-6b$

67. $3x+(-8y)-10x+4x=-3x-8y$

69. $x^2-7x+(-5x^2)+5x=-4x^2-2x$

71. $-10x-10y-10y-10x=-20x-20y$

i. 0 No

ii. -20 No

iii. $-20y$ No

iv. $-20x-20y$ Yes

v. $-20y-20x$ Yes

(iv) and (v)

Objective C Exercises

73. $12(5x)=60x$

75. $-2(5a)=-10a$

77. $-5(-6y)=30y$

79. $(6x)12=72x$

81. $(7a)(-4)=-28a$

83. $(-12b)(-9)=108b$

85. $-8(7x^2)=-56x^2$

87. $\frac{1}{6}(6x^2)=x^2$

89. $\frac{1}{8}(8x)=x$

91. $-\frac{1}{4}(-4a)=a$

93. $-\frac{1}{9}(-9b)=b$

95. $(12x)\left(\frac{1}{12}\right)=x$

97. $(-10n)\left(-\frac{1}{10}\right)=n$

99. $\frac{1}{7}(14x)=2x$

101. $-0.25(8x)=-2x$

103. $-\frac{5}{8}(24a^2)=-15a^2$

105. $-0.75(-8y)=6y$

107. $(33y)\left(\frac{1}{11}\right)=3y$

109. $(-10x)\left(\frac{1}{5}\right)=-2x$

111. $(21y)\left(-\frac{3}{7}\right)=-9y$

Objective D Exercises

113. $2(4x-3)=8x-6$

115. $-2(a+7)=-2a-14$

117. $-3(2y-8)=-6y+24$

119. $-(x+2)=-x-2$

121. $(5-3b)7=35-21b$

123. $\frac{1}{3}(6-15y)=2-5y$

125. $3(5x^2+2x)=15x^2+6x$

127. $-2(-y+9)=2y-18$

129. $(-3x-6)5=-15x-30$

131. $2(-3x^2-14)=-6x^2-28$

133. $-3(2y^2-7)=-6y^2+21$

135. $3(x^2-y^2)=3x^2-3y^2$

137. $-\frac{2}{3}(6x-18y)=-4x+12y$

139. $-(6a^2-7b^2)=-6a^2+7b^2$

141. $4(x^2-3x+5)=4x^2-12x+20$

143. $\frac{3}{4}(2x-6y+8)=\frac{3}{2}x-\frac{9}{2}y+6$

145. $4(-3a^2-5a+7)=-12a^2-20a+28$

147. $-3(-4x^2+3x-4)=12x^2-9x+12$

149. $-(3a^2+5a-4)=-3a^2-5a+4$

151. $-(8b^2-6b+9)=-8b^2+6b-9$

153. $12-7(y-9)=12-7y+63=-7y+75$

i. $5(y-9)=5y-45$ No

ii. $12-7y-63=-7y-51$ No

iii. $12-7y+63=-7y+75$ Yes

iv. $12-7y-9=-7y+3$ No

155. $6a-(5a+7)=6a-5a-7=a-7$

157. $10-(11x-3)=10-11x+3=-11x+13$

159. $8-(12+4y)=8-12-4y=-4y-4$

161. $2(x-4)-4(x+2)=2x-8-4x-8$

$=-2x-16$

163. $6(2y-7)-(3-2y)=12y-42-3+2y$

$=14y-45$

165. $2(a+2b)-(a-3b)=2a+4b-a+3b=a+7b$

167. $2[x+2(x+7)]=2[x+2x+14]=2[3x+14]$

$=6x+28$

169. $-5[2x+3(5-x)]=-5[2x+15-3x]$

$=-5[-x+15]=5x-75$

171. $-2[3x-(5x-2)]=-2[3x-5x+2]$

$=-2[-2x+2]=4x-4$

173. $-7x+3[x-(3-2x)]=-7x+3[x-3+2x]$

$=-7x+3[3x-3]$

$=-7x+9x-9=2x-9$

175. $0.12(2x+3)+x=0.24x+0.36+x$

$=1.24x+0.36$

177. $0.03x+0.04(1000-x)=0.03x+40-0.04x$

$=-0.01x+40$

Objective E Exercises

179. the unknown number: x

$\frac{x}{18}$

181. the unknown number: x

$x+20$

183. the unknown number: x

the product of eleven and the number: $11x$

$11x-8$

185. the unknown number: x

the quotient of the number and twenty: $\frac{x}{20}$

$40-\frac{x}{20}$

187. the unknown number: x
the square of the number: x^2
twice the number: $2x$
$x^2 + 2x$

189. the unknown number: x
the difference between the number and 50: $x - 50$
$10(x-50) = 10x - 500$

191. the unknown number: x
three more than the number: $x + 3$
$x-(x+3) = x - x - 3 = -3$

193. the unknown number: x
twice the number: $2x$
the difference between twice the number and four: $2x - 4$
$(2x-4)+x = 2x-4+x = 3x-4$

195. the unknown number: x
the product of three and the number: $3x$
$x + 3x = 4x$

197. the unknown number: x
the sum of the number and six: $x + 6$
$(x+6)+5 = x+6+5 = x+11$

199. the unknown number: x
the sum of the number and ten: $x + 10$
$x-(x+10) = x-x-10 = -10$

201. number of visitors to the Metropolitan Museum of Art: M
number of visitors to the Louvre: $M +$ 3,800,000

203. number of visitors to Google web sites: G
number of visitors to Microsoft web sites: $G - 63{,}000{,}000$

205. length of one piece: S
length of second piece: $12 - S$

207. distance traveled by faster car: x
distance traveled by slower car: $200 - x$

209. number of bones in your body: N
number of bones in your foot: $\frac{1}{4}N$

211. number of people surveyed: N
number of people who pay down their debt: $0.43N$

Applying the Concepts

213. length of wire: x
length of side of square: $\frac{1}{4}x$ Two examples of translation of $5x + 8$ are "eight more than the product of five and a number" and "the sum of five times a number and eight." Two examples of the translation of $5(x + 8)$ are "five times the sum of a number and eight" and "the product of five and eight more than a number.

Section 1.5

Objective A Exercises

1. Student explanations should include the idea that to find the union of two sets, we list all the elements of the first set and then list all the elements of the second set that are not elements of the first set.

3. $A = \{16, 17, 18, 19, 20, 21\}$

5. $A = \{9, 11, 13, 15, 17\}$

7. $A = \{b, c\}$

9. $A \cup B = \{3, 4, 5, 6\}$

11. $A \cup B = \{-10, -9, -8, 8, 9, 10\}$

13. $A \cup B = \{a, b, c, d, e, f\}$

15. $A \cup B = \{1, 3, 7, 9, 11, 13\}$

17. $A \cap B = \{4, 5\}$

19. $A \cap B = \varnothing$

21. $A \cap B = \{c, d, e\}$

23. Answers may vary. For example:
$A = \{1, 2, 3\}$ and $B = \{1, 2, 4, 5\}$

Objective B

25. $\{x|x > -5,\ x \in \text{negative integers}\}$

27. $\{x|x > 30,\ x \in \text{integers}\}$

29. $\{x|x > 8,\ x \in \text{real numbers}\}$

31. $(1, 2)$

33. $(3, \infty)$

35. $[-4, 5)$

37. $(-\infty, 2]$

39. $[-3,1]$

41. $\{x|-5 < x < -3\}$

43. $\{x|x \le -2\}$

45. $\{x|-3 \le x \le -2\}$

47. $\{x|x \le 6\}$

49. $[-5, 4]$

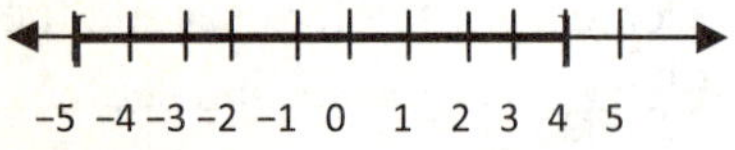

51. $\{x|x < 4\}$

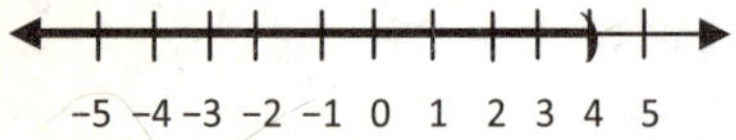

53. $\{x|x \le -4\}$

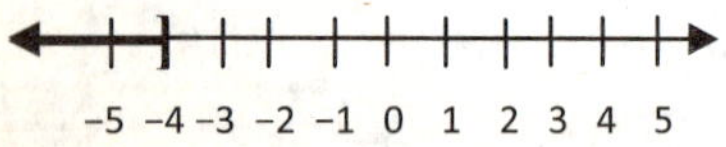

55. $(-\infty, 3]$

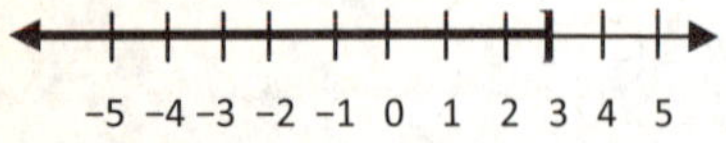

57. $[-1, 3)$

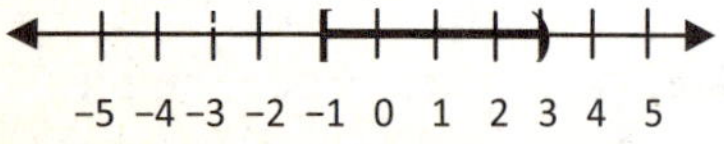

59. $\{x|-3 < x < 3\}$

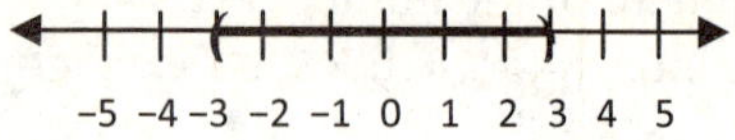

61. $\{x|2 \le x \le 4\}$

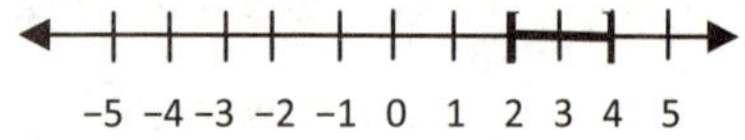

63. $\{x|-\infty < x < \infty\}$

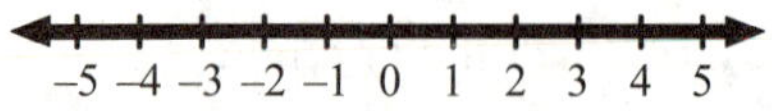

65. None

Applying the Concepts

67. $m \ge 250$

Concept Review

1. The opposite of a number is the number with its sign changed. If a number is positive, its opposite is negative. If a number is negative, its opposite is positive. The absolute value of a number is its distance from zero on the number line. Because distance is never negative, the absolute value of a number is never negative. The absolute value of a positive number is positive. The absolute value of a negative number is positive. [1.1B]

2. A minus sign indicates the operation of subtraction. A negative sign indicates the sign of a number. [1.1C]

3. The product of two nonzero numbers with the same sign is positive. The product of two nonzero numbers with different signs is negative. [1.1D]

4. The steps in the Order of Operations Agreement are:

1. Perform operations inside grouping symbols. Grouping symbols include parentheses ( ), brackets [ ], braces { }, the absolute value symbol | |, and the fraction bar.
2. Simplify exponential expressions.
3. Do multiplication and division as they occur from left to right.
4. Do addition and subtraction as they occur from left to right. [1.3A]

5. When adding fractions, you have to convert to equivalent fractions with a common denominator. One way to explain this is that you can combine like things, but you cannot combine unlike things. You can combine 4 apples and 5 apples and get 9 apples. You cannot combine 4 apples and 5 oranges and get one type of fruit. In adding whole numbers, you add like things: ones, tens, hundreds, and so on. In adding fractions, you can combine 2 *ninths* and 5 *ninths*, but you cannot add 2 *ninths* and 3 *fifths*. [1.2C]

6. No, you do not have to find a common denominator when multiplying two fractions. The product of the two fractions is the product of the numerators over the product of the denominators. [1.2D]

7. To find the reciprocal of a fraction, interchange the numerator and denominator. [1.4C]

8. In a term, the numerical coefficient is the number. The variable part consists of the variables and their exponents. [1.4A]

9. Like terms of a variable expression are terms with the same variable part. Constant terms are also considered like terms. [1.4A]

10. The empty set or null set is represented by Ø or { }. [1.5A]

11. $A \cup B$ is the union of two sets A and B. $A \cap B$ is the intersection of the two sets A and B. [1.5A]

12. The roster method encloses a list of the elements of a set in braces; for example {1, 2, 3, 4, 5}. Set-builder notation uses a rule to describe the elements of the set, for example, $\{x \mid x > 4, x \in \text{real numbers}\}$. [1.5A, 1.5B].

Chapter 1 Review Exercises

1. $-4 < 1$ True
$0 < 1$ True
$11 < 1$ False
$x < 1$ for the values -4 and 0.

2. 4

3. $-|-5| = -(5) = -5$

4. $$\begin{aligned}-3+(-12)+6+(-4) &= -15+6+(-4)\\ &= -9+(-4) = -13\end{aligned}$$

5. $16-(-3)-18 = 16+3-18 = 19-18 = 1$

6. $-6(7) = -42$

7. $-100 \div 5 = -20$

8. $$\begin{array}{r} 0.28 \\ 25\overline{)7.00} \\ \underline{50} \\ 200 \\ \underline{200} \\ 0 \end{array}$$
$\frac{7}{25} = 0.28$

9. $6.2\% = 6.2(0.01) = 0.062$

10. $\frac{5}{8} = \frac{5}{8}(100\%) = \frac{500}{8}\% = 62.5\%$

11. $\frac{1}{3} - \frac{1}{6} + \frac{5}{12} = \frac{4}{12} - \frac{2}{12} + \frac{5}{12} = \frac{4-2+5}{12} = \frac{7}{12}$

12. $5.17 - 6.238 = -1.068$

13. $-\frac{18}{35} \div \frac{17}{28} = -\frac{18}{35} \cdot \frac{28}{17} = -\frac{2 \cdot 3 \cdot 3 \cdot 2 \cdot 2 \cdot \overset{1}{\cancel{7}}}{5 \cdot \underset{1}{\cancel{7}} \cdot 17} = -\frac{72}{85}$

14. $4.32(-1.07) = -4.6224$

15. $\left(-\frac{2}{3}\right)^4 = \left(-\frac{2}{3}\right)\left(-\frac{2}{3}\right)\left(-\frac{2}{3}\right)\left(-\frac{2}{3}\right) = \frac{16}{81}$

16. $2\sqrt{36} = 2 \cdot 6 = 12$

17. $-3\sqrt{120} = -3\sqrt{4 \cdot 30} = -3 \cdot 2\sqrt{30} = -6\sqrt{30}$

18.
$$\begin{aligned} -3^2 + 4[18 + (12 - 20)] &= -3^2 + 4[18 + (-8)] \\ &= -3^2 + 4[10] \\ &= -9 + 40 = 31 \end{aligned}$$

19. $(b-a)^2 + c$

$$\begin{aligned} [3-(-2)]^2 + 4 = [3+2]^2 + 4 = [5]^2 + 4 \\ = 25 + 4 = 29 \end{aligned}$$

20.
$$\begin{aligned} 6a - 4b + 2a &= 6a + 2a - 4b \\ &= (6+2)a - 4b \\ &= 8a - 4b \end{aligned}$$

21. $-3(-12y) = -3(-12)y = 36y$

22. $5(2x-7) = 5(2x) + 5(-7) = 10x - 35$

23. $-4(2x-9) + 5(3x+2)$
$$\begin{aligned} &= -4(2x) - 4(-9) + 5(3x) + 5(2) \\ &= -8x + 36 + 15x + 10 \\ &= -8x + 15x + 36 + 10 \\ &= 7x + 46 \end{aligned}$$

24.
$$\begin{aligned} 5[2 - 3(6x-1)] &= 5[2 - 18x + 3] \\ &= 5[5 - 18x] \\ &= 25 - 90x \\ &= -90x + 25 \end{aligned}$$

25. $\{1, 3, 5, 7\}$

26. $A \cap B = \{1, 5, 9\}$

27. $\{x \mid x > 3\}$

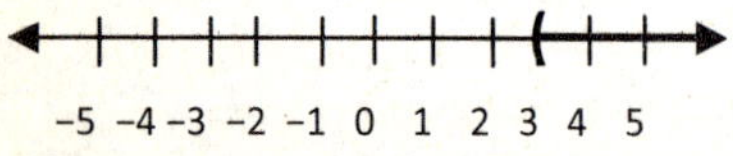

28. $[1, 4]$

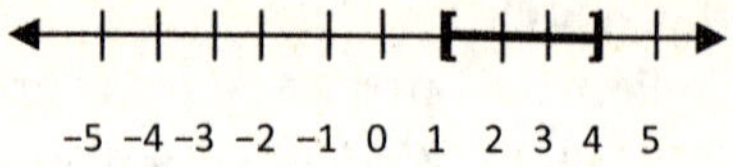

29. $(-4, \infty)$

30. **Strategy** To find the score:
- Multiply the number of correct answers by 6.
- Multiply the number of incorrect answers by −4.
- Multiply the number of blank answers by −2.
- Add the results.

Solution

$21(6) = 126$

$5(-4) = -20$

$4(-2) = -8$

$126 + (-20) + (-8) = 98$

The student's score was 98.

31. **Strategy** To find the percent
- Find the total number of pounds by adding the pounds in the three categories together.
- Divide the number of chocolate pounds (3.3) by the total number of pounds and multiply by 100%.

Solution $0.5 + 3.3 + 2.7 = 6.5$

$\left(\frac{3.3}{6.5}\right)100\% = 50.8\%$

50.8% of the candy consumed was chocolate.

32. the unknown number: x

twice the number: $2x$

one-half the number: $\frac{1}{2}x$

$2x - \frac{1}{2}x = \left(2 - \frac{1}{2}\right)x = \left(\frac{4}{2} - \frac{1}{2}\right)x = \frac{3}{2}x$

33. number of American League cards: A

number of National League cards: $5A$

Chapter 1 Test

1. $-2 > -40$

2. 4

3. $-|-4| = -(4) = -4$

4. $16 - 30 = -14$

5. $-22 + 14 + (-8) = -8 + (-8) = -16$

6. $16 - (-30) - 42 = 16 + 30 - 42 = 46 - 42 = 4$

7. $-561 \div (-33) = 17$

8. $\frac{7}{9} = 0.\overline{7}$

9. $45\% = 45\left(\frac{1}{100}\right) = \frac{45}{100} = \frac{9}{20}$

$45\% = 45(0.01) = 0.45$

10. $-\frac{2}{5} + \frac{7}{15} = -\frac{6}{15} + \frac{7}{15} = \frac{-6+7}{15} = \frac{1}{15}$

11. $6.02(-0.89) = -5.3578$

12. $\frac{5}{12} \div \left(-\frac{5}{6}\right) = \frac{5}{12} \cdot \left(-\frac{6}{5}\right) = -\frac{5 \cdot 2 \cdot 3}{2 \cdot 2 \cdot 3 \cdot 5} = -\frac{1}{2}$

13. $\frac{3}{4} \cdot (4)^2 = \frac{3}{4} \cdot 16 = \frac{3 \cdot 2 \cdot 2 \cdot 2 \cdot 2}{2 \cdot 2} = 12$

14. $-2\sqrt{45} = -2\sqrt{9 \cdot 5} = -2\sqrt{9} \cdot \sqrt{5} = -2 \cdot 3\sqrt{5} = -6\sqrt{5}$

15.
$$\begin{aligned} 16 \div 2[8 - 3(4-2)] + 1 &= 16 \div 2[8 - 3(2)] + 1 \\ &= 16 \div 2[8-6] + 1 \\ &= 16 \div 2[2] + 1 \\ &= 8[2] + 1 \\ &= 16 + 1 \\ &= 17 \end{aligned}$$

16. $b^2 - 3ab$

$(-2)^2 - 3(3)(-2) = 4 + 18 = 22$

17. $3x - 5x + 7x = (3 - 5 + 7)x = 5x$

18. $\frac{1}{5}(10x) = \frac{1}{5}(10)x = 2x$

19. $-3(2x^2 - 7y^2) = -3(2x^2) - 3(-7y^2) = -6x^2 + 21y^2$

20.
$$\begin{aligned} 2x - 3(x-2) &= 2x - 3(x) - 3(-2) \\ &= 2x - 3x + 6 \\ &= -x + 6 \end{aligned}$$

21.
$$\begin{aligned} 2x + 3[4 - (3x - 7)] &= 2x + 3[4 - 3x + 7] \\ &= 2x + 3[11 - 3x] \\ &= 2x + 33 - 9x \\ &= 2x - 9x + 33 \\ &= -7x + 33 \end{aligned}$$

22. $\{-2, -1, 0, 1, 2, 3\}$

23. $\{x \mid x < -3, x \in \text{ real numbers}\}$

24. $A \cup B = \{1, 2, 3, 4, 5, 6, 7, 8\}$

25. $\{x \mid x < 1\}$

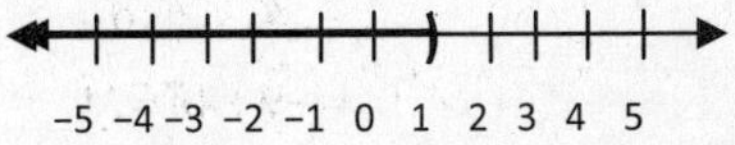

26. (0, 5)

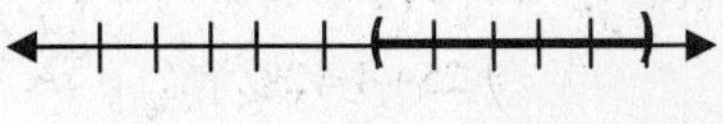

27. the number: x

the difference between a number and 3: $x - 3$

$10(x - 3) = 10x - 30$

28. catcher's throw: s

pitcher's fastball: $2s$

29. **a.** 1981, 1988, 1989, 1990, 1991, 1995

b. $-369.7 - (-81.1) = -369.7 + 81.1 = -288.6$

The difference between the trade balance in 1990 and 2000 was –\$288.6 billion.

c. The difference in trade was greatest from 1999 to 2000.

d. $\frac{-81.1}{-19.4} = 4.18 \approx 4$ times greater

e. $\frac{-369.7}{4} = -\$92.425$ billion

30. Strategy To find the difference between the highest temperature and the lowest temperature, subtract the lowest temperature (-81.4°) from the highest temperature (134.0^{0}).

Solution $134.0-(-81.4)=134.0+81.4=215.4$

The difference between the highest temperature recorded in North America and the lowest temperature recorded is 215.4° F.

Chapter 2: First-Degree Equations and Inequalities

Prep Test

1. $\frac{9}{100} = 0.09$

2. $\frac{3}{4} = \frac{3}{4}(100\%) = \frac{300}{4}\% = 75\%$

3. $3x^2 - 4x - 1$

   $3(-4)^2 - 4(-4) - 1$
   $= 3(16) - 4(-4) - 1$
   $= 48 + 16 - 1$
   $= 63$

4. $R - 0.35R = (1 - 0.35)R = 0.65R$

5. $\frac{1}{2}x + \frac{2}{3}x = \left(\frac{1}{2} + \frac{2}{3}\right)x$

   $= \left(\frac{3}{6} + \frac{4}{6}\right)x$
   $= \frac{7}{6}x$

6. $6x - 3(6 - x) = 6x - 3(6) - 3(-x)$

   $= 6x - 18 + 3x$
   $= 9x - 18$

7. $0.22(3x + 6) + x = 0.66x + 1.32 + x = 1.66x + 1.32$

8. the unknown number: n
   twice a number: $2n$
   $5 - 2n$

9. speed of old card: s
   speed of new card: $5s$

10. length of longer piece: x
    length of shorter piece: $5 - x$

Section 2.1

Objective A Exercises

1.
$$\begin{array}{c|c} 2x & = 8 \\ \hline 2(4) & 8 \end{array}$$
$8 = 8$

Yes, 4 is a solution.

3.
$$\begin{array}{c|c} 2b - 1 & = 3 \\ \hline 2(-1) - 1 & 3 \\ -2 - 1 & 3 \end{array}$$
$-3 \neq 3$

No, −1 is not a solution.

5.
$$\begin{array}{c|c} 4 - 2m & = 3 \\ \hline 4 - 2(1) & 3 \\ 4 - 2 & 3 \end{array}$$
$2 \neq 3$

No, 1 is not a solution.

7.
$$\begin{array}{c|c} 2x + 5 & = 3x \\ \hline 2(5) + 5 & 3(5) \\ 10 + 5 & 15 \end{array}$$
$15 = 15$

Yes, 5 is a solution.

9.
$$\begin{array}{c|c} 3a + 2 & = 2 - a \\ \hline 3(-2) + 2 & 2 - (-2) \\ -6 + 2 & 2 + 2 \end{array}$$
$-4 \neq 4$

No, −2 is not a solution.

11.
$$\begin{array}{c|c} 2x^2 - 1 & = 4x - 1 \\ \hline 2(2)^2 - 1 & 4(2) - 1 \\ 2(4) - 1 & 8 - 1 \\ 8 - 1 & 7 \end{array}$$
$7 = 7$

Yes, 2 is a solution.

13.
$$\begin{array}{c|c} x(x + 1) & = x^2 + 5 \\ \hline 4(4 + 1) & 4^2 + 5 \\ 4(5) & 16 + 5 \end{array}$$
$20 \neq 21$

No, 4 is not a solution.

15.
$$\begin{array}{c|c} 8t + 1 & = -1 \\ \hline 8(-1/4) + 1 & -1 \\ -2 + 1 & -1 \end{array}$$
$-1 = -1$

Yes, $-\frac{1}{4}$ is a solution.

17.

$5m+1$	$=$	$10m-3$
$5(2/5)+1$		$10(2/5)-3$
$2+1$		$4-3$

$$3 \neq 1$$

No, $\frac{2}{5}$ is not a solution.

19. Negative

Objective B Exercises

21. x will be greater than $\frac{19}{24}$ because you will add $\frac{11}{16}$ to solve the equation.

23.
$$\begin{aligned} x+5&=7 \\ x+5-5&=7-5 \\ x&=2 \end{aligned}$$
The solution is 2.

25.
$$\begin{aligned} b-4&=11 \\ b-4+4&=11+4 \\ b&=15 \end{aligned}$$
The solution is 15.

27.
$$\begin{aligned} 2+a&=8 \\ 2-2+a&=8-2 \\ a&=6 \end{aligned}$$
The solution is 6.

29.
$$\begin{aligned} n-5&=-2 \\ n-5+5&=-2+5 \\ n&=3 \end{aligned}$$
The solution is 3.

31.
$$\begin{aligned} b+7&=7 \\ b+7-7&=7-7 \\ b&=0 \end{aligned}$$
The solution is 0.

33.
$$\begin{aligned} z+9&=2 \\ z+9-9&=2-9 \\ z&=-7 \end{aligned}$$
The solution is −7.

35.
$$\begin{aligned} 10+m&=3 \\ 10-10+m&=3-10 \\ m&=-7 \end{aligned}$$
The solution is −7.

37.
$$\begin{aligned} 9+x&=-3 \\ 9-9+x&=-3-9 \\ x&=-12 \end{aligned}$$
The solution is −12.

39.
$$\begin{aligned} 2&=x+7 \\ 2-7&=x+7-7 \\ -5&=x \end{aligned}$$
The solution is −5.

41.
$$\begin{aligned} 4&=m-11 \\ 4+11&=m-11+11 \\ 15&=m \end{aligned}$$
The solution is 15.

43.
$$\begin{aligned} 12&=3+w \\ 12-3&=3-3+w \\ 9&=w \end{aligned}$$
The solution is 9.

45.
$$\begin{aligned} 4&=-10+b \\ 4+10&=-10+10+b \\ 14&=b \end{aligned}$$
The solution is 14.

47.
$$\begin{aligned} m+\frac{2}{3}&=-\frac{1}{3} \\ m+\frac{2}{3}-\frac{2}{3}&=-\frac{1}{3}-\frac{2}{3} \\ m&=-1 \end{aligned}$$
The solution is −1.

49.
$$\begin{aligned} x-\frac{1}{2}&=\frac{1}{2} \\ x-\frac{1}{2}+\frac{1}{2}&=\frac{1}{2}+\frac{1}{2} \\ x&=1 \end{aligned}$$
The solution is 1.

51.
$$\begin{aligned} \frac{5}{8}+y&=\frac{1}{8} \\ \frac{5}{8}-\frac{5}{8}+y&=\frac{1}{8}-\frac{5}{8} \\ y&=-\frac{4}{8} \\ y&=-\frac{1}{2} \end{aligned}$$
The solution is $-\frac{1}{2}$.

53.
$$\begin{aligned} m+\frac{1}{2}-\frac{1}{2}&=-\frac{1}{4}-\frac{1}{2} \\ m&=-\frac{1}{4}-\frac{2}{4} \\ m&=-\frac{3}{4} \end{aligned}$$
The solution is $-\frac{3}{4}$.

55.
$$x+\frac{2}{3}=\frac{3}{4}$$
$$x+\frac{2}{3}-\frac{2}{3}=\frac{3}{4}-\frac{2}{3}$$
$$x=\frac{9}{12}-\frac{8}{12}$$
$$x=\frac{1}{12}$$
The solution is $\frac{1}{12}$.

57.
$$-\frac{5}{6}=x-\frac{1}{4}$$
$$-\frac{5}{6}+\frac{1}{4}=x-\frac{1}{4}+\frac{1}{4}$$
$$-\frac{10}{12}+\frac{3}{12}=x$$
$$-\frac{7}{12}=x$$
The solution is $-\frac{7}{12}$.

59.
$$d+1.3619=2.0148$$
$$d+1.3619-1.3619=2.0148-1.3619$$
$$d=0.6529$$
The solution is 0.6529.

61.
$$-0.813+x=-1.096$$
$$-0.813+0.813+x=-1.096+0.813$$
$$x=-0.283$$
The solution is –0.283.

63.
$$6.149=-3.108+z$$
$$6.149+3.108=-3.108+3.108+z$$
$$9.257=z$$
The solution is 9.257.

Objective C Exercises

65.
$$5x=-15$$
$$\frac{5x}{5}=\frac{-15}{5}$$
$$x=-3$$
The solution is –3.

67.
$$3b=0$$
$$\frac{3b}{3}=\frac{0}{3}$$
$$b=0$$
The solution is 0.

69.
$$-3x=6$$
$$\frac{-3x}{-3}=\frac{6}{-3}$$
$$x=-2$$
The solution is −2.

71.
$$-3x=-27$$
$$\frac{-3x}{-3}=\frac{-27}{-3}$$
$$x=9$$
The solution is 9.

73.
$$20=\frac{1}{4}c$$
$$4(20)=4\left(\frac{1}{4}c\right)$$
$$80=c$$
The solution is 80.

75.
$$0=-5x$$
$$\frac{0}{-5}=\frac{-5x}{-5}$$
$$0=x$$
The solution is 0.

77.
$$49=-7t$$
$$\frac{49}{-7}=\frac{-7t}{-7}$$
$$-7=t$$
The solution is −7.

79.
$$\frac{x}{4}=3$$
$$4\left(\frac{1}{4}x\right)=4(3)$$
$$x=12$$
The solution is 12.

81.
$$-\frac{b}{3}=6$$
$$-3\left(-\frac{1}{3}b\right)=-3(6)$$
$$b=-18$$
The solution is −18.

83.
$$\frac{2}{5}x=6$$
$$\frac{5}{2}\left(\frac{2}{5}x\right)=\frac{5}{2}(6)$$
$$x=15$$
The solution is 15.

85.
$$-\frac{3}{5}m = 12$$
$$-\frac{5}{3}\left(-\frac{3}{5}m\right) = -\frac{5}{3}(12)$$
$$m = -20$$
The solution is −20.

87.
$$\frac{5x}{6} = 0$$
$$\frac{6}{5}\left(\frac{5}{6}x\right) = \frac{6}{5}(0)$$
$$x = 0$$
The solution is 0.

89.
$$\frac{3}{4}x = 2$$
$$\frac{4}{3}\left(\frac{3}{4}x\right) = \frac{4}{3}(2)$$
$$x = \frac{8}{3}$$
The solution is $\frac{8}{3}$.

91.
$$\frac{2}{9} = \frac{2}{3}y$$
$$\frac{3}{2}\left(\frac{2}{9}\right) = \frac{3}{2}\left(\frac{2}{3}y\right)$$
$$\frac{1}{3} = y$$
The solution is $\frac{1}{3}$.

93.
$$\frac{1}{5}x = -\frac{1}{10}$$
$$5\left(\frac{1}{5}x\right) = 5\left(-\frac{1}{10}\right)$$
$$x = -\frac{1}{2}$$
The solution is $-\frac{1}{2}$.

95.
$$-1 = \frac{2n}{3}$$
$$\frac{3}{2}(-1) = \frac{3}{2}\left(\frac{2n}{3}\right)$$
$$-\frac{3}{2} = n$$
The solution is $-\frac{3}{2}$.

97.
$$-\frac{2}{5}m = -\frac{6}{7}$$
$$-\frac{5}{2}\left(-\frac{2}{5}m\right) = -\frac{5}{2}\left(-\frac{6}{7}\right)$$
$$m = \frac{15}{7}$$
The solution is $\frac{15}{7}$.

99.
$$3n + 2n = 20$$
$$5n = 20$$
$$\frac{5n}{5} = \frac{20}{5}$$
$$n = 4$$
The solution is 4.

101.
$$10y - 3y = 21$$
$$7y = 21$$
$$\frac{7y}{7} = \frac{21}{7}$$
$$y = 3$$
The solution is 3.

103.
$$\frac{x}{1.46} = 3.25$$
$$1.46\left(\frac{1}{1.46}x\right) = 1.46(3.25)$$
$$x = 4.745$$
The solution is 4.745.

105.
$$3.47a = 7.1482$$
$$\frac{3.47a}{3.47} = \frac{7.1482}{3.47}$$
$$a = 2.06$$
The solution is 2.06.

107.
$$-3.7x = 7.881$$
$$\frac{-3.7x}{-3.7} = \frac{7.881}{-3.7}$$
$$x \approx -2.13$$
The solution is −2.13.

109. Positive

111. Negative

Objective D Exercises

113. Equal to

115.
$$P \cdot B = A$$
$$0.35(80) = A$$
$$A = 28$$
35% of 80 is 28.

117. $P \cdot B = A$

$0.012(60) = A$

$A = 0.72$

1.2% of 60 is 0.72.

119. $P \cdot B = A$

$(1.25)B = 80$

$\frac{(1.25)B}{1.25} = \frac{80}{1.25}$

$B = 64$

The number is 64.

121. $P \cdot B = A$

$P(50) = 12$

$\frac{P(50)}{50} = \frac{12}{50}$

$P = 0.24$

$P = 24\%$

The percent is 24%.

123. $P \cdot B = A$

$0.18(40) = A$

$A = 7.2$

18% of 40 is 7.2.

125. $P \cdot B = A$

$0.12(B) = 48$

$\frac{0.12(B)}{0.12} = \frac{48}{0.12}$

$B = 400$

The number is 400.

127. $\frac{1}{3}(27) = A \quad \left(33\frac{1}{3}\% = \frac{1}{3}\right)$

$9 = A$

$33\frac{1}{3}\%$ of 27 is 9.

129. $P(12) = 3$

$\frac{12P}{12} = \frac{3}{12}$

$P = 0.25$

The percent is 25%.

131. $P \cdot B = A$

$P(6) = 12$

$\frac{P(6)}{6} = \frac{12}{6}$

$P = 2$

$P = 200\%$

The percent is 200%.

133. $P \cdot B = A$

$0.0525B = 21$

$\frac{0.0525B}{0.0525} = \frac{21}{0.0525}$

$B = 400$

The number is 400.

135. $P \cdot B = A$

$0.154(50) = A$

$A = 7.7$

15.4% of 50 is 7.7.

137. $P \cdot B = A$

$0.005B = 1$

$\frac{0.005B}{0.005} = \frac{1}{0.005}$

$B = 200$

The number is 200.

139. $P \cdot B = A$

$0.0075B = 3$

$\frac{0.0075B}{0.0075} = \frac{3}{0.0075}$

$B = 400$

The number is 400.

141. $P \cdot B = A$

$2.5(12) = A$

$A = 30$

250% of 12 is 30.

143. Strategy To find the percent, solve the basic percent equation $P \cdot B = A$ using $B = 22377$ and $A = 21948$.

Solution

$P \cdot B = A$

$P \cdot 22377 = 21948$

$P = \frac{21948}{22377}$

$P = 0.981$

98.1% of those that started, finished.

145. You need to know the number of people three years old and older in the U.S that are enrolled in school.

147. (a) Strategy To find the total teen population, use the basic percent equation $P \cdot B = A$ where $P = 0.085$ and $A = 20000$.

Solution

$$P \cdot B = A$$
$$0.085B = 20000$$
$$\frac{0.085B}{0.085} = \frac{20000}{0.085}$$
$$B = 235{,}294$$
$$B \approx 240{,}000$$

The teen population in 2007 was 240,000.

147. (b) Strategy To find the teen population use the basic percent equation $P \cdot B = A$ where $P = 0.176$ and $B = 240{,}000$.

Solution

$$P \cdot B = A$$
$$0.176 \cdot 240{,}000 = A$$
$$42{,}240 = A$$
$$42{,}000 \approx A$$

The number of teen smokers in 2007, if the smoking rate were 17.6%, is 42,000.

149. Strategy To find the principal, solve the simple interest equation using $I = \$300$, $r = 8\% = 0.08$, and $t = 2$ years, for P.

Solution

$$I = Prt$$
$$300 = P(0.08)(2)$$
$$300 = 0.16P$$
$$\frac{300}{0.16} = \frac{0.16P}{0.16}$$
$$1875 = P$$

Andrea must invest $1875.

151. Strategy To determine who will earn more interest after one year, solve the simple interest equation for each account:
First, using $P = \$2500$, $r = 8\% = 0.08$, and $t = 1$ year, for I.
Second, using $P = \$3000$, $r = 7\% = 0.07$, and $t = 1$ year, for I.
Finally, compare the interest earned.

Solution

$$I = Prt$$
$$I = (2500)(0.08)(1)$$
$$I = 200 \text{ Americo's interest}$$

$$I = Prt$$
$$I = (3000)(0.07)(1)$$
$$I = 210 \text{ Octavia's interest}$$

Americo's interest was $200. Octavia's interest was $210.
Octavia earns more interest after one year.

153. Strategy To determine how much was invested at 8%, solve the simple interest equation for each account:
First, using $P = \$2000$, $r = 6\% = 0.06$, and $t = 1$ year, for I.
Second, using the amount of interest found the first step for I, $r = 8\% = 0.08$, and $t = 1$ year, for P.

Solution

$$I = Prt$$
$$I = (2000)(0.06)(1)$$
$$I = 120$$

The interest on $2000 at 6% is $120.

$$I = Prt$$
$$120 = P(0.08)(1)$$
$$\frac{120}{0.08} = P$$
$$1500 = P$$

$1500 was invested at 8%.

155. Strategy To find the amount of platinum, solve the basic percent equation using $P = 15\% = 0.15$ and $B = 12$g. The amount is unknown.

Solution

$$PB = A$$
$$0.15(12) = A$$
$$1.8 = A$$

There is 1.8 g of platinum in the necklace.

157. Strategy To find the amount of wool, solve the basic percent equation using P = 75% = 0.75 and B = 175 lb. The amount is unknown.

Solution

$$\begin{aligned} PB &= A \\ 0.75(175) &= A \\ 131.25 &= A \end{aligned}$$

There is a 131.25 lb of wool in the carpet.

159. Strategy To find the percent, solve the basic percent equation using B = 500 + 500 = 1000 and
A = 500. The percent is unknown.

Solution

$$\begin{aligned} PB &= A \\ P(1000) &= 500 \\ \frac{1000P}{1000} &= \frac{500}{1000} \\ P &= 0.5 \end{aligned}$$

The percent concentration is 50%.

161. Strategy To find the percent, solve the basic percent equation using B = 100 + 50 = 150 and A = 100(9%) = 9. The percent is unknown.

Solution

$$\begin{aligned} PB &= A \\ P(150) &= 9 \\ \frac{150P}{150} &= \frac{9}{150} \\ P &= 0.06 \end{aligned}$$

The percent concentration is 6%.

Objective E Exercises

163. (a) greater than
(b) equal to
(c) 2 mi

165. Strategy To find the number of miles traveled, solve $d = rt$ for d using r = 9 mph and $t = \frac{20}{60} = \frac{1}{3}$ h.

Solution

$$\begin{aligned} d &= rt \\ d &= 9\left(\frac{1}{3}\right) \\ d &= 3 \end{aligned}$$

The runner will travel 3 mi.

167. Strategy To find the number of miles traveled, solve $d = rt$ for d using d = 27 mi and $t = \frac{45}{60} = \frac{3}{4}$ h.

Solution

$$\begin{aligned} d &= rt \\ 27 &= r\left(\frac{3}{4}\right) \\ 36 &= r \end{aligned}$$

Marcella's average rate of speed is 36 mph.

169. Strategy To find the number of hours to walk the course:
Find the rate to run the course by solving $d = rt$ for r using d = 30 km and $t = 2$ h.

Decrease the rate by 3 km/h to find his walking rate.

Solve for $d = rt$ for t using d = 30 km and r equal to his walking rate.

Solution

$$\begin{aligned} d &= rt \\ 30 &= r(2) \\ \frac{30}{2} &= r \\ 15 &= r \quad \text{His running rate} \\ 15 - 3 &= 12 \quad \text{His walking rate} \\ d &= rt \\ 30 &= 12t \\ \frac{30}{12} &= t \\ 2.5 &= t \end{aligned}$$

It would take Palmer 2.5 h to walk the course.

171. Strategy The distance is 8 mi. Therefore $d = 8$. The joggers are running toward each other, one at 5 mph and one at 7 mph. The rate is the sum of the two rates, or 12 mph. So, $r = 12$. To find the time solve $d = rt$ for t. Convert the answer to minutes.

Solution

$$d = rt$$
$$8 = 12t$$
$$\frac{8}{12} = t$$
$$\frac{2}{3} = t$$
$$\frac{2}{3}\text{ h} = \frac{2}{3}\cdot 60\text{ min} = 40\text{ min}$$

The two joggers will meet 40 min after they start.

173. Strategy The distance is 4 mi. So, $d = 4$. The canoe is traveling against a 2 mph current. In calm water they can paddle at 10 mph. The rate is 10 mph - 2 mph = 8 mph. So $r = 8$. Solve $d = rt$ for t.

Solution

$$d = rt$$
$$4 = 8t$$
$$\frac{4}{8} = t$$
$$\frac{1}{2} = t$$

It will take them 0.5 h.

Applying the Concepts

175. Negative

177. Multiplying both sides of an equation by zero will result in the equation $0 = 0$.

179. $\frac{1}{2}x = -3$

$2\cdot\frac{1}{2}x = 2(-3)$ By the Multiplication Property of Equations, both sides of an equation can be multiplied by the same nonzero number without changing the solution of the equation. Multiply each side of the equation by the reciprocal of the coefficient of x.

$\frac{2}{1}\cdot\frac{1}{2}x = 2(-3)$ A number divided by 1 is the number. Therefore, $2 = \frac{2}{1}$.

$\left(\frac{2}{1}\cdot\frac{1}{2}\right)x = 2(-3)$ By the Associative Property of Multiplication, we can group the factors $\frac{2}{1}$ and $\frac{1}{2}$ without changing the product.

$1x = -6$ By the Inverse Property of Multiplication, the product of a nonzero number and its reciprocal is 1.

$x = -6$ By the Multiplication Property of 1, the product of a number and 1 is the number. Therefore, we can write $1x$ as x.

Section 2.2

Objective A Exercises

1.

$$3x + 1 = 10$$
$$3x + 1 - 1 = 10 - 1$$
$$3x = 9$$
$$\frac{3x}{3} = \frac{9}{3}$$
$$x = 3$$

The solution is 3.

3.

$$2a - 5 = 7$$
$$2a - 5 + 5 = 7 + 5$$
$$2a = 12$$
$$\frac{2a}{2} = \frac{12}{2}$$
$$a = 6$$

The solution is 6.

5.
$$\begin{aligned} 5 &= 4x + 9 \\ 5 - 9 &= 4x + 9 - 9 \\ -4 &= 4x \\ \frac{-4}{4} &= \frac{4x}{4} \\ -1 &= x \end{aligned}$$
The solution is −1.

7.
$$\begin{aligned} 2x - 5 &= -11 \\ 2x - 5 + 5 &= -11 + 5 \\ 2x &= -6 \\ \frac{2x}{2} &= \frac{-6}{2} \\ x &= -3 \end{aligned}$$
The solution is −3.

9.
$$\begin{aligned} 4 - 3w &= -2 \\ 4 - 4 - 3w &= -2 - 4 \\ -3w &= -6 \\ \frac{-3w}{-3} &= \frac{-6}{-3} \\ w &= 2 \end{aligned}$$
The solution is 2.

11.
$$\begin{aligned} 8 - 3t &= 2 \\ 8 - 8 - 3t &= 2 - 8 \\ -3t &= -6 \\ \frac{-3t}{-3} &= \frac{-6}{-3} \\ t &= 2 \end{aligned}$$
The solution is 2.

13.
$$\begin{aligned} 4a - 20 &= 0 \\ 4a - 20 + 20 &= 0 + 20 \\ 4a &= 20 \\ \frac{4a}{4} &= \frac{20}{4} \\ a &= 5 \end{aligned}$$
The solution is 5.

15.
$$\begin{aligned} 6 + 2b &= 0 \\ 6 - 6 + 2b &= 0 - 6 \\ 2b &= -6 \\ \frac{2b}{2} &= \frac{-6}{2} \\ b &= -3 \end{aligned}$$
The solution is −3.

17.
$$\begin{aligned} -2x + 5 &= -7 \\ -2x + 5 - 5 &= -7 - 5 \\ -2x &= -12 \\ \frac{-2x}{-2} &= \frac{-12}{-2} \\ x &= 6 \end{aligned}$$
The solution is 6.

19.
$$\begin{aligned} -1.2x + 3 &= -0.6 \\ -1.2x + 3 - 3 &= -0.6 - 3 \\ -1.2x &= -3.6 \\ \frac{-1.2x}{-1.2} &= \frac{-3.6}{-1.2} \\ x &= 3 \end{aligned}$$
The solution is 3.

21.
$$\begin{aligned} 2 &= 7 - 5a \\ 2 - 7 &= 7 - 7 - 5a \\ -5 &= -5a \\ \frac{-5}{-5} &= \frac{-5a}{-5} \\ 1 &= a \end{aligned}$$
The solution is 1.

23.
$$\begin{aligned} -35 &= -6b + 1 \\ -35 - 1 &= -6b + 1 - 1 \\ -36 &= -6b \\ \frac{-36}{-6} &= \frac{-6b}{-6} \\ 6 &= b \end{aligned}$$
The solution is 6.

25.
$$\begin{aligned} -3m - 21 &= 0 \\ -3m - 21 + 21 &= 0 + 21 \\ -3m &= 21 \\ \frac{-3m}{-3} &= \frac{21}{-3} \\ m &= -7 \end{aligned}$$
The solution is −7.

27.
$$\begin{aligned} -4y + 15 &= 15 \\ -4y + 15 - 15 &= 15 - 15 \\ -4y &= 0 \\ \frac{-4y}{-4} &= \frac{0}{-4} \\ y &= 0 \end{aligned}$$
The solution is 0.

29.
$$\begin{aligned} 9-4x &= 6 \\ 9-9-4x &= 6-9 \\ -4x &= -3 \\ \frac{-4x}{-4} &= \frac{-3}{-4} \\ x &= \frac{3}{4} \end{aligned}$$

The solution is $\frac{3}{4}$.

31.
$$\begin{aligned} 9x-4 &= 0 \\ 9x-4+4 &= 0+4 \\ 9x &= 4 \\ \frac{9x}{9} &= \frac{4}{9} \\ x &= \frac{4}{9} \end{aligned}$$

The solution is $\frac{4}{9}$.

33.
$$\begin{aligned} 1-3x &= 0 \\ 1-1-3x &= 0-1 \\ -3x &= -1 \\ \frac{-3x}{-3} &= \frac{-1}{-3} \\ x &= \frac{1}{3} \end{aligned}$$

The solution is $\frac{1}{3}$.

35.
$$\begin{aligned} 12w+11 &= 5 \\ 12w+11-11 &= 5-11 \\ 12w &= -6 \\ \frac{12w}{12} &= \frac{-6}{12} \\ w &= -\frac{6}{12} \\ w &= -\frac{1}{2} \end{aligned}$$

The solution is $-\frac{1}{2}$.

37.
$$\begin{aligned} 8b-3 &= -9 \\ 8b-3+3 &= -9+3 \\ 8b &= -6 \\ \frac{8b}{8} &= \frac{-6}{8} \\ b &= -\frac{6}{8} \\ b &= -\frac{3}{4} \end{aligned}$$

The solution is $-\frac{3}{4}$.

39.
$$\begin{aligned} 7-9a &= 4 \\ 7-7-9a &= 4-7 \\ -9a &= -3 \\ \frac{-9a}{-9} &= \frac{-3}{-9} \\ a &= \frac{3}{9} \\ a &= \frac{1}{3} \end{aligned}$$

The solution is $\frac{1}{3}$.

41.
$$\begin{aligned} 10 &= -18x+7 \\ 10-7 &= -18x+7-7 \\ 3 &= -18x \\ \frac{3}{-18} &= \frac{-18x}{-18} \\ -\frac{3}{18} &= x \\ -\frac{1}{6} &= x \end{aligned}$$

The solution is $-\frac{1}{6}$.

43.
$$\begin{aligned} 4a+\frac{3}{4} &= \frac{19}{4} \\ 4a+\frac{3}{4}-\frac{3}{4} &= \frac{19}{4}-\frac{3}{4} \\ 4a &= \frac{16}{4} \\ 4a &= 4 \\ \frac{4a}{4} &= \frac{4}{4} \\ a &= 1 \end{aligned}$$

The solution is 1.

45.
$$3x - \frac{5}{6} = \frac{13}{6}$$
$$3x - \frac{5}{6} + \frac{5}{6} = \frac{13}{6} + \frac{5}{6}$$
$$3x = \frac{18}{6}$$
$$3x = 3$$
$$\frac{3x}{3} = \frac{3}{3}$$
$$x = 1$$
The solution is 1.

47.
$$9x + \frac{4}{5} = \frac{4}{5}$$
$$9x + \frac{4}{5} - \frac{4}{5} = \frac{4}{5} - \frac{4}{5}$$
$$9x = 0$$
$$\frac{9x}{9} = \frac{0}{9}$$
$$x = 0$$
The solution is 0.

49.
$$0.9 = 10x - 0.6$$
$$0.9 + 0.6 = 10x - 0.6 + 0.6$$
$$1.5 = 10x$$
$$\frac{1.5}{10} = \frac{10x}{10}$$
$$0.15 = x$$
The solution is 0.15.

51.
$$7 = 9 - 5a$$
$$7 - 9 = 9 - 9 - 5a$$
$$-2 = -5a$$
$$\frac{-2}{-5} = \frac{-5a}{-5}$$
$$\frac{2}{5} = a$$
The solution is $\frac{2}{5}$.

53.
$$12x + 19 = 3$$
$$12x + 19 - 19 = 3 - 19$$
$$12x = -16$$
$$\frac{12x}{12} = \frac{-16}{12}$$
$$x = -\frac{16}{12}$$
$$x = -\frac{4}{3}$$
The solution is $-\frac{4}{3}$.

55.
$$-4x + 3 = 9$$
$$-4x + 3 - 3 = 9 - 3$$
$$-4x = 6$$
$$\frac{-4x}{-4} = \frac{6}{-4}$$
$$x = -\frac{6}{4}$$
$$x = -\frac{3}{2}$$
The solution is $-\frac{3}{2}$.

57.
$$\frac{1}{3}m - 1 = 5$$
$$\frac{1}{3}m - 1 + 1 = 5 + 1$$
$$\frac{1}{3}m = 6$$
$$3\left(\frac{1}{3}m\right) = 3 \cdot 6$$
$$m = 18$$
The solution is 18.

59.
$$\frac{3}{4}n + 7 = 13$$
$$\frac{3}{4}n + 7 - 7 = 13 - 7$$
$$\frac{3}{4}n = 6$$
$$\frac{4}{3}\left(\frac{3}{4}n\right) = \frac{4}{3}(6)$$
$$n = 8$$
The solution is 8.

61.
$$-\frac{3}{8}b + 4 = 10$$
$$-\frac{3}{8}b + 4 - 4 = 10 - 4$$
$$-\frac{3}{8}b = 6$$
$$-\frac{8}{3}\left(-\frac{3}{8}b\right) = -\frac{8}{3}(6)$$
$$b = -16$$
The solution is −16.

63. $\frac{y}{5}-2=3$

$\frac{y}{5}-2+2=3+2$

$\frac{y}{5}=5$

$5\left(\frac{1}{5}y\right)=5\cdot 5$

$y=25$

The solution is 25.

65. $\frac{2}{3}x-\frac{5}{6}=-\frac{1}{3}$

$6\left(\frac{2}{3}x-\frac{5}{6}\right)=6\left(-\frac{1}{3}\right)$

$4x-5=-2$

$4x=3$

$x=\frac{3}{4}$

The solution is $\frac{3}{4}$.

67. $\frac{1}{2}-\frac{2}{3}x=\frac{1}{4}$

$12\left(\frac{1}{2}-\frac{2}{3}x\right)=12\left(\frac{1}{4}\right)$

$6-8x=3$

$-8x=-3$

$x=\frac{3}{8}$

The solution is $\frac{3}{8}$.

69. $\frac{3}{2}=\frac{5}{6}+\frac{3x}{8}$

$\frac{3}{2}-\frac{5}{6}=\frac{5}{6}-\frac{5}{6}+\frac{3x}{8}$

$\frac{2}{3}=\frac{3x}{8}$

$\frac{8}{3}\left(\frac{2}{3}\right)=\frac{8}{3}\left(\frac{3x}{8}\right)$

$\frac{16}{9}=x$

The solution is $\frac{16}{9}$.

71. $\frac{11}{27}=\frac{4}{9}-\frac{2x}{3}$

$\frac{11}{27}-\frac{4}{9}=\frac{4}{9}-\frac{4}{9}-\frac{2x}{3}$

$-\frac{1}{27}=-\frac{2x}{3}$

$-\frac{3}{2}\left(-\frac{1}{27}\right)=-\frac{3}{2}\left(-\frac{2x}{3}\right)$

$\frac{1}{18}=x$

The solution is $\frac{1}{18}$.

73. $7=\frac{2x}{5}+4$

$7-4=\frac{2x}{5}+4-4$

$3=\frac{2x}{5}$

$\frac{5}{2}(3)=\frac{5}{2}\left(\frac{2}{5}x\right)$

$\frac{15}{2}=x$

The solution is $\frac{15}{2}$.

75. $7-\frac{5}{9}y=9$

$7-7-\frac{5}{9}y=9-7$

$-\frac{5}{9}y=2$

$-\frac{9}{5}\left(-\frac{5}{9}y\right)=-\frac{9}{5}(2)$

$y=-\frac{18}{5}$

The solution is $-\frac{18}{5}$.

77. $5y+9+2y=23$

$7y+9=23$

$7y+9-9=23-9$

$7y=14$

$\frac{7y}{7}=\frac{14}{7}$

$y=2$

The solution is 2.

79.
$$\begin{aligned}11z-3-7z&=9\\4z-3&=9\\4z-3+3&=9+3\\4z&=12\\\frac{4z}{4}&=\frac{12}{4}\\z&=3\end{aligned}$$
The solution is 3.

81. Negative

83. Negative

Objective B Exercises

85. Subtract $2x$ from both sides.

87.
$$\begin{aligned}6y+2&=y+17\\6y-y+2&=y-y+17\\5y+2&=17\\5y+2-2&=17-2\\5y&=15\\\frac{5y}{5}&=\frac{15}{5}\\y&=3\end{aligned}$$
The solution is 3.

89.
$$\begin{aligned}13b-1&=4b-19\\13b-4b-1&=4b-4b-19\\9b-1&=-19\\9b-1+1&=-19+1\\9b&=-18\\\frac{9b}{9}&=\frac{-18}{9}\\b&=-2\end{aligned}$$
The solution is –2.

91.
$$\begin{aligned}7a-5&=2a-20\\7a-2a-5&=2a-2a-20\\5a-5&=-20\\5a-5+5&=-20+5\\5a&=-15\\\frac{5a}{5}&=\frac{-15}{5}\\a&=-3\end{aligned}$$
The solution is –3.

93.
$$\begin{aligned}n-2&=6-3n\\n+3n-2&=6+3n+3n\\4n-2&=6\\4n-2+2&=6+2\\4n&=8\\\frac{4n}{4}&=\frac{8}{4}\\n&=2\end{aligned}$$
The solution is 2.

95.
$$\begin{aligned}4y-2&=-16-3y\\4y+3y-2&=-16-3y+3y\\7y-2&=-16\\7y-2+2&=-16+2\\7y&=-14\\\frac{7y}{7}&=\frac{-14}{7}\\y&=-2\end{aligned}$$
The solution is –2.

97.
$$\begin{aligned}m+0.4&=3m+0.8\\m-3m+0.4&=3m-3m+0.8\\-2m+0.4&=0.8\\-2m+0.4-0.4&=0.8-0.4\\-2m&=0.4\\\frac{-2m}{-2}&=\frac{0.4}{-2}\\m&=-0.2\end{aligned}$$
The solution is –0.2.

99.
$$\begin{aligned}5a+7&=2a+7\\5a-2a+7&=2a-2a+7\\3a+7&=7\\3a+7-7&=7-7\\3a&=0\\\frac{3a}{3}&=\frac{0}{3}\\a&=0\end{aligned}$$
The solution is 0.

101.
$$\begin{aligned}10-4n&=16-n\\10-4n+n&=16-n+n\\10-3n&=16\\10-10-3n&=16-10\\-3n&=6\\\frac{-3n}{-3}&=\frac{6}{-3}\\n&=-2\end{aligned}$$
The solution is –2.

103.
$$\begin{aligned} 3-2y &= 15+4y \\ 3-2y-4y &= 15+4y-4y \\ 3-6y &= 15 \\ 3-3-6y &= 15-3 \\ -6y &= 12 \\ \frac{-6y}{-6} &= \frac{12}{-6} \\ y &= -2 \end{aligned}$$
The solution is –2.

105.
$$\begin{aligned} 2b-10 &= 7b \\ 2b-2b-10 &= 7b-2b \\ -10 &= 5b \\ \frac{-10}{5} &= \frac{5b}{5} \\ -2 &= b \end{aligned}$$
The solution is –2.

107.
$$\begin{aligned} 9y &= 5y+16 \\ 9y-5y &= 5y-5y+16 \\ 4y &= 16 \\ \frac{4y}{4} &= \frac{16}{4} \\ y &= 4 \end{aligned}$$
The solution is 4.

109.
$$\begin{aligned} 6y-1 &= 2y+2 \\ 6y-2y-1 &= 2y-2y+2 \\ 4y-1 &= 2 \\ 4y-1+1 &= 2+1 \\ 4y &= 3 \\ \frac{4y}{4} &= \frac{3}{4} \\ y &= \frac{3}{4} \end{aligned}$$
The solution is $\frac{3}{4}$.

111.
$$\begin{aligned} 2y-7 &= -1-2y \\ 2y+2y-7 &= -1-2y+2y \\ 4y-7 &= -1 \\ 4y-7+7 &= -1+7 \\ 4y &= 6 \\ \frac{4y}{4} &= \frac{6}{4} \\ y &= \frac{3}{2} \end{aligned}$$
The solution is $\frac{3}{2}$.

113.
$$\begin{aligned} 5x &= 3x-8 \\ 5x-3x &= 3x-3x-8 \\ 2x &= -8 \\ \frac{2x}{2} &= \frac{-8}{2} \\ x &= -4 \end{aligned}$$

$$\begin{aligned} &4x+2 \\ &= 4(-4)+2 \\ &= -16+2 \\ &= -14 \end{aligned}$$
The answer is –14.

115.
$$\begin{aligned} 2-6a &= 5-3a \\ 2-6a+3a &= 5-3a+3a \\ 2-3a &= 5 \\ 2-2-3a &= 5-2 \\ -3a &= 3 \\ \frac{-3a}{-3} &= \frac{3}{-3} \\ a &= -1 \end{aligned}$$

$$\begin{aligned} &4a^2-2a+1 \\ &= 4(-1)^2-2(-1)+1 \\ &= 4(1)-2(-1)+1 \\ &= 4+2+1 \\ &= 6+1 \\ &= 7 \end{aligned}$$
The answer is 7.

Objective C Exercises

117. (ii)

119.
$$\begin{aligned} 6y+2(2y+3) &= 16 \\ 6y+4y+6 &= 16 \\ 10y+6 &= 16 \\ 10y+6-6 &= 16-6 \\ 10y &= 10 \\ \frac{10y}{10} &= \frac{10}{10} \\ y &= 1 \end{aligned}$$
The solution is 1.

121.
$$\begin{aligned} 12x-2(4x-6) &= 28 \\ 12x-8x+12 &= 28 \\ 4x+12 &= 28 \\ 4x+12-12 &= 28-12 \\ 4x &= 16 \\ \frac{4x}{4} &= \frac{16}{4} \\ x &= 4 \end{aligned}$$
The solution is 4.

123.
$$\begin{aligned} 9m-4(2m-3) &= 11 \\ 9m-8m+12 &= 11 \\ m+12 &= 11 \\ m+12-12 &= 11-12 \\ m &= -1 \end{aligned}$$
The solution is –1.

125.
$$\begin{aligned} 4(1-3x)+7x &= 9 \\ 4-12x+7x &= 9 \\ 4-5x &= 9 \\ 4-4-5x &= 9-4 \\ -5x &= 5 \\ \frac{-5x}{-5} &= \frac{5}{-5} \\ x &= -1 \end{aligned}$$

The solution is –1.

127.
$$\begin{aligned} 0.22(x+6) &= 0.2x+1.8 \\ 0.22x+1.32 &= 0.2x+1.8 \\ 0.22x-0.2x+1.32 &= 0.2x-0.2x+1.8 \\ 0.02x+1.32 &= 1.8 \\ 0.02x+1.32-1.32 &= 1.8-1.32 \\ 0.02x &= 0.48 \\ \frac{0.02x}{0.02} &= \frac{0.48}{0.02} \\ x &= 24 \end{aligned}$$

The solution is 24.

129.
$$\begin{aligned} 0.3x+0.3(x+10) &= 300 \\ 0.3x+0.3x+3 &= 300 \\ 0.6x+3 &= 300 \\ 0.6x+3-3 &= 300-3 \\ 0.6x &= 297 \\ \frac{0.6x}{0.6} &= \frac{297}{0.6} \\ x &= 495 \end{aligned}$$

The solution is 495.

131.
$$\begin{aligned} 5-(9-6x) &= 2x-2 \\ 5-9+6x &= 2x-2 \\ -4+6x &= 2x-2 \\ -4+6x-2x &= 2x-2x-2 \\ -4+4x &= -2 \\ -4+4+4x &= -2+4 \\ 4x &= 2 \\ \frac{4x}{4} &= \frac{2}{4} \\ x &= \frac{1}{2} \end{aligned}$$

The solution is $\frac{1}{2}$.

133.
$$\begin{aligned} 3[2-4(y-1)] &= 3(2y+8) \\ 3[2-4y+4] &= 6y+24 \\ 3[6-4y] &= 6y+24 \\ 18-12y &= 6y+24 \\ 18-12y-6y &= 6y-6y+24 \\ 18-18y &= 24 \\ 18-18-18y &= 24-18 \\ -18y &= 6 \\ \frac{-18y}{-18} &= \frac{6}{-18} \\ y &= -\frac{1}{3} \end{aligned}$$

The solution is $-\frac{1}{3}$.

135.
$$\begin{aligned} 3a+2[2+3(a-1)] &= 2(3a+4) \\ 3a+2[2+3a-3] &= 6a+8 \\ 3a+2[-1+3a] &= 6a+8 \\ 3a-2+6a &= 6a+8 \\ 9a-2 &= 6a+8 \\ 9a-6a-2 &= 6a-6a+8 \\ 3a-2 &= 8 \\ 3a-2+2 &= 8+2 \\ 3a &= 10 \\ \frac{3a}{3} &= \frac{10}{3} \\ a &= \frac{10}{3} \end{aligned}$$

The solution is $\frac{10}{3}$.

137.
$$\begin{aligned} -2[4-(3b+2)] &= 5-2(3b+6) \\ -2[4-3b-2] &= 5-6b-12 \\ -2[2-3b] &= -7-6b \\ -4+6b &= -7-6b \\ -4+6b+6b &= -7-6b+6b \\ -4+12b &= -7 \\ -4+4+12b &= -7+4 \\ 12b &= -3 \\ \frac{12b}{12} &= \frac{-3}{12} \\ b &= -\frac{1}{4} \end{aligned}$$

The solution is $-\frac{1}{4}$.

139.
$$\begin{aligned} 4-3a &= 7-2(2a+5) \\ 4-3a &= 7-4a-10 \\ 4-3a &= -3-4a \\ 4-3a+4a &= -3-4a+4a \\ 4+a &= -3 \\ 4-4+a &= -3-4 \\ a &= -7 \end{aligned}$$

$$\begin{aligned} &a^2+7a \\ &= (-7)^2+(7)(-7) \\ &= 49+(7)(-7) \\ &= 49-49 \\ &= 0 \end{aligned}$$

The answer is 0.

Objective D Exercises

141. Strategy $x = 8.5; y = 8.5, D = 3.2,$
$P = 81.60$
Unknown: z

Solution

$$\begin{aligned} P &= D(x+y+z) \\ 81.60 &= 3.2(8.5+8.5+z) \\ 81.60 &= 3.2(17+z) \\ 81.60 &= 54.4+3.2z \\ 81.60-54.4 &= 54.4-54.4+3.2z \\ 27.2 &= 3.2z \\ \frac{27.2}{3.2} &= \frac{3.2z}{3.2} \\ 8.5 &= z \end{aligned}$$

Kinzbach's score from the third judge was 8.5.

143. Strategy $x = 8; y = 8, D = 3,$
$P = 72$
Unknown: z

Solution

$$\begin{aligned} P &= D(x+y+z) \\ 72 &= 3(8+8+z) \\ 72 &= 3(16+z) \\ 72 &= 48+3z \\ 72-48 &= 48-48+3z \\ 24 &= 3z \\ \frac{24}{3} &= \frac{3z}{3} \\ 8 &= z \end{aligned}$$

Viola's score from the third judge was 8.

145. (a) $8 - 3 = 5$ ft

(b) The person who is 3 ft away.

(c) No

147. Solution To find the location of the fulcrum when the system balances, replace the variables F_1, F_2, and d in the lever system equation by the given values and solve for x.

Solution

$$\begin{aligned} F_1x &= F_2(d-x) \\ 70x &= 175(14-x) \\ 70x &= 2450-175x \\ 70x+175x &= 2450-175x+175x \\ 245x &= 2450 \\ \frac{245x}{245} &= \frac{2450}{245} \\ x &= 10 \end{aligned}$$

The fulcrum is 10 ft from the child.

149. Strategy To find the location of the fulcrum when the system balances, replace the variables F_1, F_2, and d in the lever system equation by the given values and solve for x.

Solution

$$\begin{aligned} F_1x &= F_2(d-x) \\ 90x &= 60(12-x) \\ 90x &= 720-60x \\ 90x+60x &= 720 \\ 150x &= 720 \\ \frac{150x}{150} &= \frac{720}{150} \\ x &= 4.8 \end{aligned}$$

The fulcrum is 4.8 ft from the 90-lb child.

151. Strategy To find the force when the system balances, replace the variables F_2, x, and d in the lever system equation by the given values and solve for F_1.

Solution

$$\begin{aligned} F_1x &= F_2(d-x) \\ F_1\cdot 0.15 &= 30(9-0.15) \\ F_1\cdot 0.15 &= 30(8.85) \\ 0.15F_1 &= 265.5 \\ \frac{0.15F_1}{0.15} &= \frac{265.5}{0.15} \\ F_1 &= 1770 \end{aligned}$$

A 1770-lb force is applied to the other end.

153. Strategy To find the break-even point, replace the variables P, C, and F in the cost equation by the given values and solve for x.

Solution

$$\begin{aligned} Px &= Cx + F \\ 325x &= 175x + 39{,}000 \\ 325x - 175x &= 39{,}000 \\ 150x &= 39{,}000 \\ \frac{150x}{150} &= \frac{39{,}000}{150} \\ x &= 260 \end{aligned}$$

The break-even point is 260 barbecues.

155. Strategy To find the break-even point, replace the variables P, C, and F in the cost equation by the given values and solve for x.

Solution

$$\begin{aligned} Px &= Cx + F \\ 49x &= 12x + 19{,}240 \\ 49x - 12x &= 19{,}240 \\ 37x &= 19{,}240 \\ \frac{37x}{37} &= \frac{19{,}240}{37} \\ x &= 520 \end{aligned}$$

The break-even point is 520 recorders.

157. Strategy $m = 8.3$ Unknown: C

Solution

$$\begin{aligned} m &= \frac{1}{6}(C-5) \\ 8.3 &= \frac{1}{6}(C-5) \\ 8.3 &= \frac{1}{6}C - \frac{5}{6} \\ 6 \cdot 8.3 &= 6 \cdot \frac{1}{6}C - 6 \cdot \frac{5}{6} \\ 49.8 &= C - 5 \\ 49.8 + 5 &= C - 5 + 5 \\ 54.8 &= C \end{aligned}$$

The mammal consumes 54.8 ml/min.

Section 2.3

Objective A Exercises

1. the unknown number: x

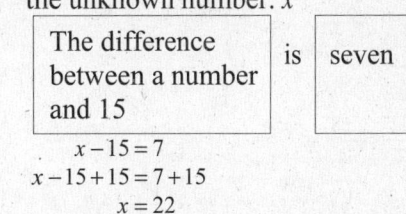

The difference between a number and 15	is	seven

$$\begin{aligned} x - 15 &= 7 \\ x - 15 + 15 &= 7 + 15 \\ x &= 22 \end{aligned}$$

The number is 22.

3. the unknown number: x

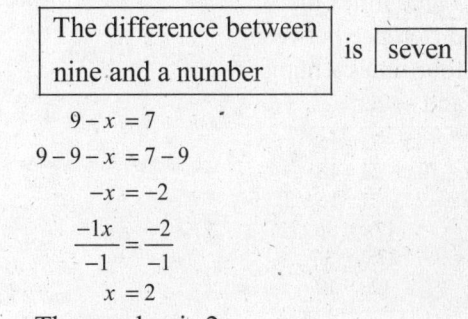

The difference between nine and a number	is	seven

$$\begin{aligned} 9 - x &= 7 \\ 9 - 9 - x &= 7 - 9 \\ -x &= -2 \\ \frac{-1x}{-1} &= \frac{-2}{-1} \\ x &= 2 \end{aligned}$$

The number is 2.

5. the unknown number: x

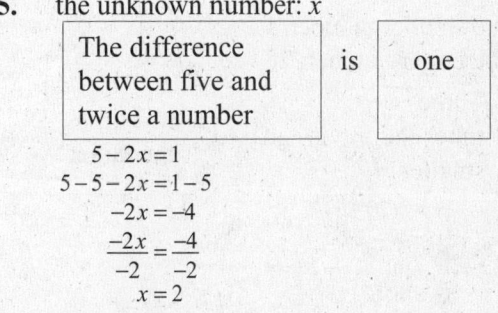

The difference between five and twice a number	is	one

$$\begin{aligned} 5 - 2x &= 1 \\ 5 - 5 - 2x &= 1 - 5 \\ -2x &= -4 \\ \frac{-2x}{-2} &= \frac{-4}{-2} \\ x &= 2 \end{aligned}$$

The number is 2.

7. the unknown number: x

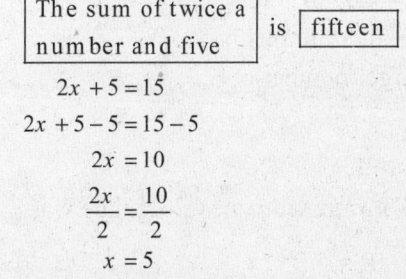

The sum of twice a number and five	is	fifteen

$$\begin{aligned} 2x + 5 &= 15 \\ 2x + 5 - 5 &= 15 - 5 \\ 2x &= 10 \\ \frac{2x}{2} &= \frac{10}{2} \\ x &= 5 \end{aligned}$$

The number is 5.

9. the unknown number: x

Six less than four times a number	is	twenty-two

$$4x-6=22$$
$$4x-6+6=22+6$$
$$4x=28$$
$$\frac{4x}{4}=\frac{28}{4}$$
$$x=7$$

The number is 7.

11. the unknown number: x

Three times the difference between four times a number and seven	is	fifteen

$$3(4x-7)=15$$
$$12x-21=15$$
$$12x-21+21=15+21$$
$$12x=36$$
$$\frac{12x}{12}=\frac{36}{12}$$
$$x=3$$

The number is 3.

13. the smaller number: x
the larger number: $20-x$

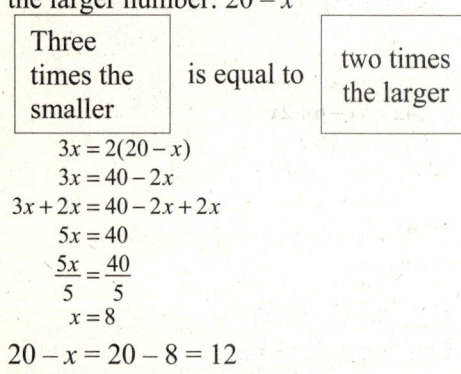

Three times the smaller	is equal to	two times the larger

$$3x=2(20-x)$$
$$3x=40-2x$$
$$3x+2x=40-2x+2x$$
$$5x=40$$
$$\frac{5x}{5}=\frac{40}{5}$$
$$x=8$$

$20-x=20-8=12$

The smaller number is 8.

The larger number is 12.

15. the smaller number: x
the larger number: $14-x$

The difference between two times the smaller and the larger	is	one

$$2x-(14-x)=1$$
$$2x-14+x=1$$
$$3x-14=1$$
$$3x=15$$
$$\frac{3x}{3}=\frac{15}{3}$$
$$x=5$$

$14-x=14-5=9$

The smaller number is 5.

The larger number is 9.

17. First odd integer: n
Second odd integer: $n+2$
Third odd integer: $n+4$
The sum of the three integers is 51.

$$n+(n+2)+(n+4)=51$$
$$3n+6=51$$
$$3n=45$$
$$n=15$$
$$n+2=15+2=17$$
$$n+4=15+4=19$$

The three integers are 15, 17, and 19.

19. First odd integer: n
Second odd integer: $n+2$
Third odd integer: $n+4$
Three times the second number is one more than the sum of the first and third numbers.

$$3(n+2)=1+n+(n+4)$$
$$3n+6=5+2n$$
$$n+6=5$$
$$n=-1$$
$$n+2=-1+2=1$$
$$n+4=-1+4=3$$

The three integers are –1, 1, and 3.

21. First even integer: n
Second even integer: $n+2$
Three times the first integer equals twice the second integer.

$$3n=2(n+2)$$
$$3n=2n+4$$
$$n=4$$
$$n+2=4+2=6$$

The integers are 4 and 6.

23. (iii)

Objective B Exercises

25.

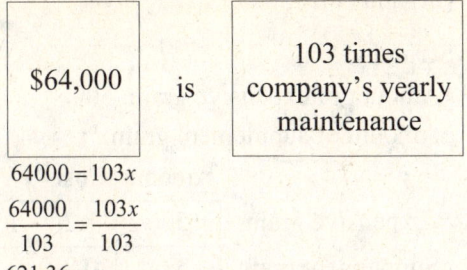

$64,000	is	103 times company's yearly maintenance

$$64000 = 103x$$
$$\frac{64000}{103} = \frac{103x}{103}$$
$$621.36 \approx x$$

The company's yearly cost for a robot was $600.

27. Strategy To find the lengths of the sides of the triangle, write and solve an equation using x to represent the length of each equal side is then $3x + 2$.

Solution

Perimeter of 46 m	is	$(3x+2)$ m + $(3x+2)$ m+ x m

$$46 = 3x + 2 + 3x + 2 + x$$
$$46 = 7x + 4$$
$$42 = 7x$$
$$\frac{42}{7} = x$$
$$6 = x$$
$$3x + 2 = 3(6) + 2 = 20$$

The lengths of the sides are 20 m, 20 m, and 6 m.

29. Strategy To find the number of minutes using the service, write and solve an equation using n to represent the number of minutes.

Solution

$15 plus $2.00 per minute	is	$37

$$15 + 2n = 37$$
$$2n = 22$$
$$n = \frac{22}{2}$$
$$n = 11$$

The customer used 11 min of hotline service.

31. Strategy To find the amount of time that the phone was used, write and solve an equation using x to represent the amount of time.

Solution

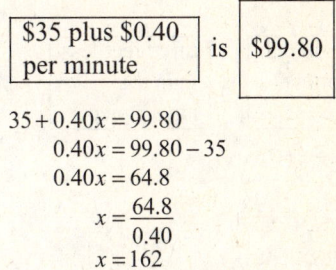

$35 plus $0.40 per minute	is	$99.80

$$35 + 0.40x = 99.80$$
$$0.40x = 99.80 - 35$$
$$0.40x = 64.8$$
$$x = \frac{64.8}{0.40}$$
$$x = 162$$

The business executive used the phone for 162 min.

33. $.15

35. Strategy To find the length and width of the path, write and solve an equation using x to represent the width. Then the length is $2x - 3$.

Solution

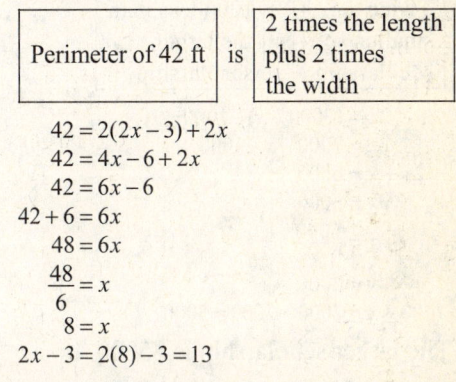

Perimeter of 42 ft	is	2 times the length plus 2 times the width

$$42 = 2(2x - 3) + 2x$$
$$42 = 4x - 6 + 2x$$
$$42 = 6x - 6$$
$$42 + 6 = 6x$$
$$48 = 6x$$
$$\frac{48}{6} = x$$
$$8 = x$$
$$2x - 3 = 2(8) - 3 = 13$$

The length is 13 m.
The width is 8 m.

37. **Strategy** To find the length of each piece, write and solve an equation using x to represent the shorter piece and $12 - x$ to represent the longer piece.

Solution

Twice the length of the shorter piece	is	three feet less than the longer piece

$$\begin{aligned} 2x &= (12 - x) - 3 \\ 2x &= 9 - x \\ 3x &= 9 \\ x &= 3 \\ 12 - x &= 12 - 3 = 9 \end{aligned}$$

The shorter piece is 3 ft.
The longer piece is 9 ft.

39. **Strategy** To find the amount of each scholarship, write and solve an equation using x to represent the smaller scholarship and $7000 - x$ to represent the larger scholarship.

Solution

Twice the smaller scholarship	is	1000 less than the larger scholarship

$$\begin{aligned} 2x &= (7000 - x) - 1000 \\ 2x &= 7000 - x - 1000 \\ 2x + x &= 6000 \\ 3x &= 6000 \\ x &= \frac{6000}{3} \\ x &= 2000 \\ 7000 - x &= 7000 - 2000 = 5000 \end{aligned}$$

The larger scholarship is \$5000.

Applying the Concepts

41. A possible problem for $6x = 123$ is "A student worked 6 hours and earned \$123. What was the student's hourly wage?" For the equation $8x + 100 = 300$, a possible problem is "A group of eight people spent \$300 at an amusement park. This included \$100 for lunch and the admission tickets for the eight people. Find the cost of each ticket."

Section 2.4

Objective A Exercises

1. (iii) and (v)

3. **Strategy**

- Amount of expensive grain: 500
  Amount of supplement grain: x

	Amount	Cost	Value
Expensive grain	500	1.2	1.2(500)
Supplement grain	x	0.8	$0.8x$
Mixture	$500 + x$	1.05	$1.05(500 + x)$

- The sum of the values before mixing equals the value after mixing.

Solution

$$\begin{aligned} 600 + 0.80x &= 1.05(500 + x) \\ 600 + 0.80x &= 525 + 1.05x \\ 600 + 0.80x - 1.05x &= 525 + 1.05x - 1.05x \\ 600 - 0.25x &= 525 \\ 600 - 600 - 0.25x &= 525 - 600 \\ -0.25x &= -75 \\ \frac{-0.25x}{-0.25} &= \frac{-75}{-0.25} \\ x &= 300 \end{aligned}$$

To make the mixture, 300 lb of supplement grain should be added to the 500 lb of the expensive grain.

5. **Strategy**

- Amount of expensive lotion: 50
  Amount of supplement lotion: 100

	Amount	Cost	Value
Expensive lotion	50	4.00	4(50)
Supplement lotion	100	2.50	2.5(100)
Mixture	150	x	$150x$

- The sum of the values before mixing equals the value after mixing.

Solution

$$\begin{aligned} 200 + 250 &= 150x \\ 450 &= 150x \\ \frac{450}{150} &= \frac{150x}{150} \\ 3 &= x \end{aligned}$$

The sunscreen mixture will cost \$3.00.

7. **Strategy**

•Amount of millet seed: 100

Amount of sunflower seed: x

	Amount	Cost	Value
Millet seed	100	0.60	0.60(100)
Sunflower seed	x	1.10	$1.10x$
Mixture	$100+x$	0.70	$0.70(100+x)$

• The sum of the values before mixing equals the value after mixing.

Solution

$$\begin{aligned} 60+1.10x &= 0.70(100+x) \\ 60+1.10x &= 70+0.70x \\ 60-60+1.10x &= 70-60+0.70x \\ 1.10x &= 10+0.70x \\ 1.10x-0.70x &= 10+0.70x-0.70x \\ 0.40x &= 10 \\ \frac{0.40x}{0.40} &= \frac{10}{0.40} \\ x &= 25 \end{aligned}$$

The mixture will need 25 lbs of sunflower seeds.

9. **Strategy**

•Amount of high-protein supplement: x

Amount of vitamin supplement: $5-x$

	Amount	Cost	Value
High-protein	x	6.75	$6.75x$
Vitamin	$5-x$	3.25	$3.25(5-x)$
Mixture	5	4.65	4.65(5)

• The sum of the values before mixing equals the value after mixing.

Solution

$$\begin{aligned} 6.75x+3.25(5-x) &= 4.65(5) \\ 6.75x+16.25-3.25x &= 23.25 \\ 3.50x+16.25 &= 23.25 \\ 3.50x+16.25-16.25 &= 23.25-16.25 \\ 3.50x &= 7.00 \\ x &= \frac{7.00}{3.50} \\ x &= 2 \\ 5-x &= 3 \end{aligned}$$

To make the mixture, 2 lb of the high protein supplement and 3 lb of the vitamin supplement were used.

11. **Strategy**

• Cost of the trail mix: x

	Amount	Cost	Value
Raisins	40	4.40	4.40(40)
Granola	100	2.30	2.30(100)
Mixture	140	x	$140x$

• The sum of the values before mixing equals the value after mixing.

Solution

$$\begin{aligned} 4.40(40)+2.30(100) &= 140x \\ 176+230 &= 140x \\ 406 &= 140x \\ \frac{406}{140} &= x \\ 2.90 &= x \end{aligned}$$

The cost of the trail mix is \$2.90 per pound.

13. **Strategy**

•Amount of alloy 1: x

Amount of alloy 2: $200-x$

	Amount	Cost	Value
Alloy 1	x	4.30	$4.30(x)$
Alloy 2	$200-x$	1.80	$1.80(200-x)$
Mixture	200	2.50	2.50(200)

• The sum of the values before mixing equals the value after mixing.

Solution

$$\begin{aligned} 4.30x+1.80(200-x) &= 2.50(200) \\ 4.30x+360.00-1.80x &= 500.00 \\ 2.50x+360 &= 500 \\ 2.50x+360-360 &= 500-360 \\ 2.50x &= 140 \\ x &= \frac{140}{2.50} \\ x &= 56 \\ 200-x &= 144 \end{aligned}$$

The amount of alloy 1 needed is 56 oz.

The amount of alloy 2 needed is 144 oz.

15. Strategy

•Cost of mixture: x

	Amount	Cost	Value
Expensive coffee	8	9.20	8(9.20)
Cheaper coffee	12	5.50	12(5.50)
Mixture	20	x	$20(x)$

• The sum of the values before mixing equals the value after mixing.

Solution

$$8(9.20)+12(5.50)=20x$$
$$73.60+66=20x$$
$$139.60=20x$$
$$\frac{139.60}{20}=\frac{20x}{20}$$
$$6.98=x$$

The cost of the coffee mixture is $6.98.

17. Strategy

•Amount of adult tickets: x

Amount of student tickets $1720-x$

	Amount	Cost	Value
Adult tickets	x	4.50	$4.50(x)$
Student tickets	$1720-x$	2.00	$2.00(1720-x)$
Mixture	1720		5980

• The sum of the values of the adult tickets and the student's tickets must equal the total collected.

Solution

$$4.50x+2.00(1720-x)=5980$$
$$4.50x+3440-2.00x=5980$$
$$2.50x+3440=5980$$
$$2.50x+3440-3440=5980-3440$$
$$2.50x=2540$$
$$\frac{2.50x}{2.50}=\frac{2540}{2.50}$$
$$x=1016$$

1016 adult tickets were sold.

Objective B Exercises

19. False

21. Strategy

• The percent concentration of tomato juice in the mixture: x

	Amount	Percent	Quantity
50% juice	100	0.50	0.50(100)
25% juice	200	0.25	0.25(200)
Mixture	300	x	$300x$

• The sum of the quantities before mixing is equal to the quantity after mixing.

Solution

$$0.50(100)+0.25(200)=300x$$
$$50+50=300x$$
$$100=300x$$
$$\frac{1}{3}=x$$

The percent concentration of tomato juice in the mixture as $33\frac{1}{3}\%$.

23. Strategy

• Amount of 50% corn: x

• Amount of mixture: $x+400$

	Amount	Percent	Quantity
50% corn	x	0.50	$0.50x$
80% corn	400	0.80	0.80(400)
Mixture	$x+400$	0.75	$0.75(x+400)$

• The sum of the quantities before mixing is equal to the quantity after mixing.

Solution

$$0.50x+0.80(400)=0.75(x+400)$$
$$0.50x+320=0.75x+300$$
$$-0.25x=-20$$
$$x=80$$

80 lbs of 50% corn must be used.

25. Strategy

• Amount of dark green paint: x

• Amount of mixture: $x+5$

	Amount	Percent	Quantity
Dark green paint	x	0.40	$0.40x$
Light green paint	5	0.20	0.20(5)
25% yellow paint	$x+5$	0.25	$0.25(x+5)$

• The sum of the quantities before mixing is equal to the quantity after mixing.

Solution

$$\begin{aligned}0.40x+0.20(5)&=0.25(x+5)\\0.40x+1&=0.25x+1.25\\0.15x&=0.25\\x&=1\frac{2}{3}\end{aligned}$$

$1\frac{2}{3}$ gallon of light green latex paint must be used.

27. Strategy

- Amount of 13% acid solution: x
- Amount of 18% acid solution: $50-x$

	Amount	Percent	Quantity
13% acid	x	0.13	$0.13x$
18% acid	$50-x$	0.18	$0.18(50-x)$
16% acid mixture	50	0.16	0.16(50)

- The sum of the quantities before mixing is equal to the quantity after mixing.

Solution

$$\begin{aligned}0.13x+0.18(50-x)&=0.16(50)\\0.13x+9.00-0.18x&=8.00\\-0.05x+9.00&=8.00\\-0.05x&=-1.00\\x&=20\\50-x=50-20&=30\end{aligned}$$

The amount of 13% solution is 20 ml.
The amount of 18% solution is 30 ml.

29. Strategy

- Percent concentration of the resulting alloy: x

	Amount	Percent	Quantity
30% gold	8	0.30	0.30(8)
25% gold	12	0.25	0.25(12)
Resulting mixture	20	x	$20x$

- The sum of the quantities before mixing is equal to the quantity after mixing.

Solution

$$\begin{aligned}0.30(8)+0.25(12)&=20x\\2.4+3.0&=20x\\5.4&=20x\\0.27&=x\end{aligned}$$

The percent concentration is 27%.

31. Strategy

- Amount of 60% mixture: x

	Amount	Percent	Quantity
Grass seed 1	x	0.60	$0.60x$
Grass seed 2	70	0.80	0.80(70)
60% mixture	$x+70$	0.74	$0.74(x+70)$

- The sum of the quantities before mixing is equal to the quantity after mixing.

Solution

$$\begin{aligned}0.60x+0.80(70)&=0.74(x+70)\\0.60x+56&=0.74x+51.8\\-0.14x&=-4.2\\x&=30\end{aligned}$$

30 lb of the 60% mixture must be used.

33. Strategy

- Amount of 20% jasmine tea: x
- Amount of 15% jasmine tea: $5-x$

	Amount	Percent	Quantity
20% jasmine tea	x	0.20	$0.20x$
15% jasmine tea	$5-x$	0.15	$0.15(5-x)$
18% mixture	5	0.18	0.18(5)

- The sum of the quantities before mixing is equal to the quantity after mixing.

Solution

$$\begin{aligned}0.20x+0.15(5-x)&=0.18(5)\\0.20x+0.75-0.15x&=0.9\\0.05x&=0.15\\x&=3\\5-x&=2\end{aligned}$$

3 lbs of 20% jasmine and 2 lbs of 15% jasmine are needed.

35. Strategy

- Amount of pure bran flakes: x

	Amount	Percent	Value
Box of cereal	50	0.40	0.40(50)
Pure bran flakes	x	1.00	$1.00x$
32% mixture	$x+50$	0.50	$0.50(x+50)$

- The sum of the quantities before mixing is equal to the quantity after mixing.

Solution

$$\begin{aligned} 0.40(50)+1.00x &= 0.50(50+x) \\ 20+1x &= 25+0.50x \\ 0.50x &= 5 \\ x &= 10 \end{aligned}$$

10 oz of pure bran flakes must be added.

Objective C Exercises

37. (i)

39. Strategy

- Speed of first plane: r
- Speed of second plane: $r + 25$

	Rate	Time	Distance
First plane	r	2	$2r$
Second plane	$r + 25$	2	$2(r + 25)$

- In 2 h, the planes are 470 miles apart.

Solution

$$\begin{aligned} 2r+2(r+25) &= 470 \\ 2r+2r+50 &= 470 \\ 4r &= 420 \\ r &= 105 \\ r+25 &= 130 \end{aligned}$$

The first plane is flying at 105 mph and the second plane is flying at 130 mph.

41. Strategy • Time: t + 8:00 AM

	Rate	Time	Distance
First plane	480	t	$480t$
Second plane	520	t	$520t$

- At what time will the planes be 3000 km apart?

Solution

$$\begin{aligned} 480t+520t &= 3000 \\ 1000t &= 3000 \\ t &= 3 \\ 3+8{:}00\text{ AM} &= 11{:}00\text{ AM} \end{aligned}$$

The planes will be 3000 km apart at 11:00 AM.

43. Strategy

- Time the motorboat travels: t
- Time the cabin cruiser travels: $t - 2$

	Rate	Time	Distance
Motorboat	9	t	$9t$
Cabin Cruiser	18	$t - 2$	$18(t - 2)$

- How many hours after the cabin cruiser leaves will the cabin cruiser meet up with the motorboat?

Solution

$$\begin{aligned} 9t &= 18(t-2) \\ 9t &= 18t-36 \\ -9t &= -36 \\ t &= 4 \\ t-2 &= 2 \end{aligned}$$

The cabin cruiser will overtake the motorboat in 2 h.

45. Strategy

- Time to airport: t
- Time in flight: $3 - t$

	Rate	Time	Distance
To airport	30	t	$30t$
In flight	60	$3 - t$	$60(3 - t)$

- The total trip is 150 mi.

Solution

$$\begin{aligned} 30t+60(3-t) &= 150 \\ 30t+180-60t &= 150 \\ 180-30t &= 150 \\ -30t &= -30 \\ t &= 1 \\ \text{Distance} &= 60(3-t) = 60(3-1) \\ &= 60(2) = 120 \end{aligned}$$

The corporate offices are 120 mi from the airport.

47. Strategy

- The speed of the car: $2r$
- The speed of the bus: r

	Rate	Time	Distance
Car	$2r$	2	$4r$
Bus	r	2	$2r$

- In 2 h the car is 68 mi ahead of the bus.

Solution

$$\begin{aligned} 2r+68&=4r\\ 68&=2r\\ 34&=r\\ 2r&=68 \end{aligned}$$

The car is traveling at 68 mph.

49. Strategy

- The travel time to the airport: t
- The travel time from the airport: $5-t$

	Rate	Time	Distance
To the airport	100	t	$100t$
From the airport	150	$5-t$	$150(5-t)$

- The distance traveled both ways is the same.

Solution

$$\begin{aligned} 100t&=150(5-t)\\ 100t&=750-150t\\ 250t&=750\\ t&=3\\ 100t&=300 \end{aligned}$$

The distance between airports is 300 mi.

51. Strategy

- Time the first plane traveled: t
- Time the second plane traveled: $t-1$

	Rate	Time	Distance
First plane	500	t	$500t$
Second plane	500	$t-1$	$500(t-1)$

- The planes pass each other.

Solution

$$\begin{aligned} 500t+500(t-1)&=3000\\ 500t+500t-500&=3000\\ 1000t&=2500\\ t&=2\frac{1}{2} \end{aligned}$$

The planes will pass each other after 2.5 h.

53. Strategy

- Time: t

	Rate	Time	Distance
First cyclist	16	t	$16t$
Second cyclist	18	t	$18t$

- The course is 51 mi.

Solution

$$\begin{aligned} 16t+18t&=51\\ 34t&=51\\ t&=1.5 \end{aligned}$$

The two cyclists will meet after 1.5 h.

55. Strategy

- Time for car: t
- Time for cyclist: $t+3$

	Rate	Time	Distance
Car	48	t	$48t$
Cyclist	12	$t+3$	$12(t+3)$

- The two vehicles travel the same distance.

Solution

$$\begin{aligned} 48t&=12(t+3)\\ 48t&=12t+36\\ 36t&=36\\ t&=1 \end{aligned}$$

$$\text{Distance} = 48t = 48(1) = 48$$

The car overtakes the cyclist 48 mi from the starting point.

Applying the Concepts

57. Strategy

- Amount of water evaporated: x

	Amount	Percent	Quantity
Water	x	0	$0(x)$
12% salt	50	0.12	$0.12(50)$
15% salt	$50-x$	0.15	$0.15(50-x)$

- The difference of the quantities after is equal to the quantity after evaporation.

Solution

$$\begin{aligned} 0.12(50)-0x&=0.15(50-x)\\ 6&=7.5-0.15x\\ -1.5&=-0.15x\\ 10&=x \end{aligned}$$

10 oz of water should be evaporated.

59. Strategy • The rate for 2nd mi: r

	Rate	Time	Distance
First mile	30	$\frac{1}{30}$	1
Second mile	r	$\frac{1}{r}$	1
Both miles	60	$\frac{2}{60}=\frac{1}{30}$	2

• The time traveled during the first mile plus the time traveled during the second mile is equal to the total time traveled during both miles.

Solution

$$\frac{1}{30}+\frac{1}{r}=\frac{1}{30}$$
$$\frac{1}{r}=0$$
$$r=0$$

There is no solution to this problem. It is not possible to increase the speed enough to average 60 mph.

Section 2.5

Objective A Exercises

1. The Addition Property of Inequalities states that the same number can be added to each side of an inequality without changing the solution set of the inequality.
Examples will vary. For instance:

$8>6$ and $-5<-1$
$8+7>6+7$ and $-5+(-2)<-1+(-2)$
$15>13$ and $-7<-3$

3. Replace x with each value to determine if the inequality holds.
i) $-17+7\le-3$; $-10\le-3$; solution
ii) $8+7\le-3$; $15\le-3$; not a solution
iii) $-10+7\le-3$; $-3\le-3$; solution
iv) $0+7\le-3$; $7\le-3$; not a solution

5. $x-3<2$
$x<5$
$\{x|\,x<5\}$

7. $4x\le8$
$$\frac{4x}{4}\le\frac{8}{4}$$
$x\le2$
$\{x|\,x\le2\}$

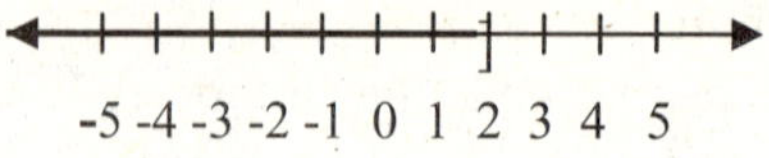

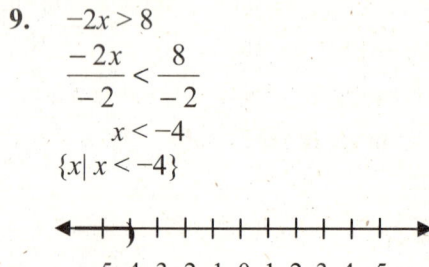

9. $-2x>8$
$$\frac{-2x}{-2}<\frac{8}{-2}$$
$x<-4$
$\{x|\,x<-4\}$

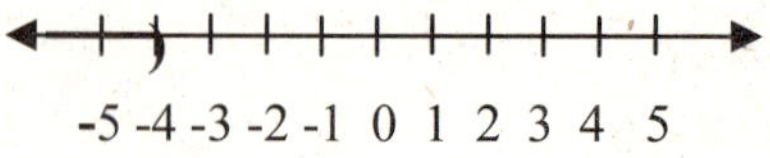

11. $3x-1>2x+2$
$x-1>2$
$x>3$
The solution set is $\{x|\,x>3\}$.

13. $2x-1>7$
$2x>8$
$$\frac{2x}{2}>\frac{8}{2}$$
$x>4$
The solution set is $\{x|\,x>4\}$.

15. $5x-2\le8$
$5x\le10$
$$\frac{5x}{5}\le\frac{10}{5}$$
$x\le2$
The solution set is $\{x|\,x\le2\}$.

17. $6x+3>4x-1$
$6x>4x-4$
$2x>-4$
$$\frac{2x}{2}>\frac{-4}{2}$$
$x>-2$
The solution set is $\{x|\,x>-2\}$.

19. $8x + 1 \geq 2x + 13$

$$6x + 1 \geq 13$$
$$6x \geq 12$$
$$\frac{6x}{6} \geq \frac{12}{6}$$
$$x \geq 2$$

The solution set is $\{x | x \geq 2\}$.

21. $4 - 3x < 10$

$$-3x < 6$$
$$\frac{-3x}{-3} > \frac{6}{-3}$$
$$x > -2$$

The solution set is $\{x | x > -2\}$.

23. $7 - 2x \geq 1$

$$-2x \geq -6$$
$$\frac{-2x}{-2} \leq \frac{-6}{-2}$$
$$x \leq 3$$

The solution set is $\{x | x \leq 3\}$.

25. $-3 - 4x > -11$

$$-4x > -8$$
$$\frac{-4x}{-4} < \frac{-8}{-4}$$
$$x < 2$$

The solution set is $\{x | x < 2\}$.

27. $4x - 2 < x - 11$

$$3x - 2 < -11$$
$$3x < -9$$
$$\frac{3x}{3} < \frac{-9}{3}$$
$$x < -3$$

The solution set is $\{x | x < -3\}$.

29. $x + 7 \geq 4x - 8$

$$-3x + 7 \geq -8$$
$$-3x \geq -15$$
$$\frac{-3x}{-3} \leq \frac{-15}{-3}$$
$$x \leq 5$$

The solution set is $\{x | x \leq 5\}$.

31. $3x + 2 \leq 7x + 4$

$$-4x + 2 \leq 4$$
$$-4x \leq 2$$
$$\frac{-4x}{-4} \geq \frac{2}{-4}$$
$$x \geq -\frac{1}{2}$$

The solution set is $\{x | x \geq -\frac{1}{2}\}$.

33. The solution to the inequality $nx > a$, where both n and a are negative contains both positive and negative numbers.

35. The solution to the inequality $x - n > -a$, where both n and a are positive and $n < a$ contains both positive and negative numbers.

37.

$$\frac{3}{5}x - 2 < \frac{3}{10} - x$$
$$10\left(\frac{3}{5}x - 2\right) < 10\left(\frac{3}{10} - x\right)$$
$$6x - 20 < 3 - 10x$$
$$16x - 20 < 3$$
$$16x < 23$$
$$\frac{16x}{16} < \frac{23}{16}$$
$$x < \frac{23}{16}$$

The solution is $\left(-\infty, \frac{23}{16}\right)$.

39. $\frac{2}{3}x-\frac{3}{2}<\frac{7}{6}-\frac{1}{3}x$

$6\left(\frac{2}{3}x-\frac{3}{2}\right)<6\left(\frac{7}{6}-\frac{1}{3}x\right)$

$4x-9<7-2x$

$6x-9<7$

$6x<16$

$\frac{6x}{6}<\frac{16}{6}$

$x<\frac{8}{3}$

The solution is $\left(-\infty, \frac{8}{3}\right)$.

41. $\frac{1}{2}x-\frac{3}{4}<\frac{7}{4}x-2$

$4\left(\frac{1}{2}x-\frac{3}{4}\right)<4\left(\frac{7}{4}x-2\right)$

$2x-3<7x-8$

$-5x-3<-8$

$-5x<-5$

$\frac{-5x}{-5}>\frac{-5}{-5}$

$x>1$

The solution is $(1, \infty)$.

43. $4(2x-1)>3x-2(3x-5)$

$8x-4>3x-6x+10$

$8x-4>-3x+10$

$11x-4>10$

$11x>14$

$\frac{11x}{11}>\frac{14}{11}$

$x>\frac{14}{11}$

The solution is $\left(\frac{14}{11}, \infty\right)$.

45. $2-5(x+1)\geq 3(x-1)-8$

$2-5x-5\geq 3x-3-8$

$-3-5x\geq 3x-11$

$-5x\geq 3x-8$

$-8x\geq -8$

$\frac{-8x}{-8}\leq\frac{-8}{-8}$

$x\leq 1$

The solution is $(-\infty, 1]$.

47. $3+2(x+5)\geq x+5(x+1)+1$

$3+2x+10\geq x+5x+5+1$

$2x+13\geq 6x+6$

$-4x+13\geq 6$

$-4x\geq -7$

$\frac{-4x}{-4}\leq\frac{-7}{-4}$

$x\leq\frac{7}{4}$

The solution is $\left(-\infty, \frac{7}{4}\right]$.

49. $3-4(x+2)\leq 6+4(2x+1)$

$3-4x-8\leq 6+8x+4$

$-4x-5\leq 10+8x$

$-12x-5\leq 10$

$-12x\leq 15$

$\frac{-12x}{-12}\geq\frac{15}{-12}$

$x\geq -\frac{5}{4}$

The solution is $\left[-\frac{5}{4}, \infty\right)$.

51. $12-2(3x-2)\geq 5x-2(5-x)$

$12-6x+4\geq 5x-10+2x$

$16-6x\geq 7x-10$

$-6x\geq 7x-26$

$-13x\geq -26$

$\frac{-13x}{-13}\leq\frac{-26}{-13}$

$x\leq 2$

The solution is $(-\infty, 2]$.

Objective B Exercises

53. Writing $-3 > x > 4$ does not make sense because there is no number that is less than -3 *and* greater than 4.

55. $x - 3 \le 1$ and $2x \ge -4$

$x \le 4$ $\quad x \ge -2$

$\{x \mid x \le 4\}$ $\quad \{x \mid x \ge -2\}$

$\{x \mid x \le 4\} \cap \{x \mid x \ge -2\} = [-2, 4]$

57. $2x < 6$ or $x - 4 > 1$

$x < 3$ $\quad x > 5$

$\{x \mid x < 3\}$ $\quad \{x \mid x > 5\}$

$\{x \mid x < 3\} \cup \{x \mid x > 5\} = (-\infty, 3) \cup (5, \infty)$

59. $\frac{1}{2}x > -2$ and $5x < 10$

$x > -4$ $\quad x < 2$

$\{x \mid x > -4\}$ $\quad \{x \mid x < 2\}$

$\{x \mid x > -4\} \cap \{x \mid x < 2\} = (-4, 2)$

61. $\frac{2}{3}x > 4$ or $2x < -8$

$x > 6$ $\quad x < -4$

$\{x \mid x > 6\}$ $\quad \{x \mid x < -4\}$

$\{x \mid x > 6\} \cup \{x \mid x < -4\} = (-\infty, -4) \cup (6, \infty)$

63. $3x < -9$ and $x - 2 < 2$

$x < -3$ $\quad x < 4$

$\{x \mid x < -3\}$ $\quad \{x \mid x < 4\}$

$\{x \mid x < -3\} \cap \{x \mid x < 4\} = (-\infty, -3)$

65. $2x - 3 > 1$ and $3x - 1 < 2$

$2x > 4$ $\quad 3x < 3$

$x > 2$ $\quad x < 1$

$\{x \mid x > 2\}$ $\quad \{x \mid x < 1\}$

$\{x \mid x > 2\} \cap \{x \mid x < 1\} = \emptyset$

67. $4x + 1 < 5$ and $4x + 7 > -1$

$4x < 4$ $\quad 4x > -8$

$x < 1$ $\quad x > -2$

$\{x \mid x < 1\}$ $\quad \{x \mid x > -2\}$

$\{x \mid x < 1\} \cap \{x \mid x > -2\} = (-2, 1)$

69. The inequality $x > -3$ or $x < 2$ describes all real numbers.

71. The inequality $x < -3$ or $x > 2$ describes two intervals of real numbers.

72. $6x - 2 < -14$ or $5x + 1 > 11$

$6x < -12$ $\quad 5x > 10$

$x < -2$ $\quad x > 2$

$\{x \mid x < -2\}$ $\quad \{x \mid x > 2\}$

$\{x \mid x < -2\} \cup \{x \mid x > 2\} = \{x \mid x < -2 \text{ or } x > 2\}$

75. $5 < 4x - 3 < 21$

$5 + 3 < 4x - 3 + 3 < 21 + 3$

$8 < 4x < 24$

$\frac{8}{4} < \frac{4x}{4} < \frac{24}{4}$

$2 < x < 6$

$\{x \mid 2 < x < 6\}$

77. $-2 < 3x + 7 < 1$

$-2 + (-7) < 3x + 7 + (-7) < 1 + (-7)$

$-9 < 3x < -6$

$\frac{-9}{3} < \frac{3x}{3} < \frac{-6}{3}$

$-3 < x < -2$

$\{x \mid -3 < x < -2\}$

79. $3x - 5 > 10$ or $3x - 5 < -10$

$3x > 15$ $\quad 3x < -5$

$x > 5$ $\quad x < -\frac{5}{3}$

$\{x \mid x > 5\}$ $\quad \{x \mid x < -\frac{5}{3}\}$

$\{x \mid x > 5\} \cup \{x \mid x < -\frac{5}{3}\} = \{x \mid x > 5 \text{ or } x < -\frac{5}{3}\}$

81. $8x + 2 \le -14$ and $4x - 2 > 10$

$8x \le -16$ $\quad 4x > 12$

$x \le -2$ $\quad x > 3$

$\{x \mid x \le -2\}$ $\quad \{x \mid x > 3\}$

$\{x \mid x \le -2\} \cap \{x \mid x > 3\} = \emptyset$

83. $5x + 12 \ge 2$ or $7x - 1 \le 13$

$5x \ge -10$ $\quad 7x \le 14$

$x \ge -2$ $\quad x \le 2$

$\{x \mid x \ge -2\}$ $\quad \{x \mid x \le 2\}$

$\{x \mid x \ge -2\} \cup \{x \mid x \le 2\} =$ the set of real

numbers

85. $3 \le 7x - 14 \le 31$

$3 + 14 \le 7x - 14 + 14 \le 31 + 14$

$17 \le 7x \le 45$

$\frac{17}{7} \le \frac{7x}{7} \le \frac{45}{7}$

$\frac{17}{7} \le x \le \frac{45}{7}$

$\{x \mid \frac{17}{7} \le x \le \frac{45}{7}\}$

87. $1 - 3x < 16$ and $1 - 3x > -16$

$-3x < 15$ $\quad -3x > -17$

$x > -5$ $\quad x < \frac{-17}{-3}$

$\{x \mid x > -5\}$ $\quad \{x \mid x < \frac{17}{3}\}$

$\{x \mid x > -5\} \cap \{x \mid x < \frac{17}{3}\} = \{x \mid -5 < x < \frac{17}{3}\}$

89. $6x + 5 < -1$ or $1 - 2x < 7$

$6x < -6$ $\quad -2x < 6$

$x < -1$ $\quad x > -3$

$\{x \mid x < -1\}$ $\quad \{x \mid x > -3\}$

$\{x \mid x < -1\} \cup \{x \mid x > -3\} =$ The set of real

numbers.

91. $9 - x \ge 7$ and $9 - 2x < 3$

$-x \ge -2$ $\quad -2x < -6$

$x \le 2$ $\quad x > 3$

$\{x \mid x \le 2\}$ $\quad \{x \mid x > 3\}$

$\{x \mid x \le 2\} \cap \{x \mid x > 3\} = \emptyset$

Objective C Exercises

93. The temperature did not go above 42°F can be written as $t \le 42$.

95. The high temperature was 42°F can be written as $t \le 42$.

97. Strategy: Let x represent the smallest number.

Solution:

Two times the difference between a number and eight	$\leq$	Five times the sum of the number and four.

$$2(x-8) \leq 5(x+4)$$
$$2x-16 \leq 5x+20$$
$$-3x-16 \leq 20$$
$$-3x \leq 36$$
$$x \geq -12$$

The smallest number is −12.

99. Strategy: Let x represent the width of the rectangle.
The length of the rectangle is $2x-5$.
To find the maximum width solve the inequality $2L+2W<60$.

Solution: $2L+2W<60$
$$2(2x-5)+2x<60$$
$$4x-10+2x<60$$
$$6x-10<60$$
$$6x<70$$
$$x<\frac{70}{6}=11\frac{2}{3}$$

The maximum width of the rectangle is 11 cm.

101. Strategy: Let d represent the number of days to run advertisement.
To find the maximum number of days the advertisement can run on the website solve the inequality $250+12d \leq 1500$.

Solution: $250+12d \leq 1500$
$$12d \leq 1250$$
$$d \leq \frac{1250}{12}$$
$$d \leq 104\frac{1}{6}$$

You can run the advertisement for 104 days.

103. Strategy: Let x represent the cost of a gallon of paint.
Since a gallon of paint covers 100 square feet and the room is 320 square feet the homeowner will need to buy 4 gallons of paint.
To find the maximum cost per gallon solve the inequality $24+4x \leq 100$.

Solution: $24+4x \leq 100$
$$4x \leq 76$$
$$x \leq 19$$

The maximum that the homeowner can pay for a gallon of paint is $19.

105. Strategy: To find the temperature range in Fahrenheit degrees solve the compound inequality $0<\frac{5(F-32)}{9}<30$.

Solution: $0<\frac{5(F-32)}{9}<30$
$$\frac{9}{5}(0)<\frac{9}{5}\left(\frac{5(F-32)}{9}\right)<\frac{9}{5}(30)$$
$$0<F-32<54$$
$$0+32<F-32+32<54+32$$
$$32^{\circ}<F<86^{\circ}$$

107. Strategy: Let N represent the amount of sales.
To find the minimum amount of sales solve the inequality $1000+0.05N \geq 3200$.

Solution: $1000+0.05N \geq 3200$
$$0.05N \geq 2200$$
$$N \geq 44{,}000$$

George's amount of sales must be $44,000 or more per month.

109. Strategy: Let N represent the number of checks.
To find the maximum number of checks solve the inequality
$8 + 0.12(N - 100) < 5 + 0.15(N - 100)$.

Solution:

$$8 + 0.12(N - 100) < 5 + 0.15(N - 100)$$
$$8 + 0.12N - 12 < 5 + 0.15N - 15$$
$$0.12N - 4 < 0.15N - 10$$
$$0.12N + 6 < 0.15N$$
$$6 < 0.03N$$
$$200 < N$$

The first account is less expensive when you write more than 200 checks.

111. Strategy: Let n represent the score on the last test.
To find the range of scores solve the inequality
$70 \leq \frac{56+91+83+62+n}{5} \leq 79$.

Solution:

$$70 \leq \frac{56+91+83+62+n}{5} \leq 79$$
$$70 \leq \frac{292+n}{5} \leq 79$$
$$5(70) \leq 5 \cdot \frac{292+n}{5} \leq 5(79)$$
$$350 \leq 292 + n \leq 395$$
$$350 - 292 \leq 292 - 292 + n \leq 395 - 292$$
$$58 \leq n \leq 103$$

Since 100 is the maximum core, the range of scores needed to receive an C grade is $58 \leq n \leq 100$.

113. Strategy: Let x represent the first even integer.
To find the three consecutive even integers solve the inequality
$30 < x + (x + 2) + (x + 4) < 5$.

Solution:

$$30 < x + (x + 2) + (x + 4) < 51$$
$$30 < 3x + 6 < 51$$
$$30 - 6 < 3x + 6 - 6 < 51 - 6$$
$$24 < 3x < 45$$
$$\frac{24}{3} < \frac{3x}{3} < \frac{45}{3}$$
$$8 < x < 15$$

The four consecutive integers are 10, 12 and 14; 12, 14 and 16; 14, 16 and 18.

Applying the Concepts

115. a) Always true
b) Sometimes true
c) Sometimes true
d) Sometimes true
e) Always true

Section 2.6

Objective A Exercises

1. $|2 - 8| = 6$
$|-6| = 6$
$6 = 6$
Yes, 2 is a solution.

3. $|3(-1) - 4| = 7$
$|-3 - 4| = 7$
$|-7| = 7$
$7 = 7$
Yes, −1 is a solution.

5. $|x| = 7$
$x = 7$ or $x = -7$
The solutions are 7 and – 7.

7. $|b| = 4$
$b = 4$ or $b = -4$
The solutions are 4 and – 4.

9. $|-y| = 6$
$-y = 6$ or $-y = -6$
$y = -6$ or $y = 6$
The solutions are 6 and – 6.

11. $|-a| = 7$
$-a = 7$ or $-a = -7$
$a = -7$ or $a = 7$
The solutions are 7 and – 7.

13. $|x| = -4$
There is no solution to this equation because the absolute value of a number must be nonnegative.

15. $|-t| = -3$
There is no solution to this equation because the absolute value of a number must be nonnegative.

17. $|x + 2| = 3$
$x + 2 = 3$ or $x + 2 = -3$
$x = 1$ $\quad$ $x = -5$
The solutions are 1 and – 5.

19. $|y - 5| = 3$
$y - 5 = 3$ or $y - 5 = -3$
$y = 8$ $\quad$ $y = 2$
The solutions are 2 and 8.

21. $|a - 2| = 0$
$a - 2 = 0$
$a = 2$
The solution is 2.

23. $|x - 2| = -4$
There is no solution to this equation because the absolute value of a number must be nonnegative.

25. $|3 - 4x| = 9$
$3 - 4x = 9$ or $3 - 4x = -9$
$-4x = 6$ $\quad$ $-4x = -12$
$x = -\frac{3}{2}$ $\quad$ $x = 3$

The solutions are 3 and $-\frac{3}{2}$.

27. $|2x - 3| = 0$
$2x - 3 = 0$
$2x = 3$
$x = \frac{3}{2}$

The solution is $\frac{3}{2}$.

29. $|3x - 2| = -4$
There is no solution to this equation because the absolute value of a number must be nonnegative.

31. $|x - 2| - 2 = 3$
$|x - 2| = 5$
$x - 2 = 5$ or $x - 2 = -5$
$x = 7$ $\quad$ $x = -3$
The solutions are 7 and – 3.

33. $||3a + 2| - 4 = 4$
$|3a + 2| = 8$
$3a + 2 = 8$ or $3a + 2 = -8$
$3a = 6$ $\quad$ $3a = -10$
$a = 2$ $\quad$ $a = -\frac{10}{3}$

The solutions are 2 and $-\frac{10}{3}$.

35. $|2 - y| + 3 = 4$
$|2 - y| = 1$
$2 - y = 1$ or $2 - y = -1$
$-y = -1$ $\quad$ $-y = -3$
$y = 1$ $\quad$ $y = 3$
The solutions are 1 and 3.

37. $|2x - 3| + 3 = 3$
$|2x - 3| = 0$
$2x - 3 = 0$
$2x = 3$
$x = \frac{3}{2}$

The solution is $\frac{3}{2}$.

39. $|2x - 3| + 4 = -4$
$|2x - 3| = -8$
There is no solution to this equation because the absolute value of a number must be nonnegative.

41. $|6x-5|-2=4$

$|6x-5|=6$

$6x-5=6$ or $6x-5=-6$

$6x=11$ $6x=-1$

$x=\frac{11}{6}$ $x=-\frac{1}{6}$

The solutions are $\frac{11}{6}$ and $-\frac{1}{6}$.

43. $|3t+2|+3=4$

$|3t+2|=1$

$3t+2=1$ or $3t+2=-1$

$3t=-1$ $3t=-3$

$t=-\frac{1}{3}$ $t=-1$

The solutions are $-\frac{1}{3}$ and -1.

45. $3-|x-4|=5$

$-|x-4|=2$

$|x-4|=-2$

There is no solution to this equation because the absolute value of a number must be nonnegative.

47. $8-|2x-3|=5$

$-|2x-3|=-3$

$|2x-3|=3$

$2x-3=3$ or $2x-3=-3$

$2x=6$ $2x=0$

$x=3$ $x=0$

The solutions are 3 and 0.

49. $|2-3x|+7=2$

$|2-3x|=-5$

There is no solution to this equation because the absolute value of a number must be nonnegative.

51. $|8-3x|-3=2$

$|8-3x|=5$

$8-3x=5$ or $8-3x=-5$

$-3x=-3$ $-3x=-13$

$x=1$ $x=\frac{13}{3}$

The solutions are 1 and $\frac{13}{3}$.

53. $|2x-8|+12=2$

$|2x-8|=-10$

There is no solution to this equation because the absolute value of a number must be nonnegative.

55. $2+|3x-4|=5$

$|3x-4|=3$

$3x-4=3$ or $3x-4=-3$

$3x=7$ $3x=1$

$x=\frac{7}{3}$ $x=\frac{1}{3}$

The solutions are $\frac{7}{3}$ and $\frac{1}{3}$.

57. $5-|2x+1|=5$

$-|2x+1|=0$

$2x+1=0$

$2x=-1$

$x=-\frac{1}{2}$

The solution is $-\frac{1}{2}$.

59. $6-|2x+4|=3$

$-|2x+4|=-3$

$|2x+4|=3$

$2x+4=3$ or $2x+4=-3$

$2x=-1$ $2x=-7$

$x=-\frac{1}{2}$ $x=-\frac{7}{2}$

The solutions are $-\frac{1}{2}$ and $-\frac{7}{2}$.

61. $8-|1-3x|=-1$

$-|1-3x|=-9$

$|1-3x|=9$

$1-3x=9$ $1-3x=-9$

$-3x=8$ or $-3x=-10$

$x=-\frac{8}{3}$ $x=\frac{10}{3}$

The solutions are $-\frac{8}{3}$ and $\frac{10}{3}$.

63. $5 + |2 - x| = 3$
$|2 - x| = -2$
There is no solution to this equation because the absolute value of a number must be nonnegative.

65. Two positive solutions.

67. Two negative solutions.

Objective B Exercises

69. $|x| > 3$
$x > 3$ or $x < -3$
$\{x \mid x > 3\}$ $\{x \mid x < -3\}$

$\{x \mid x > 3\} \cup \{x \mid x < -3\} = \{x \mid x > 3 \text{ or } x <$

$-3\}$

71. $|x + 1| > 2$
$x + 1 > 2$ or $x + 1 < -2$
$x > 1$ $x < -3$
$\{x \mid x > 1\}$ $\{x \mid x < -3\}$

$\{x \mid x > 1\} \cup \{x \mid x < -3\} = \{x \mid x > 1 \text{ or } x <$

$-3\}$

73. $|x - 5| \leq 1$
$-1 \leq x - 5 \leq 1$
$-1 + 5 \leq x - 5 + 5 \leq 1 + 5$
$4 \leq x \leq 6$
$\{x \mid 4 \leq x \leq 6\}$

75. $|2 - x| \geq 3$
$2 - x \geq 3$ or $2 - x \leq -3$
$-x \geq 1$ $-x \leq -5$
$x \leq -1$ $x \geq 5$
$\{x \mid x \leq -1\}$ $\{x \mid x \geq 5\}$

$\{x \mid x \leq -1\} \cup \{x \mid x \geq 5\} = \{x \mid x \leq -1 \text{ or } x \geq 5\}$

77. $|2x + 1| < 5$
$-5 < 2x + 1 < 5$
$-5 - 1 < 2x + 1 - 1 < 5 - 1$
$-6 < 2x < 4$
$-3 < x < 2$
$\{x \mid -3 < x < 2\}$

79. $|5x + 2| > 12$
$5x + 2 > 12$ or $5x + 2 < -12$
$5x > 10$ $5x < -14$
$x > 2$ $x < -\frac{14}{5}$
$\{x \mid x > 2\}$ $\{x \mid x < -\frac{14}{5}\}$

$\{x \mid x > 2\} \cup \{x \mid x < -\frac{14}{5}\}$

$= \{x \mid x > 2 \text{ or } x < -\frac{14}{5}\}$

81. $|4x - 3| \leq -2$
The absolute value of a number must be nonnegative. The solution set is the empty

set ∅.

83. $|2x + 7| > -5$
$2x + 7 > -5$ or $2x + 7 < 5$
$2x > -12$ $2x < -2$
$x > -6$ $x < -1$
$\{x \mid x > -6\}$ $\{x \mid x < -1\}$

$\{x \mid x > -6\} \cup \{x \mid x < -1\}$= The set of all real

numbers.

85. $|4 - 3x| \geq 5$
$4 - 3x \geq 5$ or $4 - 3x \leq -5$
$-3x \geq 1$ $-3x \leq -9$
$x \leq -\frac{1}{3}$ $x \geq 3$

$\{x \mid x \le -\frac{1}{3}\}$ $\{x \mid x \ge 3\}$

$\{x \mid x \le -\frac{1}{3}\} \cup \{x \mid x \ge 3\}$

$= \{x \mid x \le -\frac{1}{3} \text{ or } x \ge 3\}$

87. $|5 - 4x| \le 13$

$-13 \le 5 - 4x \le 13$

$-13 + (-5) \le 5 + (-5) - 4x \le 13 + (-5)$

$-18 \le -4x \le 8$

$\frac{18}{4} \ge x \ge -2$

$\{x \mid -2 \le x \le \frac{9}{2}\}$

89. $|6 - 3x| \le 0$

$0 \le 6 - 3x \le 0$

$-6 \le -3x \le -6$

$2 \le x \le 2$

$2 \le x \le 2 = \{x \mid x = 2\}$

91. $|2 - 9x| > 20$

$2 - 9x > 20$ or $2 - 9x < -20$

$-9x > 18$ $-9x < -22$

$x < -2$ $x > \frac{22}{9}$

$\{x \mid x < -2\}$ $\{x \mid x > \frac{22}{9}\}$

$\{x \mid x < -2\} \cup \{x \mid x > \frac{22}{9}\}$

$= \{x \mid x < -2 \text{ or } x > \frac{22}{9}\}$

93. $|2x - 3| + 2 < 8$

$|2x - 3| < 6$

$-6 < 2x - 3 < 6$

$-6 + 3 < 2x - 3 + 3 < 6 + 3$

$-3 < 2x < 9$

$-\frac{3}{2} < x < \frac{9}{2}$

$\{x \mid -\frac{3}{2} < x < \frac{9}{2}\}$

95. $|2 - 5x| - 4 > -2$

$|2 - 5x| > 2$

$2 - 5x > 2$ or $2 - 5x < -2$

$-5x > 0$ $-5x < -4$

$x < 0$ $x > \frac{4}{5}$

$\{x \mid x < 0\}$ $\{x \mid x > \frac{4}{5}\}$

$\{x \mid x < 0\} \cup \{x \mid x > \frac{4}{5}\} = \{x \mid x < 0 \text{ or } x > \frac{4}{5}\}$

97. $8 - |2x - 5| < 3$

$-|2x - 5| < -5$

$|2x - 5| > 5$

$2x - 5 < -5$ or $2x - 5 > 5$

$2x < 0$ $2x > 10$

$x < 0$ $x > 5$

$\{x \mid x < 0\}$ $\{x \mid x > 5\}$

$\{x \mid x < 0\} \cup \{x \mid x > 5\} = \{x \mid x < 0 \text{ or } x > 5\}$

99. All negative solutions.

Objective C Exercises

101. The desired dosage is 3 ml. The tolerance is 0.2 ml.

103. Strategy: Let d represent the diameter of the bushing, T the tolerance and x the lower and upper limits of the diameter.
Solve the absolute value inequality $|x - d| \le T$.

Solution: $|x - d| \le T$

$|x - 1.75| \le 0.008$

$-0.008 \le x - 1.75 \le 0.008$

$-0.008 + 1.75 \le x - 1.75 + 1.75 \le 0.008 + 1.75$

$1.742 \le x \le 1.758$

The lower and upper limits of the diameter of the bushing are 1.742 in. and 1.758 in.

105. Strategy: Let x represent the percent of American voters who felt the economy is an important issue.
Solve the absolute value inequality $|x-41| \le 3$.

Solution: $|x-41| \le 3$

$$-3 \le x-41 \le 3$$
$$-3+41 \le x-41+41 \le 3+41$$
$$38 \le x \le 44$$

The lower and upper limits of American voters who felt the economy is an important issue 38% and 44%.

107. Strategy: Let L represent the length of the piston.
Solve the absolute value inequality $|L-9\frac{5}{8}| \le \frac{1}{32}$.

Solution: $|L-9\frac{5}{8}| \le \frac{1}{32}$

$$-\frac{1}{32} \le L-9\frac{5}{8} \le \frac{1}{32}$$
$$-\frac{1}{32}+9\frac{5}{8} \le L-9\frac{5}{8}+9\frac{5}{8} \le \frac{1}{32}+9\frac{5}{8}$$
$$9\frac{19}{32} \le L \le 9\frac{21}{32}$$

The upper and lower limits of the length of the piston are $9\frac{19}{32}$ in. and $9\frac{21}{32}$ in.

109. Strategy: Let M represent the range, in ohms, for a resistor.
Let T represent the tolerance of the resistor.
Solve the absolute value inequality $|M-29{,}000| \le T$.

Solution: $T=(0.02)(29{,}000)$
$= 580$ ohm

$$|M-29{,}000| \le 580$$
$$-580 \le M-29{,}000 \le 580$$
$$-580+29{,}000 \le M-29{,}000+29{,}000 \le 580+29{,}000$$
$$28{,}420 \le M \le 29{,}580$$

The upper and lower limits of the resistor are 28,420 ohms and 29,580 ohms.

Applying the Concepts

111. a) The equation $|x+3| = x+3$ is true for all x for which $x+3 \ge 0$.

$$x+3 \ge 0$$
$$x \ge -3$$
$$\{x \mid x \ge -3\}$$

b) The equation $|a-4| = 4-a$ is true for all a for which $4-a \ge 0$.

$$4-a \ge 0$$
$$-a \ge -4$$
$$a \le 4$$
$$\{a \mid a \le 4\}$$

113. a) $|x+y| \le |x|+|y|$
b) $|x-y| \ge |x|-|y|$
c) $||x|-|y|| \ge |x|-|y|$
d) $|xy| = |x||y|$

Concept Review

1. The Addition Property of Equations is used to remove a term from one side of an equation by adding the opposite of that term to each side of the equation. [2.1B]

2. To check the solution of an equation, substitute the proposed solution for the variable in the original equation. Simplify the resulting numerical expressions. If the left and right sides are equal, the proposed solution is a solution of the equation. If the left and right sides are not equal, the proposed solution is not a solution of the equation. [2.1B]

3. To solve an equation containing parentheses, first use the Distributive Property to remove the parentheses. Combine like terms on each side of the equation. Use the Addition Property of Equations to rewrite the equation with only one variable term. Then use the Addition Property of Equations to rewrite the equation with only one constant term. Use the Multiplication Property of Equations to rewrite the equation in the form *variable* = *constant.* The constant is the solution. [2.2C]

4. Consecutive integers differ by 1. For example, 8 and 9 are consecutive integers. Consecutive even integers are even integers that differ by 2. For example, 8 and 10 are consecutive even integers. [2.3A]

5. When mixing a 15% solution with a 20% solution, the percent concentration of the resulting solution must be between 15% and 20%. [2.4B]

6. In solving a percent mixture problem use the formula $Ar = Q$, where A is the amount of the solution or alloy, r is the percent of concentration, and Q is the quantity of substance in the solution or alloy. [2.4B]

7. In solving a uniform motion problem, use the formula $rt = d$, where r is the rate of travel, t is the time spent traveling, and d is the distance traveled. [2.4C]

8. The Multiplication Property of Equations states that both sides of an equation can be multiplied by the same nonzero number without changing the solution to the equation. The Multiplication Property of Inequalities consists of two rules: (1) Each side of an inequality can be multiplied by the same positive number, without changing the solution set; (2) If each side of an inequality is multiplied by the same negative number, the inequality must be reversed in order to keep the solution set of the inequality unchanged. [2.5A]

9. The compound inequality $-c < ax + b < c$ is equivalent to an absolute value inequality in the form $|ax + b| < c,\ c > 0$ [2.6B]

10. The compound inequality $ax + b < -c$ or $ax + b > c$ is equivalent to an absolute value inequality in of the form $|ax + b| > c.$ [2.6B]

11. To check the solution to an absolute value equation, substitute the proposed solution for the variable in the original equation. Simplify the resulting numerical expressions. If the left and right sides are equal, the proposed solution is a solution of the equation. If the left and right sides are not equal, the proposed solution is not a solution. [2.6A]

Chapter 2 Review Exercises

1.
$$\begin{aligned} 3t - 3 + 2t &= 7t - 15 \\ 5t - 3 &= 7t - 15 \\ 5t - 7t - 3 &= 7t - 7t - 15 \\ -2t - 3 &= -15 \\ -2t - 3 + 3 &= -15 + 3 \\ -2t &= -12 \\ \frac{-2t}{-2} &= \frac{-12}{-2} \\ t &= 6 \end{aligned}$$

The solution is 6.

2.
$$3x-7>-2$$
$$3x-7+7>-2+7$$
$$3x>5$$
$$\frac{3x}{3}>\frac{5}{3}$$
$$x>\frac{5}{3}$$

The solution set is $\left\{x \mid x>\frac{5}{3}\right\}$.

3.

$5x-2=4x+5$	
$5(3)-2$	$4(3)+5$
$15-2$	$12+5$

$$13\neq 17$$

No, 3 is not a solution.

4.
$$x+4=-5$$
$$x+4-4=-5-4$$
$$x=-9$$

The solution is –9.

5.
$$3x<4 \qquad\qquad x+2>-1$$
$$\frac{3x}{3}<\frac{4}{3} \qquad\qquad x+2-2>-1-2$$
$$x<\frac{4}{3} \quad \text{and} \quad x>-3$$
$$\left\{x \mid x<\frac{4}{3}\right\}\cup\{x \mid x>-3\}=\left\{x \mid -3<x<\frac{4}{3}\right\}$$

The solution is set is $\left\{x \mid -3<x<\frac{4}{3}\right\}$.

6.
$$\frac{3}{5}x-3=2x+5$$
$$5\left(\frac{3}{5}x-3\right)=5(2x+5)$$
$$3x-15=10x+25$$
$$3x-10x-15=10x-10x+25$$
$$-7x-15=25$$
$$-7x-15+15=25+15$$
$$-7x=40$$
$$\frac{-7x}{-7}=\frac{40}{-7}$$
$$x=-\frac{40}{7}$$

The solution is $-\frac{40}{7}$.

7.
$$-\frac{2}{3}x=\frac{4}{9}$$
$$-\frac{3}{2}\left(-\frac{2}{3}x\right)=-\frac{3}{2}\left(\frac{4}{9}\right)$$
$$x=-\frac{2}{3}$$

The solution is $-\frac{2}{3}$.

8.
$$|x-4|-8=-3$$
$$|x-4|-8+8=-3+8$$
$$|x-4|=5$$
$$x-4=5 \qquad\qquad x-4=-5$$
$$x-4+4=5+4 \qquad\qquad x-4+4=-5+4$$
$$x=9 \qquad\qquad x=-1$$

The solutions are 9 and –1.

9.
$$|2x-5|<3$$
$$-3<2x-5<3$$
$$-3+5<2x-5+5<3+5$$
$$2<2x<8$$
$$\frac{2}{2}<\frac{2x}{2}<\frac{8}{2}$$
$$1<x<4$$

The solution set is $\{x \mid 1<x<4\}$.

10.
$$\frac{2x-3}{3}+2=\frac{2-3x}{5}$$
$$15\left(\frac{2x-3}{3}+2\right)=15\left(\frac{2-3x}{5}\right)$$
$$5(2x-3)+30=3(2-3x)$$
$$10x-15+30=6-9x$$
$$10x+15=6-9x$$
$$10x+9x+15=6-9x+9x$$
$$19x+15=6$$
$$19x+15-15=6-15$$
$$19x=-9$$
$$\frac{19x}{19}=\frac{-9}{19}$$
$$x=-\frac{9}{19}$$

The solution is $-\frac{9}{19}$.

11. $2(a-3)=5(4-3a)$

$2a-6=20-15a$

$2a+15a-6=20-15a+15a$

$17a-6=20$

$17a-6+6=20+6$

$17a=26$

$\frac{17a}{17}=\frac{26}{17}$

$a=\frac{26}{17}$

The solution is $\frac{26}{17}$.

12. $5x-2>8$ or $3x+2<-4$

$5x-2+2>8+2$ $\quad$ $3x+2-2<-4-2$

$5x>10$ or $3x<-6$

$\frac{5x}{5}>\frac{10}{5}$ $\quad$ $\frac{3x}{3}<\frac{-6}{3}$

$x>2$ $\quad$ $x<-2$

$\{x\,|\,x>2\}\cup\{x\,|\,x<-2\}=\{x\,|\,x>2 \text{ or } x<-2\}$

The solution set is $\{x\,|\,x>2 \text{ or } x<-2\}$,

13. $|4x-5|\ge 3$

$4x-5\ge 3$ $\quad$ $4x-5\le -3$

$4x-5+5\ge 3+5$ $\quad$ $4x-5+5\le -3+5$

$4x\ge 8$ or $4x\le 2$

$\frac{4x}{4}\ge\frac{8}{4}$ $\quad$ $\frac{4x}{4}\le\frac{2}{4}$

$x\ge 2$ $\quad$ $x\le\frac{1}{2}$

$\{x\,|\,x\ge 2\}\cup\left\{x\,|\,x\le\frac{1}{2}\right\}=\left\{x\,|\,x\ge 2 \text{ or } x\le\frac{1}{2}\right\}$

The solution set is $\left\{x\,|\,x\ge 2 \text{ or } x\le\frac{1}{2}\right\}$.

14. $P(12)=30$

$\frac{P(12)}{12}=\frac{30}{12}$

$P=2.5$

The percent is 250%.

15. $\frac{1}{2}x-\frac{5}{8}=\frac{3}{4}x+\frac{3}{2}$

$8\left(\frac{1}{2}x-\frac{5}{8}\right)=8\left(\frac{3}{4}x+\frac{3}{2}\right)$

$4x-5=6x+12$

$4x-6x-5=6x-6x+12$

$-2x-5=12$

$-2x-5+5=12+5$

$-2x=17$

$\frac{-2x}{-2}=\frac{17}{-2}$

$x=-\frac{17}{2}$

The solution is $-\frac{17}{2}$.

16. $6+|3x-3|=2$

$6-6+|3x-3|=2-6$

$|3x-3|=-4$

There is no solution to this equation because the absolute value of a number must be non-negative.

17. $3x-2>x-4$ or $7x-5<3x+3$

$3x-x-2>x-x-4$ $\quad$ $7x-3x-5<3x-3x+3$

$2x-2>-4$ $\quad$ $4x-5<3$

$2x-2+2>-4+2$ or $4x-5+5<3+5$

$2x>-2$ $\quad$ $4x<8$

$\frac{2x}{2}>\frac{-2}{2}$ $\quad$ $\frac{4x}{4}<\frac{8}{4}$

$x>-1$ $\quad$ $x<2$

$\{x\,|\,x>-1\}\cup\{x\,|\,x<2\}$

$=\{x\,|\,x \text{ is any real number}\}$

The solution set is $\{x\,|\,x \text{ is any real number}\}$.

18. $2x-(3-2x)=4-3(4-2x)$

$2x-3+2x=4-12+6x$

$4x-3=6x-8$

$4x-6x-3=6x-6x-8$

$-2x-3=-8$

$-2x-3+3=-8+3$

$-2x=-5$

$\frac{-2x}{-2}=\frac{-5}{-2}$

$x=\frac{5}{2}$

The solution is $\frac{5}{2}$.

19.
$$\begin{aligned} x+9&=-6\\ x+9-9&=-6-9\\ x&=-15 \end{aligned}$$
The solution is –15.

20.
$$\begin{aligned} \frac{2}{3}&=x+\frac{3}{4}\\ \frac{2}{3}-\frac{3}{4}&=x+\frac{3}{4}-\frac{3}{4}\\ \frac{8}{12}-\frac{9}{12}&=x\\ -\frac{1}{12}&=x \end{aligned}$$
The solution is $-\frac{1}{12}$.

21.
$$\begin{aligned} -3x&=-21\\ \frac{-3x}{-3}&=\frac{-21}{-3}\\ x&=7 \end{aligned}$$
The solution is 7.

22.
$$\begin{aligned} \frac{2}{3}a&=\frac{4}{9}\\ \frac{3}{2}\left(\frac{2}{3}a\right)&=\frac{3}{2}\left(\frac{4}{9}\right)\\ a&=\frac{2}{3} \end{aligned}$$
The solution is $\frac{2}{3}$.

23.
$$\begin{aligned} 3y-5&=3-2y\\ 3y+2y-5&=3-2y+2y\\ 5y-5&=3\\ 5y-5+5&=3+5\\ 5y&=8\\ \frac{5y}{5}&=\frac{8}{5}\\ y&=\frac{8}{5} \end{aligned}$$
The solution is $\frac{8}{5}$.

24.
$$\begin{aligned} 4x-5+x&=6x-8\\ 5x-5&=6x-8\\ 5x-6x-5&=6x-6x-8\\ -x-5&=-8\\ -x-5+5&=-8+5\\ -x&=-3\\ \frac{-x}{-1}&=\frac{-3}{-1}\\ x&=3 \end{aligned}$$
The solution is 3.

25.
$$\begin{aligned} 3(x-4)&=-5(6-x)\\ 3x-12&=-30+5x\\ 3x-5x-12&=-30+5x-5x\\ -2x-12&=-30\\ -2x-12+12&=-30+12\\ -2x&=-18\\ \frac{-2x}{-2}&=\frac{-18}{-2}\\ x&=9 \end{aligned}$$
The solution is 9.

26.
$$\begin{aligned} \frac{3x-2}{4}+1&=\frac{2x-3}{2}\\ 4\left(\frac{3x-2}{4}+1\right)&=4\left(\frac{2x-3}{2}\right)\\ 3x-2+4&=4x-6\\ 3x+2&=4x-6\\ 3x-4x+2&=4x-4x-6\\ -x+2&=-6\\ -x+2-2&=-6-2\\ -x&=-8\\ \frac{-x}{-1}&=\frac{-8}{-1}\\ x&=8 \end{aligned}$$
The solution is 8.

27.
$$\begin{aligned} 5x-8&<-3\\ 5x-8+8&<-3+8\\ 5x&<5\\ \frac{5x}{5}&<\frac{5}{5}\\ x&<1 \end{aligned}$$
The solution set is $(-\infty,1)$.

28.
$$\begin{aligned} 2x-9 &\le 8x+15 \\ 2x-8x-9 &\le 8x-8x+15 \\ -6x-9 &\le 15 \\ -6x-9+9 &\le 15+9 \\ -6x &\le 24 \\ \frac{-6x}{-6} &\ge \frac{24}{-6} \\ x &\ge -4 \end{aligned}$$

The solution set is $[-4,\infty)$.

29.
$$\begin{aligned} \frac{2}{3}x-\frac{5}{8} &\ge \frac{3}{4}x+1 \\ 24\left(\frac{2}{3}x-\frac{5}{8}\right) &\ge 24\left(\frac{3}{4}x+1\right) \\ 16x-15 &\ge 18x+24 \\ 16x-18x-15 &\ge 18x-18x+24 \\ -2x-15 &\ge 24 \\ -2x-15+15 &\ge 24+15 \\ -2x &\ge 39 \\ \frac{-2x}{-2} &\le \frac{39}{-2} \\ x &\le -\frac{39}{2} \end{aligned}$$

The solution set is $\left\{x \mid x \le -\frac{39}{2}\right\}$.

30.
$$\begin{aligned} 2-3(2x-4) &\le 4x-2(1-3x) \\ 2-6x+12 &\le 4x-2+6x \\ -6x+14 &\le 10x-2 \\ -6x-10x+14 &\le 10x-10x-2 \\ -16x+14 &\le -2 \\ -16x+14-14 &\le -2-14 \\ -16x &\le -16 \\ \frac{-16x}{-16} &\ge \frac{-16}{-16} \\ x &\ge 1 \end{aligned}$$

The solution set is $[1,\infty)$.

31.
$$\begin{aligned} -5 < 4x-1 &< 7 \\ -5+1 < 4x-1+1 &< 7+1 \\ -4 < 4x &< 8 \\ \frac{-4}{4} < \frac{4x}{4} &< \frac{8}{4} \\ -1 < x &< 2 \end{aligned}$$

The solution set is $\{x \mid -1 < x < 2\}$.

32. $|2x-3| = 8$

$$\begin{aligned} 2x-3 &= 8 & \quad\text{or}\quad 2x-3 &= -8 \\ 2x-3+3 &= 8+3 & 2x-3+3 &= -8+3 \\ 2x &= 11 & 2x &= -5 \\ \frac{2x}{2} &= \frac{11}{2} & \frac{2x}{2} &= \frac{-5}{2} \\ x &= \frac{11}{2} & x &= -\frac{5}{2} \end{aligned}$$

The solutions are $\frac{11}{2}$ and $-\frac{5}{2}$.

33.
$$\begin{aligned} |5x+8| &= 0 \\ 5x+8 &= 0 \\ 5x+8-8 &= 0-8 \\ 5x &= -8 \\ \frac{5x}{5} &= \frac{-8}{5} \\ x &= -\frac{8}{5} \end{aligned}$$

The solution is $-\frac{8}{5}$.

34. $|5x-4| < -2$

There is no solution to this equation because the absolute value of a number must be non-negative.

35. **Strategy** • Time to travel to the island: t

•Time to return to dock: $2\frac{1}{3}-t$

	Rate	Time	Distance
To island	16	t	$16t$
Back to dock	12	$\frac{7}{3}-t$	$12\left(\frac{7}{3}-t\right)$

• The distance to the island equals the distance back to the dock.

Solution

$$16t = 12\left(\frac{7}{3} - t\right)$$
$$16t = 28 - 12t$$
$$16t + 12t = 28 - 12t + 12t$$
$$28t = 28$$
$$\frac{28t}{28} = \frac{28}{28}$$
$$t = 1$$

The distance is $16t = 16(1) = 16$.
The island is 16 mi from the dock.

36. Strategy • Gallons of apple juice: x

	Amount	Cost	Value
Apple	x	4.20	$4.20x$
Cranberry	40	6.50	40(6.50)
Mixture	$40 + x$	5.20	$5.20(40 + x)$

• The sum of the values before mixing equals the value after mixing.

Solution

$$4.20x + 40(6.50) = 5.20(40 + x)$$
$$4.20x + 260 = 208 + 5.20x$$
$$4.20x - 5.20x + 260 = 208$$
$$-x + 260 = 208$$
$$-x + 260 - 260 = 208 - 260$$
$$-x = -52$$
$$\frac{-x}{-1} = \frac{-52}{-1}$$
$$x = 52$$

The mixture must contain 52 gal of apple juice.

37. Strategy • To find the minimum amount of sales, write and solve an inequality using N to represent the amount of sales.

Solution

$$800 + 0.04N \geq 3000$$
$$800 - 800 + 0.04x \geq 3000 - 800$$
$$0.04x \geq 2200$$
$$\frac{0.04x}{0.04} \geq \frac{2200}{0.04}$$
$$x \geq 55{,}000$$

The executive's amount of sales must be \$55,000 or more.

38. The unknown number is x.

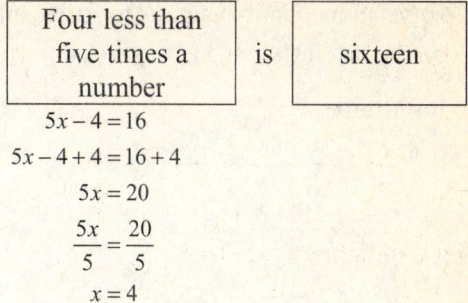

Four less than five times a number	is	sixteen

$$5x - 4 = 16$$
$$5x - 4 + 4 = 16 + 4$$
$$5x = 20$$
$$\frac{5x}{5} = \frac{20}{5}$$
$$x = 4$$

The number is 4.

39. Strategy • Let b represent the diameter of the bushing, T the tolerance, and d the lower and upper limits of diameter. Solve the absolute value inequality $|d - b| \leq T$ for d.

Solution

$$|d - b| \leq T$$
$$-0.003 < d - 2.75 \leq 0.003$$
$$-0.003 < d - 2.75 \leq 0.003$$
$$-0.003 + 2.75 \leq d - 2.75 + 2.75 \leq 0.003 + 2.75$$
$$2.747 \leq d \leq 2.753$$

The lower limit of the bushing is 2.747 in. and the upper limit is 2.753 in.

40. Strategy • The smaller integer: n
The larger integer: $20 - n$
•Five times the smaller integer is two more than twice the larger integer.

Solution

$$5n = 2 + 2(20 - n)$$
$$5n = 2 + 40 - 2n$$
$$5n = 42 - 2n$$
$$5n + 2n = 42 - 2n + 5n$$
$$7n = 42$$
$$\frac{7n}{7} = \frac{42}{7}$$
$$n = 6$$
$$20 - n = 10 - 6 = 14$$

The integers are 6 and 14.

41. **Strategy** To find the range of scores, write and solve an inequality using N to represent the score on the last test.

Solution

$$80 \le \frac{92+66+72+88+N}{5} \le 90$$
$$80 \le \frac{318+N}{5} \le 90$$
$$5(80) \le 5\left(\frac{318+N}{5}\right) \le 5(90)$$
$$400 \le 318+N \le 450$$
$$400-318 \le 318-318+N \le 450-318$$
$$82 \le N \le 132$$

Since 100 is the maximum score, the range of scores on the last text that will yield a B grade is $82 \le N \le 100$.

42. **Strategy** • Rate of the first plane: r
•Rate of the second plane: $r + 80$

	Rate	Time	Distance
1st plane	r	1.75	$1.75r$
2nd plane	$r+80$	1.75	$1.75(r+80)$

• The total distance traveled by the two planes was 1680 mi.

Solution

$$1.75r+1.75(r+80)=1680$$
$$1.75r+1.75r+140=1680$$
$$3.5r+140=1680$$
$$3.5r+140-140=1680-140$$
$$3.5r=1540$$
$$\frac{3.5r}{3.5}=\frac{1540}{3.5}$$
$$r=440$$

$r + 80 = 440 + 80 = 520$

The speed of the first plane is 440 mph. The speed of the second plane is 520 mph.

43. **Strategy** • Pounds of 30% tin: x
Pounds of 70% tin: $500 - x$

	Amount	Percent	Quantity
30%	x	0.30	$0.30x$
70%	$500-x$	0.70	$0.70(500-x)$
40%	500	0.40	$500(0.40)$

• The sum of the quantities before mixing equals the quantity after mixing.

Solution

$$0.30x+0.70(500-x)=0.40(500)$$
$$0.30x+350-0.70x=200$$
$$-0.40x+350=200$$
$$-0.40x+350-350=200-350$$
$$-0.40x=-150$$
$$\frac{-0.40x}{-0.40}=\frac{-150}{-0.40}$$
$$x=375$$
$$500-x=500-375=125$$

375 lb of the 30% tin alloy and 125 lb of the 70% tin alloy were used.

44. **Strategy** Let r represent the length of the piston rod, T the tolerance and L the lower and upper limits of the length. Solve the absolute value of $|L-r| \le T$ for L.

Solution

$$|L-r| \le T$$
$$\left|L-10\frac{3}{8}\right| \le \frac{1}{32}$$
$$-\frac{1}{32} \le L-10\frac{3}{8} \le \frac{1}{32}$$
$$-\frac{1}{32}+10\frac{3}{8} \le L-10\frac{3}{8}+10\frac{3}{8} \le \frac{1}{32}+10\frac{3}{8}$$
$$-\frac{1}{32}+10\frac{12}{32} \le L \le \frac{1}{32}+10\frac{12}{32}$$
$$10\frac{11}{32} \le L \le 10\frac{13}{32}$$

The lower and upper limits of the length of the piston are $10\frac{11}{32}$ and $10\frac{13}{32}$ in.

Chapter 2 Test

1.

$$\begin{aligned} x-2&=-4\\ x-2+2&=-4+2\\ x&=-2 \end{aligned}$$

The solution is -2.

2.

$$\begin{aligned} b+\frac{3}{4}&=\frac{5}{8}\\ b+\frac{3}{4}-\frac{3}{4}&=\frac{5}{8}-\frac{3}{4}\\ b&=\frac{5}{8}-\frac{6}{8}\\ b&=-\frac{1}{8} \end{aligned}$$

The solution is $-\frac{1}{8}$.

3.

$$\begin{aligned} -\frac{3}{4}y&=-\frac{5}{8}\\ -\frac{4}{3}\left(-\frac{3}{4}y\right)&=-\frac{4}{3}\left(-\frac{5}{8}\right)\\ y&=\frac{5}{6} \end{aligned}$$

The solution is $\frac{5}{6}$.

4.

$$\begin{aligned} 3x-5&=7\\ 3x-5+5&=7+5\\ 3x&=12\\ \frac{3x}{3}&=\frac{12}{3}\\ x&=4 \end{aligned}$$

The solution is 4.

5.

$$\begin{aligned} \frac{3}{4}y-2&=6\\ \frac{3}{4}y-2+2&=6+2\\ \frac{3}{4}y&=8\\ \frac{4}{3}\cdot\frac{3}{4}y&=\frac{4}{3}\cdot 8\\ y&=\frac{32}{2} \end{aligned}$$

The solution is $\frac{32}{3}$.

6.

$$\begin{aligned} 2x-3-5x&=8+2x-10\\ -3x-3&=2x-2\\ -3x-2x-3&=2x-2x-2\\ -5x-3&=-2\\ -5x-3+3&=-2+3\\ -5x&=1\\ \frac{-5x}{-5}&=\frac{1}{-5}\\ x&=-\frac{1}{5} \end{aligned}$$

The solution is $-\frac{1}{5}$.

7.

$$\begin{aligned} 2[a-(2-3a)-4]&=a-5\\ 2[a-2+3a-4]&=a-5\\ 2[4a-6]&=a-5\\ 8a-12&=a-5\\ 8a-a-12&=a-a-5\\ 7a-12&=-5\\ 7a-12+12&=-5+12\\ 7a&=7\\ \frac{7a}{7}&=\frac{7}{7}\\ a&=1 \end{aligned}$$

The solution is 1.

8.

$$\begin{array}{r|l} \multicolumn{2}{c}{x^2-3x=2x-6}\\ \hline (-2)^2-3(-2) & 2(-2)-6\\ 4-3(-2) & -4-6\\ 4+6 & -10\\ & \\ \multicolumn{2}{c}{10\neq-10} \end{array}$$

No, -2 is not a solution.

9.
$$\frac{2x+1}{3}-\frac{3x+4}{6}=\frac{5x-9}{9}$$
$$18\left(\frac{2x+1}{3}-\frac{3x+4}{6}\right)=18\left(\frac{5x-9}{9}\right)$$
$$18\left(\frac{2x+1}{3}\right)-18\left(\frac{3x+4}{6}\right)=2(5x-9)$$
$$6(2x+1)-3(3x+4)=10x-18$$
$$12x+6-9x-12=10x-18$$
$$3x-6=10x-18$$
$$3x-10x-6=10x-10x-18$$
$$-7x-6=-18$$
$$-7x-6+6=-18+6$$
$$-7x=-12$$
$$\frac{-7x}{-7}=\frac{-12}{-7}$$
$$x=\frac{12}{7}$$

The solution is $\frac{12}{7}$.

10.
$$3x-2\ge 6x+7$$
$$3x-6x-2\ge 6x-6x+7$$
$$-3x-2\ge 7$$
$$-3x-2+2\ge 7+2$$
$$-3x\ge 9$$
$$\frac{-3x}{-3}\le\frac{9}{-3}$$
$$x\le -3$$

The solution set is $\{x\mid x\le -3\}$.

11.
$$P\cdot B=A$$
$$0.005(8)=A$$
$$0.04=A$$

0.5% of 8 is 0.04.

12.
$$4x-1>5 \qquad\qquad 2-3x<8$$
$$4x-1+1>5+1 \qquad\qquad 2-2-3x<8-2$$
$$4x>6 \qquad \text{or} \qquad -3x<6$$
$$\frac{4x}{4}>\frac{6}{4} \qquad\qquad \frac{-3x}{-3}>\frac{6}{-3}$$
$$x>\frac{3}{2} \qquad\qquad x>-2$$
$$\left\{x\mid x>\frac{3}{2}\right\}\cup\{x\mid x>-2\}=\{x\mid x>-2\}$$

The solution set is $\{x\mid x>-2\}$.

13.
$$4-3x\ge 7 \qquad\qquad 2x+3\ge 7$$
$$4-4-3x\ge 7-4 \qquad\qquad 2x+3-3\ge 7-3$$
$$-3x\ge 3 \qquad \text{and} \qquad 2x\ge 4$$
$$\frac{-3x}{-3}\le\frac{3}{-3} \qquad\qquad \frac{2x}{2}\ge\frac{4}{2}$$
$$x\le -1 \qquad\qquad x\ge 2$$
$$\{x\mid x\le -1\}\cap\{x\mid x\ge 2\}=\varnothing$$

There is no solution.

14. $|3-5x|=12$
$$3-5x=12 \qquad\qquad 3-5x=-12$$
$$3-3-5x=12-3 \qquad\qquad 3-3-5x=-12-3$$
$$-5x=9 \qquad\qquad -5x=-15$$
$$\frac{-5x}{-5}=\frac{9}{-5} \qquad\qquad \frac{-5x}{-5}=\frac{-15}{-5}$$
$$x=-\frac{9}{5} \qquad\qquad x=3$$

The solutions are $-\frac{9}{5}$ and 3.

15.
$$2-|2x-5|=-7$$
$$2-2-|2x-5|=-7-2$$
$$-|2x-5|=-9$$
$$\frac{-|2x-5|}{-1}=\frac{-9}{-1}$$
$$|2x-5|=9$$
$$2x-5=9 \qquad\qquad 2x-5=-9$$
$$2x-5+5=9+5 \qquad\qquad 2x-5+5=-9+5$$
$$2x=14 \qquad\qquad 2x=-4$$
$$\frac{2x}{2}=\frac{14}{2} \qquad\qquad \frac{2x}{2}=\frac{-4}{2}$$
$$x=7 \qquad\qquad x=-2$$

The solutions are –2 and 7.

16. $|3x-5|\le 4$
$$-4\le 3x-5\le 4$$
$$-4+5\le 3x-5+5\le 4+5$$
$$1\le 3x\le 9$$
$$\frac{1}{3}\le\frac{3x}{3}\le\frac{9}{3}$$
$$\frac{1}{3}\le x\le 3$$

The solution set is $\left\{x\mid \frac{1}{3}\le x\le 3\right\}$.

17. $|4x-3|>5$

$$\begin{array}{lcl} 4x-3>5 & & 4x-3<-5 \\ 4x-3+3>5+3 & & 4x-3+3<-5+3 \\ 4x>8 & \text{or} & 4x<-2 \\ \frac{4x}{4}>\frac{8}{4} & & \frac{4x}{4}<\frac{-2}{4} \\ x>2 & & x<-\frac{1}{2} \end{array}$$

The solution set is $\left\{x \mid x>2 \text{ or } x<-\frac{1}{2}\right\}$.

18. **Strategy** To find the number of miles, write and solve an inequality using N to represent the number of miles.

Solution

cost of Gambelli < cost of McDougal

$$12+0.10N<24$$
$$12-12+0.10N<24-12$$
$$0.10N<12$$
$$\frac{0.10N}{0.10}<\frac{12}{0.10}$$
$$N<120$$

Gambelli will cost less if you drive less 120 mi.

19. **Strategy** • Let b represent the diameter of the bushing, T the tolerance, and d the lower and upper limits of diameter. Solve the absolute value inequality $|d-b|\le T$ for d.

Solution

$$|d-b|\le T$$
$$|d-2.65|<0.002$$
$$-0.002<d-2.65\le 0.002$$
$$-0.002+2.65\le d-2.65+2.65\le 0.002+2.65$$
$$2.648\le d\le 2.652$$

The lower limit of the bushing is 2.648 in. and the upper limit is 2.652 in.

20. The smaller number: x

The larger number: $15-x$

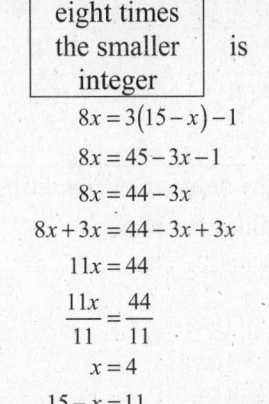

eight times the smaller integer	is	one less than three times the larger integer

$$8x=3(15-x)-1$$
$$8x=45-3x-1$$
$$8x=44-3x$$
$$8x+3x=44-3x+3x$$
$$11x=44$$
$$\frac{11x}{11}=\frac{44}{11}$$
$$x=4$$
$$15-x=11$$

The smaller number is 4.
The larger number is 11.

21. **Strategy** • Amount of water: x

20% salt solution is 80% water.
16% salt solution is 84% water.

	Amount	Percent	Quantity
Water	x	1.00	$1.00x$
20% salt solution	5	0.80	0.80(5)
16% salt solution	$x+5$	0.84	$0.84(x+5)$

• The sum of the quantities before mixing equals the quantity after mixing.

Solution

$$1.00x+0.80(5)=0.84(x+5)$$
$$x+4=0.84x+4.2$$
$$1.00x-0.84x+4=4.2$$
$$0.16x+4=4.2$$
$$0.16x+4-4=4.2-4$$
$$0.16x=0.2$$
$$\frac{0.16x}{0.16}=\frac{0.2}{0.16}$$
$$x=1.25$$

1.25 gal of water must be used.

22. Strategy • Cost of hamburger mixture: x

	Amount	Cost	Value
\$2.10 hamburger	100	2.10	2.10(100)
\$3.70 hamburger	60	3.70	60(3.70)
Mixture	160	x	$160x$

• The sum of the values before mixing equals the value after mixing.

Solution

$$2.10(100)+3.70(60)=160x$$
$$210+222=160x$$
$$432=160x$$
$$\frac{432}{160}=\frac{160x}{160}$$
$$2.70=x$$

The hamburger mixture costs \$2.70 per lb.

23. Strategy

• Time jogger runs a distance: t

•Time jogger returns the same distance: $1\frac{45}{60}-t$

	Rate	Time	Distance
run out	8	t	$8t$
return	6	$\frac{7}{4}-t$	$6\left(\frac{7}{4}-t\right)$

• The distance traveled is the same.

Solution

$$8t=6\left(\frac{7}{4}-t\right)$$
$$8t=\frac{21}{2}-6t$$
$$8t+6t=\frac{21}{2}-6t+6t$$
$$14t=\frac{21}{2}$$
$$\frac{14t}{14}=\frac{1}{14}\left(\frac{21}{2}\right)$$
$$t=\frac{3}{4}$$

The jogger ran for $\frac{3}{4}$ h on the way there.

$$8\left(\frac{3}{4}\right)=6$$

The jogger ran a distance of 6 mi one way and a total distance of 12 mi.

24. Strategy • Rate of the slower train: r

•Rate of the faster train: $r+5$

	Rate	Time	Distance
Slow train	r	2	$2r$
Fast train	$r+5$	2	$2(r+5)$

• The total distance traveled by the two planes was 250 mi.

Solution

$$2r+2(r+5)=250$$
$$2r+2r+10=250$$
$$4r+10=250$$
$$4r+10-10=250-10$$
$$4r=240$$
$$\frac{4r}{4}=\frac{240}{4}$$
$$r=60$$

$r+5=60+5=65$

The speed of the slow train is 60 mph. The speed of the fast train is 65 mph.

25. Strategy • Ounces of pure water: x

	Amount	Percent	Amount
Water	x	0	0
8% salt solution	60	0.08	0.08(60)
16% salt solution	$60+x$	0.03	$0.03(60+x)$

• The sum of the values before mixing equals the value after mixing.

Solution

$$0+0.08(60)=0.03(60+x)$$
$$4.8=1.8+0.03x$$
$$4.8-1.8=1.8-1.8+0.03x$$
$$3=0.03x$$
$$\frac{3}{0.03}=\frac{0.03x}{0.03}$$
$$100=x$$

There are 100 oz of pure water.

Cumulative Review Exercises

1. $-6-(-20)-8=-6+20-8=14-8=6$

2. $(-2)(-6)(-4)=12(-4)=-48$

3. $-\frac{5}{6}-\left(-\frac{7}{16}\right)=-\frac{40}{48}-\left(-\frac{21}{48}\right)=\frac{-40-(-21)}{48}=\frac{-40+21}{48}$
$=-\frac{19}{48}$

4. $-4^2\cdot\left(-\frac{3}{2}\right)^3=-(4)(4)\cdot\left(-\frac{3}{2}\right)\left(-\frac{3}{2}\right)\left(-\frac{3}{2}\right)=-16\cdot\left(-\frac{27}{8}\right)$
$=16\cdot\frac{27}{8}=\frac{16\cdot 27}{8}=\frac{\overset{1}{2}\cdot\overset{1}{2}\cdot\overset{1}{2}\cdot 2\cdot 3\cdot 3\cdot 3}{\underset{1}{2}\cdot\underset{1}{2}\cdot\underset{1}{2}}$
$=\frac{2\cdot 3\cdot 3\cdot 3}{1}=\frac{54}{1}=54$

5. $\frac{5}{8}-\left(-\frac{1}{2}\right)^2\div\left(\frac{1}{3}-\frac{3}{4}\right)=\frac{5}{8}-\frac{1}{4}\div\left(\frac{1}{3}-\frac{3}{4}\right)$
$=\frac{5}{8}-\frac{1}{4}\div\left(\frac{4}{12}-\frac{9}{12}\right)$
$=\frac{5}{8}-\frac{1}{4}\div\left(-\frac{5}{12}\right)$
$=\frac{5}{8}-\frac{1}{4}\times\left(-\frac{12}{5}\right)$
$=\frac{5}{8}+\frac{3}{5}=\frac{25}{40}+\frac{24}{40}=\frac{49}{40}$

6. $3(a-c)-2ab$
$=3[2-(-4)]-2(2)(3)=3[2+4]-2(2)(3)$
$=3[6]-2(2)(3)=18-2(2)(3)=18-4(3)$
$=18-12=6$

7. $3x-8x+(-12x)=-5x+(-12x)$
$=-5x-12x=-17x$

8. $2a-(-b)-7a-5b=2a+b-7a-5b$
$=(2a-7a)+(b-5b)$
$=-5a+(-4b)=-5a-4b$

9. $(16x)\left(\frac{1}{8}\right)=\frac{1}{8}(16x)=\left(\frac{1}{8}\cdot 16\right)x=2x$

10. $-4(-9y)=4(9y)=(4\cdot 9)y=36y$

11. $-2(-x^2-3x+2)=-2(-x^2)+(-2)(-3x)+(-2)(2)$
$=2x^2+6x-4$

12. $-2(x-3)+2(4-x)=-2x+6+8-2x$
$=(-2x-2x)+(6+8)=-4x+14$

13. $-3[2x-4(x-3)]+2=-3[2x-4x+12]+2$
$=-3[-2x+12]+2=6x-36+2$
$=6x-34$

14. $A\cap B=\{-4,\ -2,\ 0,\ 2\}\cap\{-4,\ 0,\ 4,\ 8\}=\{-4,\ 0\}$

15. $\{x\mid x<3\}\cap\{x\mid x>-2\}$

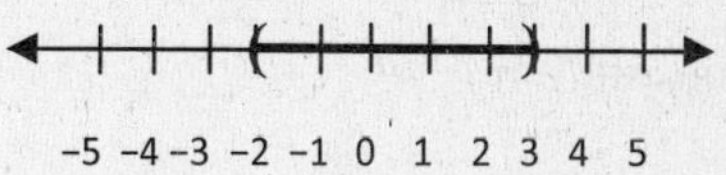

16.
$$\begin{array}{r|l} x^2+6x+9 & = \quad x+3 \\ \hline (-3)^2+6(-3)+9 & -3+3 \\ 9-18+9 & 0 \\ -9+9 & 0 \end{array}$$
$$0=0$$
Yes, –3 is a solution.

17.
$$\begin{aligned} \frac{3}{5}x &= -15 \\ \frac{5}{3}\cdot\frac{3}{5}x &= \frac{5}{3}\cdot(-15) \\ 1\cdot x &= -25 \\ x &= -25 \end{aligned}$$
The solution is −25.

18.
$$\begin{aligned} 7x-8 &= -29 \\ 7x-8+8 &= -29+8 \\ 7x &= -21 \\ \frac{7x}{7} &= \frac{-21}{7} \\ x &= -3 \end{aligned}$$
The solution is −3.

19.
$$\begin{aligned} 13-9x &= -14 \\ 13-13-9x &= -14-13 \\ -9x &= -27 \\ \frac{-9x}{-9} &= \frac{-27}{-9} \\ x &= 3 \end{aligned}$$
The solution is 3.

20.
$$\begin{aligned} 5x-8 &= 12x+13 \\ 5x-12x-8 &= 5x-5x+13 \\ -7x-8 &= 13 \\ -7x-8+8 &= 13+8 \\ -7x &= 21 \\ \frac{-7x}{-7} &= \frac{21}{-7} \\ x &= -3 \end{aligned}$$
The solution is –3.

21.
$$\begin{aligned} 11-4x &= 2x+8 \\ 11-4x-2x &= 2x-2x+8 \\ 11-6x &= 8 \\ 11-11-6x &= 8-11 \\ -6x &= -3 \\ \frac{-6x}{-6} &= \frac{-3}{-6} \\ x &= \frac{1}{2} \end{aligned}$$
The solution is $\frac{1}{2}$.

22.
$$\begin{aligned} 8x-3(4x-5) &= -2x-11 \\ 8x-12x+15 &= -2x-11 \\ -4x+15 &= -2x-11 \\ -2x &= -26 \\ x &= 13 \end{aligned}$$
The solution is 13.

23.
$$\begin{aligned} 3-2(2x-1) &\ge 3(2x-2)+1 \\ 3-4x+2 &\ge 6x-6+1 \\ -4x+5 &\ge 6x-5 \\ -4x-6x+5 &\ge -5 \\ -10x+5 &\ge -5 \\ -10x+5-5 &\ge -5-5 \\ -10x &\ge -10 \\ \frac{-10x}{-10} &\le \frac{-10}{-10} \\ x &\le 1 \end{aligned}$$
The solution set is $\{x\,|\,x\le 1\}$.

24.
$$\begin{aligned} 3x+2 &\le 5 \\ 3x+2-2 &\le 5-2 \\ 3x &\le 3 \\ \frac{3x}{3} &\le \frac{3}{3} \\ x &\le 1 \end{aligned} \quad \text{and} \quad \begin{aligned} x+5 &\ge 1 \\ x+5-5 &\ge 1-5 \\ x &\ge -4 \end{aligned}$$
$$\{x\,|\,x\le 1\}\cap\{x\,|\,x\ge -4\}=\{x\,|\,-4\le x\le 1\}$$
The solution set is $\{x\,|\,-4\le x\le 1\}$.

25. $|3-2x|=5$

$$\begin{aligned} 3-2x&=5 & 3-2x&=-5\\ 3-3-2x&=5-3 & 3-3-2x&=-5-3\\ -2x&=2 & -2x&=-8\\ \frac{-2x}{-2}&=\frac{2}{-2} & \frac{-2x}{-2}&=\frac{-8}{-2}\\ x&=-1 & x&=4 \end{aligned}$$

The solutions are –1 and 4.

26. $|3x-1|>5$

$$\begin{aligned} 3x-1&<-5 & & 3x-1&>5\\ 3x-1+1&<-5+1 & & 3x-1+1&>5+1\\ 3x&<-4 & \text{or} & 3x&>6\\ \frac{3x}{3}&<\frac{-4}{3} & & \frac{3x}{3}&>\frac{6}{3}\\ x&<-\frac{4}{3} & & x&>2 \end{aligned}$$

$$\left\{x \mid x<-\frac{4}{3}\right\}\cup\{x \mid x>2\}=\left\{x \mid x>2 \text{ or } x<-\frac{4}{3}\right\}$$

The solution set is $\left\{x \mid x>2 \text{ or } x<-\frac{4}{3}\right\}$.

27. $55\%=55\left(\frac{1}{100}\right)=\frac{55}{100}=\frac{11}{20}$

28. $1.03=1.03(100\%)=103\%$

29. Percent · Base = Amount

$$\begin{aligned} 25\%\cdot B&=30\\ 0.25B&=30\\ \frac{0.25B}{0.25}&=\frac{30}{0.25}\\ B&=120 \end{aligned}$$

30. The unknown number: x

$$\begin{aligned} 6x+13&=3x-5\\ 6x-3x+13&=3x-3x-5\\ 3x+13&=-5\\ 3x+13-13&=-5-13\\ 3x&=-18\\ \frac{3x}{3}&=\frac{-18}{3}\\ x&=-6 \end{aligned}$$

The solution is –6.

31. Strategy • Amount of oat flour: x

	Amount	Cost	Quantity
Oat	x	0.80	$0.80x$
Wheat	40	0.50	0.50(40)
Mixture	$x+40$	0.60	$0.60(x+40)$

Solution

$$\begin{aligned} 0.80x+0.50(40)&=0.60(x+40)\\ 0.80x+20&=0.60x+24\\ 0.20x&=4\\ x&=20 \end{aligned}$$

20 lb of oat flour are needed for the mixture.

32. Strategy • Amount pure gold: x

	Amount	Percent	Quantity
Pure gold	x	1.00	$1.00x$
Alloy	100	0.20	0.20(100)
Mixture	$x+100$	0.36	$0.36(x+100)$

• The sum of the quantities before mixing is equal to the quantity after mixing.

Solution

$$\begin{aligned} 1.00x+0.20(100)&=0.36(x+100)\\ 1.00x+20&=0.36x+36\\ 0.64x&=16\\ x&=25 \end{aligned}$$

25 g of pure gold must be added.

33. Strategy

- Time running: t
- Time jogging: $55 - t$

	Rate	Time	Distance
Running	8	t	$8t$
Jogging	3	$55 - t$	$3(55 - t)$

- The distance traveled is the same.

Solution

$$8t = 3(55 - t)$$
$$8t = 165 - 3t$$
$$11t = 165$$
$$t = 15$$

Distance = 8t = 8(15) = 120

The length of the track is 120 m.

Chapter 3: Geometry

Prep Test

1. $$\begin{aligned} x+47&=90 \\ x&=43 \end{aligned}$$

2. $$\begin{aligned} 32+97+x&=180 \\ 129+x&=180 \\ x&=51 \end{aligned}$$

3. $2(18)+2(10)=36+20=56$

4. abc

$(2)(3.14)(9)=6.28(9)=56.52$

5. xyz^3

$\left(\frac{4}{3}\right)(3.14)\left(3^3\right)=\frac{4}{3}(3.14)27=113.04$

6. $\frac{1}{2}a(b+c)$

$=\frac{1}{2}(6)(25+15)=\frac{1}{2}(6)(40)=3(40)=120$

Section 3.1

Objective A Exercises

1. The measure of the given angle is 40°. The measure of the angle is between 0° and 90°, so the angle is acute.

3. The measure of the given angle is 115°. The measure of the angle is between 90° and 180°, so the angle is obtuse.

5. The measure of the given angle is 90°. The angle is right.

7. **Strategy** Complementary angles are two angles whose sum is 90°. To find the complement, let x represent the complement of a 62° angle. Write an equation and solve for x.

Solution

$$\begin{aligned} x+62^\circ&=90^\circ \\ x&=28^\circ \end{aligned}$$

The complement of a 62° angle is a 28° angle.

9. **Strategy** Supplementary angles are two angles whose sum is 180°. To find the supplement, let x represent the supplement of a 162° angle. Write an equation and solve for x.

Solution

$$\begin{aligned} x+162^\circ&=180^\circ \\ x&=18^\circ \end{aligned}$$

The supplement of a 162° angle is an 18° angle.

11. $$\begin{aligned} AB+BC+CD&=AD \\ 12+BC+9&=35 \\ 21+BC&=35 \\ BC&=14 \end{aligned}$$

$BC=14$ cm

13. $$\begin{aligned} QR+RS&=QS \\ QR+3(QR)&=QS \\ 7+3\cdot 7&=QS \\ 7+21&=QS \\ 28&=QS \end{aligned}$$

$QS=28$ ft

15. $$\begin{aligned} EF+FG&=EG \\ EF+\frac{1}{2}(EF)&=EG \\ 20+\frac{1}{2}(20)&=EG \\ 20+10&=EG \\ 30&=EG \end{aligned}$$

$EG=30$ m

17. $$\begin{aligned} \angle LOM+\angle MON&=\angle LON \\ 53^\circ+\angle MON&=139^\circ \\ \angle MON&=139^\circ-53^\circ=86^\circ \end{aligned}$$

The measure of $\angle MON$ is 86°.

19. **Strategy** To find the measure of $\angle x$, write an equation using the fact that the sum of the measures of $\angle x$ and 74° is 145°. Solve for $\angle x$.

Solution

$$\begin{aligned} \angle x+74^\circ&=145^\circ \\ \angle x&=71^\circ \end{aligned}$$

The measure of $\angle x$ is 71°.

21. **Strategy** To find the measure of $\angle x$, write an equation using the fact that the sum of the measures of $\angle x$ and $\angle 2x$ is 90°. Solve for $\angle x$.

Solution

$$x + 2x = 90°$$
$$3x = 90°$$
$$x = 30°$$

The measure of $\angle x$ is 30°.

23. **Strategy** To find the measure of $\angle x$, write an equation using the fact that the sum of x and $x + 18°$ is 90°. Solve for x.

Solution

$$x + x + 18° = 90°$$
$$2x + 18° = 90°$$
$$2x = 72°$$
$$x = 36°$$

The measure of $\angle x$ is 36°.

25. **Strategy** To find the measure of $\angle a$, write an equation using the fact that the sum of $\angle a$ and 53° is 180°. Solve for $\angle a$.

Solution

$$\angle a + 53° = 180°$$
$$\angle a = 127°$$

The measure of $\angle a$ is 127°.

27. **Strategy** The sum of the measures of the three angles shown is 360°. To find $\angle a$, write an equation and solve for $\angle a$.

Solution

$$\angle a + 76° + 168° = 360°$$
$$\angle a + 244° = 360°$$
$$\angle a = 116°$$

The measure of $\angle a$ is 116°.

29. **Strategy** The sum of the measures of the three angles shown is 180°. To find x, write an equation and solve for x.

Solution

$$3x + 4x + 2x = 180°$$
$$9x = 180°$$
$$x = 20°$$

The measure of x is 20°.

31. **Strategy** The sum of the measures of the three angles shown is 180°. To find x, write an equation and solve for x.

Solution

$$5x + (x + 20°) + 2x = 180°$$
$$8x + 20° = 180°$$
$$8x = 160°$$
$$x = 20°$$

The measure of x is 20°.

33. **Strategy** The sum of the measures of the four angles shown is 360°. To find x, write an equation and solve for x.

Solution

$$3x + 4x + 6x + 5x = 360°$$
$$18x = 360°$$
$$x = 20°$$

The measure of x is 20°.

35. **Strategy**

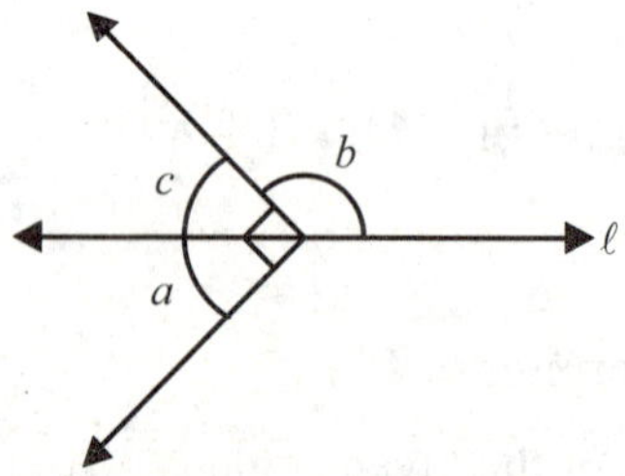

To find the measure of $\angle b$:
- Use the fact that $\angle a$ and $\angle c$ are complementary angles.
- Find $\angle b$ by using the fact that $\angle c$ and $\angle b$ are supplementary angles.

Solution

$$\angle a + \angle c = 90°$$
$$51° + \angle c = 90°$$
$$\angle c = 39°$$

$$\angle b + \angle c = 180°$$
$$\angle b + 39° = 180°$$
$$\angle b = 141°$$

The measure of $\angle b$ is 141°.

37. An acute angle.

39. An obtuse angle.

Objective B Exercises

41. **Strategy** The angles labeled are adjacent angles of intersecting lines and are, therefore, supplementary angles. To find x, write an equation and solve for x.

Solution

$x + 74° = 180°$

$x = 106°$

The measure of x is 106°.

43. **Strategy** The angles labeled are vertical angles and are, therefore, equal. To find x, write an equation and solve for x.

Solution

$5x = 3x + 22°$

$2x = 22°$

$x = 11°$

The measure of x is 11°.

45. **Strategy** • To find the measure of $\angle a$, use the fact that corresponding angles of parallel lines are equal.
• To find the measure of $\angle b$, use the fact that adjacent angles of intersecting lines are supplementary.

Solution

$\angle a = 38°$

$\angle b + \angle a = 180°$

$\angle b + 38° = 180°$

$\angle b = 142°$

The measure of $\angle a$ is 38°.
The measure of $\angle b$ is 142°.

47. **Strategy** • To find the measure of $\angle a$, use the fact that alternate interior angles of parallel lines are equal.
• To find the measure of $\angle b$, use the fact that adjacent angles of intersecting lines are supplementary.

Solution

$\angle a = 47°$

$\angle a + \angle b = 180°$

$47° + \angle b = 180°$

$\angle b = 133°$

The measure of $\angle a$ is 47°.
The measure of $\angle b$ is 133°.

49. **Strategy**

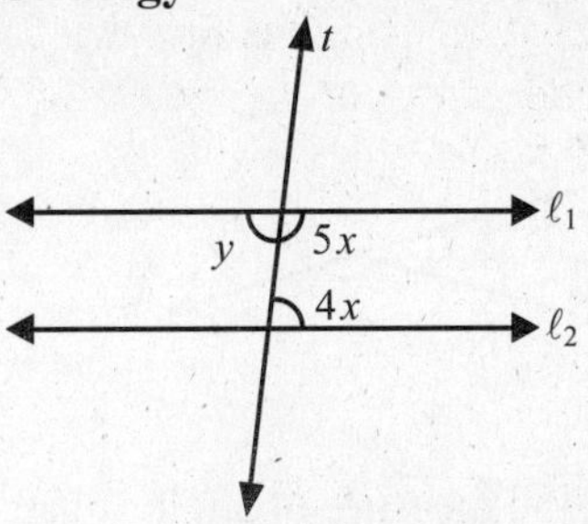

$4x = y$ because alternate interior angles have the same measure. $y + 5x = 180°$ because adjacent angles of intersecting lines are supplementary. Substitute $4x$ for y and solve for x.

Solution

$4x + 5x = 180°$

$9x = 180°$

$x = 20°$

The measure of x is 20°.

51. **Strategy**

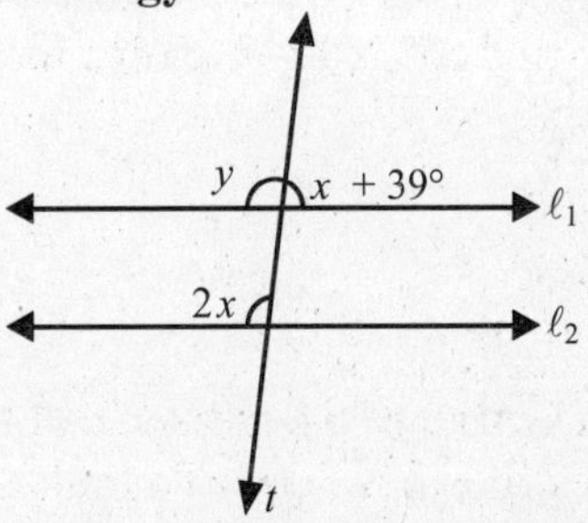

$y = 2x$ because corresponding angles have the same measure. $y + x + 39° = 180°$ because adjacent angles of intersecting lines are supplementary angles. Substitute $2x$ for y and solve for x.

Solution

$2x + x + 39° = 180°$

$3x + 39° = 180°$

$3x = 141°$

$x = 47°$

The measure of x is 47°.

53. True, $\angle a$ and $\angle b$ are vertical angles.

Objective C Exercises

55. Strategy

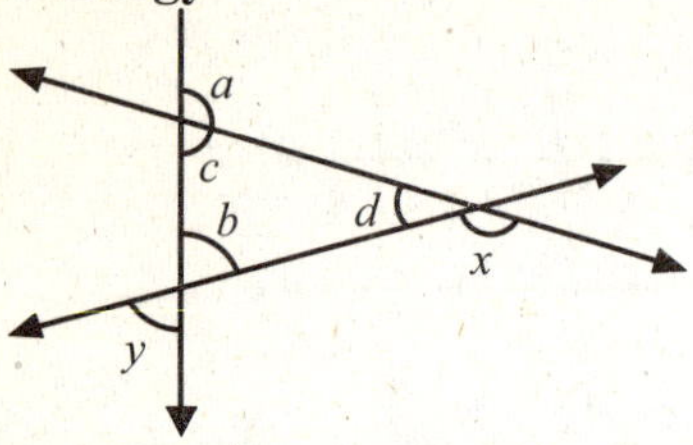

- To find the measure of angle y, use the fact that $\angle b$ and $\angle y$ are vertical angles.
- To find the measure of angle x:
Find the measure of angle c by using the fact that the sum of an interior and an exterior angle is 180°.
Find the measure of angle d by using the fact that the sum of the interior angles of triangle is 180°.
Find the measure of angle x by using the fact that the sum of an interior and an exterior angle is 180°.

Solution

$\angle y = \angle b = 70°$

$\angle a + \angle c = 180°$

$95° + \angle c = 180°$

$\angle c = 85°$

$\angle b + \angle c + \angle d = 180°$

$70° + 85° + \angle d = 180°$

$155° + \angle d = 180°$

$\angle d = 25°$

$\angle d + \angle x = 180°$

$25° + \angle x = 180°$

$\angle x = 155°$

The measure of $\angle x$ is 155°.
The measure of $\angle y$ is 70°.

57. Strategy

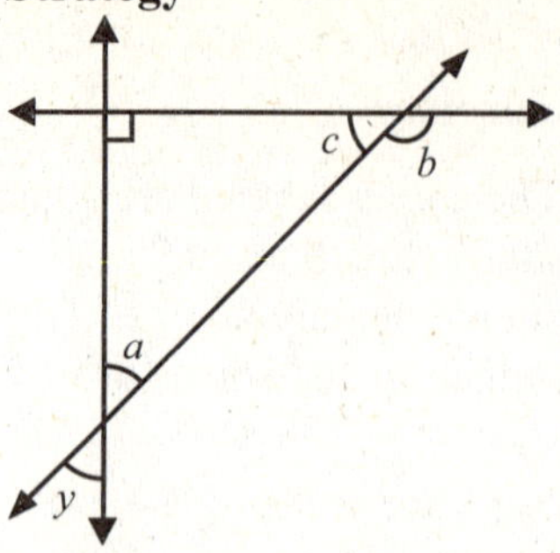

- To find the measure of angle a, use the fact that $\angle a$ and $\angle y$ are vertical angles.
- To find the measure of angle b:
Find the measure of angle c by using the fact that the sum of the interior angles of a triangle is 180°.
Find the measure of angle b by using the fact that the sum of an interior and an exterior angle is 180°.

Solution

$\angle a = \angle y = 45°$

$\angle a + \angle c + 90° = 180°$

$45° + \angle c + 90° = 180°$

$\angle c + 135° = 180°$

$\angle c = 45°$

$\angle c + \angle b = 180°$

$45° + \angle b = 180°$

$\angle b = 135°$

The measure of $\angle a$ is 45°.
The measure of $\angle b$ is 135°.

59. Strategy To find the measure of $\angle BOC$, use the fact that the sum of the measures of the angles x, $\angle AOB$, and $\angle BOC$ is 180°. Since $\overline{AO} \perp \overline{OB}$, $\angle AOB = 90°$.

Solution

$x + \angle AOB + \angle BOC = 180°$

$x + 90° + \angle BOC = 180°$

$\angle BOC = 90° - x$

The measure of $\angle BOC$ is $90° - x$.

61. **Strategy** To find the measure of the third angle, use the fact that the sum of the measures of the interior angles of a triangle is 180°. Write and solve an equation using x to represent the measure of the third angle.

Solution

$$x+90°+30°=180°$$
$$x+120°=180°$$
$$x=60°$$

The measure of the third angle is 60°.

63. **Strategy** To find the measure of the third angle, use the fact that the sum of the measures of the interior angles of a triangle is 180°. Write an equation using x to represent the measure of the third angle. Solve the equation for x.

Solution

$$x+42°+103°=180°$$
$$x+145°=180°$$
$$x=35°$$

The measure of the third angle is 35°.

65. **Strategy** To find the measure of the third angle, use the fact that the sum of the measures of the interior angles of a triangle is 180°. Write an equation using x to represent the measure of the third angle. Solve the equation for x.

Solution

$$x+13°+65°=180°$$
$$x+78°=180°$$
$$x=102°$$

The measure of the third angle is 102°.

67. False

Applying the Concepts

69. The three angles, a, b, and c, lie along a straight line; they form a straight angle. A straight angle measures 180°. The sum of the measures of the three angles of a triangle is 180°.

71.

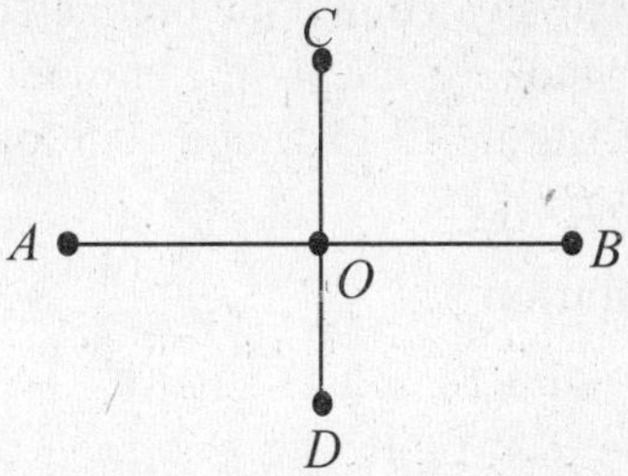

$\angle AOC$ and $\angle BOC$ are supplementary angles; therefore, $\angle AOC+\angle BOC=180°$. Because $\angle AOC$ and $\angle BOC$ are equal, by substitution $\angle AOC+\angle AOC=180°$. Therefore, $2(\angle AOC)=180°$ and $\angle AOC=90°$. Therefore, $\overline{AB}\perp\overline{CD}$.

Section 3.2

Objective A Exercises

1. The polygon has 6 sides so the polygon is a hexagon.

3. The polygon has 5 sides so the polygon is a pentagon.

5. The triangle has no sides equal so the triangle is a scalene triangle.

7. The triangle has three sides equal so the triangle is an equilateral triangle.

9. The triangle has one obtuse angle so the triangle is an obtuse triangle.

11. The triangle has three acute angle so the triangle is an acute triangle.

13. **Strategy** To find the perimeter, use the formula for the perimeter of a triangle. Substitute 12 for a, 20 for b, and 24 for c. Solve for P.

Solution

$P=a+b+c=12+20+24=56$

The perimeter is 56 in.

15. **Strategy** To find the perimeter, use the formula for the perimeter of a square. Substitute 3.5 for s and solve for P.

Solution

$P=4s=4\cdot 3.5=14$

The perimeter is 14 ft.

17. **Strategy** To find the perimeter, use the formula for the perimeter of a rectangle. Substitute 13 for L and 10.5 for W. Solve for P.

Solution

$$P = 2L + 2W = 2\cdot 13 + 2\cdot 10.5$$
$$= 26 + 21 = 47$$

The perimeter is 47 mi.

19. **Strategy** To find the circumference, use the circumference formula that involves the radius. For the exact answer, leave the answer in terms of π. For an approximation, use the π key on a calculator. $r = 4$.

Solution

$$C = 2\pi r = 2\pi(4) = 8\pi \approx 25.13$$

The circumference is 8π cm. The circumference is approximately 25.13 cm.

21. **Strategy** To find the circumference, use the circumference formula that involves the radius. For the exact answer, leave the answer in terms of π. For an approximation, use the π key on a calculator. $r = 5.5$.

Solution

$$C = 2\pi r = 2\pi(5.5) = 11\pi \approx 34.56$$

The circumference is 11π mi. The circumference is approximately 34.56 mi.

23. **Strategy** To find the circumference, use the circumference formula that involves the diameter. For the exact answer, leave the answer in terms of π. For an approximation, use the π key on a calculator. $d = 17$.

Solution

$$C = \pi d = \pi(17) = 17\pi \approx 53.41$$

The circumference is 17π ft. The circumference is approximately 53.41 ft.

25. **Strategy** To find the perimeter, use the formula for the perimeter of a triangle. Substitute 3.8 for a, 5.2 for b, and 8.4 for c. Solve for P.

Solution

$$P = a + b + c = 3.8 + 5.2 + 8.4 = 17.4$$

The perimeter is 17.4 cm.

27. **Strategy** To find the perimeter, use the formula for the perimeter of a triangle. Substitute $2\frac{1}{2}$ for a and b, and 3 for c. Solve for P.

Solution

$$P = a + b + c = 2\frac{1}{2} + 2\frac{1}{2} + 3 = 8$$

The perimeter is 8 cm.

29. **Strategy** To find the perimeter, use the formula for the perimeter of a rectangle. Substitute 8.5 for L and 3.5 for W. Solve for P.

Solution

$$P = 2L + 2W = 2(8.5) + 2(3.5) = 17 + 7 = 24$$

The perimeter is 24 m.

31. **Strategy** To find the perimeter, use the formula for the perimeter of a square. Substitute 12.2 for s. Solve for P.

Solution

$$P = 4s = 4(12.2) = 48.8$$

The perimeter is 48.8 cm.

33. **Strategy** To find the perimeter, multiply the measure of one of the equal sides (3.5) by 5.

Solution

$$P = 5(3.5) = 17.5$$

The perimeter is 17.5 in.

35. **Strategy** To find the diameter, substitute 4.2 for r in the formula $d = 2r$ and solve for d.

Solution

$$d = 2r = 2(4.2) = 8.4$$

The diameter is 8.4 cm.

37. **Strategy** To find the circumference, use the circumference formula that involves the diameter. Leave the answer in terms of π. $d = 1.5$.

Solution

$$C = \pi d = \pi(1.5) = 1.5\pi$$

The circumference is 1.5π in.

39. **Strategy** To find the circumference, use the circumference formula that involves the radius. An approximation is asked for; use the π key on a calculator.
$r = 36$.

Solution
$C = 2\pi r = 2\pi(36) = 72\pi \approx 226.19$

The circumference is approximately 226.19 cm.

41. **Strategy** To find the amount of fencing, use the formula for the perimeter of a rectangle. Substitute 18 for L and 12 for W. Solve for P.

Solution
$P = 2L + 2W = 2(18) + 2(12)$
$= 36 + 24 = 60$

The 60 ft of fencing is needed.

43. **Strategy** To find the amount to be nailed down, use the formula for the perimeter of a rectangle. Substitute 12 for L and 10 for W. Solve for P.

Solution
$P = 2L + 2W = 2(12) + 2(10)$
$= 24 + 20 = 44$

44 ft of carpet must be nailed down.

45. **Strategy** To find the length, use the formula for the perimeter of a rectangle. Substitute 440 for P and 100 for W. Solve for L.

Solution
$P = 2L + 2W$
$440 = 2L + 2(100)$
$440 = 2L + 200$
$240 = 2L$
$120 = L$

The length is 120 ft.

47. **Strategy** To find the third side of the banner, use the formula for the perimeter of a triangle. Substitute 46 for P, 18 for a, and 18 for b. Solve for c.

Solution
$P = a + b + c$
$46 = 18 + 18 + c$
$46 = 36 + c$
$10 = c$

The third side of the banner is 10 in.

49. **Strategy** To find the length of each side, use the formula for the perimeter of a square. Substitute 48 for P. Solve for s.

Solution
$P = 4s$
$48 = 4s$
$12 = s$

The length of each side is 12 in.

51. **Strategy** To find the length of the diameter, use the circumference formula that involves the diameter. An approximation is asked for; use the π key on a calculator. $C = 8$.

Solution
$C = \pi d$
$8 = \pi d$
$\frac{8}{\pi} = d$
$2.55 \approx d$

The diameter is approximately 2.55 cm.

53. **Strategy** To find the length of molding, use the circumference formula that involves the diameter. An approximation is asked for; use the π key on a calculator. $d = 4.2$.

Solution
$C = \pi d = \pi(4.2) \approx 13.19$

The length of molding is approximately 13.19 ft.

55. **Strategy** To find the distance:
- Convert the diameter to feet.
- Multiply the circumference by 8.

An approximation is asked for; use the π key on a calculator.

Solution
24 in. = 2 ft
$\text{distance} = 8C = 8\pi d = 8\pi(2) = 16\pi \approx 50.27$

The bicycle travels approximately 50.27 ft.

57. **Strategy** To find the circumference of the earth, use the circumference formula that involves the radius. An approximation is asked for; use the π key on a calculator. $r = 6356$.

Solution
$C = 2\pi r = 2\pi(6356) = 12,712\pi \approx 39,935.93$

The circumference of the earth is approximately 39,935.93 km.

59. The perimeter of a square is $P = 4s$, and the circumference of a circle is $C = \pi d$. If the side of the square is equal to the diameter

of the circle, then $s = d$. The perimeter of

the square can be written $P = 4d$.

$4d > \pi d$, since $4 > \pi$. The perimeter of

the square is greater than the

circumference of the circle.

Objective B Exercises

61. **Strategy** To find the area, use the formula for the area of a rectangle. Substitute 12 for L and 5 for W. Solve for A.

Solution
$A = LW = 12(5) = 60$

The area is $60\ \text{ft}^2$.

63. **Strategy** To find the area, use the formula for the area of a square. Substitute 4.5 for s. Solve for A.

Solution

$A = s^2 = (4.5)^2 = 20.25$

The area is $20.25\ \text{in}^2$.

65. **Strategy** To find the area, use the formula for the area of a triangle. Substitute 42 for b and 26 for h. Solve for A.

Solution
$A = \frac{1}{2}bh = \frac{1}{2}(42)(26) = 546$

The area is $546\ \text{ft}^2$.

67. **Strategy** To find the area, use the formula for the area of a circle. Substitute 4 for r. Solve for A. For the exact answer, leave the answer in terms of π. For an approximation, use the π key on a calculator.

Solution
$A = \pi r^2 = \pi(4)^2 = 16\pi \approx 50.27$

The area is $16\pi\ \text{cm}^2$. The area is approximately $50.27\ \text{cm}^2$.

69. **Strategy** To find the area, use the formula for the area of a circle. Substitute 5.5 for r. Solve for A. For the exact answer, leave the answer in terms of π. For an approximation, use the π key on a calculator.

Solution

$A = \pi r^2 = \pi(5.5)^2 = 30.25\pi \approx 95.03$

The area is $30.25\pi\ \text{mi}^2$. The area is approximately $95.03\ \text{mi}^2$.

71. **Strategy** To find the area:
- Find the radius of the circle.
- Use the formula for the area of a circle.

For the exact answer, leave the answer in terms of π. For an approximation, use the π key on a calculator.

Solution
$r = \frac{1}{2}d = \frac{1}{2}(17) = 8.5$

$A = \pi r^2 = \pi(8.5)^2 = 72.25\pi \approx 226.98$

The area is $72.25\pi\ \text{ft}^2$. The area is approximately $226.98\ \text{ft}^2$.

73. **Strategy** To find the area, use the formula for the area of a square. Substitute 12.5 for s. Solve for A.

Solution
$A = s^2 = (12.5)^2 = 156.25$

The area is 156.25 cm^2.

75. **Strategy** To find the area, use the formula for the area of a rectangle. Substitute 38 for L and 15 for W. Solve for A.

Solution

$A = LW = 38(15) = 570$

The area is 570 in^2.

77. **Strategy** To find the area, use the formula for the area of a parallelogram. Substitute 16 for b and 12 for h. Solve for A.

Solution

$A = bh = 16(12) = 192$

The area is 192 in^2.

79. **Strategy** To find the area, use the formula for the area of a triangle. Substitute 6 for b and 4.5 for h. Solve for A.

Solution

$A = \frac{1}{2}bh = \frac{1}{2}(6)(4.5) = 13.5$

The area is 13.5 ft^2.

81. **Strategy** To find the area, use the formula for the area of a trapezoid. Substitute 35 for b_1, 20 for b_2, and 12 for h. Solve for A.

Solution

$A = \frac{1}{2}h(b_1 + b_2) = \frac{1}{2} \cdot 12(35 + 20) = 330$

The area is 330 cm^2.

83. **Strategy** To find the area, use the formula for the area of a circle. Leave the answer in terms of π. $r = 5$.

Solution

$A = \pi r^2 = \pi(5)^2 = 25\pi$

The area is $25\pi \text{ in}^2$.

85. **Strategy** To find the area:

- Find the radius of the circle. $d = 3.4$
- Use the formula for the area of a circle.

An approximation is asked for; use the π key on a calculator.

Solution

$r = \frac{1}{2}d = \frac{1}{2}(3.4) = 1.7$

$A = \pi r^2 = \pi(1.7)^2 = 2.89\pi \approx 9.08$

The area is approximately 9.08 ft^2.

87. **Strategy** To find the area:

- Find the radius of the telescope. $d = 200$
- Use the formula for the area of a circle.

Leave the answer in terms of π.

Solution

$r = \frac{1}{2}d = \frac{1}{2}(200) = 100$

$A = \pi r^2 = \pi(100)^2 = 10{,}000\pi$

The area is $10{,}000\pi \text{ in}^2$.

89. **Strategy** To find the area, use the formula for the area of a rectangle. Substitute 14 for L and 9 for W. Solve for A.

Solution

$A = LW = 14(9) = 126$

The area of the flower garden is 126 ft^2.

91. **Strategy** To find the amount of turf, use the formula for the area of a rectangle. Substitute 100 for L and 75 for W. Solve for A.

Solution

$A = LW = 100(75) = 7500$

7500 yd^2 of artificial turf must be purchased.

93. **Strategy** To find the width, use the formula for the area of a rectangle. Substitute 300 for A and 30 for L. Solve for W.

Solution

$$A = LW$$
$$300 = 30W$$
$$10 = W$$

The width of the rectangle is 10 in.

95. **Strategy** To find the length of the base, use the formula for the area of a triangle. Substitute 50 for A and 5 for h. Solve for b.

Solution

$A = \frac{1}{2}bh$

$50 = \frac{1}{2}b(5)$

$50 = \frac{5}{2}b$

$20 = b$

The base of the triangle is 20 m.

97. **Strategy** To find the number of quarts of stain:

- Use the formula for the area of a rectangle to find the area of the deck.
- Divide the area of the deck by the area one quart will cover (50).

Solution

$A = LW$

$A = 10(8)$

$A = 80$

$80 \div 50 = 1.6$

Because a portion of a second quart is needed, 2 qt of stain should be purchased.

99. **Strategy** To find the cost of the wallpaper:

- Use the formula for the area of a rectangle to find the areas of the two walls.
- Add the areas of the two walls.
- Divide the total area by the area in one roll (40) to find the total number of rolls.
- Multiply the number of rolls by 24.50.

Solution

$A_1 = LW = 9(8) = 72$

$A_2 = LW = 11(8) = 88$

$A = A_1 + A_2 = 72 + 88 = 160$

$160 \div 40 = 4$

$4 \cdot 24.50 = 98$

The cost to wallpaper the two walls is $98.

101. **Strategy** To find the increase in area:

- Use the formula for the area of a circle to find the area of a circle with $r = 8$.
- Use the formula for the area of a circle to find the area of a circle with radius $r = 8 + 2 = 10$.
- Subtract the area of the smaller circle from the area of the larger circle. An approximation is asked for; use the π key on a calculator.

Solution

$A_1 = \pi r^2 = \pi(8)^2 = 64\pi$

$A_2 = \pi(10)^2 = 100\pi$

$A_2 - A_1 = 100\pi - 64\pi = 36\pi \approx 113.10$

The area is increased by 113.10 in^2.

103. **Strategy** To find the area of the walkway:

- Use the formula for the area of a rectangle to find the plot of grass. Substitute 30 for L and 20 for W.
- Use the formula for a rectangle to find the area of the total area (walkway + grass). Substitute 34 $(30 + 2 + 2)$ for L and 24 $(20 + 2 + 2)$ for W.
- Subtract the area of the grass from the total area.

Solution

$A_1 = LW = 30(20) = 600$

$A_2 = LW = 34(24) = 816$

$A_2 - A_1 = 816 - 600 = 216$

The area of the walkway is 216 m^2.

105. **Strategy** To find the cost of the carpet:

- Use the formula for the area of a rectangle to find the area of the carpet.
- Use the conversion factor $\frac{1 \text{ yd}^2}{9 \text{ ft}^2}$.
- Multiply the area measured in square yards by 21.95.

Solution

$A = LW = 24(15) = 360$

$360 \text{ ft}^2 = 360 \text{ ft}^2 \cdot \frac{1 \text{ yd}^2}{9 \text{ ft}^2} = 40 \text{ yd}^2$

$40(21.95) = 878$

The cost of the carpet is $878.

107. Figure 1:

$A = \frac{1}{2}bh$

$A = \frac{1}{2}xy$

Figure 2:

$A = \frac{1}{2}bh$

$A = \frac{1}{2}xy$

The area of the first triangle is equal to the area of the second triangle.

Applying the Concepts

109. $A = LW$

Double the length and double the width:

$A = (2L)(2W) = 4LW$

$4LW$ is four times the quantity LW.

The area of the resulting rectangle is 4 times larger.

Section 3.3

Objective A Exercises

1. **Strategy** To find the volume, use the formula for the volume of a rectangular solid. $L = 14$, $W = 10$, $H = 6$.

Solution $V = LWH = 14(10)(6) = 840$

The volume is 840 in^3.

3. **Strategy** To find the volume, use the formula for the volume of a pyramid. $s = 3$, $h = 5$.

Solution

$V = \frac{1}{3}s^2h = \frac{1}{3}(3^2)(5) = \frac{1}{3}(9)(5) = 15$

The volume is 15 ft^3.

5. **Strategy** To find the volume:

- Find the radius of the sphere. $d = 3$.
- Use the formula for the volume of a sphere.

Solution

$r = \frac{1}{2}d = \frac{1}{2}(3) = 1.5$

$V = \frac{4}{3}\pi r^3 = \frac{4}{3}\pi(1.5)^3 = \frac{4}{3}\pi(3.375)$

$V = 4.5\pi \approx 14.14$

The volume is 4.5π cm^3. The volume is approximately 14.14 cm^3.

7. **Strategy** To find the volume, use the formula for the volume of a rectangular solid. $L = 6.8$, $W = 2.5$, $H = 2$.

Solution

$V = LWH = 6.8(2.5)(2) = 34$

The volume of the storage unit is 34 m^3.

9. **Strategy** To find the volume, use the formula for the volume of a cube. $s = 3.5$.

Solution

$V = s^3 = (3.5)^3 = 42.875$

The volume of the cube is 42.875 in^3.

11. **Strategy** To find the volume:

- Find the radius of the sphere. $d = 6$.
- Use the formula for the volume of a sphere.

Solution

$r = \frac{1}{2}d = \frac{1}{2}(6) = 3$

$V = \frac{4}{3}\pi r^3 = \frac{4}{3}\pi(3)^3 = \frac{4}{3}\pi(27) = 36\pi$

The volume is 36π ft^3.

13. **Strategy** To find the volume:

- Find the radius of the cylinder. $d = 24$.
- Use the formula for the volume of a cylinder. $h = 18$.

Solution

$r = \frac{1}{2}d = \frac{1}{2}(24) = 12$

$V = \pi r^2 h = \pi(12^2)(18) = \pi(144)(18)$

$V = 2592\pi \approx 8143.01$

The volume of the cylinder is approximately 8143.01 cm^3.

15. **Strategy** To find the volume, use the formula for the volume of a cone. $r = 5$, $h = 9$.

Solution

$V = \frac{1}{3}\pi r^2 h = \frac{1}{3}\pi(5)^2(9) = \frac{1}{3}\pi(25)(9) = 75\pi$

The volume of the cone is 75π in^3.

17. **Strategy** To find the volume, use the formula for the volume of a pyramid. $s = 6$, $h = 10$.

Solution

$V = \frac{1}{3}s^2h = \frac{1}{3}(6^2)(10) = \frac{1}{3}(36)(10) = 120$

The volume of the pyramid is 120 in^3.

19. **Strategy** To find the width, use the formula for the volume of a rectangular solid. $V = 52.5$, $L = 7$, $H = 3$.

Solution

$V = LWH$

$52.5 = 7(W)(3)$

$52.5 = 21W$

$2.5 = W$

The width of the freezer is 2.5 ft.

21. **Strategy** To find the height, use the formula for the volume of a cylinder. $V=502.4$, $d=10$, $r=5$.

Solution

$$V=\pi r^2 h$$
$$502.4=\pi(5)^2 h$$
$$502.4=25\pi h$$
$$6.40\approx h$$

The height of the cylinder is approximately 6.40 in.

23. **Strategy** To find the volume, use the formula for the volume of a rectangular solid. $L = 360$, $w = 160$, and $h = 3$.

Solution

$$V=LWH$$
$$V=360(160)(3)$$
$$V=172{,}800$$

The volume of the guacamole would be 172,800 ft^3.

25. **Strategy** Compare the volume of a cube with side = r to the volume of a sphere.

Solution

Cube

$V=s^3$

$V=r^3$

Sphere

$V=\frac{4}{3}\pi r^3$

$\frac{4}{3}\pi r^3 > r^3$, so the volume of the sphere is greater than the volume of the cube.

Objective B Exercises

27. **Strategy** To find the surface area, use the formula for the surface area of a rectangular solid. $L=5$, $W=4$, $H=3$.

Solution

$$SA=2LW+2LH+2WH$$
$$SA=2(5)(4)+2(5)(3)+2(4)(3)$$
$$SA=40+30+24=94$$

The surface area of the rectangular solid is 94 m^2.

29. **Strategy** To find the surface area, use the formula for the surface area of a pyramid. $s=4$, $l=5$.

Solution

$$SA=s^2+2sl$$
$$SA=4^2+2(4)(5)=16+40=56$$

The surface area of the pyramid is 56 m^2.

31. **Strategy** To find the surface area, use the formula for the surface area of a cylinder. $r=6$, $h=2$.

Solution $SA=2\pi r^2+2\pi rh$

$$SA=2\pi(6^2)+2\pi(6)(2)=2\pi(36)+24\pi$$
$$SA=72\pi+24\pi=96\pi\approx 301.59$$

The surface area of the cylinder is 96π in^2. The surface area of the cylinder is approximately 301.59 in^2.

33. **Strategy** To find the surface area, use the formula for the surface area of a rectangular solid. $H=5$, $L=8$, $W=4$.

Solution

$$SA=2LW+2LH+2WH$$
$$SA=2(8)(4)+2(8)(5)+2(4)(5)$$
$$SA=64+80+40=184$$

The surface area of the rectangular solid is 184 ft^2.

35. **Strategy** To find the surface area, use the formula for the surface area of a cube. $s=$ 3.4

Solution

$$SA=6s^2=6(3.4)^2=6(11.56)=69.36$$

The surface area of the cube is 69.36 m^2.

37. **Strategy** To find the surface area:

- Find the radius of the sphere. $d=15$.
- Use the formula for the surface area of a sphere.

Solution

$$r=\frac{1}{2}d=\frac{1}{2}(15)=7.5$$
$$SA=4\pi r^2=4\pi(7.5)^2=4\pi(56.25)=225\pi$$

The surface area of the sphere is 225π cm^2.

39. **Strategy** To find the surface area, use the formula for the surface area of a cylinder. $r=4$, $h=12$.

Solution

$SA=2\pi r^2+2\pi rh=2\pi(4)^2+2\pi(4)(12)$

$SA=2\pi(16)+96\pi=32\pi+96\pi=128\pi$

$SA\approx 402.12$

The surface area of the cylinder is approximately 402.12 in^2.

41. **Strategy** To find the surface area, use the formula for the surface area of a cone. $r=1.5$, $l=2.5$.

Solution

$SA=\pi r^2+\pi rl=\pi(1.5)^2+\pi(1.5)(2.5)$

$SA=\pi(2.25)+3.75\pi=6\pi$

The surface area of the cone is 6π ft^2.

43. **Strategy** To find the surface area, use the formula for the surface area of a pyramid. $s=9$, $l=12$.

Solution

$SA=s^2+2sl=9^2+2(9)(12)$

$SA=81+216=297$

The surface area of the pyramid is 297 in^2.

45. **Strategy** To find the width, use the formula for the surface area of a rectangular solid. $SA=108$, $L=6$, and $H=4$.

Solution

$$\begin{aligned}SA&=2LW+2LH+2WH\\108&=2(6)W+2(6)(4)+2W(4)\\108&=12W+48+8W\\108&=20W+48\\60&=20W\\3&=W\end{aligned}$$

The width of the rectangular solid is 3 cm.

47. **Strategy** To find the number of cans of paint:

• Find the formula for the surface area of a cylinder. $r=12$, $h=30$.

• Divide the surface area by 300.

Solution

$SA=2\pi r^2+2\pi rh=2\pi(12)^2+2\pi(12)(30)$

$SA=2\pi(144)+720\pi=288\pi+720\pi=1008\pi$

$1008\pi\div 300\approx 10.56$

Because a portion of an eleventh can is needed, 11 cans of paint should be purchased.

49. **Strategy** To find the amount of glass, use the formula for the surface area of a rectangular solid. Omit the top of the fish tank. The formula becomes $SA=LW+2LH+2WH$. $L=12$, $W=8$, $H=9$.

Solution

$SA=LW+2LH+2WH$

$SA=12(8)+2(12)(9)+2(8)(9)$

$SA=96+216+144=456$

The fish tank requires 456 in^2 of glass.

51. **Strategy** To find the difference in area:

• Use the formula for the surface area of a pyramid. $s=5$, $l=8$.

• Use the formula for the surface area of a cone. $r=\frac{1}{2}d=\frac{1}{2}(5)=2.5$, $l=8$.

• Subtract the surface area of the cone from the surface area of the pyramid.

Solution

$SA=s^2+2sl=5^2+2(5)(8)=25+80=105$

$$\begin{aligned}SA&=\pi r^2+\pi rl=\pi(2.5)^2+\pi(2.5)(8)\\&=\pi(6.25)+20\pi\\&=26.25\pi\approx 82.47\end{aligned}$$

$105-82.47=22.53$

The surface area of the pyramid is approximately 22.53 cm^2 larger than the surface area of the cone.

53. **a.** The distance from the edge of the base to the vertex of a regular pyramid is longer than the distance, perpendicular to the base, from the base to the vertex. The statement is always true.

b. The distance from the edge of the base of a cone to the vertex is longer than the distance, perpendicular to the base, from the base to the vertex. The statement is never true.

c. The four triangular faces of a regular pyramid could be equilateral triangles, but they could be isosceles triangles that are not equilateral. The statement is sometimes true.

Applying the Concepts

55. **a.** For example, cut perpendicular to the top and bottom faces and parallel to two of the sides.

b. For example, beginning at an edge that is perpendicular to the bottom face, cut at an angle through to the bottom face.

c. For example, beginning at the top face, at a distance *d* from a vertex, cut at an angle to the bottom face, ending at a distance greater than *d* from the opposite vertex.

d. For example, beginning on the top face, at a distance *d* from a vertex, cut across the cube to a point just above the opposite vertex.

Chapter 3 Concept Review

1. Perpendicular lines are intersecting lines that form right angles. [3.1A]
2. Two angles are complementary when the sum of their measures is 90°. [3.1A]
3. The obtuse angles $\angle a$, $\angle c$, $\angle w$, and $\angle y$ have the same measure. [3.1B]
4. The sum of the measures of the three angles in a triangle is 180°. [3.1C]
5. To find the radius of a circle when you know the diameter, multiply the diameter by $\frac{1}{2}$. [3.2A]
6. The formula for the perimeter of a rectangle is $P = 2L + 2W$, where P is the perimeter, L is the length, and W is the width. [3.2A]
7. To find the circumference of a circle, multiply π times the diameter or multiply 2π times the radius. [3.2A]
8. The formula for the area of a triangle is $A = \frac{1}{2}bh$, where A is the area, b is the base, and h is the height. [3.2B]
9. To find the volume of a rectangular solid, you need to know the length, the width, and the height. [3.3A]
10. The perimeter of a plane geometric figure is the measure of the distance around the figure. Perimeter is measured in linear units. Area is a measure of the amount of space in a region. Area is measured in square units. [3.2A]
11. The surface area of a rectangular solid is the sum of the areas of six rectangles. [3.3B]

Chapter 3 Review Exercises

1. **Strategy** • To find the measure of angle *c*, use the fact that the sum of an interior and an exterior angle is 180°.

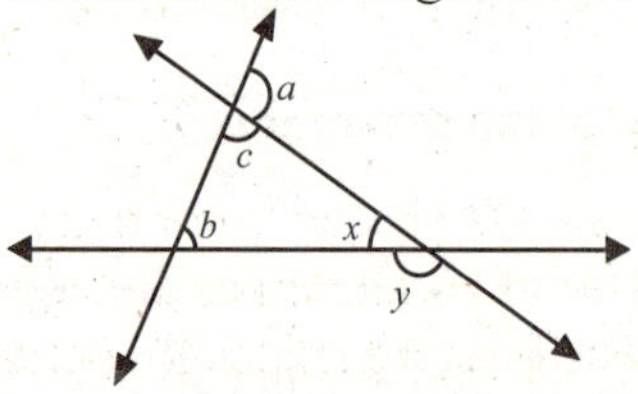

• To find the measure of angle *x*, use the fact that the sum of the measurements of the interior angles of a triangle is 180°.
• To find the measure of angle *y*, use the fact that the sum of an interior and an exterior angle is 180°.

Solution

$$\begin{aligned} \angle a + \angle c &= 180^\circ \\ 74^\circ + \angle c &= 180^\circ \\ \angle c &= 106^\circ \end{aligned}$$

$$\begin{aligned} \angle b + \angle c + \angle x &= 180^\circ \\ 52^\circ + 106^\circ + \angle x &= 180^\circ \\ 158^\circ + \angle x &= 180^\circ \\ \angle x &= 22^\circ \end{aligned}$$

$\angle x + \angle y = 180°$
$22° + \angle y = 180°$,
$\angle y = 158°$

The measure of $\angle x$ is 22° and the measure of $\angle y$ is 158°.

2. **Strategy** To find x, use the fact that adjacent angles of intersecting lines are supplementary.

Solution

$112° + x = 180°$

$x = 68°$

The measure of x is 68°.

3. $AC = AB + BC = 3(BC) + BC$

$= 4(BC) = 4(11) = 44$

The length of AC is 44 cm.

4. **Strategy** The sum of the measures of the three angles shown is 180°. To find x, write an equation and solve for x.

Solution

$4x + 3x + (x + 28°) = 180°$

$8x + 28° = 180°$

$8x = 152°$

$x = 19°$

The measure of x is 19°.

5. **Strategy** To find the volume, use the formula for the volume of a pyramid. $s = 6$, $h = 8$.

Solution

$V = \frac{1}{3}s^2h = \frac{1}{3}(6)^2(8) = \frac{1}{3}(36)(8) = 96 \text{ cm}^3$

The volume of the pyramid is 96 cm^3.

6. **Strategy** $\angle a = 138°$ because alternate interior angles of parallel lines are equal. $\angle a + \angle b = 180°$ because adjacent angles of intersecting lines are supplementary.

Solution

$\angle a + \angle b = 180°$

$\angle a = 138°$

$138° + \angle b = 180°$

$\angle b = 42°$

The measure of $\angle a$ is 138°. The measure of $\angle b$ is 42°.

7. **Strategy** To find the surface area, use the formula for the surface area of a rectangular solid. $L = 10$, $W = 5$, $H = 4$.

Solution

$SA = 2LW + 2LH + 2WH$

$SA = 2(10)(5) + 2(10)(4) + 2(5)(4)$

$SA = 100 + 80 + 40 = 220$

The surface area of the solid is 220 ft^2.

8. **Strategy** Supplementary angles are two angles whose sum is 180°. To find the supplement, let x represent the supplement of a 32° angle. Write and equation and solve for x.

Solution

$32° + x = 180°$

$x = 148°$

The supplement of a 32° angle is a 148° angle.

9. **Strategy** To find the area, use the formula for the area of a rectangle. Substitute 12 for L and 6.5 for W. Solve for A.

Solution

$A = LW = 12(6.5) = 78$

The area is 78 cm^2.

10. **Strategy** To find the area, use the formula for the area of a rectangle. Substitute 9 for b and 14 for h. Solve for A.

Solution

$A = \frac{1}{2}bh = \frac{1}{2}(9)(14) = 63$

The area is 63 cm^2.

11. **Strategy** To find the volume, use the formula for the volume of a rectangular solid. $L = 6.5$, $W = 2$, $H = 3$.

Solution

$V = LWH = (6.5)(2)(3) = 39$

The area is 39 ft^3.

12. Strategy To find the third angle, use the fact that the sum of the measures of the interior angles of a triangle is 180°. Let $x =$ the third angle.

Solution

$$37°+48°+x = 180°$$
$$85°+x = 180°$$
$$x = 95°$$

The third angle is 95°.

13. Strategy To find the base, use the formula for the area of a triangle. Substitute 7 for *h* and 28 for *A*. Solve for *b*.

Solution

$$A = \frac{1}{2}bh$$
$$28 = \frac{1}{2}b(7)$$
$$56 = 7b$$
$$8 = b$$

The base of the triangle is 8 cm.

14. Strategy To find the volume, use the formula for the volume of a sphere. The radius of the sphere is one-half the diameter.

Solution

$$r = \frac{1}{2}d = \frac{1}{2}(12) = 6$$
$$V = \frac{4}{3}\pi r^3 = \frac{4}{3}\pi(6^3) = \frac{4}{3}\pi(216) = 288\pi$$

The volume is $288\pi\ \text{mm}^3$.

15. Strategy To find the volume, use the formula for the volume of a right circular cone. Substitute 7 for *r* and 16 for *h*. Solve for *V*. Leave the answer in terms of π.

Solution

$$V = \frac{1}{3}\pi r^2 h = \frac{1}{3}\pi(7)^2(16) = \frac{1}{3}\pi(49)(16)$$
$$= \frac{784\pi}{3}$$

The volume is $\frac{784\pi}{3}\ \text{cm}^3$.

16. Strategy To find the length of each side, use the formula for the perimeter of a square. $P = 86$.

Solution

$$P = 4s$$
$$86 = 4s$$
$$21.5 = s$$

The side of the square is 21.5 cm.

17. Strategy To find the number of cans of paint:

- Find the surface area by using the formula for the surface area of a cylinder.
- Divide the surface area by 200.

Solution

$$SA = 2\pi r^2 + 2\pi rh$$
$$SA = 2\pi(6)^2 + 2\pi(6)(15)$$
$$SA = 2\pi(36) + 180\pi = 72\pi + 180\pi$$
$$SA = 252\pi$$
$$252\pi \div 200 \approx 3.96$$

Because a portion of a fourth can is needed, 4 cans of paint should be purchased.

18. Strategy To find the amount of fencing, use the formula for the perimeter of a rectangle.

Solution

$$P = 2L + 2W = 2(56) + 2(48) = 112 + 96$$
$$= 208$$

208 yd of fencing are needed to fence the park.

19. a. Strategy To find the area, use the formula for the volume of a rectangular solid. $L = 9$, and $W = 5$.

Solution

$$A = LW = 9 \cdot 5 = 45$$

The area of the cell is 45 ft ^{2}.

b. Strategy To find the volume, use the formula for the area of a rectangular solid. $L = 9$, $W = 5$, $H = 7$.

Solution

$$V = LWH = 9 \cdot 5 \cdot 7 = 315$$

The volume of the cell is 315 ft ^{3}.

20. Strategy To find the area of the walkway:
- Find the length and width of the total area.
- Find the total area.
- Subtract the area of the plot of grass from the total area.

Solution

$L = 40 + 4 = 44$

$W = 25 + 4 = 29$

$A = 44 \times 29 = 1276$

Area of the grass:

$A = 40 \times 25 = 1000$

Area of the walkway is 1276 – 1000 = 276

The area of the walkway is 276 m^2.

Chapter 3 Test

1. $r = \frac{1}{2}d = \frac{1}{2}(1.5) = 0.75$

The radius is 0.75 m.

2. $C = \pi d = 2\pi r = 2 \cdot \pi \cdot 5 = 31.42$

The circumference is 31.42 cm.

3. $P = 2L + 2W = 2(8) + 2(5) = 26$

The perimeter is 26 ft.

4.
$$\begin{aligned} AB + BC + CD &= AD \\ 15 + BC + 6 &= 24 \\ 21 + BC &= 24 \\ BC &= 3 \end{aligned}$$

BC is 3.

5. $r = \frac{1}{2}d = \frac{1}{2}(8) = 4$

$V = \frac{4}{3}\pi r^3 = \frac{4}{3}\pi(4)^3 = \frac{4}{3}\pi(64) = 268.08$

The volume of the sphere is 268.08 ft^3.

6. $r = \frac{1}{2}d = \frac{1}{2}(9) = 4.5$

$A = \pi r^2 = \pi(4.5)^2 = 63.62$

The area is 63.62 cm^2.

7.
$$\begin{aligned} \angle b &= 80^\circ \\ \angle a + \angle b &= 180^\circ \\ \angle a + 80^\circ &= 180^\circ \\ \angle a &= 100^\circ \end{aligned}$$

The measure of angle a is 100° and the measures of angle b is 80°.

8.
$$\begin{aligned} x + 105^\circ &= 180^\circ \\ x &= 75^\circ \end{aligned}$$

The supplement of a 105° angle is a 75° angle.

9.
$$\begin{aligned} a + 45^\circ &= 180^\circ \\ a &= 135^\circ \\ b &= 45^\circ \end{aligned}$$

The measure of angle a is 135° and the measure of angle b is 45°.

10. $A = LW = (11)(5) = 55$

The area is 55 m^2.

11. $V = \pi r^2 h = \pi(3)^2(6) = 169.65$

The volume is 169.65 m^3.

12. $P = 2L + 2W = 2(2) + 2(1.4) = 6.8$

The perimeter is 6.8 m.

13.
$$\begin{aligned} x + 32^\circ &= 90^\circ \\ x &= 58^\circ \end{aligned}$$

The complement of a 32° angle is a 58^0 angle.

14.
$$\begin{aligned} SA &= 2\pi r^2 + 2\pi rh \\ &= 2\pi(2.5)^2 + 2\pi(2.5)(8) \\ &= 164.93 \text{ ft}^2 \end{aligned}$$

The surface area is 164.93 ft^2.

15.
$$\begin{aligned} \pi r_2^2 - \pi r_1^2 &= \pi(10^2) - \pi(8^2) \\ &= 100\pi - 64\pi = 36\pi = 113.10 \text{ in}^2 \end{aligned}$$

There is 113.10 in^2 more in a 10 inch pizza than in an 8 inch pizza.

16. Right triangles have a 90° angle and two acute angles.

$$\begin{aligned} 32^\circ + 90^\circ + x &= 180^\circ \\ x &= 58^\circ \end{aligned}$$

The measure of the other two angles in the triangle are 90° and 58°.

17. Distance $= 10C$

$C = \pi D = 3.14 \cdot 28 \div 12 \text{ in} \approx 7.33 \text{ ft}$

Distance $= 10 \cdot 7.33 = 73.3$ ft

The bicycle traveled 73.3 ft.

18. $A = LW = 18 \cdot 14 \div 9 = 28 \text{ yd}^2$

The area of the room is 28 yd^2.

19. $r=\frac{1}{2}d=\frac{1}{2}(9)=4.5$

$V=\pi r^2h=\pi\cdot4.5^2\cdot18=1145.11\text{ ft}^3$

The volume of the silo is 1145.11 ft^2.

20. $A=\frac{1}{2}bh=\frac{1}{2}(8)(2.75)=11\text{ m}^2$

The area is 11 m^2.

Cumulative Review Exercises

1. $x\le1$

$-3\le1$ True

$0\le1$ True

$1\le1$ True

The inequality is true for –3, 0, and 1.

2. $8.9\%=8.9(0.01)=0.089$

3. $\frac{7}{20}=\frac{7}{20}(100\%)=\frac{700}{20}\%=35\%$

4. $-\frac{4}{9}\div\frac{2}{3}=-\frac{4}{9}\cdot\frac{3}{2}=-\left(\frac{4\cdot3}{9\cdot2}\right)=-\frac{2}{3}$

5. $5.7(-4.3)=-24.51$

6. $-\sqrt{125}=-\sqrt{25\cdot5}=-\sqrt{25}\sqrt{5}=-5\sqrt{5}$

7. $5-3\left[10+(5-6)^2\right]=5-3\left[10+(-1)^2\right]=5-3[10+1]$

$=5-3[11]=5-33=-28$

8. $a(b-c)^3=-1(-2-[-4])^3=-1(-2+4)^3=-1(2)^3$

$=-1(8)=-8$

9. $5m+3n-8m=(5m-8m)+3n$

$=-3m+3n$

10. $-7(-3y)=[-7(-3)]y=21y$

11. $4(3x+2)-(5x-1)=12x+8-5x+1$

$=7x+9$

12. $\{-2,-1\}$

13. $C\cup D=\{-10,0,10,20,30\}$

14.

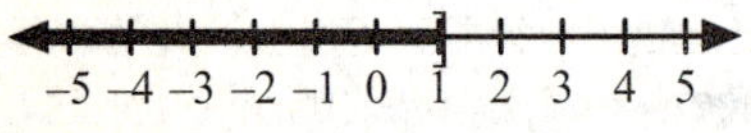

15. $4x+2=6x-8$

$4x-6x+2=6x-6x-8$

$-2x+2=-8$

$-2x+2-2=-8-2$

$-2x=-10$

$\frac{-2x}{-2}=\frac{-10}{-2}$

$x=5$

The solution is 5.

16. $3(2x+5)=18$

$6x+15=18$

$6x+15-15=18-15$

$6x=3$

$\frac{6x}{6}=\frac{3}{6}$

$x=\frac{1}{2}$

The solution is $\frac{1}{2}$.

17. $4y-3\ge6y+5$

$-2y-3\ge5$

$-2y\ge8$

$\frac{-2y}{-2}\le\frac{8}{-2}$

$y\le-4$

The solution set is $\{y|y\le-4\}$.

18. $8-4(3x+5)\le6(x-8)$

$8-12x-20\le6x-48$

$-12-12x\le6x-48$

$-12-18x\le-48$

$-18x\le-36$

$\frac{-18x}{-18}\ge\frac{-36}{-18}$

$x\ge2$

The solution set is $\{x|x\ge2\}$.

19. $2x-3>5$ or $x+4<1$

$2x>8$ $\quad x<-3$

$x>4$

$\{x|x>4\}$ $\quad \{x|x<-3\}$

$\{x|x>4\}\cup\{x|x<-3\}=\{x|x<-3\text{ or }x>4\}$

20. $-3\le2x-7\le5$

$-3+7\le2x-7+7\le5+7$

$4\le2x\le12$

$\frac{4}{2}\le\frac{2x}{2}\le\frac{12}{2}$

$2\le x\le6$

$\{x|2\le x\le6\}$

21. $|3x-1|=2$

$3x-1=2$ or $3x-1=-2$

$3x=3$ $\quad 3x=-1$

$x=1$ $\quad x=-\frac{1}{3}$

The solutions are 1 and $-\frac{1}{3}$.

22. $|x-8|\le 2$

$-2\le x-8\le 2$

$-2+8\le x-8+8\le 2+8$

$6\le x\le 10$

$\{x|6\le x\le 10\}$

23. Strategy To find x, use the fact that adjacent angles of intersecting lines are supplementary.

Solution

$49°+x=180°$

$x=131°$

The measure of x is 131°.

24. Strategy The unknown number: x

the difference between four times a number and ten	is	two

$4x-10=2$

$4x=12$

$x=3$

The number is 3.

25. Strategy To find the third angle, use the fact that the sum of the measures of the interior angles of a triangle is 180°. Let $x=$ the third angle.

Solution

$37°+21°+x=180°$

$58°+x=180°$

$x=122°$

The third angle is 122°.

26. Strategy: Total amount of interest earned: x

	Principal	Rate	Interest
Amount at 4.5%	5000	0.045	0.045(5000)
Amount at 3.5 %	2500	0.035	0.035(2500)

Solution:

$0.045(5000)+0.035(2500)=x$

$225+87.5=x$

$312.5=x$

The total amount of interest earned was $312.50.

27. Strategy To find the third side of the triangle, use the formula for the perimeter of a triangle. In an isosceles triangle, two sides are of equal measure.

Solution

$P=a+b+c$

$19.5=7.5+7.5+c$

$19.5=15+c$

$4.5=c$

The third side measures 4.5 m.

28. Strategy To find the percent, use the basic percent equation.
Percent $=P$, base = 49,982, amount = 35,408.

Percent · Base = Amount

Solution

$P\cdot 49{,}982=35{,}408$

$P=\frac{35{,}408}{49{,}982}=0.708=70.8\%$

70.8% of the men's median annual earnings is the women's median annual earnings.

29. Strategy To find the area:

- Find the radius of the circle.
- Use the formula for the area of a circle.

Leave the answer in terms of π.

Solution

$r=\frac{1}{2}d=\frac{1}{2}(9)=4.5$

$A=\pi r^2=\pi(4.5)^2=20.25\pi$

The area is 20.25π cm².

30. Strategy Use the formula for the volume of a rectangular solid. Substitute 144 for V, 12 for L, and 4 for W. Solve for H.

Solution

$V=LWH$

$144=12(4)H$

$144=48H$

$3=H$

The height is 3 ft.

Chapter 4: Linear Functions and Inequalities in Two Variables

Prep Test

1. $-4(x-3) = -4x + 12$

2. $\sqrt{(-6)^2 + (-8)^2} = \sqrt{36+64} = \sqrt{100} = 10$

3. $\dfrac{3-(-5)}{2-6} = \dfrac{3+5}{2-6} = \dfrac{8}{-4} = -2$

4. $-2x + 5$
   $-2(-3) + 5 = 6 + 5 = 11$

5. $\dfrac{2r}{r-1}$
   $\dfrac{2(5)}{5-1} = \dfrac{10}{4} = 2.5$

6. $2p^3 - 3p + 4$
   $$\begin{aligned} 2(-1)^3 - 3(-1) + 4 &= 2(-1) - 3(-1) + 4 \\ &= -2 + 3 + 4 \\ &= 5 \end{aligned}$$

7. $\dfrac{x_1 + x_2}{2}$
   $\dfrac{7+(-5)}{2} = \dfrac{2}{2} = 1$

8. $$\begin{aligned} 3x - 4y &= 12 \\ 3x - 4(0) &= 12 \\ 3x &= 12 \\ x &= 4 \end{aligned}$$

Section 4.1

Objective A Exercises

1. 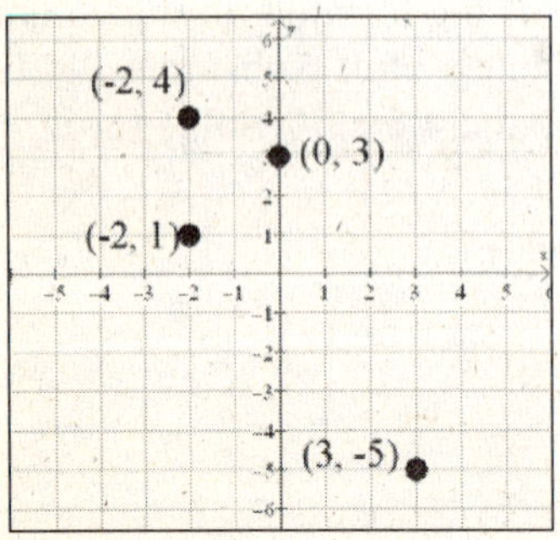

3. 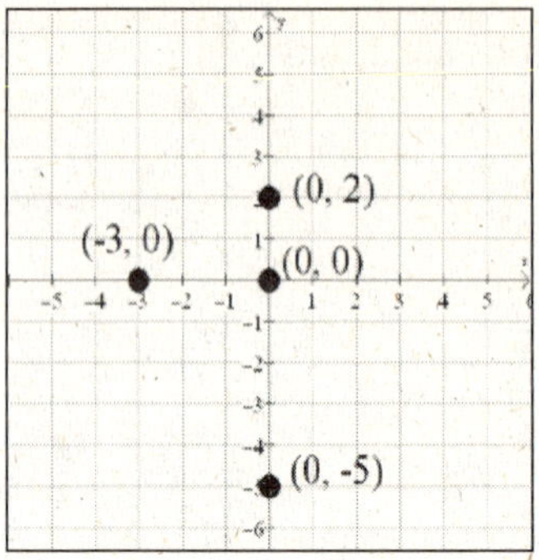

5. 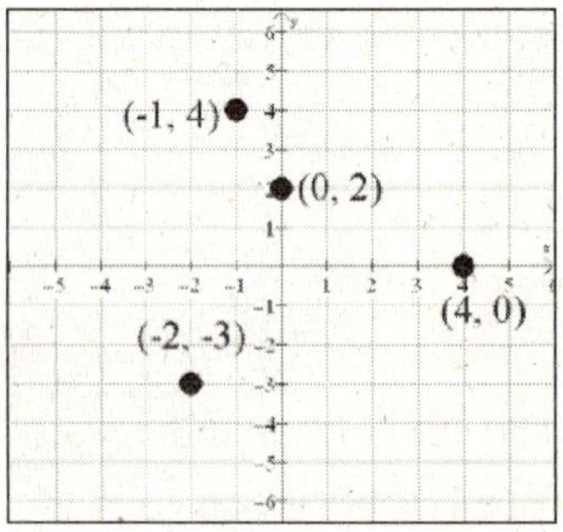

7. *A* (2, 3)
   B (4, 0)
   C (−4, 1)
   D (−2, −2)

9. *A* (−2, 5)
   B (3, 4)
   C (0, 0)
   D (−3, −2)

11. **a.** Abscissa of point *A*: 2
    Abscissa of point *C*: −4
    b. Ordinate of point *B*: 1
    Ordinate of point *D*: −3

13. **a.** *y*-axis
    b. *x*-axis

Objective B Exercises

15.

$y=-x+7$	
4	$-(3)+7$
	$-3+7$
	4

$4=4$

Yes, (3, 4) is a solution of $y=-x+7$.

17.

$y=\frac{1}{2}x-1$	
2	$\frac{1}{2}(-1)-1$
	$-\frac{1}{2}-1$
	$-\frac{3}{2}$

$2\neq-\frac{3}{2}$

No, (2, −3) is not a solution of $y=\frac{1}{2}x-1$

19.

$2x-5y=4$	
$2(4)-5(1)$	4
$8-5$	
3	

$3\neq4$

No, (4, 1) is a not solution of $2x-5y=4$

21. If $x>2$, $-3x<-6$, then $-3x+6<0$.
y is negative.

23.
$$\begin{aligned}y&=3x-2\\&=3(3)-2\\&=9-2=7\end{aligned}$$
The ordered-pair solution is (3,7)

25.
$$\begin{aligned}y&=\frac{2}{3}x-1\\&=\frac{2}{3}(6)-1\\&=4-1=3\end{aligned}$$
The ordered-pair solution is (6, 3).

27.
$$\begin{aligned}y&=-3x+1\\&=-3(0)+1\\&=0+1=1\end{aligned}$$
The ordered-pair solution is (0, 1).

29.
$$\begin{aligned}y&=\frac{2}{5}x+2\\&=\frac{2}{5}(-5)+2\\&=-2+2=0\end{aligned}$$
The ordered-pair solution is (−5, 0).

31.

x	$y=2x$	(x,y)
−2	$2(-2)=-4$	(−2, −4)
−1	$2(-1)=-2$	(−1, −2)
0	$2(0)=0$	(0, 0)
2	$2(2)=4$	(2, 4)

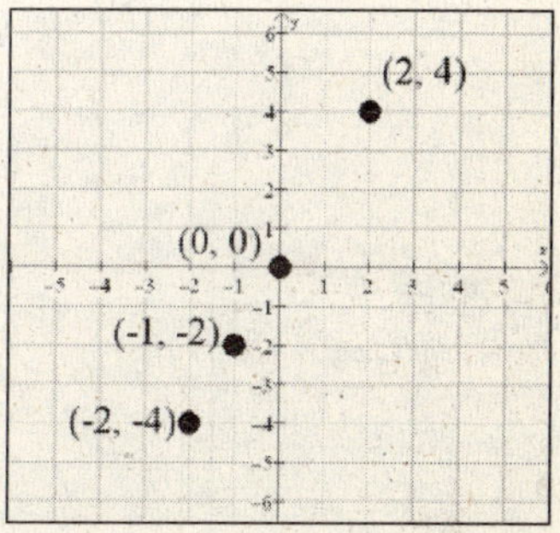

33.

x	$y=\frac{2}{3}x+1$	(x, y)
-3	$\frac{2}{3}(-3)+1=-1$	$(-3, -1)$
0	$\frac{2}{3}(0)+1=1$	$(0, 1)$
3	$\frac{2}{3}(3)+1=3$	$(3, 3)$

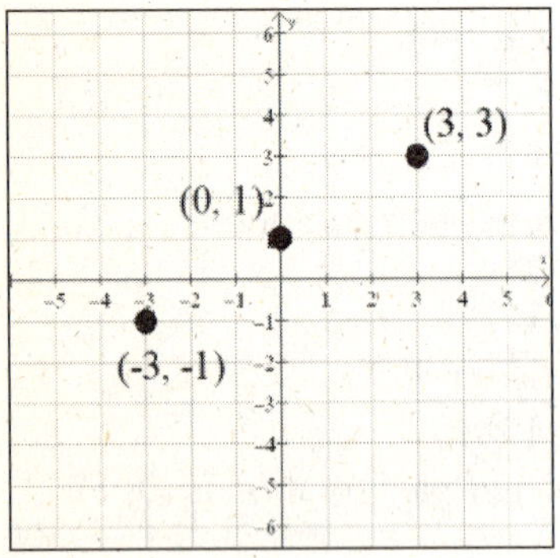

35. Solve $2x+3y=6$ for y.

$$2x+3y=6$$
$$3y=-2x+6$$
$$y=-\frac{2}{3}x+2$$

x	$y=-\frac{2}{3}x+2$	(x, y)
-3	$-\frac{2}{3}(-3)+2=4$	$(-3, 4)$
0	$-\frac{2}{3}(0)+2=2$	$(0, 2)$
3	$-\frac{2}{3}(3)+2=0$	$(3, 0)$

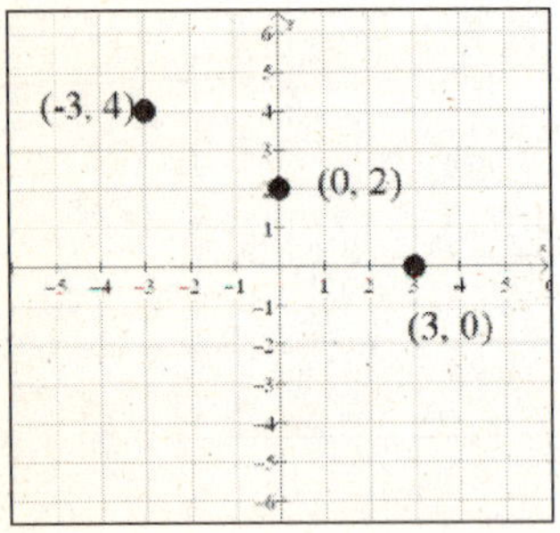

Objective C Exercises

37. a) 25 g of cerium selenate will dissolve at 50°C.

b) 5 g of cerium selenate will dissolve at 80°C.

39.

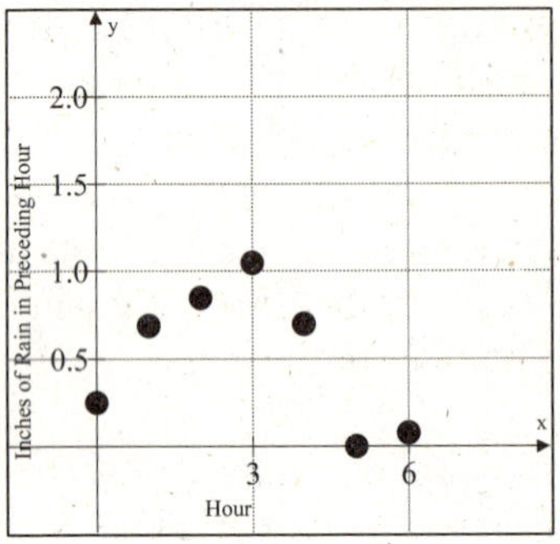

Applying the Concepts

41. Ordered pairs: $(-2,4)$, $(-1, 1)$, $(0,0)$, $(1, 1)$, $(2, 4)$

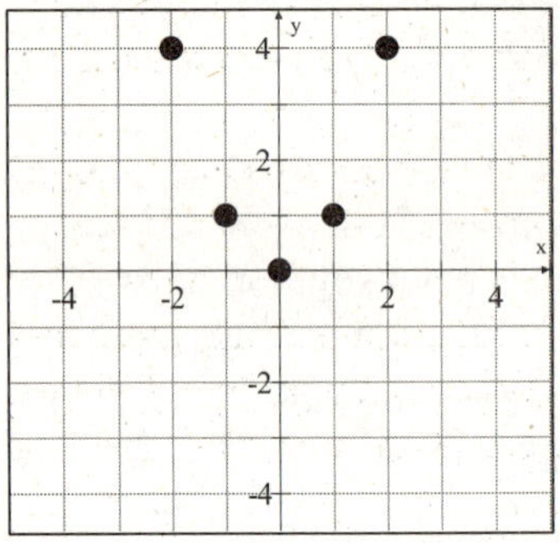

43. The graph of all ordered pairs (x, y) that are 5 units from the origin is a circle of radius 5 that has its center at $(0,0)$.

45.

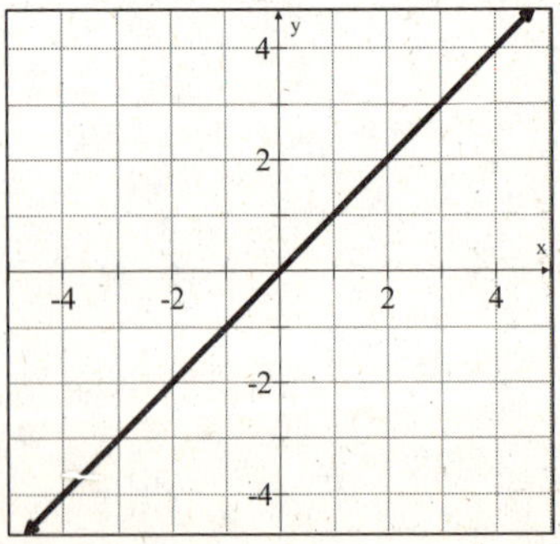

Section 4.2

Objective A Exercises

1. A function is a set of ordered pairs in which no two ordered pairs can have the same x-coordinate and different y-coordinates.

3. The diagram does represent a function because each number in the domain is paired with one number in the range.

5. The diagram does represent a function because each number in the domain is paired with one number in the range.

7. No, the diagram does not represent a function. The 6 in the domain is paired with two different values in the range.

9. This is a function because each x-coordinate is paired with only one y-coordinate.

11. This is a function because each x-coordinate is paired with only one y-coordinate.

13. This is a function because each x-coordinate is paired with only one y-coordinate.

15. This is not a function because there are x-coordinates paired with two different y-coordinates. They are $(1,1)$, $(1,-1)$ and $(4,2)$, $(4,-2)$.

17. a) Yes, this table defines a function because there is only one fine for any given number of miles per hour over the speed limit.
b) If $x = 16$, then $y = \$125$.

19. True

21. $f(x) = 5x - 4$
$f(3) = 5(3) - 4$
$f(3) = 15 - 4$
$f(3) = 11$

23. $f(x) = 5x - 4$
$f(0) = 5(0) - 4$
$f(0) = -4$

25. $G(t) = 4 - 3t$
$G(0) = 4 - 3(0)$
$G(0) = 4$

27. $G(t) = 4 - 3t$
$G(-2) = 4 - 3(-2)$
$G(-2) = 4 + 6$
$G(-2) = 10$

29. $q(r) = r^2 - 4$
$q(3) = 3^2 - 4$
$q(3) = 9 - 4$
$q(3) = 5$

31. $q(r) = r^2 - 4$
$q(-2) = (-2)^2 - 4$
$q(-2) = 4 - 4$
$q(-2) = 0$

33. $F(x) = x^2 + 3x - 4$
$F(4) = 4^2 + 3(4) - 4$
$F(4) = 16 + 12 - 4$
$F(4) = 24$

35. $F(x) = x^2 + 3x - 4$
$F(-3) = (-3)^2 + 3(-3) - 4$
$F(-3) = 9 - 9 - 4$
$F(-3) = -4$

37. $H(p) = \dfrac{3p}{p+2}$

$H(1) = \dfrac{3(1)}{1+2}$

$H(1) = \dfrac{3}{3}$

$H(1) = 1$

39. $H(p) = \dfrac{3p}{p+2}$

$H(t) = \dfrac{3t}{t+2}$

41. $s(t) = t^3 - 3t + 4$
$s(-1) = (-1)^3 - 3(-1) + 4$
$s(-1) = -1 + 3 + 4$
$s(-1) = 6$

43. $s(t) = t^3 - 3t + 4$
$s(a) = a^3 - 3a + 4$

45. $P(x) = 4x + 7$
$P(-2 + h) - P(-2)$
$= 4(-2 + h) + 7 - [4(-2) + 7]$
$= -8 + 4h + 7 + 8 - 7$
$= 4h$

47. \$114.29

49. a) \$4.75 per game
b) \$4.00 per game

51. Domain = {1, 2, 3, 4, 5}
Range = {1, 4, 7, 10, 13}

53. Domain = {0, 2, 4, 6}
Range = {1, 2, 3, 4}

55. Domain = {1, 3, 5, 7, 9}
Range = {0}

57. Domain = {−2, −1, 0, 1, 2}
Range = {0, 1, 2}

59. Domain = {−2, −1, 0, 1, 2}
Range = {−3, 3, 6, 7, 9}

61. Values of x for which $x - 1 = 0$ are excluded from the domain of the function.
$x - 1 = 0$
$x = 1$

63. Values of x for which $x + 8 = 0$ are excluded from the domain of the function.
$x + 8 = 0$
$x = -8$

65. No values are excluded.

67. No values are excluded.

69. $x = 0$

71. No values are excluded.

73. No values are excluded.

75. Values of x for which $x + 2 = 0$ are excluded from the domain of the function.
$x + 2 = 0$
$x = -2$

77. No values are excluded.

79. $f(x) = 4x - 3$
$f(0) = 4(0) - 3 = -3$
$f(1) = 4(1) - 3 = 1$
$f(2) = 4(2) - 3 = 5$
$f(3) = 4(3) - 3 = 9$
Range = {−3, 1, 5, 9}

81. $g(x) = 5x - 8$
$g(-3) = 5(-3) - 8 = -23$
$g(-1) = 5(-1) - 8 = -13$
$g(0) = 5(0) - 8 = -8$
$g(1) = 5(1) - 8 = -3$
$g(3) = 5(3) - 8 = 7$
Range = {−23, −13, −8, −3, 7}

83. $h(x) = x^2$
$h(-2) = (-2)^2 = 4$
$h(-1) = (-1)^2 = 1$
$h(0) = 0$
$h(1) = (1)^2 = 1$
$h(2) = (2)^2 = 4$
Range = {0, 1, 4}

85. $f(x) = 2x^2 - 2x + 2$
$f(-4) = 2(-4)^2 - 2(-4) + 2 = 42$
$f(-2) = 2(-2)^2 - 2(-2) + 2 = 14$
$f(0) = 2(0)^2 - 2(0) + 2 = 2$
$f(4) = 2(4)^2 - 2(4) + 2 = 26$
Range = {2, 14, 26, 42}

87. $H(x) = \dfrac{5}{1 - x}$

$H(-2) = \dfrac{5}{1-(-2)} = \dfrac{5}{3}$

$H(0) = \dfrac{5}{1-0} = 5$

$H(2) = \dfrac{5}{1-2} = -5$

Range = $\{-5, \dfrac{5}{3}, 5\}$

89. $f(x) = \dfrac{2}{x-4}$

$f(-2) = \dfrac{2}{-2-4} = -\dfrac{1}{3}$

$f(0) = \dfrac{2}{0-4} = -\dfrac{1}{2}$

$f(2) = \dfrac{2}{2-4} = -1$

$f(6) = \dfrac{2}{6-4} = 1$

Range = $\{-1, -\dfrac{1}{2}, -\dfrac{1}{3}, 1\}$

91. $H(x) = 2 - 3x - x^2$

$H(-5) = 2 - 3(-5) - (-5)^2 = -8$

$H(0) = 2$

$H(5) = 2 - 3(5) - 5^2 = -38$

Range = $\{-38, -8, 2\}$

Applying the Concepts

93. A relation and a function are similar in that both are sets of ordered pairs. A function is a specific type of relation. A function is a relation in which there are no two ordered pairs with the same first element.

95. a) $\{(-2,-8), (-1,-1), (0,0), (1,1), (2,8)\}$
b) Yes, this set of ordered pairs defines a function because each member of the domain is assigned exactly one member of the range.

97. Evaluate the function $P = f(v) = 0.015v^3$ when $v = 15$.
$P = f(v) = 0.015(15)^3 = 50.625$
The power produced will be 50.625 watts.

99. a) The speed of the paratrooper 11.5 s after the beginning of the jump is 36.3 ft/s.
b) 30 ft/s

101. a) 275,000 malware attacks
b) 700,000 malware attacks

Section 4.3

Objective A Exercises

1. If the three ordered pairs appear to lie on the same line, then it is more likely that your calculations are correct.

3. $y = x - 3$

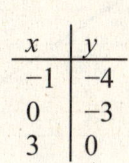

x	y
−1	−4
0	−3
3	0

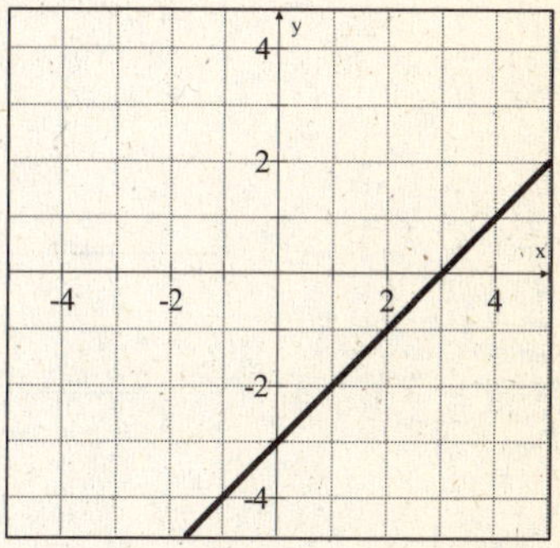

5. $y = -3x + 2$

x	y
0	2
1	−1
2	−4

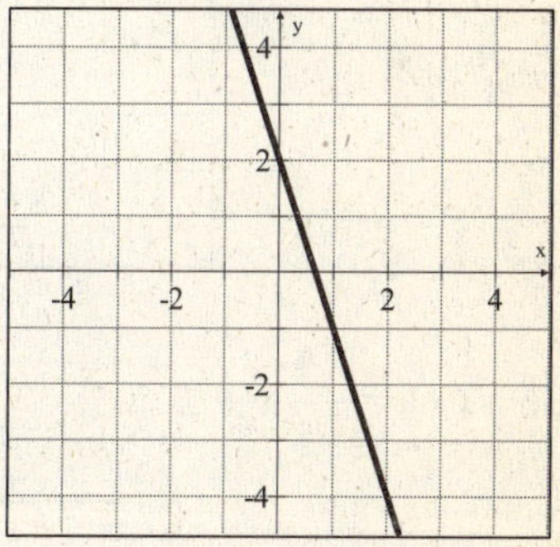

7. $f(x) = 3x - 4$

x	$f(x)$
0	−4
1	−1
2	2

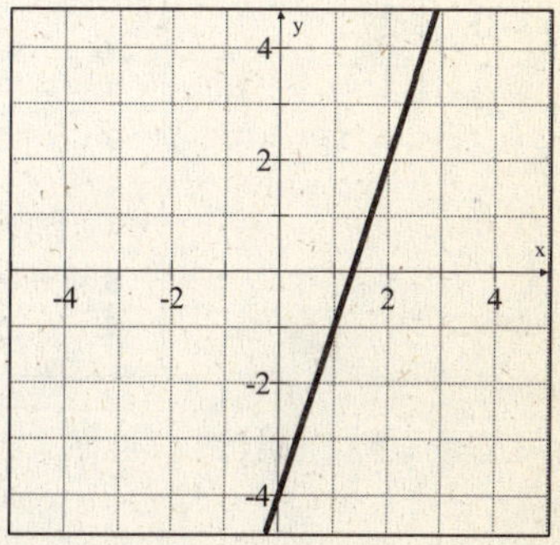

9. $f(x) = -\frac{2}{3}x$

x	$f(x)$
−3	2
0	0
3	−2

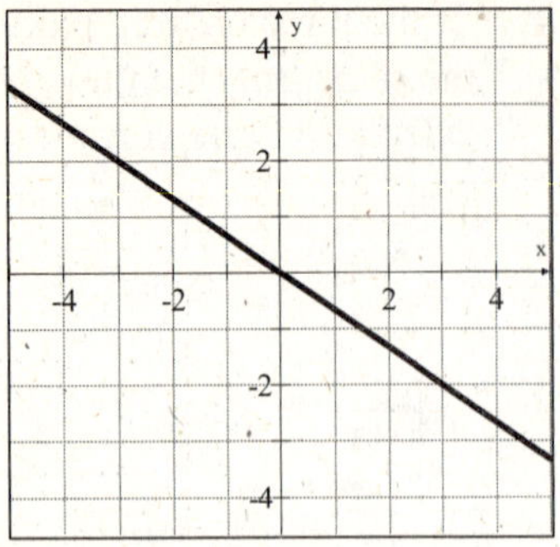

11. $y = \frac{2}{3}x - 4$

x	y
0	−4
3	−2
6	0

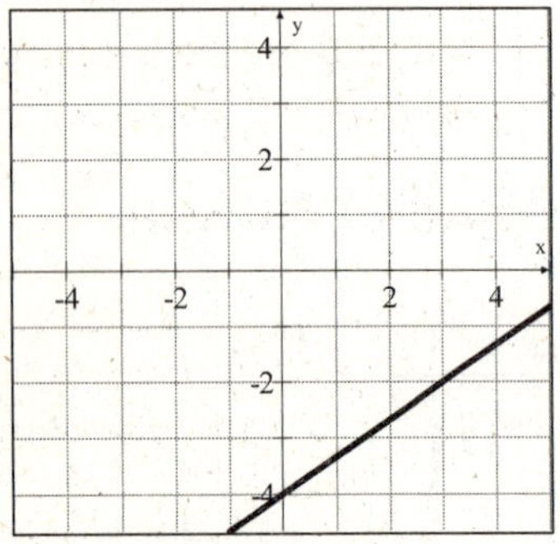

13. $f(x) = -\frac{1}{3}x + 2$

x	$f(x)$
−3	3
0	2
3	1

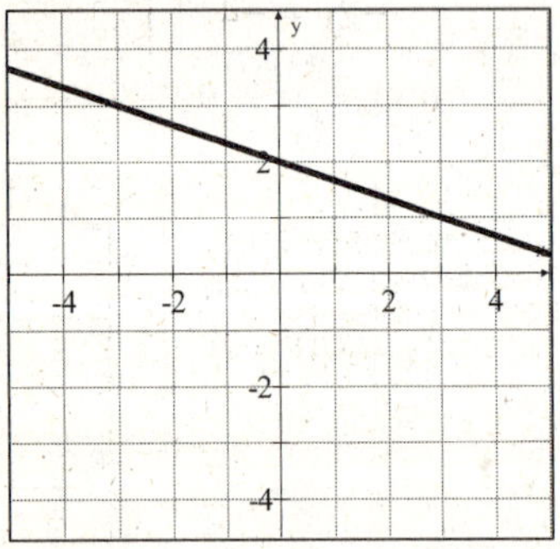

Objective B Exercises

15. $2x + y = -3$

$y = -2x - 3$

x	y
0	−3
1	−5
−1	−1

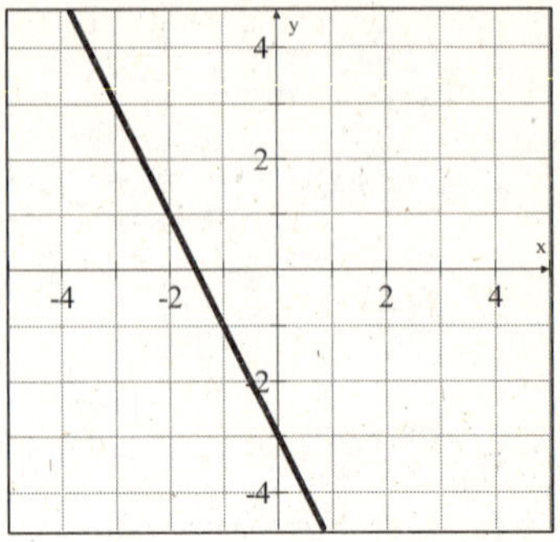

17. $x - 4y = 8$

$-4y = -x + 8$

$y = \frac{1}{4}x - 2$

x	y
0	−2
4	−1
8	0

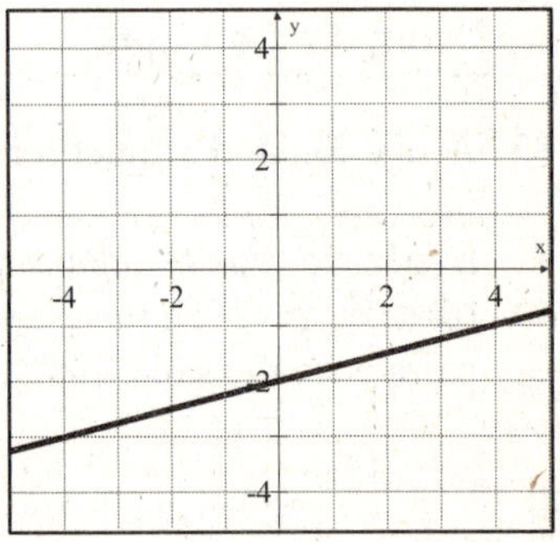

19. $4x + 3y = 12$

$3y = -4x + 12$

$y = -\frac{4}{3}x + 4$

x	y
0	4
3	0
6	−4

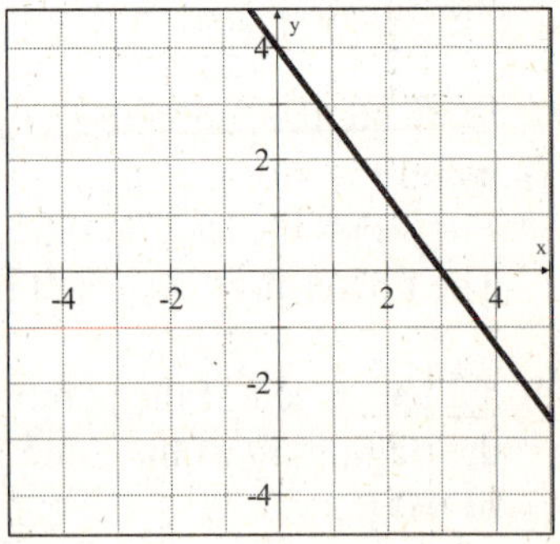

21. $x - 3y = 0$

$-3y = -x$

$y = \dfrac{1}{3}x$

x	y
0	0
3	1
−3	−1

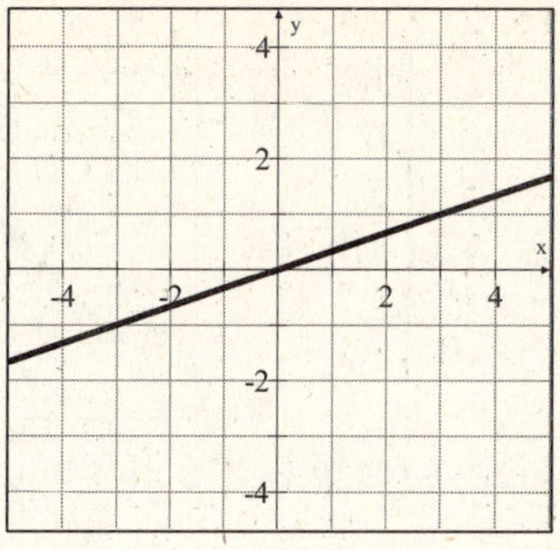

23. $y = -2$

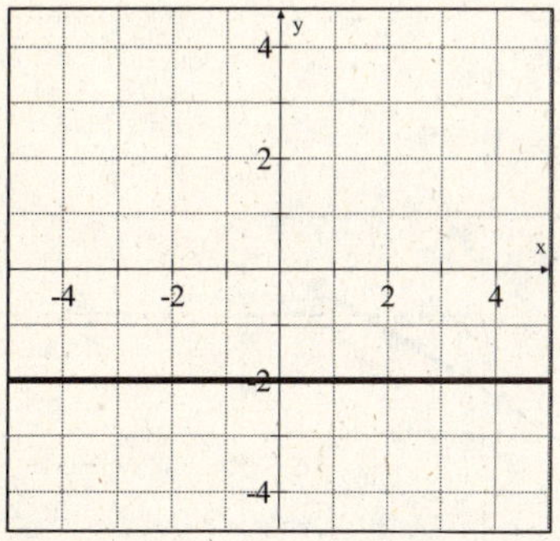

25. $x = -3$

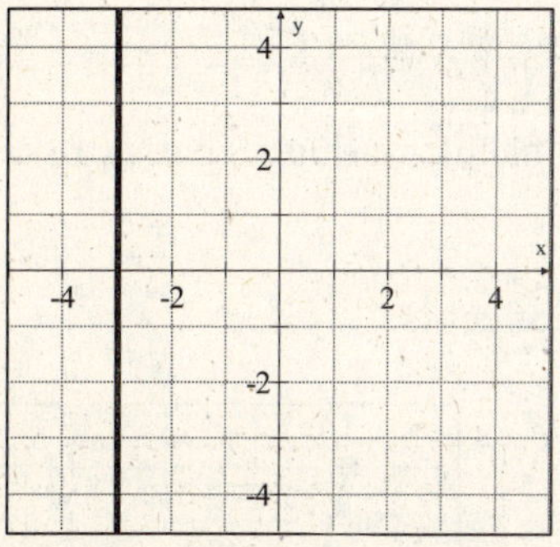

27. $3x - y = -2$

$-y = -3x - 2$

$y = 3x + 2$

x	y
0	2
1	5
−1	−1

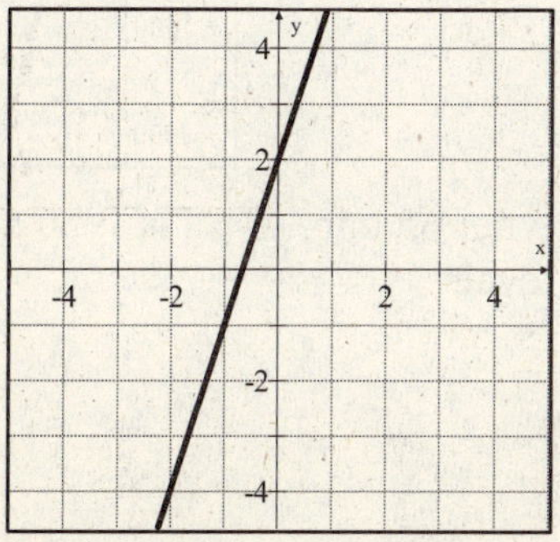

29. $3x - 2y = 8$

$-2y = -3x + 8$

$y = \dfrac{3}{2}x - 4$

x	y
0	−4
2	−1
4	2

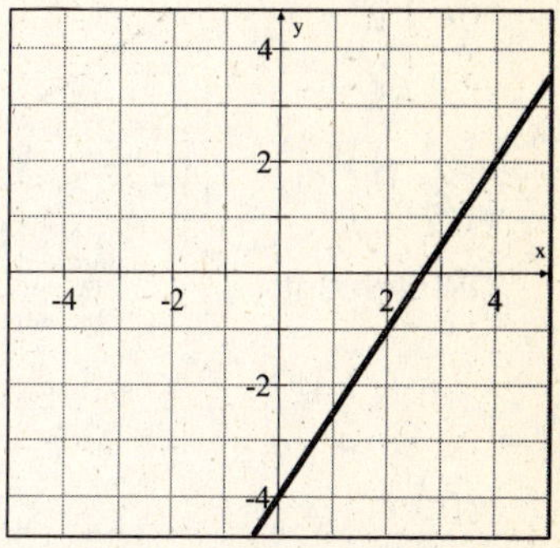

31. No. If $B = 0$, then it is not possible to solve $Ax + By = C$ for y.

Objective C Exercises

33. x- intercept:

$x - 2y = -4$

$x - 2(0) = -4$

$x = -4$

$(-4,0)$

y-intercept:

$x - 2y = -4$

$0 - 2y = -4$

$-2y = -4$

$y = 2$

$(0,2)$

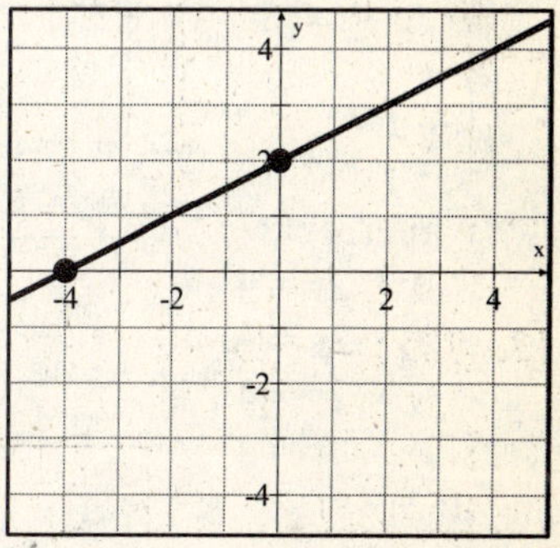

35. x- intercept:

$2x - 3y = 9$

$2x - 3(0) = 9$

$2x = 9$

$x = \dfrac{9}{2}$

$\left(\dfrac{9}{2},0\right)$

y-intercept:

$2x - 3y = 9$

$2(0) - 3y = 9$

$-3y = 9$

$y = -3$

$(0,-3)$

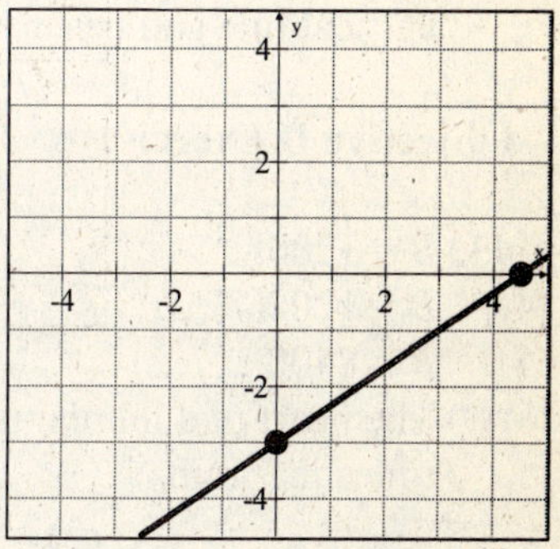

37. x- intercept:

$$2x + y = 3$$
$$2x + 0 = 3$$
$$2x = 3$$
$$x = \frac{3}{2}$$
$$\left(\frac{3}{2}, 0\right)$$

y-intercept:

$$2x + y = 3$$
$$2(0) + y = 3$$
$$y = 3$$
$$(0,3)$$

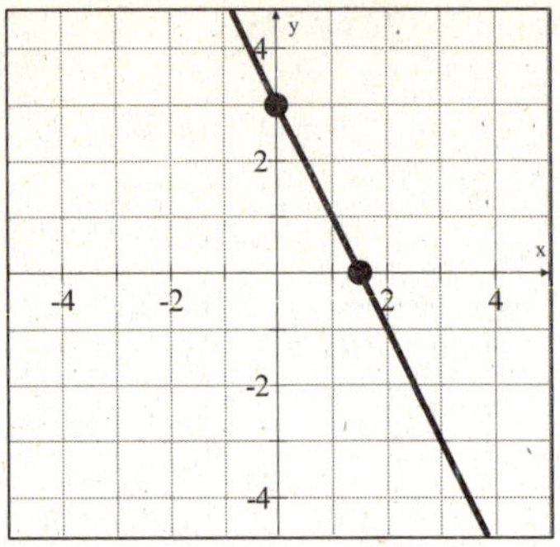

39. x- intercept:

$$3x + 2y = 4$$
$$3x + 2(0) = 4$$
$$3x = 4$$
$$x = \frac{4}{3}$$
$$\left(\frac{4}{3}, 0\right)$$

y-intercept:

$$3x + 2y = 4$$
$$3(0) + 2y = 4$$
$$2y = 4$$
$$y = 2$$
$$(0,2)$$

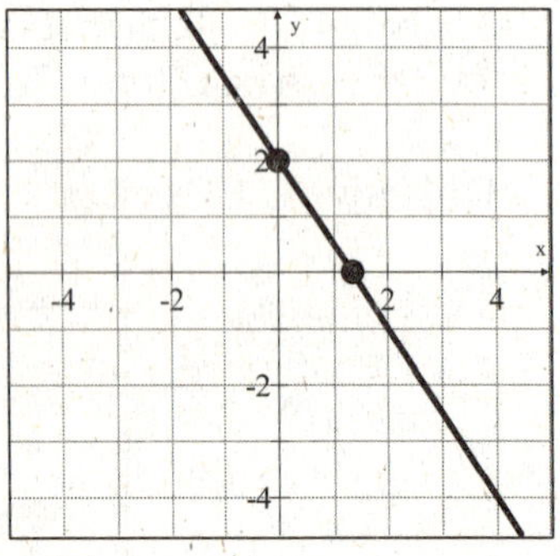

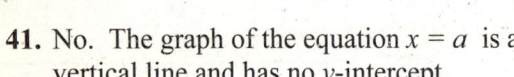

41. No. The graph of the equation $x = a$ is a vertical line and has no y-intercept.

Objective D Exercises

43. $B = 1200t$
$B = 1200(7)$
$B = 8400$
The heart of a hummingbird will beat 8400 times in 7 min.

45. $W = 11t$

t	W
0	0
5	55
15	165

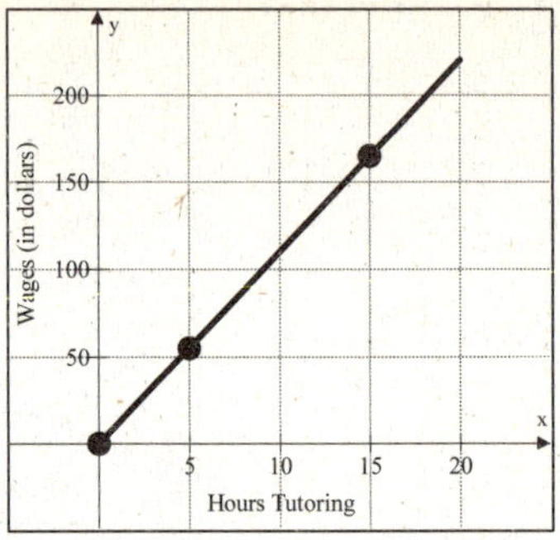

Marlys receives $165 for tutoring 15 h.

47. $C = 80n + 5000$

n	C
0	5000
50	9000
100	13000

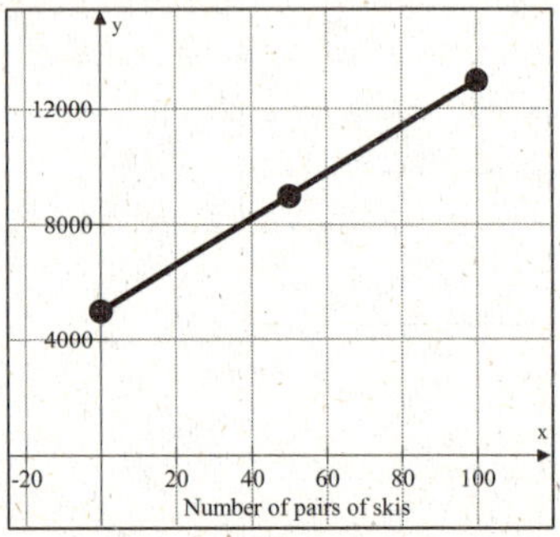

The cost of manufacturing 50 pairs of skis is $9000.

Applying the Concepts

49. a) $D = -30t$

t	D
0	0
20	−600
65	−1950

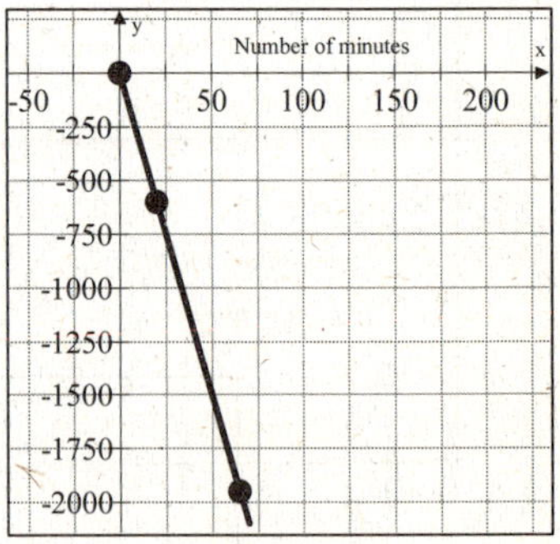

After 65 min, Alvin is 1950 m below sea level.

b) $D = -48t$

t	D
0	0
20	−960
65	−3120

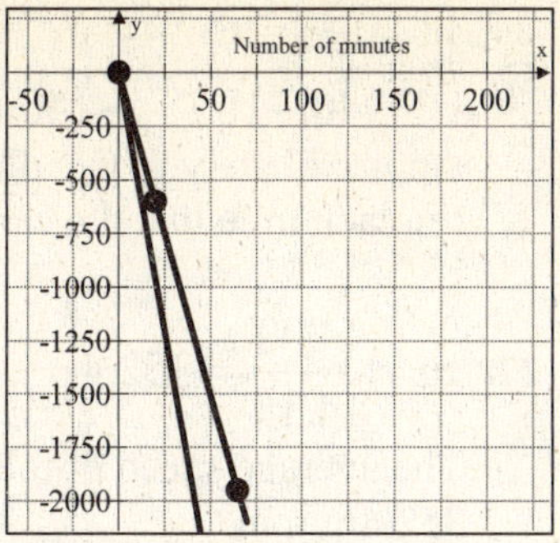

The point is below (65, −1950).

51. Yes, the points with y-coordinates less than −6500.

53. To graph the equation of a straight line by plotting points, find three ordered pair solutions of the equation. Plot these ordered pairs in a rectangular coordinate system. Draw a straight line through the points.

55. The x and y intercepts of the graph of the equation $4x + 3y = 0$ are both (0,0). A straight line is determined by two points, so we need to find another point on the line in order to graph this equation.

57. $\frac{x}{2} + \frac{y}{3} = 1$

x-intercept (2,0)
y-intercept (0,3)

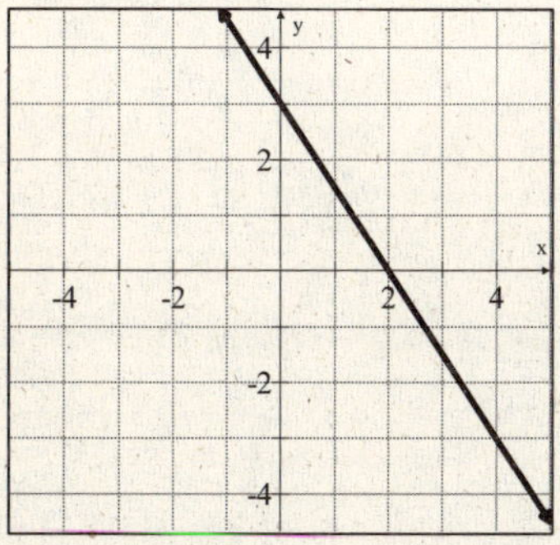

Section 4.4

Objective A Exercises

1. (1,3), (3,1)

$$m = \frac{y_2 - y_1}{x_2 - x_1} = \frac{1-3}{3-1} = \frac{-2}{2} = -1$$

The slope is −1.

3. (−1,4), (2,5)

$$m = \frac{y_2 - y_1}{x_2 - x_1} = \frac{5-4}{2-(-1)} = \frac{1}{3}$$

The slope is $\frac{1}{3}$.

5. (−1,3), (−4, 5)

$$m = \frac{y_2 - y_1}{x_2 - x_1} = \frac{5-3}{-4-(-1)} = \frac{2}{-3} = -\frac{2}{3}$$

The slope is $-\frac{2}{3}$.

7. (0,3), (4,0)

$$m = \frac{y_2 - y_1}{x_2 - x_1} = \frac{3-0}{0-4} = \frac{3}{-4} = -\frac{3}{4}$$

The slope is $-\frac{3}{4}$.

9. (2,4), (2,−2)

$$m = \frac{y_2 - y_1}{x_2 - x_1} = \frac{4-(-2)}{2-2} = \frac{6}{0}$$

The slope is undefined.

11. (2,5), (−3,−2)

$$m = \frac{y_2 - y_1}{x_2 - x_1} = \frac{5-(-2)}{2-(-3)} = \frac{7}{5}$$

The slope is $\frac{7}{5}$.

13. (2,3), (−1,3)

$$m = \frac{y_2 - y_1}{x_2 - x_1} = \frac{3-3}{2-(-1)} = \frac{0}{3} = 0$$

The slope is 0.

15. (0,4), (−2, 5)

$$m = \frac{y_2 - y_1}{x_2 - x_1} = \frac{5-4}{-2-0} = \frac{1}{-2} = -\frac{1}{2}$$

The slope is $-\frac{1}{2}$.

17. (−3,−1), (−3,4)

$$m = \frac{y_2 - y_1}{x_2 - x_1} = \frac{4-(-1)}{-3-(-3)} = \frac{5}{0}$$

The slope is undefined.

19. If a and c are equal the slope of l is undefined.

21. $m = \frac{240-80}{6-2} = \frac{160}{4} = 40$

The average speed of the motorist is 40 mph.

23. $m = \frac{275-125}{20-50} = \frac{150}{-30} = -5$

The temperature of the oven decreases 5°/min.

25. $m = \frac{13-6}{40-180} = \frac{7}{-140} = -0.05$

Approximately 0.05 gallon of fuel is used for each mile that the car is driven.

27. $m = \frac{5000}{14.19} = 352.4$

The average speed of the runner was 352.4m/min.

29. a) $\frac{6\ in}{5\ ft} = \frac{6\ in}{60\ in} = \frac{1}{10} > \frac{1}{12}$

No it does not meet the requirements for ANSI.

b) $\frac{12}{170} = \frac{6}{85} < \frac{1}{12}$

Yes, it does meet the requirements for ANSI.

Objective B Exercises

	Equation	*Value of m*	*Value of b*	*Slope*	*y-intercept*
31.	$y = -3x + 5$	−3	5	-3	(0,5)
33.	$y = 4x$	4	0	4	(0,0)

35. $y = \frac{1}{2}x + 2$

$m = \frac{1}{2}$

y – intercept (0,2)

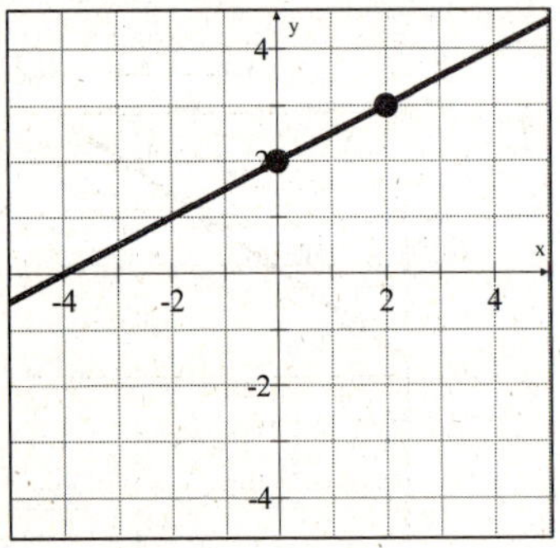

37. $y = -\dfrac{3}{2}x$

$m = -\dfrac{3}{2}$

y – intercept
(0,0)

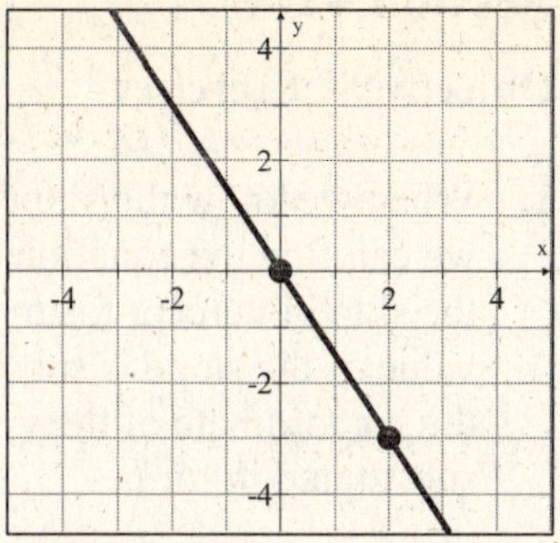

39. $y = -\dfrac{1}{2}x + 2$

$m = -\dfrac{1}{2}$

y – intercept
(0,2)

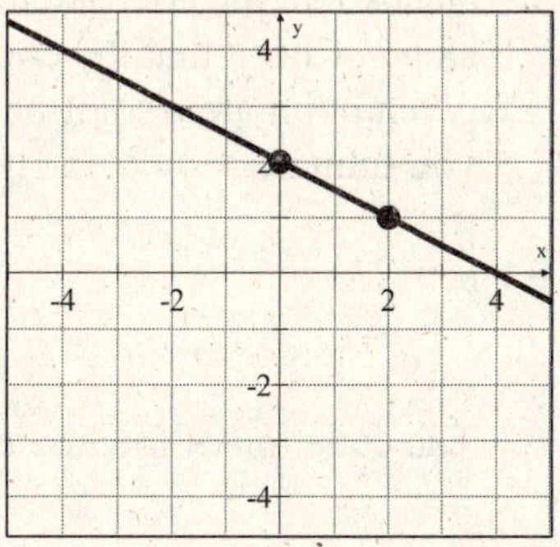

41. $y = 2x - 4$

$m = 2$

y – intercept
(0,–4)

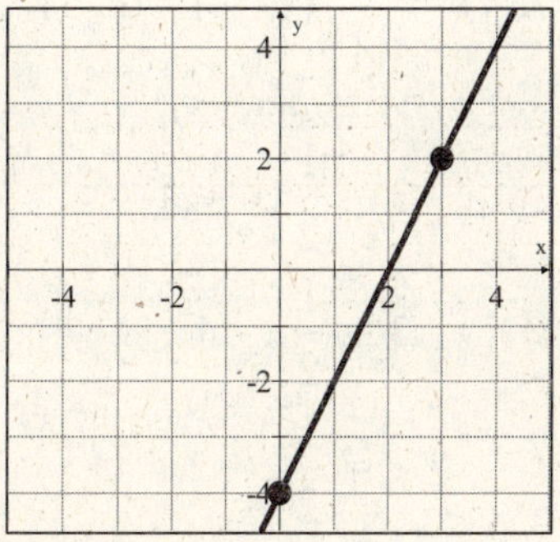

43. $4x - y = 1$

$-y = -4x + 1$

$y = 4x - 1$

$m = 4$

y – intercept
(0,–1)

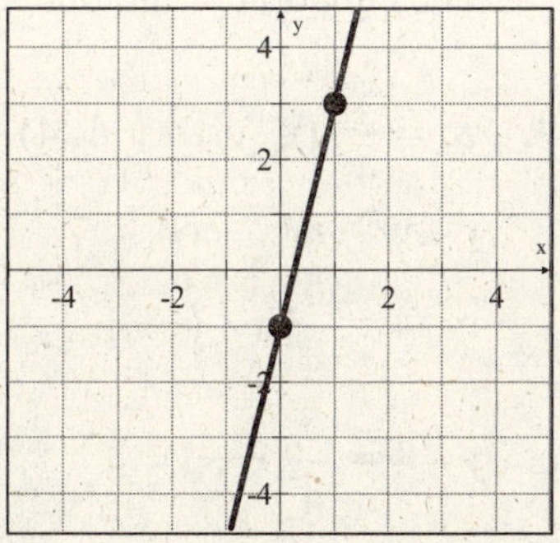

45. $x - 3y = 3$

$-3y = -x + 3$

$y = \dfrac{1}{3}x - 1$

$m = \dfrac{1}{3}$

y-intercept
(0,–1)

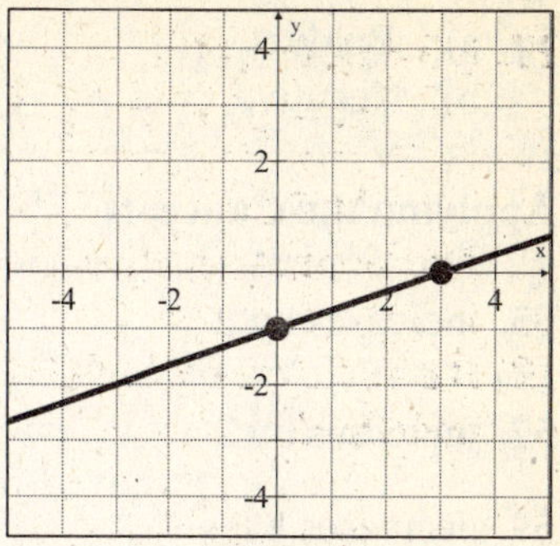

47.

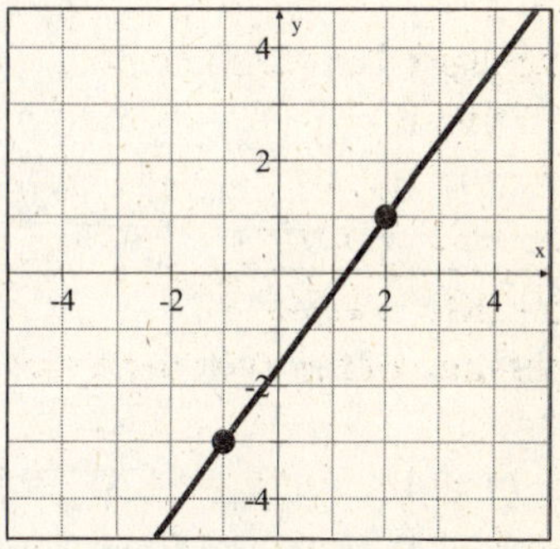

49.

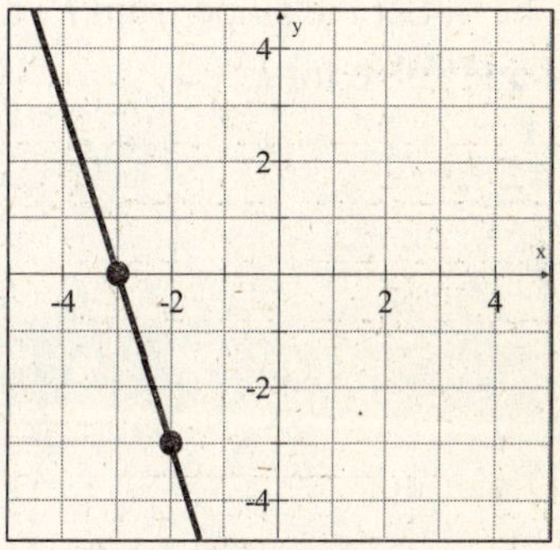

51.

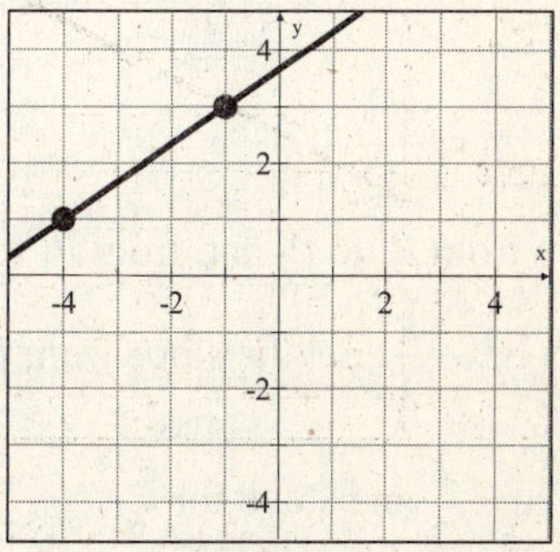

53. a) Below
b) Negative

Applying the Concepts

55. increases by 2

57. increases by 2

59. decreases by $\frac{2}{3}$

61. i. D ii. C
iii. B iv. F
v. E vi. A

63. $P_1 = (3,2)$
$P_2 = (4,6)$
$P_3 = (5,k)$
P_1 to P_2: $m = 4$
The slope from P_2 to P_3 and from P_1 to P_3 must also be 4. Set the slope from P_1 to P_3 equal to 4 and solve for k.

$$\frac{2-k}{-2} = 4$$
$$2-k = -8$$
$$k = -10$$

65. $P_1 = (k, 1)$
$P_2 = (0,-1)$
$P_3 = (2,-2)$

P_2 to P_3: $m = -\frac{1}{2}$

The slope from P_1 to P_2 and from P_1 to P_3 must also be $-\frac{1}{2}$. Set the slope from P_1 to P_2 equal to $-\frac{1}{2}$ and solve for k.

$$\frac{2}{k-0} = -\frac{1}{2}$$
$$k = -4$$

Section 4.5

Objective A Exercises

1. When we know the slope and the y-intercept we can find the equation of the line using the slope-intercept form, $y = mx + b$. The value of the slope is substituted in for m, and the y-coordinate of the y-intercept is substituted in for b.

3. Check that the coefficient of x is the given slope. Check that the coordinates of the given point are a solution of the equation you found.

5. $m = 2, b = 5$
$y = mx + b$
$y = 2x + 5$
The equation of the line is $y = 2x + 5$.

7. $m = \frac{1}{2}, (x_1, y_1) = (2, 3)$

$$y - y_1 = m(x - x_1)$$
$$y - 3 = \frac{1}{2}(x - 2)$$
$$y - 3 = \frac{1}{2}x - 1$$
$$y = \frac{1}{2}x + 2$$

The equation of the line is $y = \frac{1}{2}x + 2$.

9. $m = \frac{5}{4}, (x_1, y_1) = (-1, 4)$

$$y - y_1 = m(x - x_1)$$
$$y - 4 = \frac{5}{4}[x - (-1)]$$
$$y - 4 = \frac{5}{4}(x + 1)$$
$$y - 4 = \frac{5}{4}x + \frac{5}{4}$$
$$y = \frac{5}{4}x + \frac{21}{4}$$

The equation of the line is $y = \frac{5}{4}x + \frac{21}{4}$.

11. $m = -\frac{5}{3}$, $(x_1, y_1) = (3, 0)$

$y - y_1 = m(x - x_1)$

$y - 0 = -\frac{5}{3}(x - 3)$

$y = -\frac{5}{3}(x - 3)$

$y = -\frac{5}{3}x + 5$

The equation of the line is $y = -\frac{5}{3}x + 5$.

13. $m = -3$, $(x_1, y_1) = (2, 3)$

$y - y_1 = m(x - x_1)$

$y - 3 = -3(x - 2)$

$y - 3 = -3x + 6$

$y = -3x + 9$

The equation of the line is $y = -3x + 9$.

15. $m = -3$, $(x_1, y_1) = (-1, 7)$

$y - y_1 = m(x - x_1)$

$y - 7 = -3[x - (-1)]$

$y - 7 = -3(x + 1)$

$y - 7 = -3x - 3$

$y = -3x + 4$

The equation of the line is $y = -3x + 4$.

17. $m = \frac{2}{3}$, $(x_1, y_1) = (-1, -3)$

$y - y_1 = m(x - x_1)$

$y - (-3) = \frac{2}{3}[(x - (-1)]$

$y + 3 = \frac{2}{3}(x + 1)$

$y + 3 = \frac{2}{3}x + \frac{2}{3}$

$y = \frac{2}{3}x - \frac{7}{3}$

The equation of the line is $y = \frac{2}{3}x - \frac{7}{3}$.

19. $m = \frac{1}{2}$, $(x_1, y_1) = (0, 0)$

$y - y_1 = m(x - x_1)$

$y - 0 = \frac{1}{2}(x - 0)$

$y = \frac{1}{2}x$

The equation of the line is $y = \frac{1}{2}x$.

21. $m = 3$, $(x_1, y_1) = (2, -3)$

$y - y_1 = m(x - x_1)$

$y - (-3) = 3(x - 2)$

$y + 3 = 3x - 6$

$y = 3x - 9$

The equation of the line is $y = 3x - 9$.

23. $m = -\frac{2}{3}$, $(x_1, y_1) = (3, 5)$

$y - y_1 = m(x - x_1)$

$y - 5 = -\frac{2}{3}(x - 3)$

$y - 5 = -\frac{2}{3}x + 2$

$y = -\frac{2}{3}x + 7$

The equation of the line is $y = -\frac{2}{3}x + 7$.

25. $m = -1$, $b = -3$

$y = -x - 3$

The equation of the line is $y = -x - 3$.

27. $m = \frac{7}{5}$, $(x_1, y_1) = (1, -4)$

$y - y_1 = m(x - x_1)$

$y - (-4) = \frac{7}{5}(x - 1)$

$y + 4 = \frac{7}{5}x - \frac{7}{5}$

$y = \frac{7}{5}x - \frac{27}{5}$

The equation of the line is $y = \frac{7}{5}x - \frac{27}{5}$.

29. $m = -\frac{2}{5}$, $(x_1, y_1) = (4, -1)$

$$y - y_1 = m(x - x_1)$$
$$y - (-1) = -\frac{2}{5}(x - 4)$$
$$y + 1 = -\frac{2}{5}x + \frac{8}{5}$$
$$y = -\frac{2}{5}x + \frac{3}{5}$$

The equation of the line is $y = -\frac{2}{5}x + \frac{3}{5}$.

31. Slope is undefined, $(x_1, y_1) = (3, -4)$
The line is a vertical line. All points on the line have an abscissa of 3.
The equation of the line is $x = 3$.

33. $m = -\frac{5}{4}$, $(x_1, y_1) = (-2, -5)$

$$y - y_1 = m(x - x_1)$$
$$y - (-5) = -\frac{5}{4}[x - (-2)]$$
$$y + 5 = -\frac{5}{4}(x + 2)$$
$$y + 5 = -\frac{5}{4}x - \frac{10}{4}$$
$$y = -\frac{5}{4}x - \frac{15}{2}$$

The equation of the line is $y = -\frac{5}{4}x - \frac{15}{2}$.

35. $m = 0$, $(x_1, y_1) = (-2, -3)$

$$y - y_1 = m(x - x_1)$$
$$y - (-3) = 0[x - (-2)]$$
$$y + 3 =$$
$$y = -3$$

The equation of the line is $y = -3$.

37. $m = -2$, $(x_1, y_1) = (4, -5)$

$$y - y_1 = m(x - x_1)$$
$$y - (-5) = -2(x - 4)$$
$$y + 5 = -2x + 8$$
$$y = -2x + 3$$

The equation of the line is $y = -2x + 3$.

39. Slope is undefined, $(x_1, y_1) = (-5, -1)$
The line is a vertical line. All points on the line have an abscissa of −5.
The equation of the line is $x = -5$.

Objective B Exercises

41. Check that the coordinates of each given point are a solution of your equation.

43. (0,2), (3,5)

$$m = \frac{y_2 - y_1}{x_2 - x_1} = \frac{5-2}{3-0} = \frac{3}{3} = 1$$
$$y - y_1 = m(x - x_1)$$
$$y - 2 = 1(x - 0)$$
$$y - 2 = x$$
$$y = x + 2$$

The equation of the line is $y = x + 2$.

45. (0,−3), (−4,5)

$$m = \frac{y_2 - y_1}{x_2 - x_1} = \frac{5-(-3)}{-4-0} = \frac{8}{-4} = -2$$
$$y - y_1 = m(x - x_1)$$
$$y - (-3) = -2(x - 0)$$
$$y + 3 = -2x$$
$$y = -2x - 3$$
$$y = -2x - 3$$

The equation of the line is $y = -2x - 3$.

47. (2,3), (5,5)

$$m = \frac{y_2 - y_1}{x_2 - x_1} = \frac{5-3}{5-2} = \frac{2}{3}$$
$$y - y_1 = m(x - x_1)$$
$$y - 3 = \frac{2}{3}(x - 2)$$
$$y - 3 = \frac{2}{3}x - \frac{4}{3}$$
$$y = \frac{2}{3}x + \frac{5}{3}$$

The equation of the line is $y = \frac{2}{3}x + \frac{5}{3}$.

49. (−1,3), (2,4)

$$m = \frac{y_2 - y_1}{x_2 - x_1} = \frac{4-3}{2-(-1)} = \frac{1}{3}$$
$$y - y_1 = m(x - x_1)$$

$$y-3=\frac{1}{3}[x-(-1)]$$
$$y-3=\frac{1}{3}(x+1)$$
$$y-3=\frac{1}{3}x+\frac{1}{3}$$
$$y=\frac{1}{3}x+\frac{10}{3}$$

The equation of the line is $y=\frac{1}{3}x+\frac{10}{3}$.

51. (−1,−2), (3,4)

$$m=\frac{y_2-y_1}{x_2-x_1}=\frac{4-(-2)}{3-(-1)}=\frac{6}{4}=\frac{3}{2}$$
$$y-y_1=m(x-x_1)$$
$$y-4=\frac{3}{2}(x-3)$$
$$y-4=\frac{3}{2}x-\frac{9}{2}$$
$$y=\frac{3}{2}x-\frac{1}{2}$$

The equation of the line is $y=\frac{3}{2}x-\frac{1}{2}$.

53. (0,3), (2,0)

$$m=\frac{y_2-y_1}{x_2-x_1}=\frac{3-0}{0-2}=-\frac{3}{2}$$
$$y-y_1=m(x-x_1)$$
$$y=-\frac{3}{2}(x-2)$$
$$y=-\frac{3}{2}x+3$$

The equation of the line is $y=-\frac{3}{2}x+3$.

55. (−3,−1), (2,−1)

$$m=\frac{y_2-y_1}{x_2-x_1}=\frac{-1-(-1)}{2-(-3)}=\frac{0}{5}=0$$
$$y-y_1=m(x-x_1)$$
$$y-(-1)=0[(x-(-3)]$$
$$y+1=0$$
$$y=-1$$

The equation of the line is $y=-1$.

57. (−2,−3), (−1,−2)

$$m=\frac{y_2-y_1}{x_2-x_1}=\frac{-3-(-2)}{-2-(-1)}=\frac{-1}{-1}=1$$
$$y-y_1=m(x-x_1)$$
$$y-(-3)=1[x-(-2)]$$
$$y+3=1(x+2)$$
$$y=x-1$$

The equation of the line is $y=x-1$.

59. (−2,3), (2,−1)

$$m=\frac{y_2-y_1}{x_2-x_1}=\frac{3-(-1)}{-2-2}=\frac{4}{-4}=-1$$
$$y-y_1=m(x-x_1)$$
$$y-(-1)=-1(x-2)$$
$$y+1=-x+2$$
$$y=-x+1$$

The equation of the line is $y=-x+1$.

61. (2,3), (5,−5)

$$m=\frac{y_2-y_1}{x_2-x_1}=\frac{3-(-5)}{2-5}=\frac{8}{-3}=-\frac{8}{3}$$
$$y-y_1=m(x-x_1)$$
$$y-3=-\frac{8}{3}(x-2)$$
$$y-3=-\frac{8}{3}x+\frac{16}{3}$$
$$y=-\frac{8}{3}x+\frac{25}{3}$$

The equation of the line is $y=-\frac{8}{3}x+\frac{25}{3}$.

63. (2,0), (0,−1)

$$m=\frac{y_2-y_1}{x_2-x_1}=\frac{0-(-1)}{2-0}=\frac{1}{2}$$
$$y-y_1=m(x-x_1)$$
$$y-0=\frac{1}{2}(x-2)$$
$$y=\frac{1}{2}x-1$$

The equation of the line is $y=\frac{1}{2}x-1$.

65. (3,−4), (−2,−4)

$$m = \frac{y_2 - y_1}{x_2 - x_1} = \frac{-4-(-4)}{3-(-2)} = \frac{0}{5} = 0$$

$$y - y_1 = m(x - x_1)$$
$$y - (-4) = 0(x - 3)$$
$$y + 4 = 0$$
$$y = -4$$

The equation of the line is $y = -4$.

67. (0,0), (4,3)

$$m = \frac{y_2 - y_1}{x_2 - x_1} = \frac{3-0}{4-0} = \frac{3}{4}$$

$$y - y_1 = m(x - x_1)$$
$$y - 0 = \frac{3}{4}(x - 0)$$
$$y = \frac{3}{4}x$$

The equation of the line is $y = \frac{3}{4}x$.

69. (2,−1), (−1, 3)

$$m = \frac{y_2 - y_1}{x_2 - x_1} = \frac{3-(-1)}{-1-2} = \frac{4}{-3} = -\frac{4}{3}$$

$$y - y_1 = m(x - x_1)$$
$$y - (-1) = -\frac{4}{3}(x - 2)$$
$$y + 1 = -\frac{4}{3}x + \frac{8}{3}$$
$$y = -\frac{4}{3}x + \frac{5}{3}$$

The equation of the line is $y = -\frac{4}{3}x + \frac{5}{3}$.

71. (−2,5), (−2,−5)

$$m = \frac{y_2 - y_1}{x_2 - x_1} = \frac{5-(-5)}{-2-(-2)} = \frac{10}{0}$$

The slope is undefined. The line is a vertical line. All points on the line have an abscissa of −2. The equation of the line is $x = -2$.

73. (2,1), (−2,−3)

$$m = \frac{y_2 - y_1}{x_2 - x_1} = \frac{1-(-3)}{2-(-2)} = \frac{4}{4} = 1$$

$$y - y_1 = m(x - x_1)$$
$$y - 1 = 1(x - 2)$$
$$y = x - 1$$

The equation of the line is $y = x - 1$.

75. (−4, −3), (2,5)

$$m = \frac{y_2 - y_1}{x_2 - x_1} = \frac{5-(-3)}{2-(-4)} = \frac{8}{6} = \frac{4}{3}$$

$$y - y_1 = m(x - x_1)$$
$$y - 5 = \frac{4}{3}(x - 2)$$
$$y - 5 = \frac{4}{3}x - \frac{8}{3}$$
$$y = \frac{4}{3}x + \frac{7}{3}$$

The equation of the line is $y = \frac{4}{3}x + \frac{7}{3}$.

77. (0,3), (3,0)

$$m = \frac{y_2 - y_1}{x_2 - x_1} = \frac{3-0}{0-3} = \frac{3}{-3} = -1$$

$$y - y_1 = m(x - x_1)$$
$$y - 0 = -1(x - 3)$$
$$y = -x + 3$$

The equation of the line is $y = -x + 3$.

Objective C Exercises

79. Strategy: Let x represent the number of minutes after takeoff.
Let y represent the height of the plane in feet.
Use the slope-intercept form of an equation to find the equation of the line.

Solution:
a) y-intercept (0,0); slope is 1200
$y = mx + b$
$y = 1200x + 0$
The linear function is $f(x) = 1200x$.
$$0 \le x \le 26\frac{2}{3}$$

b) Find the height of the plane 11 minutes after takeoff.
$y = 1200(11) = 13,200$
Eleven minutes after takeoff the height of the plane will be 13,200 ft.

81. Strategy: Let x represent the year.
Let y represent the percent of trees that are hardwoods.
Use the point – slope formula to find the equation of the line.

Solution:
a) (1964, 57), (2004, 82)
$$m = \frac{82-57}{2004-1964} = \frac{25}{40} = 0.625$$
$$y - y_1 = m(x - x_1)$$
$$y - 57 = 0.625(x - 1964)$$
$$y - 57 = 0.625x - 1227.5$$
$$y = 0.625x - 1170.5$$
The linear equation is $f(x) = 0.625x - 1170.5$
b) Predict the percent of trees that will be hardwoods in 2012.
$y = 0.625(2012) - 1170.5 = 87$
In 2012 it is predicted that 87% of the trees will be hardwoods.

83. Strategy: Let x represent the number of miles driven.
Let y represent the number of gallons of gas in the tank.
Use the slope-intercept form of an equation to find the equation of the line.

Solution:
a) y-intercept (0,16); slope is −0.032
$y = -0.032x + 16$
Since $0 \le y \le 16$, we have
$$0 \le -0.032x + 16 \le 16$$
$$-16 \le -0.032x \le 0$$
$$500 \ge x \ge 0$$
The linear function is $f(x) = -0.032x + 16$, for $500 \ge x \ge 0$.
b) Find the number of gallons of gas left in the tank after driving 150 mi.
$y = -0.032(150) + 16 = 11.2$
After driving 150 mi there are 11.2 gal of gas left in the tank.

85. Strategy: Let x represent the price of a motorcycle.
Let y represent the number of motorcycles sold.

Use the point – slope formula to find the equation of the line.

Solution:
a) (9000, 50,000), (8750, 55,000)
$$m = \frac{55,000-50,000}{8750-9000} = \frac{5000}{-250} = -20$$
$$y - y_1 = m(x - x_1)$$
$$y - 50,000 = -20(x - 9000)$$
$$y - 50,000 = -20x + 180,000)$$
$$y = -20x + 230,000$$
The linear function is $f(x) = -20x + 230,000$.
b) Find the number of motorcycles sold when the price is \$8500.
$y = -20(8500) + 230,000 = 60,000$
When the price of a motorcycle is \$8500, 60,000 will be sold.

87. Strategy: Let x represent the number of ounces of lean hamburger.
Let y represent the number of calories.
Use the point – slope formula to find the equation of the line.

Solution:

a) (2, 126), (3, 189)

$$m = \frac{189-126}{3-2} = 63$$

$$y - y_1 = m(x - x_1)$$
$$y - 126 = 63(x-2)$$
$$y - 126 = 63x - 126$$
$$y = 63x$$

The linear function is $f(x) = 63x$.

b) Find the number of calories in a 5 oz serving.

$y = 63(5) = 315$

There are 315 calories in a 5 oz serving of lean hamburger.

89. Substitute 15,000 in for $f(x)$ and solve the equation for x.

91. (2,5), (0,3)

$$m = \frac{5-3}{2-0} = \frac{2}{2} = 1$$
$$y - y_1 = m(x - x_1)$$
$$y - 3 = 1(x-0)$$
$$y = x + 3$$
$$f(x) = x + 3$$

93. (1,3), (−1,5)

$$m = \frac{5-3}{-1-1} = \frac{2}{-2} = -1$$
$$y - y_1 = m(x - x_1)$$
$$y - 3 = -1(x-1)$$
$$y - 3 = -x + 1$$
$$y = -x + 4$$
$$f(x) = -x + 4$$
$$f(4) = -4 + 4 = 0$$

95. Given $m = \frac{4}{3}$ and a point (3,2)

a) $y - y_1 = m(x - x_1)$

$$y - 2 = \frac{4}{3}(x-3)$$
$$y - 2 = \frac{4}{3}x - 4$$
$$y = \frac{4}{3}x - 2$$

For $x = -6$ we have

$$y = \frac{4}{3}(-6) - 2 = -8 - 2 = -10$$

b) For $y = 6$ we have

$$6 = \frac{4}{3}x - 2$$
$$8 = \frac{4}{3}x$$
$$\frac{3}{4}\cdot 8 = \frac{3}{4}\cdot\frac{4}{3}x$$
$$6 = x$$

Applying the Concepts

97. The slope of any line parallel to the y-axis is undefined. In order to use the point-slope formula, we must be able to substitute the slope of the line for m. In other words, in order to use the point-slope formula the slope of the line must be defined.

99. Student solutions will vary.
Find the equation of the line:

$$m = \frac{0-6}{6-(-3)} = \frac{-6}{9} = -\frac{2}{3}$$
$$y - 0 = -\frac{2}{3}(x-6)$$
$$y = -\frac{2}{3}x + 4$$

Possible answers are:

If $x = 0$, $y = -\frac{2}{3}(0) + 4 = 4$

$x = 3$, $y = -\frac{2}{3}(3) + 4 = 2$

$x = 6$, $y = -\frac{2}{3}(6) + 4 = 0$

(0,4), (3,2), (6,0)

101. Find the x and y coordinates for the midpoint of the line segment:

$$x_m = \frac{2+(-4)}{2} = \frac{-2}{2} = -1$$

$$y_m = \frac{5+1}{2} = \frac{6}{2} = 3$$

The midpoint is (-1, 3).
Use the point-slope formula to find the equation of the line.

$$y - y_1 = m(x - x_1)$$
$$y - 3 = -2[x - (-1)]$$
$$y - 3 = -2(x+1)$$
$$y - 3 = -2x - 2$$
$$y = -2x + 1$$

Section 4.6

Objective A Exercises

1. Two lines are parallel if they have the same slope and different y-intercepts.

3. No. If two nonvertical lines are perpendicular, the product of their slopes is -1. Therefore one line must have a positive slope and the other line must have a negative slope.

5. $m = -5$

7. $m = -\frac{1}{4}$

9. Yes, the lines are perpendicular. $x = -2$ is a vertical line and $y = 3$ is a horizontal line.

11. No, the lines are not parallel. $x = -3$ is a vertical line and $y = \frac{1}{3}$ is a horizontal line.

13. No, the lines are not parallel because their slopes are not equal.

15. Yes, the lines are perpendicular. Their slopes are negative reciprocals of each other.

17. $2x + 3y = 2$

$$3y = -2x + 2$$
$$y = -\frac{2}{3}x + \frac{2}{3}$$
$$m_1 = -\frac{2}{3}$$
$$2x + 3y = -4$$
$$3y = -2x - 4$$
$$y = -\frac{2}{3}x - \frac{4}{3}$$
$$m_2 = -\frac{2}{3}$$

Since $m_1 = m_2 = -\frac{2}{3}$ the lines are parallel.

19. $x - 4y = 2$

$$-4y = -x + 2$$
$$y = \frac{1}{4}x - \frac{1}{2}$$
$$m_1 = \frac{1}{4}$$
$$4x + y = 8$$
$$y = -4x + 8$$
$$m_2 = -4$$

Since $m_1 \cdot m_2 = \frac{1}{4} \cdot (-4) = -1$ the lines are perpendicular.

21. $m_1 = \frac{6-2}{1-3} = \frac{4}{-2} = -2$

$$m_2 = \frac{-1-3}{-1-(-1)} = \frac{-4}{0}$$
$$m_1 \neq m_2$$

The lines are not parallel.

23. $m_1 = \frac{-1-2}{4-(-3)} = \frac{-3}{7} = -\frac{3}{7}$

$$m_2 = \frac{-4-3}{-2-1} = \frac{-7}{-3} = \frac{7}{3}$$
$$m_1 \cdot m_2 = -\frac{3}{7}\left(\frac{7}{3}\right) = -1$$

The lines are perpendicular.

25. Since the new line is parallel to $y = 2x + 1$ both lines will have the same slope. The slope of the new line is $m = 2$.

Use the point-slope formula to find the equation of the line.
$m = 2$ and $(3, -2)$

$$y - y_1 = m(x - x_1)$$
$$y - (-2) = 2(x - 3)$$
$$y + 2 = 2x - 6$$
$$y = 2x - 8$$

The equation of the line is $y = 2x - 8$.

27. Since the new line is perpendicular to $y = -\frac{2}{3}x - 2$ the slope of the new line must be the negative reciprocal of the slope of the given line. The slope of the new line is $m = \frac{3}{2}$.

Use the point-slope formula to find the equation of the line.

$m = \frac{3}{2}$ and $(-2, -1)$

$$y - y_1 = m(x - x_1)$$
$$y - (-1) = \frac{3}{2}[x - (-2)]$$
$$y + 1 = \frac{3}{2}(x + 2)$$
$$y + 1 = \frac{3}{2}x + 3$$
$$y = \frac{3}{2}x + 2$$

The equation of the line is $y = \frac{3}{2}x + 2$.

29. Since the new line is parallel to $2x - 3y = 2$ both lines will have the same slope.

$$2x - 3y = 2$$
$$-3y = -2x + 2$$
$$y = \frac{2}{3}x - \frac{2}{3}$$

The slope of the new line is $m = \frac{2}{3}$.

Use the point-slope formula to find the equation of the line.

$m = \frac{2}{3}$ and $(-2, -4)$

$$y - y_1 = m(x - x_1)$$
$$y - (-4) = \frac{2}{3}[x - (-2)]$$
$$y + 4 = \frac{2}{3}(x + 2)$$
$$y + 4 = \frac{2}{3}x + \frac{4}{3}$$
$$y = \frac{2}{3}x - \frac{8}{3}$$

The equation of the line is $y = \frac{2}{3}x - \frac{8}{3}$.

31. Since the new line is perpendicular to $y = -3x + 4$ the slope of the new line must be the negative reciprocal of the slope of the given line.

$m_1 = -3$

$-3 \cdot m_2 = -1$ therefore $m_2 = \frac{1}{3}$

The slope of the new line is $m_2 = \frac{1}{3}$.

Use the point-slope formula to find the equation of the line.

$m_2 = \frac{1}{3}$ and $(4, 1)$

$$y - y_1 = m(x - x_1)$$
$$y - 1 = \frac{1}{3}(x - 4)$$
$$y - 1 = \frac{1}{3}x - \frac{4}{3}$$
$$y = \frac{1}{3}x - \frac{1}{3}$$

The equation of the line is $y = \frac{1}{3}x - \frac{1}{3}$.

33. Since the new line is perpendicular to $3x-5y=2$ the slope of the new line must be the negative reciprocal of the slope of the given line.

$$3x-5y=2$$
$$-5y=-3x+2$$
$$y=\frac{3}{5}x-\frac{2}{5}$$
$$m_1=\frac{3}{5}$$
$$\frac{3}{5}\cdot m_2=-1$$
$$m_2=-\frac{5}{3}$$

The slope of the new line is $m_2=-\frac{5}{3}$.

Use the point-slope formula to find the equation of the line.

$m_2=-\frac{5}{3}$ and $(-1,-3)$

$$y-y_1=m(x-x_1)$$
$$y-(-3)=-\frac{5}{3}[x-(-1)]$$
$$y+3=-\frac{5}{3}(x+1)$$
$$y+3=-\frac{5}{3}x-\frac{5}{3}$$
$$y=-\frac{5}{3}x-\frac{14}{3}$$

The equation of the line is $y=-\frac{5}{3}x-\frac{14}{3}$.

Applying the Concepts

35. Use the points (0,0) and (6, 3).

$$m_1=\frac{3-0}{6-0}=\frac{3}{6}=\frac{1}{2}$$
$$m_1\cdot m_2=-1$$
$$\frac{1}{2}\cdot m_2=-1$$
$$m_2=-2$$

Using the point (6, 3)

$$y-3=-2(x-6)$$
$$y-3=-2x+12$$
$$y=-2x+15$$

The equation of the line is $y=-2x+15$.

37. Write the equations of the line in slope-intercept form.

$$A_1x+B_1y=C_1$$
$$B_1y=-A_1x+C_1$$
$$y=-\frac{A_1x}{B_1}+\frac{C_1}{B_1}$$
$$A_2x+B_2y=C_2$$
$$B_2y=-A_2x+C_2$$
$$y=-\frac{A_2x}{B_2}+\frac{C_2}{B_2}$$

If the two lines are parallel their slopes must be equal.

$-\frac{A_1}{B_1}=-\frac{A_2}{B_2}$ so $\frac{A_1}{B_1}=\frac{A_2}{B_2}$.

Section 4.7

Objective A Exercises

1. A half-plane is the set of points on one side of a line in the plane.

3. $y>2x-7$

$$0>2(0)-7$$
$$0>-7$$

Yes, (0,0) is a solution.

5. $y\le -\frac{2}{3}x-8$

$$0\le -\frac{2}{3}(0)-8$$
$$0\le -8$$

No, (0,0) is not a solution.

7. $y \le \frac{3}{2}x - 3$

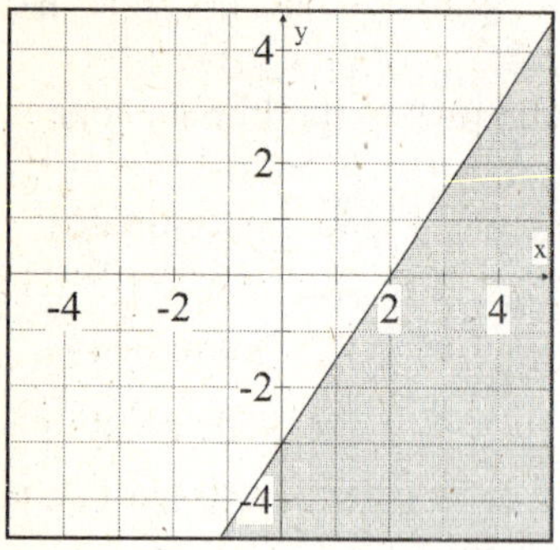

9. $y < -\frac{1}{3}x + 1$

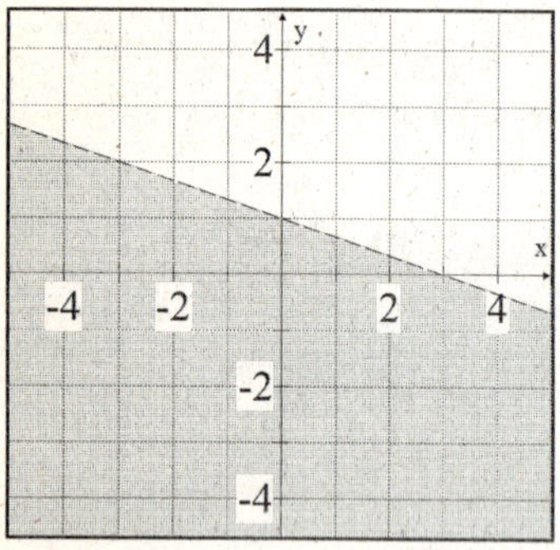

11. $4x - 5y > 10$

$-5y > -4x + 10$

$y < \frac{4}{5}x - 2$

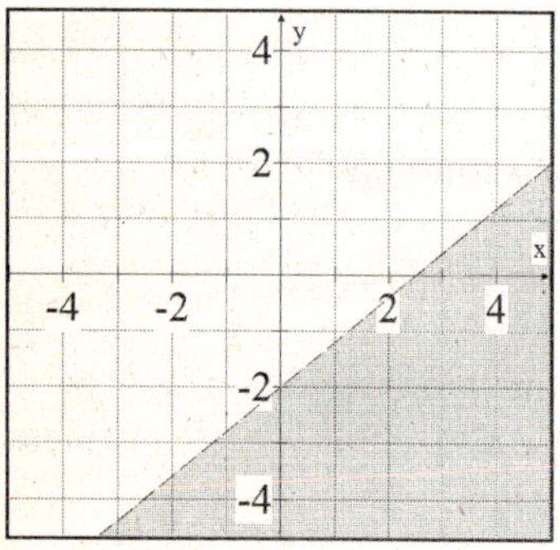

13. $x + 3y < 6$

$3y < -x + 6$

$y < -\frac{1}{3}x + 2$

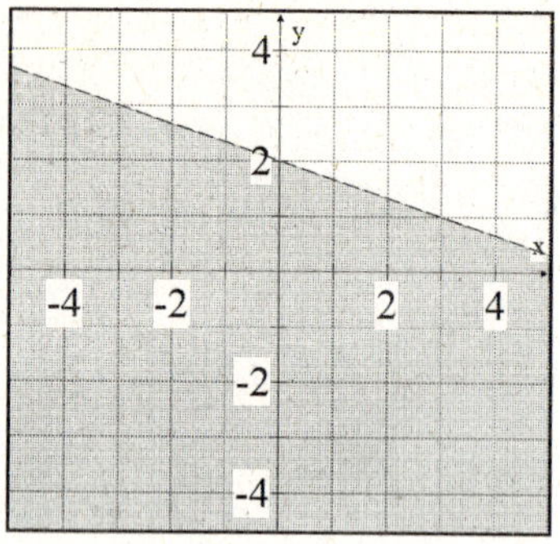

15. $2x + 3y \ge 6$

$3y \ge -2x + 6$

$y \ge -\frac{2}{3}x + 2$

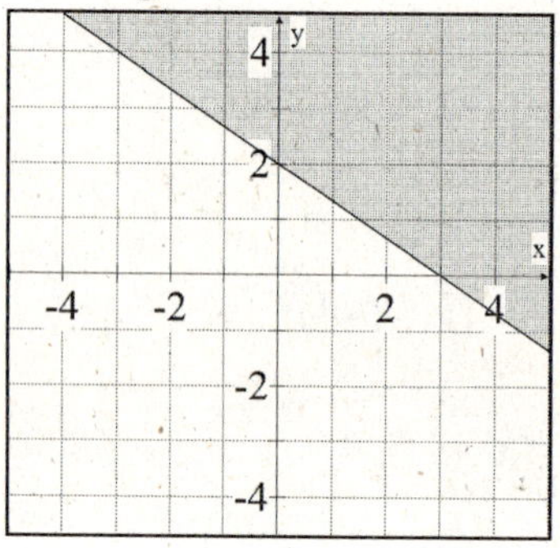

17. $-x + 2y > -8$

$2y > x - 8$

$y > \frac{1}{2}x - 4$

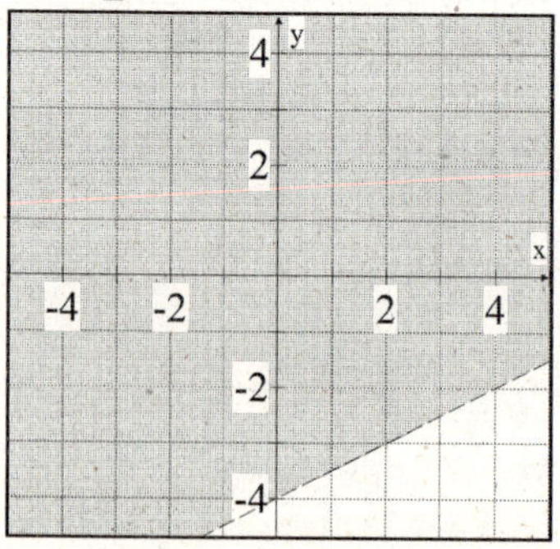

19. $y - 4 < 0$

$y < 4$

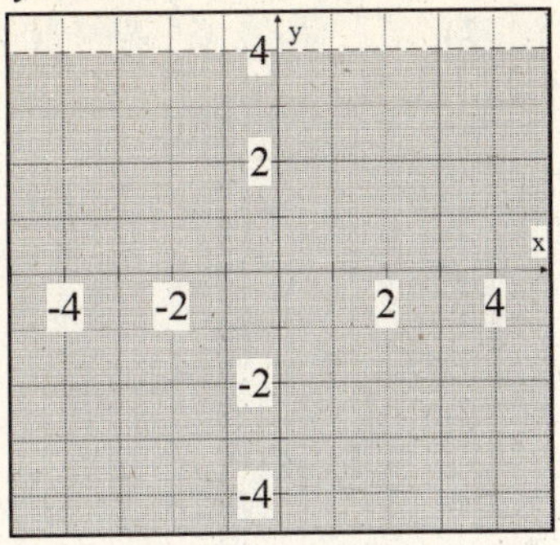

21. $6x + 5y < 15$

$5y < -6x + 15$

$y < -\frac{6}{5}x + 3$

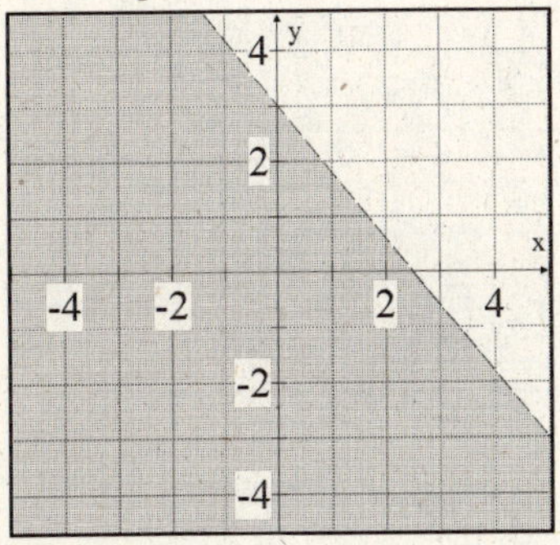

23. $-5x + 3y \geq -12$

$3y \geq 5x - 12$

$y \geq \frac{5}{3}x - 4$

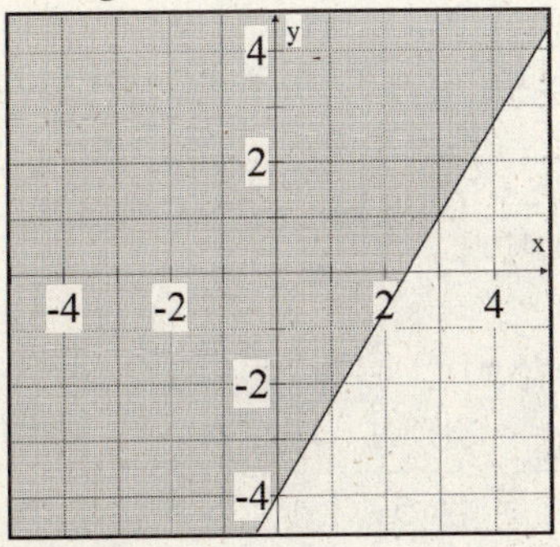

25. Quadrant I

Applying the Concepts

27. The inequality $y < 3x - 1$ is not a function because given a value of x there is more than one corresponding value of y. For example, both (3,2) and (3,−1) are ordered pairs that satisfy the inequality. This contradicts the definition of a function because there are two ordered pairs with the same first coordinate and different second coordinates.

29. There are no points whose coordinates satisfy both $y \leq x - 1$ and $y \geq x + 2$. The solution set of $y \leq x - 1$ is all points on or below the line $y = x - 1$. The solution set of $y \geq x + 2$ is all points on or above the line $y = x + 2$. Since the lines $y = x - 1$ and $y = x + 2$ are parallel lines and $y = x + 2$ is above the line $y = x - 1$ there are no points that lie both below $y = x - 1$ and above $y = x + 2$.

Chapter 4 Concept Review

1. The ordinate is the second number in an ordered pair. The abscissa is the first number in the ordered pair.

2. A linear equation in two variables has an infinite number of solutions.

3. A relation is a function when no two ordered pairs of the relation have the same first coordinate.

4. The value of a dependent variable y depends on the value of the independent variable x. The value of the independent x is not dependent on the value of any other variable.

5. In the general equation $y = mx + b$, m represents the slope and b is the ordinate of the y-intercept.

6. A straight line is determined by two points. However, to ensure the accuracy of the graph, find three ordered-pair solutions. If the three solutions do not lie on a straight line, there has been an error in calculating an ordered-pair solution or in plotting the points.

7. The equation of a vertical line is of the form $x = a$. The equation of a horizontal line is of the form $y = b$.

8. In the ordered pair for the x-intercept the y coordinate is 0. In the ordered pair for the y-intercept, the x coordinate is 0.

9. When the slope of a line is undefined, the equation of the line is of the form $x = a$.

10. Given two ordered pairs on a line, use the slope formula to find the slope of the line.

The slope formula is $m = \dfrac{y_2 - y_1}{x_2 - x_1}$, where m is the slope, and (x_1, y_1) and (x_2, y_2) are the two points on the line.

11. Parallel lines never meet. The distance between them is always the same. Perpendicular lines are lines that intersect at right angles.

12. The point-slope formula states the if (x_1, y_1) is a point on a line with slope m then $y - y_1 = m(x - x_1)$.

Chapter 4 Review Exercises

1. $y = \dfrac{x}{x-2}$

$y = \dfrac{4}{4-2} = \dfrac{4}{2} = 2$

The ordered pair is (4,2).

2. $P(x) = 3x + 4$

$P(-2) = 3(-2) + 4 = -6 + 4 = -2$

$P(a) = 3(a) + 4 = 3a + 4$

3. $y = 2x^2 - 5$

Ordered pairs: (−2, 3), (−1, −3), (0, −5), (1, −3) and (2, 3).

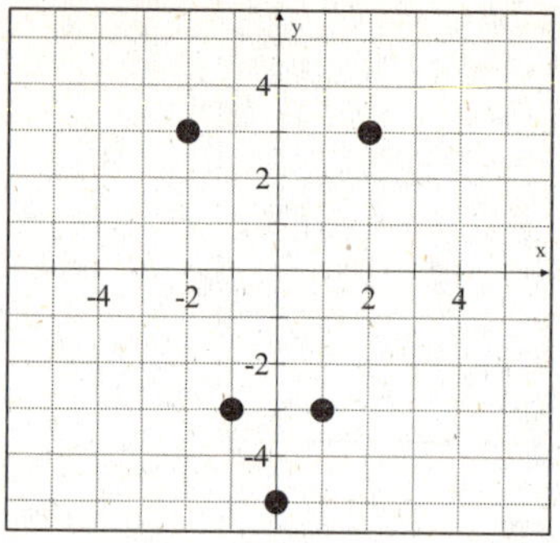

4.

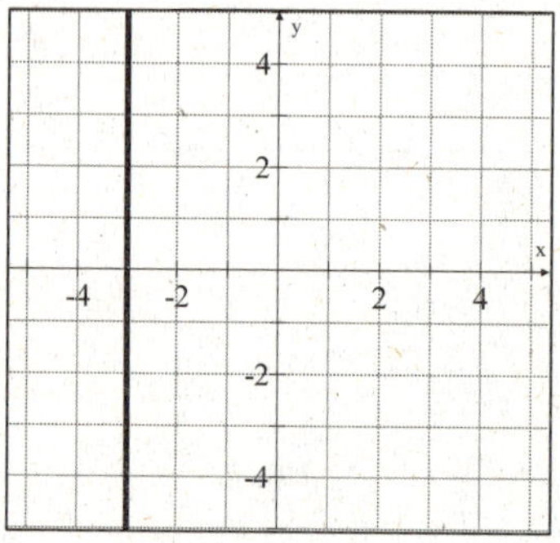

5. $f(x) = x^2 + x - 1$

$f(-2) = (-2)^2 + (-2) - 1 = 1$

$f(-1) = (-1)^2 + (-1) - 1 = -1$

$f(0) = (0)^2 + 0 - 1 = -1$

$f(1) = (1)^2 + 1 - 1 = 1$

$f(2) = (2)^2 + 2 - 1 = 5$

Range = {−1, 1, 5}

6. Domain = {−1, 0, 1, 5}
Range = {0, 2, 4}

7. (1, −4) and (6, 0)

$m = \dfrac{y_2 - y_1}{x_2 - x_1} = \dfrac{0-(-4)}{6-1} = \dfrac{4}{5}$

(−2, 3) and (3, 7)

$m = \dfrac{y_2 - y_1}{x_2 - x_1} = \dfrac{7-3}{3-(-2)} = \dfrac{4}{5}$

Yes the lines are parallel.

8. $f(x) = \dfrac{x}{x+4}$

The function is not defined for zero in the denominator.

$x + 4 = 0$

$x = -4$

$f(x)$ is not defined for $x = -4$.

9. $y = -\dfrac{2}{3}x - 2$

x-intercept:

$0 = -\dfrac{2}{3}x - 2$

$2 = -\dfrac{2}{3}x$

$x = -3$

$(-3, 0)$

y-intercept:

$y = -\dfrac{2}{3}(0) - 2$

$y = -2$

$(0, -2)$

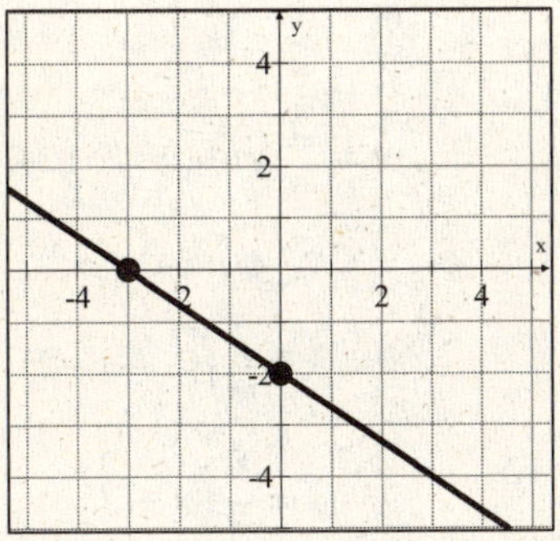

10. $3x + 2y = -6$

x-intercept:

$3x + 2(0) = -6$

$3x = -6$

$x = -2$

$(-2, 0)$

y-intercept:

$3(0) + 2y = -6$

$2y = -6$

$y = -3$

$(0, -3)$

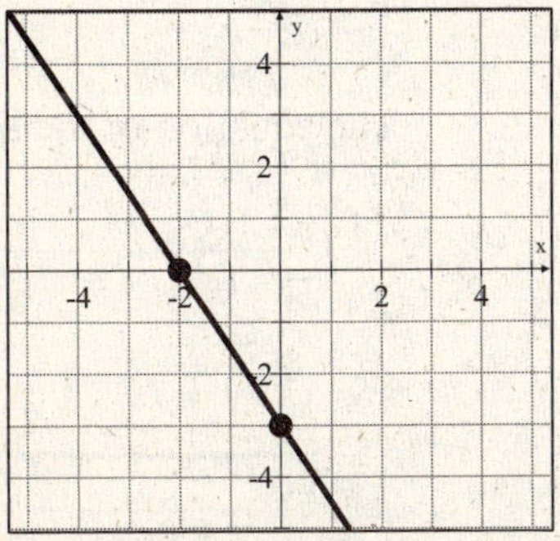

11. $y = -2x + 2$

x	y
0	2
1	0
-1	4

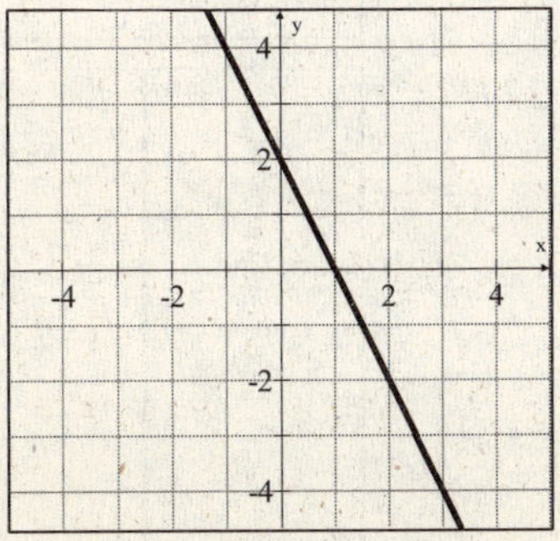

12. $4x - 3y = 12$

$-3y = -4x + 12$

$y = \dfrac{4}{3}x - 4$

x	y
0	−4
3	0
6	4

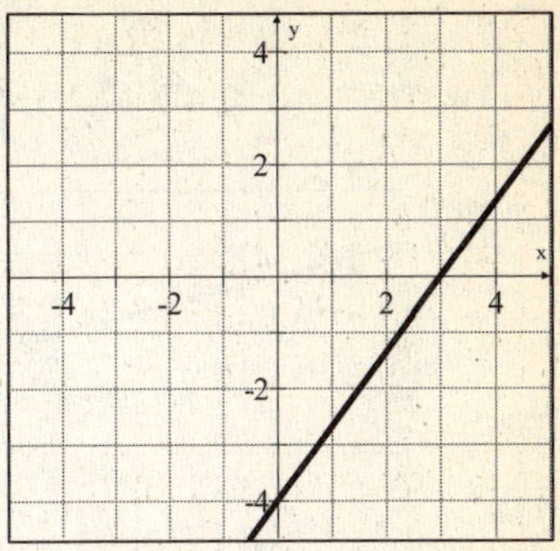

13. $(3, -2)$ and $(-1, 2)$

$m = \dfrac{y_2 - y_1}{x_2 - x_1}$

$m = \dfrac{-2-2}{3-(-1)} = \dfrac{-4}{4} = -1$

14. Use the point-slope formula to find the equation of the line.

$m = \dfrac{5}{2}$ and $(-3, 4)$

$y - y_1 = m(x - x_1)$

$y - 4 = \dfrac{5}{2}[x - (-3)]$

$y - 4 = \dfrac{5}{2}(x + 3)$

$y - 4 = \dfrac{5}{2}x + \dfrac{15}{2}$

$y = \dfrac{5}{2}x + \dfrac{23}{2}$

15.

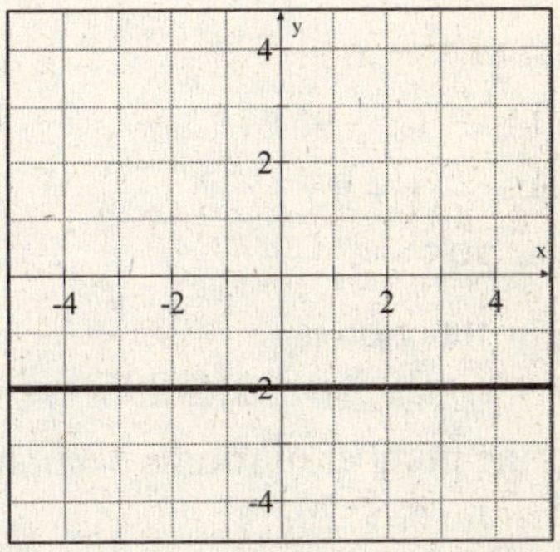

16. $m = -\dfrac{1}{4}$ and $(-2, 3)$

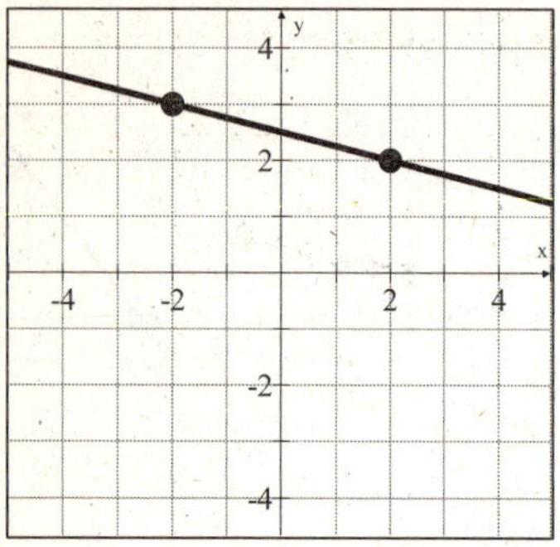

17. $f(x) = x^2 - 2$

$f(-2) = (-2)^2 - 2 = 2$

$f(-1) = (-1)^2 - 2 = -1$

$f(0) = (0)^2 - 2 = -2$

$f(1) = (1)^2 - 2 = -1$

$f(2) = (2)^2 - 2 = 2$

Range = $\{-2, -1, 2\}$

18. Strategy: Let x represent the room rate.
Let y represent the number of rooms occupied.
Use the point – slope formula to find the equation of the line.

Solution:

a) (95, 200), (105, 190)

$m = \dfrac{200-190}{95-105} = \dfrac{10}{-10} = -1$

$y - y_1 = m(x - x_1)$

$y - 200 = -1(x - 95)$

$y - 200 = -1x + 95$

$y = -x + 295$

The linear function is

$f(x) = -x + 295,\ 0 \le x \le 295$

b) Find the number of rooms occupied when the rate is $120.

$f(120) = -1(120) + 295 = 175$

When the room rate is $125, 175 rooms will be occupied.

19. The slope for the parallel line is $m = -4$.

$m = -4$ and $(-2, 3)$

$y - y_1 = m(x - x_1)$

$y - 3 = -4[(x - (-2)]$

$y - 3 = -4(x + 2)$

$y - 3 = -4x - 8$

$y = -4x - 5$

The equation of the line is $y = -4x - 5$.

20. The slope for the perpendicular line is $m = \dfrac{5}{2}$.

$m = \dfrac{5}{2}$ and $(-2, 3)$

$y - y_1 = m(x - x_1)$

$y - 3 = \dfrac{5}{2}[(x - (-2)]$

$y - 3 = \dfrac{5}{2}(x + 2)$

$y - 3 = \dfrac{5}{2}x + 5$

$y = \dfrac{5}{2}x + 8$

The equation of the line is $y = \dfrac{5}{2}x + 8$.

21.

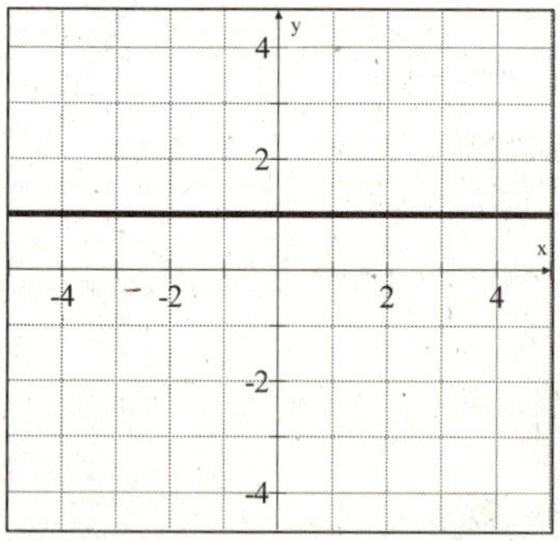

22.

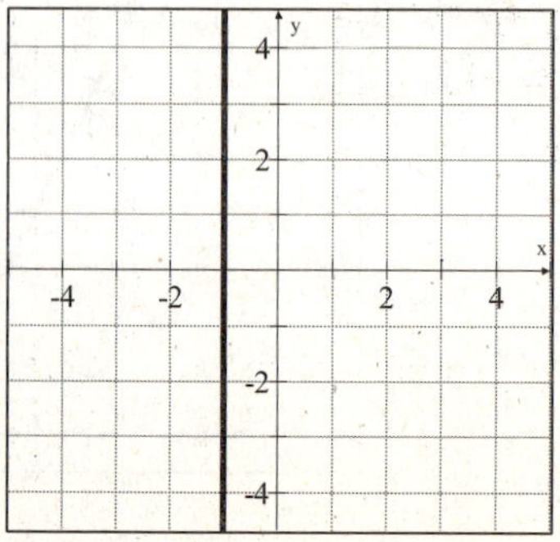

23. $m = -\frac{2}{3}$ and $(-3, 3)$

$$y - y_1 = m(x - x_1)$$
$$y - 3 = -\frac{2}{3}[(x - (-3)]$$
$$y - 3 = -\frac{2}{3}(x + 3)$$
$$y - 3 = -\frac{2}{3}x - 2$$
$$y = -\frac{2}{3}x + 1$$

The equation of the line is $y = -\frac{2}{3}x + 1$.

24. $(-8, 2)$ and $(4, 5)$

$$m = \frac{5-2}{4-(-8)} = \frac{3}{12} = \frac{1}{4}$$
$$y - y_1 = m(x - x_1)$$
$$y - 5 = \frac{1}{4}(x - 4)$$
$$y - 5 = \frac{1}{4}x - 1$$
$$y = \frac{1}{4}x + 4$$

The equation of the line is $y = \frac{1}{4}x + 4$.

25. $3x - 2y = -12$

x-intercept:

$$3x - 2(0) = -12$$
$$3x = -12$$
$$x = -4$$

$(-4, 0)$

y-intercept:

$$3(0) - 2y = -12$$
$$-2y = -12$$
$$y = 6$$

$(0, 6)$

26. $y = 3x - 4$

$$y = 3(-1) - 4$$
$$y = -7$$

The ordered-pair solution is $(-1, -7)$.

27.

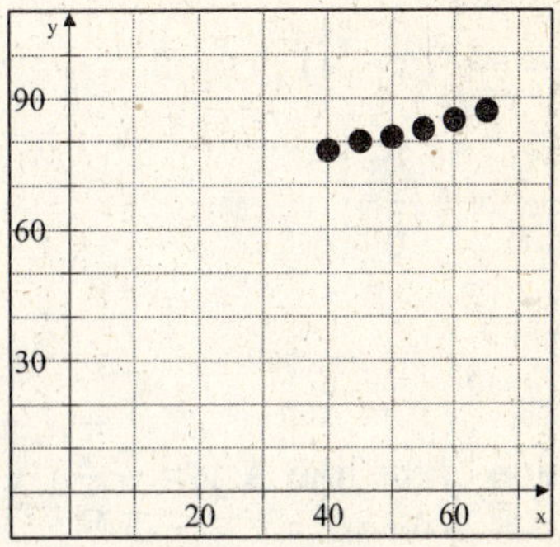

28. $y \geq 2x - 3$

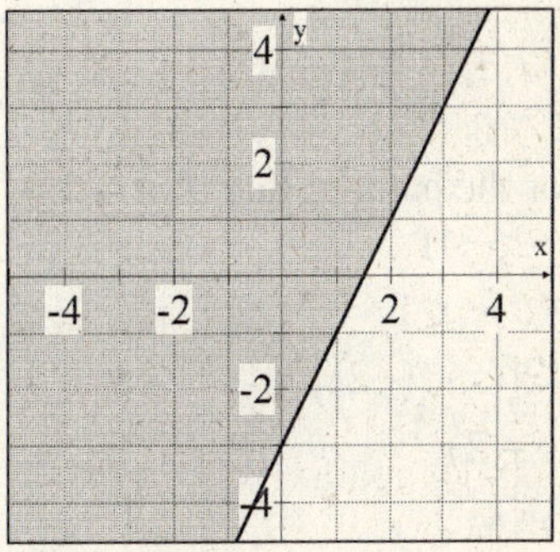

29. $3x - 2y < 6$

$$-2y < -3x + 6$$
$$y > \frac{3}{2}x - 3$$

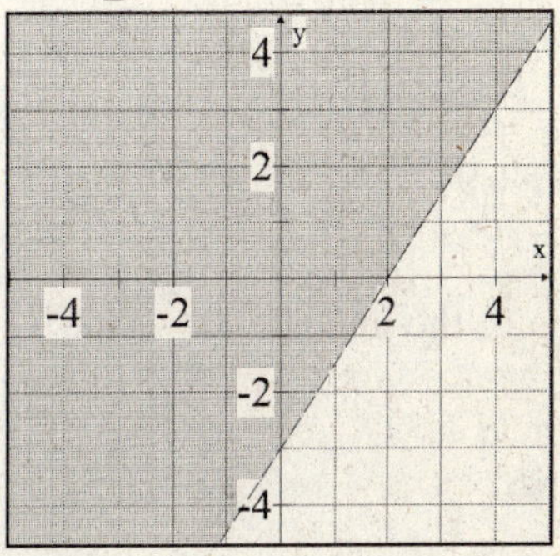

30. (−2,4) and (4,−3)

$$m=\frac{4-(-3)}{-2-4}=\frac{7}{-6}=-\frac{7}{6}$$

$$y-y_1=m(x-x_1)$$
$$y-(-3)=-\frac{7}{6}(x-4)$$
$$y+3=-\frac{7}{6}x+\frac{28}{6}$$
$$y=-\frac{7}{6}x+\frac{5}{3}$$

The equation of the line is $y=-\frac{7}{6}x+\frac{5}{3}$.

31. $4x-2y=7$

$$-2y=-4x+7$$
$$y=2x-\frac{7}{2}$$

The slope for the parallel line is $m=2$.

$m=2$ and $(-2,-4)$

$$y-y_1=m(x-x_1)$$
$$y-(-4)=2[(x-(-2)]$$
$$y+4=2(x+2)$$
$$y+4=2x+4$$
$$y=2x$$

The equation of the line is $y=2x$.

32. The slope for the parallel line is $m=-3$.

$m=-3$ and $(3,-2)$

$$y-y_1=m(x-x_1)$$
$$y-(-2)=-3(x-3)$$
$$y+2=-3x+9$$
$$y=-3x+7$$

The equation of the line is $y=-3x+7$.

33. $y=-\frac{2}{3}x+6$

$$m_1=-\frac{2}{3}$$
$$m_1\cdot m_2=-1$$
$$-\frac{2}{3}\cdot m_2=-1$$
$$m_2=\frac{3}{2}$$

The slope for the perpendicular line is $m=\frac{3}{2}$.

$m=\frac{3}{2}$ and $(2,5)$

$$y-y_1=m(x-x_1)$$
$$y-5=\frac{3}{2}(x-2)$$
$$y-5=\frac{3}{2}x-3$$
$$y=\frac{3}{2}x+2$$

The equation of the line is $y=\frac{3}{2}x+2$.

34. $m=-\frac{1}{3}$ and $(-1,4)$

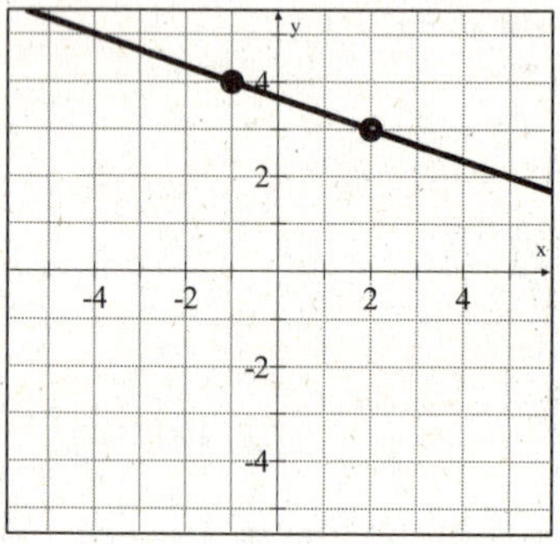

35.

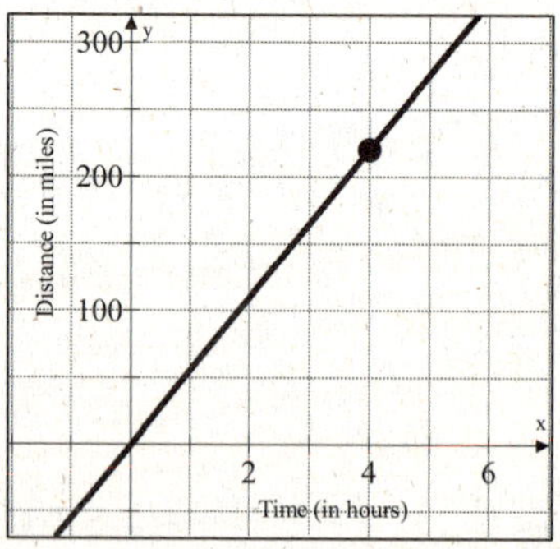

After 4 h the car will travel 220 mi.

36. (500, 12,000) and (200, 6000)

$$m=\frac{12{,}000-6000}{500-200}=\frac{6000}{300}=20$$

The slope is 20. The manufacturing cost is \$20 per calculator.

37. a) The y-intercept is (0, 25,000).
The slope is 80.
$y = mx + b$
$y = 80x + 25{,}000$
The linear function is $f(x) = 80x + 25{,}000$.
b) Predict the cost of building a house with 2000 square feet.
$f(2000) = 80(2000) + 25{,}000$
$f(x) = 185{,}000$
The house will cost $185,000 to build.

Chapter 4 Test

1. $P(x) = 2 - x^2$
ordered pairs: (−2,−2), (−1,1), (0,2), (1,1), (2, −2)

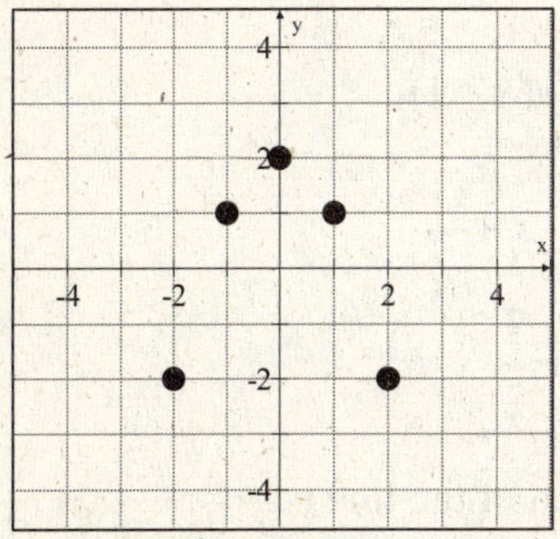

2. $y = 2x + 6$
$y = 2(-3) + 6$
$y = -6 + 6$
$y = 0$
The ordered pair is (−3, 0).

3. $y = \frac{2}{3}x - 4$

x	y
0	−4
3	−2
6	0

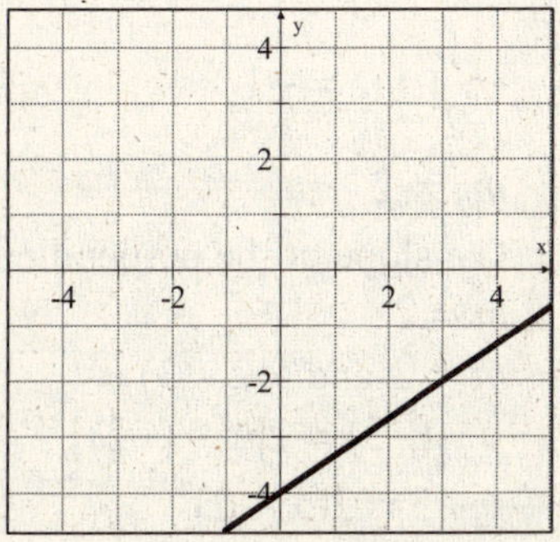

4. $2x + 3y = -3$

x	y
0	−1
3	−3
-3	1

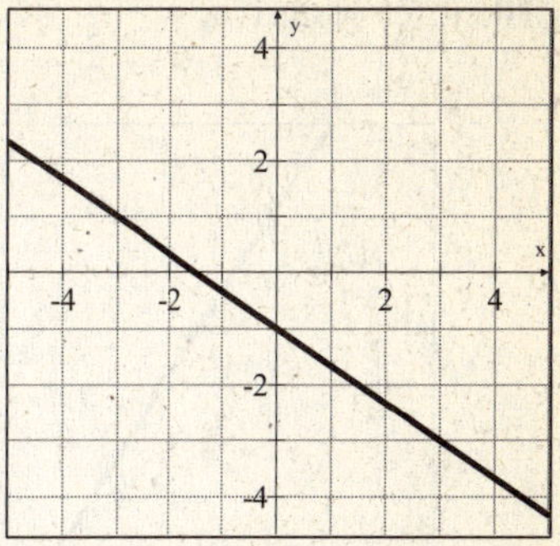

5. The equation of a vertical line that contains (−2, 3) is $x = -2$.

6. $m_1 = \dfrac{5-1}{2-(-3)} = \dfrac{4}{5}$

$m_2 = \dfrac{6-1}{0-(-4)} = \dfrac{5}{4}$

$m_1 \cdot m_2 = \dfrac{4}{5} \cdot \dfrac{5}{4} = 1$

The lines are not perpendicular.

7. (−2,3) and (4,2)

$m = \dfrac{y_2 - y_1}{x_2 - x_1} = \dfrac{3-2}{-2-4} = \dfrac{1}{-6} = -\dfrac{1}{6}$

The slope of the line is $-\dfrac{1}{6}$.

8. $P(x) = 3x^2 - 2x + 1$
$P(2) = 3(2)^2 - 2(2) + 1$
$P(2) = 9$

9. $2x - 3y = 6$
x-intercept:
$2x - 3(0) = 6$
$2x = 6$
$x = 3$
(3,0)
y-intercept:
$2(0) - 3y = 6$
$-3y = 6$
$y = -2$
(0, −2)

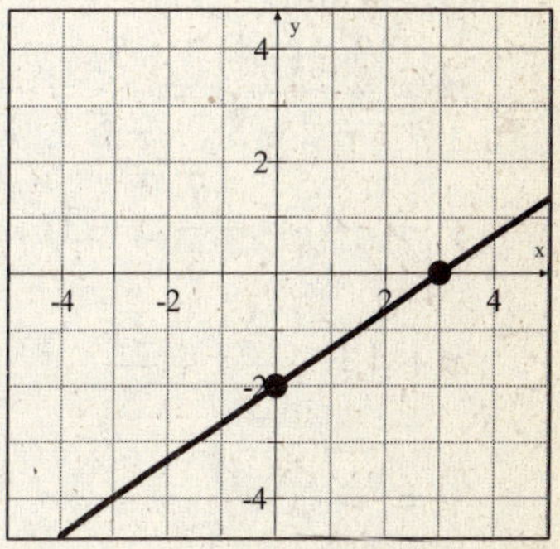

10. $(-2, 3)$ and $m = -\frac{3}{2}$

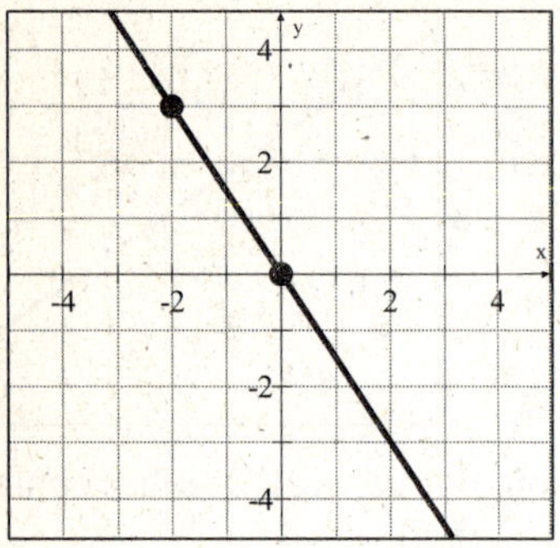

11. $m = \frac{2}{5}$ and $(-5, 2)$

$$y - 2 = \frac{2}{5}[x - (-5)]$$
$$y - 2 = \frac{2}{5}(x + 5)$$
$$y - 2 = \frac{2}{5}x + 2$$
$$y = \frac{2}{5}x + 4$$

The equation of the line is $y = \frac{2}{5}x + 4$.

12. $f(x) = \frac{2x + 1}{x}$

The function is not defined for zero in the denominator.

$x = 0$ is excluded from the domain of $f(x)$.

13. $(3, -4)$ and $(-2, 3)$

$$m = \frac{3 - (-4)}{-2 - 3} = \frac{7}{-5} = -\frac{7}{5}$$
$$y - (-4) = -\frac{7}{5}(x - 3)$$
$$y + 4 = -\frac{7}{5}x + \frac{21}{5}$$
$$y = -\frac{7}{5}x + \frac{1}{5}$$

The equation of the line is $y = -\frac{7}{5}x + \frac{1}{5}$.

14. A horizontal line has a slope of 0.

$m = 0$ and $(4, -3)$

$$y - (-3) = 0(x - 4)$$
$$y + 3 = 0$$
$$y = -3$$

The equation of the line is $y = -3$.

15. Domain = $\{-4, -2, 0, 3\}$

Range = $\{0, 2, 5\}$

16. A line parallel to $y = -\frac{3}{2}x - 6$ has a slope of $m = -\frac{3}{2}$.

$m = -\frac{3}{2}$ and $(1, 2)$

$$y - 2 = -\frac{3}{2}(x - 1)$$
$$y - 2 = -\frac{3}{2}x + \frac{3}{2}$$
$$y = -\frac{3}{2}x + \frac{7}{2}$$

The equation of the line is $y = -\frac{3}{2}x + \frac{7}{2}$.

17. $y = -\frac{1}{2}x - 3$

$$m_1 = -\frac{1}{2}$$
$$m_1 \cdot m_2 = -1$$
$$-\frac{1}{2} \cdot m_2 = -1$$
$$m_2 = 2$$

The slope of the perpendicular line is $m = 2$.

$m = 2$ and $(-2, -3)$

$$y - (-3) = 2[x - (-2)]$$
$$y + 3 = 2(x + 2)$$
$$y + 3 = 2x + 4$$
$$y = 2x + 1$$

The equation of the line is $y = 2x + 1$.

18. $3x - 4y > 8$

$-4y > -3x + 8$

$y < \frac{3}{4}x - 2$

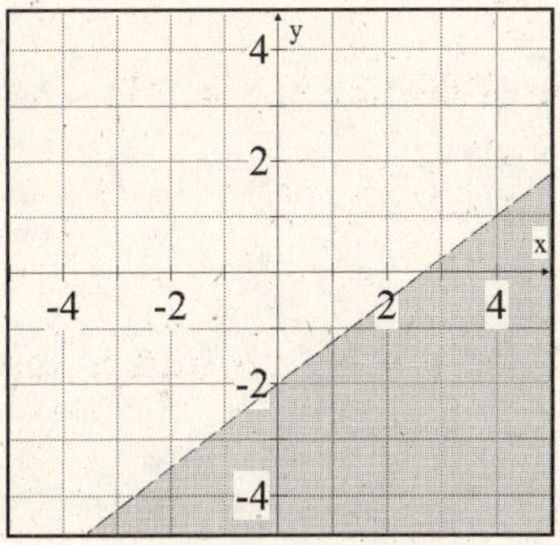

19. Strategy: Use two points on the graph to find the slope of the line.

Solution: (3, 120,000) and (12, 30,000)

$m = \frac{120{,}000 - 30{,}000}{3 - 12} = \frac{90{,}000}{-9} = -10{,}000$

The value of the house decreases by $10,000 each year.

20. Strategy: Let x represent the tuition cost. Let y represent the number of students. Use the point – slope formula to find the equation of the line.

Solution:

a) $m = \frac{-6}{20} = -\frac{3}{10}$ and (250, 100)

$y - y_1 = m(x - x_1)$

$y - 100 = -\frac{3}{10}(x - 250)$

$y - 100 = -\frac{3}{10}x + 75$

$y = -\frac{3}{10}x + 175$

The linear function that will predict enrollment based on the cost of tuition is

$f(x) = -\frac{3}{10}x + 175$

b) Find the number of students who enroll when tuition is $300.

$f(300) = -\frac{3}{10}(300) + 175$

$f(300) = 85$

When tuition is $300, 85 students will enroll.

Cumulative Review Exercises

1. Replace x with each element of the set and determine whether the inequality is true.

$x \le -3$

$-5 \le -3$ True

$-3 \le -3$ True

$-1 \le -3$ False

The inequality is true for -5 and -3.

2. $\frac{17}{20} = 17 \div 20 = 0.85$

3. $3\sqrt{45} = 3\sqrt{3^2 \cdot 5} = 3\sqrt{3^2}\sqrt{5}$

$= 3^2\sqrt{5} = 9\sqrt{5}$

4. $12 - 18 \div 3(-2)^2 = 12 - 18 \div 3(4)$

$= 12 - 6(4) = 12 - 24$

$= -12$

5. $\frac{a-b}{a^2 - c} = \frac{-2-3}{(-2)^2 - (-4)} = \frac{-2-3}{(-2)^2 + 4}$

$= \frac{-2-3}{4+4} = \frac{-5}{8} = -\frac{5}{8}$

6. $3d - 9 - 7d = (3d - 7d) - 9 = -4d - 9$

7. $4(-8z) = (4)(-8)z = -32z$

8. $2(x + y) - 5(3x - y) = 2x + 2y - 15x + 5y$

$= -13x + 7y$

9. $\{x \mid x < -2\} \cup \{x \mid x > 0\}$

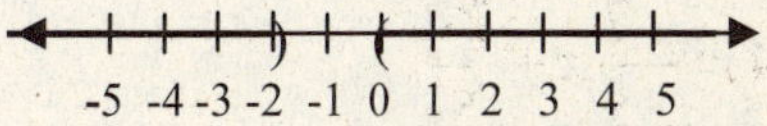

10. $2x - \frac{2}{3} = \frac{7}{3}$

$2x = \frac{9}{3}$

$2x = 3$

$x = \frac{3}{2}$

11. $3x - 2(10x - 6) = x - 6$

$3x - 20x + 12 = x - 6$

$-17x + 12 = x - 6$

$-17x - x + 12 = x - x - 6$

$-18x + 12 = -6$

$-18x + 12 - 12 = -6 - 12$

$-18x = -18$

$\frac{-18x}{-18} = \frac{-18}{-18}$

$x = 1$

The solution is 1.

12. $4x - 3 < 9x + 2$

$4x - 9x - 3 < 9x - 9x + 2$

$-5x - 3 < 2$

$-5x - 3 + 3 < 2 + 3$

$-5x < 5$

$\frac{-5x}{-5} > \frac{5}{-5}$

$x > -1$

$\{x | x > -1\}$

13. $3x - 1 < 4$ and $x - 2 > 2$

$3x < 5$ $\quad x > 4$

$x < \frac{5}{3}$

$\{x | \ x < \frac{5}{3}\} \cap \{x | \ x > 4\} = \emptyset$

14. $|3x - 5| < 5$

$-5 < 3x - 5 < 5$

$-5 + 5 < 3x - 5 + 5 < 5 + 5$

$0 < 3x < 10$

$\frac{0}{3} < \frac{3x}{3} < \frac{10}{3}$

$0 < x < \frac{10}{3}$

$\{x | \ 0 < x < \frac{10}{3}\}$

15. $f(t) = t^2 + t$

$f(2) = 2^2 + 2 = 4 + 2$

$f(2) = 6$

16. $(2, -3)$ and $(4, 1)$

$m = \frac{y_2 - y_1}{x_2 - x_1} = \frac{1-(-3)}{4-2} = \frac{4}{2} = 2$

17.

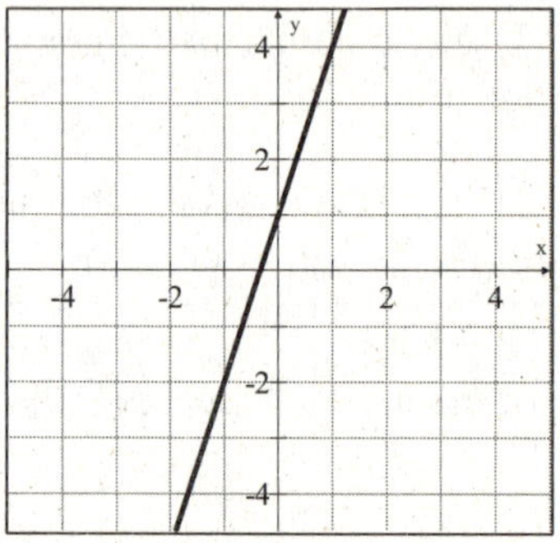

18.

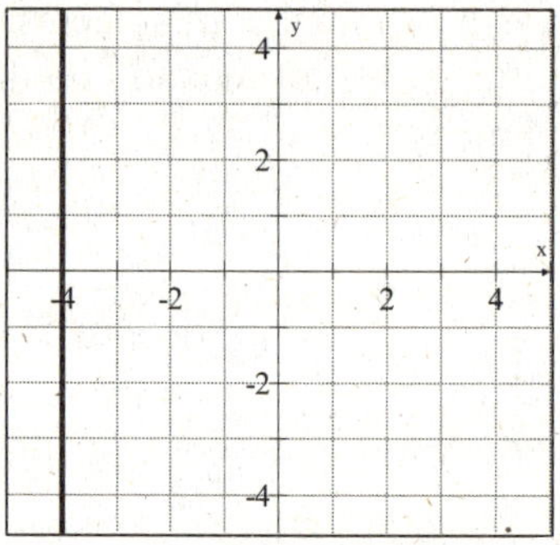

19.

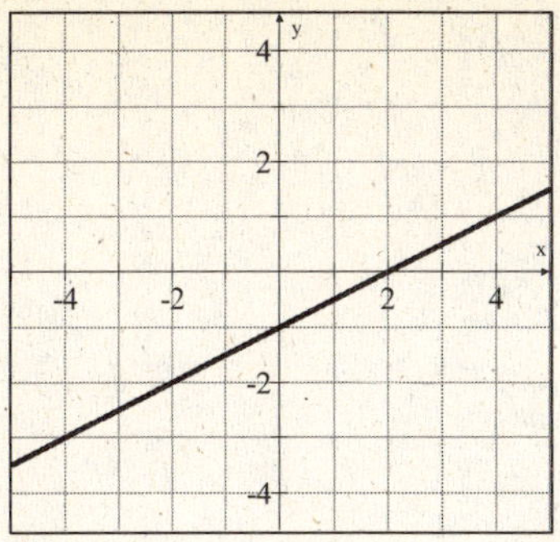

20. $3x - 2y \ge 6$

$-2y \ge -3x + 6$

$y \le \frac{3}{2}x - 3$

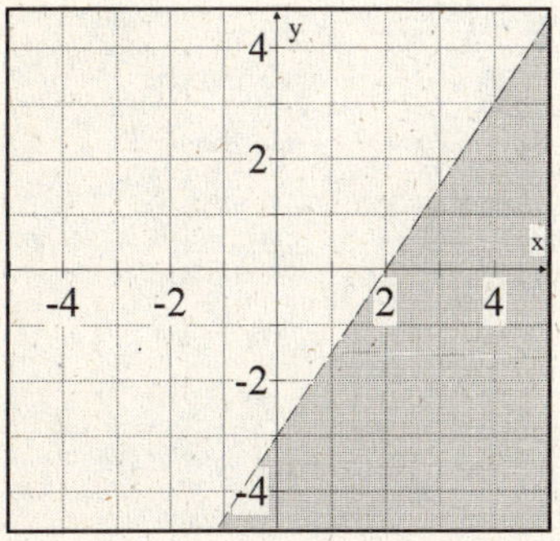

21. (6,–4) and (–3,–1)

$m = \frac{-1-(-4)}{-3-6} = \frac{3}{-9} = -\frac{1}{3}$

$y-(-4) = -\frac{1}{3}(x-6)$

$y+4 = -\frac{1}{3}x + 2$

$y = -\frac{1}{3}x - 2$

The equation of the line is $y = -\frac{1}{3}x - 2$.

22. A line parallel to $y = -\frac{3}{2}x + 2$ has a slope of $m = -\frac{3}{2}$.

$m = -\frac{3}{2}$ and (2, 4)

$y - 4 = -\frac{3}{2}(x-2)$

$y - 4 = -\frac{3}{2}x + 3$

$y = -\frac{3}{2}x + 7$

The equation of the line is $y = -\frac{3}{2}x + 7$.

23. Strategy: Let r represent the speed of 1st plane.
The speed of the 2nd plane is $2r$.

	Rate	Time	Distance
1st plane	r	3	$3r$
2nd plane	$2r$	3	$3(2r)$

The total distance traveled by the two planes is 1800 mi.

Solution: $3r + 3(2r) = 1800$

$3r + 6r = 1800$

$9r = 1800$

$r = 200$

$2r = 2(200) = 400$

The first plane is traveling at 200 mph and the second plane is traveling at 400 mph.

24. Strategy: Let x represent the pounds of coffee costing \$9.00.
Pounds of coffee costing \$6.00 is $60 - x$

	Amount	Cost	Value
\$9 coffee	x	9	$9x$
\$6 coffee	$60 - x$	6	$6(60 - x)$
Mixture	60	8	$8(60)$

The sum of the values before mixing is equal to the quantity after mixing.

Solution:

$9x + 6(60 - x) = 8(60)$

$9x + 360 - 6x = 480$

$3x + 360 = 480$

$3x = 120$

$x = 40$

$60 - x = 60 - 40 = 20$

The mixture contains 40 lb of \$9.00 coffee and 20 lb of \$6.00 coffee.

25. Strategy: Use two points on the graph to find the slope of the line.

Solution: (3, 40,000) and (12, 10,000)

$$m = \frac{40{,}000 - 10{,}000}{3 - 12} = \frac{30{,}000}{-9} = -\frac{10{,}000}{3}$$

The value of the truck decreases by $3333.33 per year.

Chapter 5: Systems of Linear Equations and Inequalities

Prep Test

1. $10\left(\frac{3}{5}x+\frac{1}{2}y\right)=10\left(\frac{3}{5}x\right)+10\left(\frac{1}{2}y\right)=6x+5y$

2. $3x+2y-z$
$3(-1)+2(4)-(-2)$
$=-3+8+2$
$=7$

3. $3x-2z=4$
$3x-2(-2)=4$
$3x+4=4$
$3x=0$
$x=0$

4. $3x+4(-2x-5)=-5$
$3x-8x-20=-5$
$-5x-20=-5$
$-5x=15$
$x=-3$
The solution is −3.

5. $0.45x+0.06(-x+4000)=630$
$0.45x-0.06x+240=630$
$0.39x+240=630$
$0.39x=390$
$x=1000$
The solution is 1000.

6. $3x-2y=6$
$-2y=-3x+6$

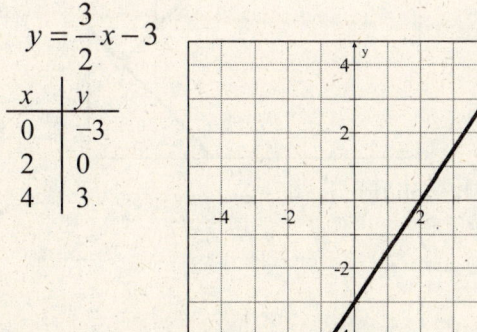

$y=\frac{3}{2}x-3$

x	y
0	−3
2	0
4	3

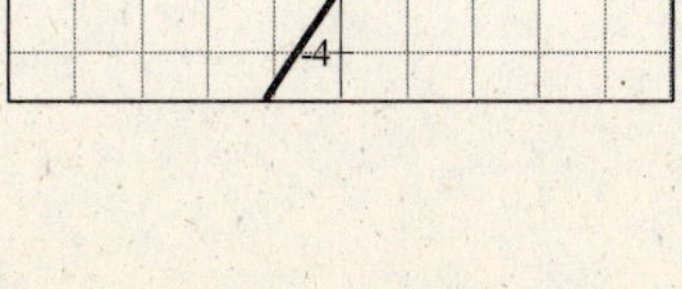

7. $y>-\frac{3}{5}x+1$

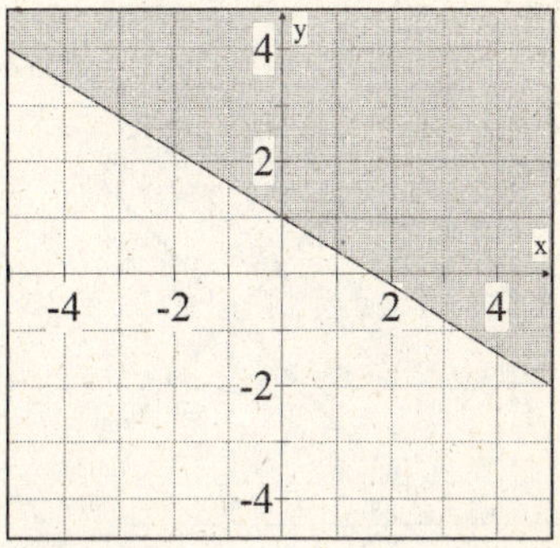

Section 5.1

Objective A Exercises

1. $3x-2y=2$
$3(0)-2(-1)=2$
$0+2=2$
$2=2$
$x+2y=6$
$0+2(-1)=6$
$-2\neq 6$
No, (0,−1) is not a solution of the system of equations.

3. $x+y=-8$
$-3+(-5)=-8$
$-8=-8$
$2x+5y=-31$
$2(-3)+5(-5)=-31$
$-6-25=-31$
$-31=-31$
Yes, (−3,−5) is a solution of the system of equations.

5. The graphs intersect at only one point. The system is an independent system of equations.

7. The graphs are parallel so the system of equations has no solution. The system is an inconsistent system of equations.

9. The system is an independent system of equations.

11. $x + y = 2$
$x - y = 4$

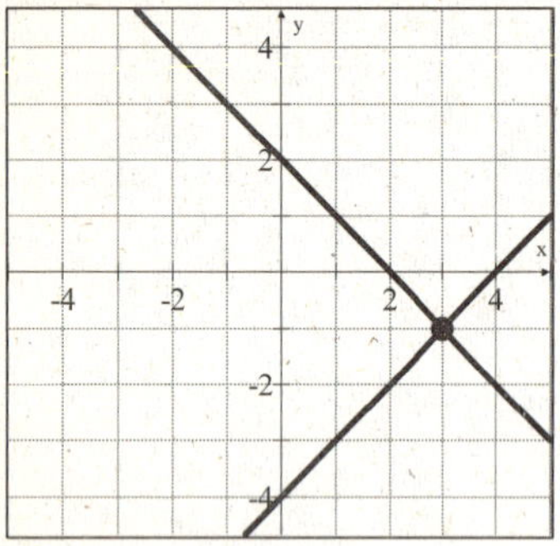

The solution is (3, −1).

13. $x - y = -2$
$x + 2y = 10$

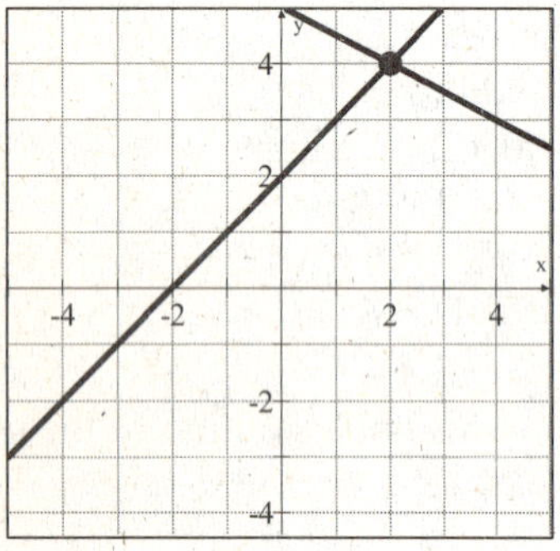

The solution is (2, 4).

15. $3x - 2y = 6$
$y = 3$

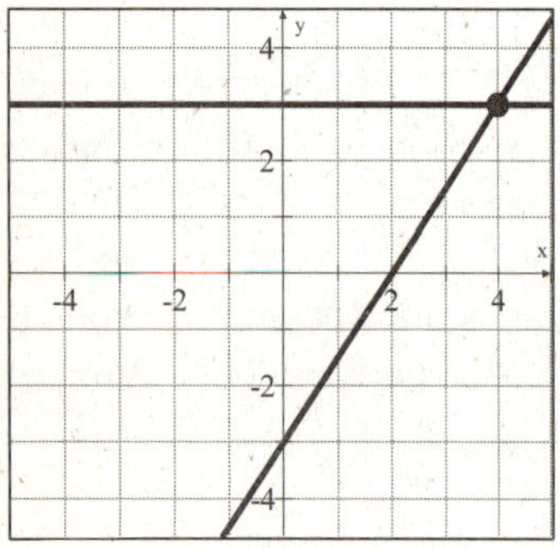

The solution is (4, 3).

17. $x = 4$
$y = -1$

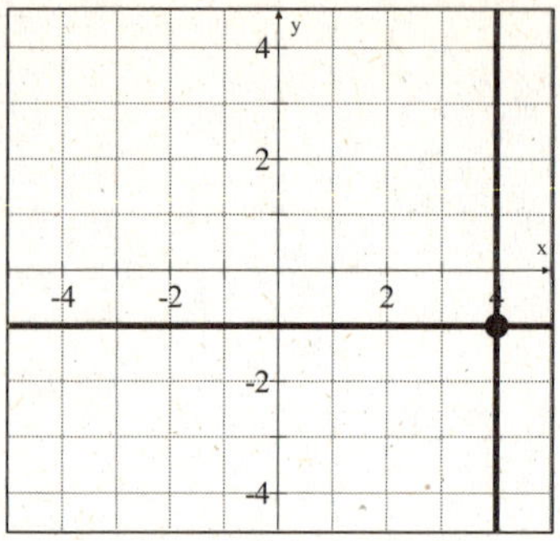

The solution is (4, −1).

19. $2x + y = 3$
$x - 2 = 0$

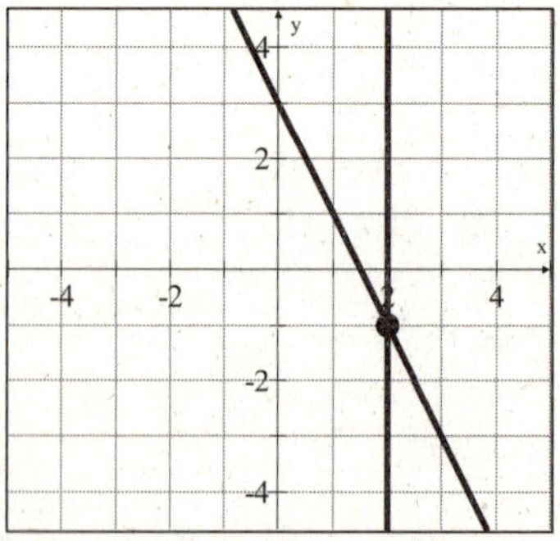

The solution is (2, −1).

21. $x - y = 6$
$x + y = 2$

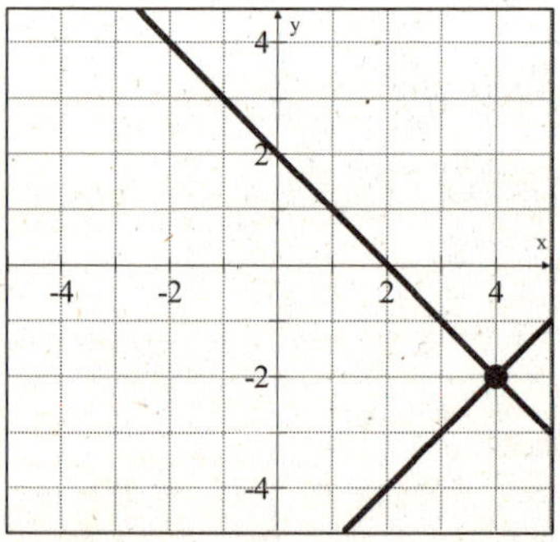

The solution is (4, −2).

23. $y = x - 5$
$2x + y = 4$

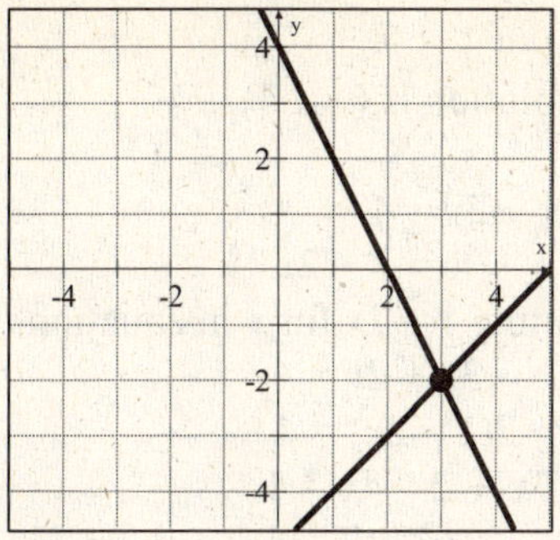

The solution is (3, −2).

25. $y = \frac{1}{2}x - 2$
$x - 2y = 8$

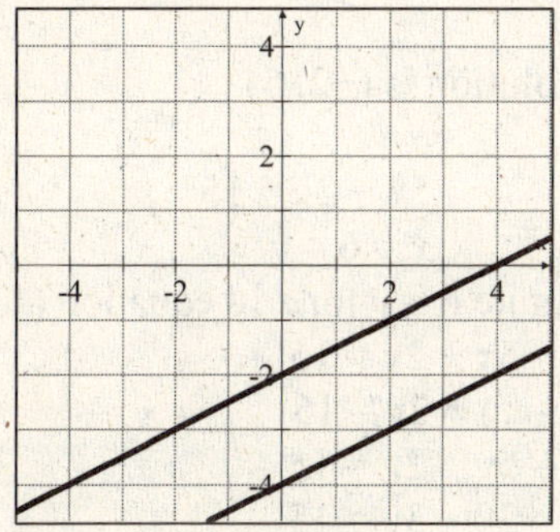

The lines are parallel and do not intersect so there is no solution.

27. $2x - 5y = 10$
$y = \frac{2}{5}x - 2$

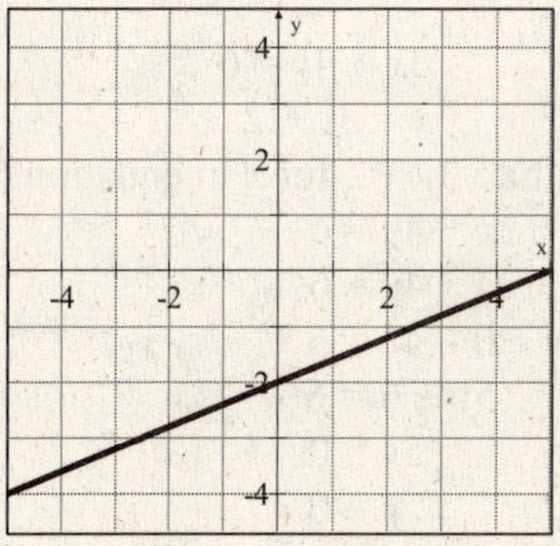

The two equations represent the same line. The system of equations is dependent. The solutions are the ordered pairs $\left(x, \frac{2}{5}x - 2\right)$.

Objective B Exercises

29. (1) $y = -x + 1$
(2) $2x - y = 5$
Substitute $-x + 1$ for y in equation (2).
$$2x - y = 5$$
$$2x - (-x + 1) = 5$$
$$2x + x - 1 = 5$$
$$3x - 1 = 5$$
$$3x = 6$$
$$x = 2$$
Substitute 2 for x in equation (1).
$y = -x + 1$
$y = -2 + 1$
$y = -1$
The solution is (2, −1).

31. (1) $x = 2y - 3$
(2) $3x + y = 5$
Substitute $2y - 3$ for x in equation (2).
$$3x + y = 5$$
$$3(2y - 3) + y = 5$$
$$6y - 9 + y = 5$$
$$7y - 9 = 5$$
$$7y = 14$$
$$y = 2$$
Substitute 2 for y in equation (1).
$x = 2y - 3$
$x = 2(2) - 3$
$x = 4 - 3$
$x = 1$
The solution is (1, 2).

33. (1) $3x + 5y = -1$
(2) $y = 2x - 8$
Substitute $2x - 8$ for y in equation (1).
$$3x + 5y = -1$$
$$3x + 5(2x - 8) = -1$$
$$3x + 10x - 40 = -1$$
$$13x - 40 = -1$$
$$13x = 39$$
$$x = 3$$
Substitute 3 for x in equation (2).
$y = 2x - 8$
$y = 2(3) - 8$
$y = 6 - 8$
$y = -2$
The solution is $(3, -2)$.

35. (1) $4x - 3y = 2$
(2) $y = 2x + 1$
Substitute $2x + 1$ for y in equation (1).
$$4x - 3y = 2$$
$$4x - 3(2x + 1) = 2$$
$$4x - 6x - 3 = 2$$
$$-2x - 3 = 2$$
$$-2x = 5$$
$$x = -\frac{5}{2}$$
Substitute $-\frac{5}{2}$ for x in equation (2).
$y = 2x + 1$
$y = 2\left(-\frac{5}{2}\right) + 1$
$y = -5 + 1$
$y = -4$
The solution is $\left(-\frac{5}{2}, -4\right)$.

37. (1) $3x - 2y = -11$
(2) $x = 2y - 9$
Substitute $2y - 9$ for x in equation (1).
$$3x - 2y = -11$$
$$3(2y - 9) - 2y = -11$$
$$6y - 27 - 2y = -11$$
$$4y - 27 = -11$$
$$4y = 16$$
$$y = 4$$
Substitute 4 for y in equation (2).
$x = 2y - 9$
$x = 2(4) - 9$
$x = 8 - 9$
$x = -1$
The solution is $(-1, 4)$.

39. (1) $3x + 2y = 4$
(2) $y = 1 - 2x$
Substitute $1 - 2x$ for y in equation (1).
$$3x + 2y = 4$$
$$3x + 2(1 - 2x) = 4$$
$$3x + 2 - 4x = 4$$
$$-x + 2 = 4$$
$$-x = 2$$
$$x = -2$$
Substitute -2 for x in equation (2).
$y = 1 - 2x$
$y = 1 - 2(-2)$
$y = 1 + 4$
$y = 5$
The solution is $(-2, 5)$.

41. (1) $5x + 2y = 15$
(2) $x = 6 - y$
Substitute $6 - y$ for x in equation (1).
$$5x + 2y = 15$$
$$5(6 - y) + 2y = 15$$
$$30 - 5y + 2y = 15$$
$$30 - 3y = 15$$
$$-3y = -15$$
$$y = 5$$
Substitute 5 for y in equation (2).
$x = 6 - y$
$x = 6 - 5$
$x = 1$
The solution is $(1, 5)$.

43. (1) $3x - 4y = 6$
(2) $x = 3y + 2$
Substitute $3y + 2$ for x in equation (1).
$$3x - 4y = 6$$
$$3(3y + 2) - 4y = 6$$
$$9y + 6 - 4y = 6$$
$$5y + 6 = 6$$
$$5y = 0$$
$$y = 0$$
Substitute 0 for y in equation (2).

$x = 3y + 2$
$x = 2(0) + 2$
$x = 0 + 2$
$x = 2$
The solution is (2, 0).

45. (1) $3x + 7y = -5$
(2) $y = 6x - 5$
Substitute $6x - 5$ for y in equation (1).
$3x + 7y = -5$
$3x + 7(6x - 5) = -5$
$3x + 42x - 35 = -5$
$45x - 35 = -5$
$45x = 30$
$x = \frac{2}{3}$
Substitute $\frac{2}{3}$ for x in equation (2).
$y = 6x - 5$
$y = 6 \cdot \frac{2}{3} - 5$
$y = 4 - 5$
$y = -1$
The solution is $(\frac{2}{3}, -1)$.

47. (1) $3x - y = 10$
(2) $6x - 2y = 5$
Solve equation (1) for y.
$3x - y = 10$
$-y = -3x + 10$
$y = 3x - 10$
Substitute $3x - 10$ for y in equation (2).
$6x - 2y = 5$
$6x - 2(3x - 10) = 5$
$6x - 6x + 20 = 5$
$20 = 5$
No solution. This is not a true equation. The lines are parallel and the system is inconsistent.

49. (1) $3x + 4y = 14$
(2) $2x + y = 1$
Solve equation (2) for y.
$2x + y = 1$
$y = -2x + 1$
Substitute $-2x + 1$ for y in equation (1).
$3x + 4y = 14$
$3x + 4(-2x + 1) = 14$
$3x - 8x + 4 = 14$
$-5x + 4 = 14$
$-5x = 10$
$x = -2$
Substitute -2 for x in equation (2).
$2x + y = 1$
$2(-2) + y = 1$
$-4 + y = 1$
$y = 5$
The solution is $(-2, 5)$.

51. (1) $3x + 5y = 0$
(2) $x - 4y = 0$
Solve equation (2) for x.
$x - 4y = 0$
$x = 4y$
Substitute $4y$ for x in equation (1).
$3x + 5y = 0$
$3(4y) + 5y = 0$
$12y + 5y = 0$
$17y = 0$
$y = 0$
Substitute 0 for y in equation (2).
$x - 4y = 0$
$x - 4(0) = 0$
$x - 0 = 0$
$x = 0$
The solution is (0, 0).

53. (1) $2x - 4y = 16$
(2) $-x + 2y = -8$
Solve equation (2) for x.
$-x + 2y = -8$
$x = 2y + 8$
Substitute $2y + 8$ for x in equation (1).
$2x - 4y = 16$
$2(2y + 8) - 4y = 16$
$4y + 16 - 4y = 16$
$16 = 16$
This is a true equation. The equations are dependent. The solutions are the ordered pairs $\left(x, \frac{1}{2}x - 4\right)$.

55. (1) $y = 3x + 2$
(2) $y = 2x + 3$
Substitute $2x + 3$ for y in equation (1).
$y = 3x + 2$
$2x + 3 = 3x + 2$
$3 = x + 2$
$x = 1$
Substitute 1 for x in equation (2).
$y = 2x + 3$
$y = 2(1) + 3$
$y = 5$
The solution is (1, 5).

57. (1) $y = 3x + 1$
(2) $y = 6x - 1$
Substitute $6x - 1$ for y in equation (1).
$y = 3x + 1$
$6x - 1 = 3x + 1$
$3x - 1 = 1$
$3x = 2$
$x = \frac{2}{3}$
Substitute $\frac{2}{3}$ for x in equation (2).
$y = 6x - 1$
$y = 6 \cdot \frac{2}{3} - 1$
$y = 3$
The solution is $(\frac{2}{3}, 3)$.

59. The value of $\frac{a}{b}$ is $\frac{2}{3}$.

Objective C Exercises

61. The interest rates on the two accounts are 5.5% and 7.2%.

63. Strategy: Let x represent the amount invested at 4.2%.
$2800 is invested at 3.5%.

	Principal	Rate	Interest
Amount at 3.5%	2800	0.035	0.035(2800)
Amount at 4.2%	x	0.042	$0.042x$

The sum of the interest earned is $329.

Solution: $0.035(2800) + 0.042x = 329$
$98 + 0.042x = 329$
$0.042x = 231$
$x = 5500$
$5500 is invested at 4.2%.

65. Strategy: Let x represent the amount invested at 6.5%.
Let y represent the total amount invested at 5%.
$6000 is invested at 4%.

	Principal	Rate	Interest
Amount at 4%	6000	0.04	0.04(6000)
Amount at 6.5%	x	0.065	$0.065x$
Total invested	y	0.05	$0.05y$

The total amount invested is y.
$y = 6000 + x$
The total interest earned is equal to 5% of the total investment.
$0.04(6000) + 0.065x = 0.05y$

Solution: (1) $y = 6000 + x$
(2) $0.04(6000) + 0.065x = 0.05y$
Substitute $6000 + x$ for y in equation (2).
$0.04(6000) + 0.065x = 0.05(6000 + x)$
$240 + 0.065x = 300 + 0.05x$
$240 + 0.015x = 300$
$0.015x = 60$
$x = 4000$
$4000 must be invested at 6.5%.

67. Strategy: Let x represent the amount invested at 3.5%.
Let y represent the amount invested at 4.5%.

	Principal	Rate	Interest
Amount at 3.5%	x	0.035	$0.035x$
Amount at 4.5%	y	0.045	$0.045y$

The total amount invested is $42,000.
$x + y = 42{,}000$
The interest earned from the 3.5% investment is equal to the interest earned from the 4.5% investment.
$0.035x = 0.045y$

Solution: (1) $x + y = 42{,}000$
(2) $0.035x = 0.045y$
Solve equation (1) for y and substitute for y in equation (2).
$y = 42{,}000 - x$
$0.035x = 0.045(42{,}000 - x)$
$0.035x = 1890 - 0.045x$
$0.080x = 1890$
$x = 23{,}625$
$y = 42{,}000 - x = 42{,}000 - 23{,}625 = 18{,}375$

\$23,625 is invested at 3.5% and \$18,375 is invested at 4.5%.

69. Strategy: Let x represent the amount invested at 4.5%.
Let y represent the amount invested at 8%.

	Principal	Rate	Interest
Amount at 4.5%	x	0.045	$0.045x$
Amount at 8%	y	0.08	$0.08y$

The total amount invested is \$16,000.
$x + y = 16{,}000$
The total interest earned is \$1070.
$0.045x + 0.08y = 1070$

Solution: (1) $x + y = 16{,}000$
(2) $0.045x + 0.08y = 1070$
Solve equation (1) for y and substitute for y in equation (2).
$y = 16{,}000 - x$
$0.045x + 0.08(16{,}000 - x) = 1070$
$0.045x + 1280 - 0.08x = 1070$
$-0.035x + 1280 = 1070$
$-0.035x = -210$
$x = 6000$
\$6000 is invested at 4.5% .

Applying the Concepts

71.

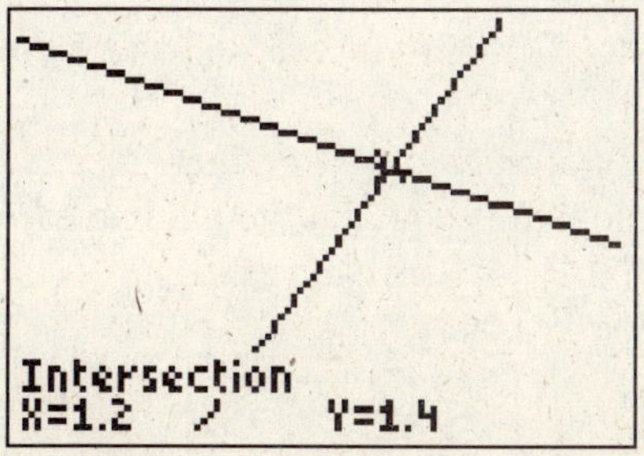

The solution is (1.20, 1.40).

73.

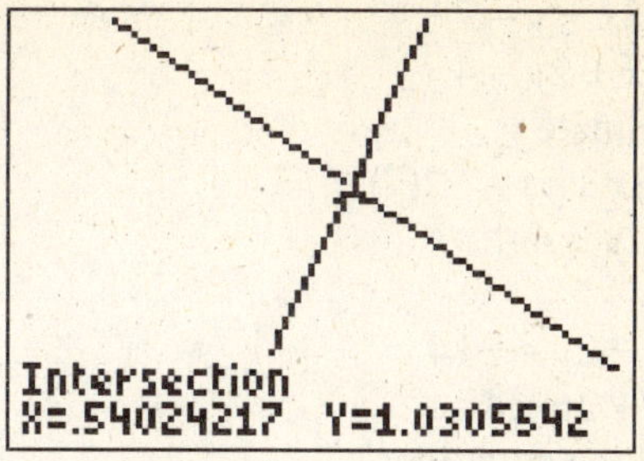

The solution is (0.54, 1.03).

Section 5.2

Objective A Exercises

1. Student answers may vary. Possible answers are 6 and −5.

3. (1) $x - y = 5$
(2) $x + y = 7$
Eliminate y. Add the two equations.
$2x = 12$
$x = 6$
Replace x with 6 in equation (1).
$x - y = 5$
$6 - y = 5$
$-y = -1$
$y = 1$
The solution is (6,1).

5. (1) $3x + y = 4$
(2) $x + y = 2$
Eliminate y.
$3x + y = 4$
$-1(x + y) = -1(2)$

$3x + y = 4$
$-x - y = -2$
Add the equations.
$2x = 2$
$x = 1$
Replace x with 1 in equation (2).
$x + y = 2$
$1 + y = 2$
$y = 1$
The solution is (1,1).

7. (1) $3x + y = 7$
(2) $x + 2y = 4$
Eliminate y.
$-2(3x + y) = -2(7)$
$x + 2y = 4$

$-6x - 2y = -14$
$x + 2y = 4$
Add the equations.
$-5x = -10$
$x = 2$
Replace x with 2 in equation (2).
$x + 2y = 4$
$2 + 2y = 4$
$2y = 2$
$y = 1$
The solution is (2,1).

9. (1) $2x + 3y = -1$
(2) $x + 5y = 3$
Eliminate x.
$2x + 3y = -1$
$-2(x + 5y) = -2(3)$

$2x + 3y = -1$
$-2x - 10y = -6$
Add the equations.
$-7y = -7$
$y = 1$
Replace y with 1 in equation (2).
$x + 5y = 3$
$x + 5(1) = 3$
$x + 5 = 3$
$x = -2$
The solution is (−2,1).

11. (1) $3x - y = 4$
(2) $6x - 2y = 8$
Eliminate y.
$-2(3x - y) = -2(4)$
$-6x + 2y = -8$

$6x - 2y = 8$
$6x - 2y = 8$
Add the equations.
$0 = 0$
This is a true equation. The equations are dependent. The solutions are the ordered pairs $(x, 3x - 4)$.

13. (1) $2x + 5y = 9$
(2) $4x - 7y = -16$
Eliminate x.
$-2(2x + 5y) = -2(9)$
$4x - 7y = -16$

$-4x - 10y = -18$
$4x - 7y = -16$
Add the equations.
$-17y = -34$
$y = 2$
Replace y with 2 in equation (1).
$2x + 5y = 9$
$2x + 5(2) = 9$
$2x + 10 = 9$
$2x = -1$
$x = -\frac{1}{2}$

The solution is $\left(-\frac{1}{2}, 2\right)$.

15. (1) $4x - 6y = 5$
(2) $2x - 3y = 7$
Eliminate y.
$4x - 6y = 5$
$-2(2x - 3y) = -2(7)$

$4x - 6y = 5$
$-4x + 6y = -14$
Add the equations.
$0 = -9$
This is not a true equation. The system of equations is inconsistent and therefore has no solution.

17. (1) $3x - 5y = 7$
(2) $x - 2y = 3$
Eliminate x.
$3x - 5y = 7$
$-3(x - 2y) = -3(3)$

$3x - 5y = 7$
$-3x + 6y = -9$
Add the equations.
$y = -2$
Replace y with -2 in equation (2).
$x - 2y = 3$
$x - 2(-2) = 3$
$x + 4 = 3$
$x = -1$
The solution is $(-1,-2)$.

19. (1) $x + 3y = 7$
(2) $-2x + 3y = 22$
Eliminate x.
$2(x + 3y) = 2(7)$
$-2x + 3y = 22$

$2x + 6y = 14$
$-2x + 3y = 22$
Add the equations.
$9y = 36$
$y = 4$
Replace y with 4 in equation (1).
$x + 3y = 7$
$x + 3(4) = 7$
$x + 12 = 7$
$x = -5$
The solution is $(-5,4)$.

21. (1) $3x + 2y = 16$
(2) $2x - 3y = -11$
Eliminate x.
$-2(3x + 2y) = -2(16)$
$3(2x - 3y) = 3(-11)$

$-6x - 4y = -32$
$6x - 9y = -33$
Add the equations.
$-13y = -65$
$y = 5$
Replace y with 5 in equation (1).
$3x + 2y = 16$
$3x + 2(5) = 16$
$3x + 10 = 16$
$3x = 6$
$x = 2$
The solution is $(2,5)$.

23. (1) $4x + 4y = 5$
(2) $2x - 8y = -5$
Eliminate x.
$4x + 4y = 5$
$-2(2x - 8y) = -2(-5)$

$4x + 4y = 5$
$-4x + 16y = 10$
Add the equations.
$20y = 15$
$y = \frac{3}{4}$

Replace y with $\frac{3}{4}$ in equation (1).
$4x + 4y = 5$
$4x + 4\left(\frac{3}{4}\right) = 5$
$4x + 3 = 5$
$4x = 2$
$x = \frac{1}{2}$

The solution is $\left(\frac{1}{2},\frac{3}{4}\right)$.

25. (1) $5x + 4y = 0$
(2) $3x + 7y = 0$
Eliminate x.
$-3(5x + 4y) = -3(0)$
$5(3x + 7y) = 5(0)$
$-15x - 12y = 0$
$15x + 35y = 0$
Add the equations.
$23y = 0$
$y = 0$
Replace y with 0 in equation (1).

$5x+4y=0$
$5x+4(0)=0$
$5x+0=0$
$5x=0$
$x=0$
The solution is (0,0).

27. (1) $5x+2y=1$
(2) $2x+3y=7$
Eliminate x.
$-2(5x+2y)=-2(1)$
$5(2x+3y)=5(7)$

$-10x-4y=-2$
$10x+15y=35$
Add the equations.
$11y=33$
$y=3$
Replace y with 3 in equation (1).
$5x+2y=1$
$5x+2(3)=1$
$5x+6=1$
$5x=-5$
$x=-1$
The solution is (−1,3).

29. (1) $3x-6y=6$
(2) $9x-3y=8$
Eliminate y.
$3x-6y=6$
$-2(9x-3y)=-2(8)$

$3x-6y=6$
$-18x+6y=-16$
Add the equations.
$-15x=-10$
$x=\frac{2}{3}$

Replace x with $\frac{2}{3}$ in equation (1).

$3x-6y=6$
$3\left(\frac{2}{3}\right)-6y=6$
$2-6y=6$
$-6y=4$
$y=-\frac{2}{3}$

The solution is $\left(\frac{2}{3},-\frac{2}{3}\right)$.

31. (1) $\frac{3}{4}x+\frac{1}{3}y=-\frac{1}{2}$
(2) $\frac{1}{2}x-\frac{5}{6}y=-\frac{7}{2}$
Clear the fractions.
$12\left(\frac{3}{4}x+\frac{1}{3}y\right)=12\left(-\frac{1}{2}\right)$
$6\left(\frac{1}{2}x-\frac{5}{6}y\right)=6\left(-\frac{7}{2}\right)$

$9x+4y=-6$
$3x-5y=-21$
Eliminate x.
$9x+4y=-6$
$-3(3x-5y)=-3(-21)$

$9x+4y=-6$
$-9x+15y=63$
Add the equations.
$19y=57$
$y=3$
Replace y with 3 in equation (1).
$\frac{3}{4}x+\frac{1}{3}y=-\frac{1}{2}$
$\frac{3}{4}x+\frac{1}{3}(3)=-\frac{1}{2}$
$\frac{3}{4}x+1=-\frac{1}{2}$
$\frac{3}{4}x=-\frac{3}{2}$
$x=-2$
The solution is (−2,3).

33. (1) $\frac{5x}{6}+\frac{y}{3}=\frac{4}{3}$
(2) $\frac{2x}{3}-\frac{y}{2}=\frac{11}{6}$
Clear the fractions.
$6\left(\frac{5x}{6}+\frac{y}{3}\right)=6\left(\frac{4}{3}\right)$
$6\left(\frac{2x}{3}-\frac{y}{2}\right)=6\left(\frac{11}{6}\right)$

$5x + 2y = 8$
$4x - 3y = 11$
Eliminate y.
$3(5x + 2y) = 3(8)$
$2(4x - 3y) = 2(11)$

$15x + 6y = 24$
$8x - 6y = 22$
Add the equations.
$23x = 46$
$x = 2$
Replace x with 2 in equation (1).
$\frac{5x}{6} + \frac{y}{3} = \frac{4}{3}$
$\frac{5(2)}{6} + \frac{y}{3} = \frac{4}{3}$
$\frac{5}{3} + \frac{y}{3} = \frac{4}{3}$
$\frac{y}{3} = -\frac{1}{3}$
$y = -1$
The solution is (2,−1).

35. (1) $\frac{2x}{5} - \frac{y}{2} = \frac{13}{2}$
(2) $\frac{3x}{4} - \frac{y}{5} = \frac{17}{2}$
Clear the fractions.
$10\left(\frac{2x}{5} - \frac{y}{2}\right) = 10\left(\frac{13}{2}\right)$
$20\left(\frac{3x}{4} - \frac{y}{5}\right) = 20\left(\frac{17}{2}\right)$

$4x - 5y = 65$
$15x - 4y = 170$
Eliminate y.
$4(4x - 5y) = 4(65)$
$-5(15x - 4y) = -5(170)$

$16x - 20y = 260$
$-75x + 20y = -850$
Add the equations.
$-59x = -590$
$x = 10$
Replace x with 10 in equation (1).
$\frac{2x}{5} - \frac{y}{2} = \frac{13}{2}$
$\frac{2(10)}{5} - \frac{y}{2} = \frac{13}{2}$
$4 - \frac{y}{2} = \frac{13}{2}$
$-\frac{y}{2} = \frac{5}{2}$
$y = -5$
The solution is (10,−5).

37. (1) $\frac{3x}{2} - \frac{y}{4} = -\frac{11}{12}$
(2) $\frac{x}{3} - y = -\frac{5}{6}$
Clear the fractions.
$12\left(\frac{3x}{2} - \frac{y}{4}\right) = 12\left(-\frac{11}{12}\right)$
$6\left(\frac{x}{3} - y\right) = 6\left(-\frac{5}{6}\right)$

$18x - 3y = -11$
$2x - 6y = -5$
Eliminate y.
$-2(18x - 3y) = -2(-11)$
$2x - 6y = -5$

$-36x + 6y = 22$
$2x - 6y = -5$
Add the equations.
$-34x = 17$
$x = -\frac{1}{2}$

Replace x with $-\frac{1}{2}$ in equation (1).
$\frac{3x}{2} - \frac{y}{4} = -\frac{11}{12}$
$\frac{3}{2}\left(-\frac{1}{2}\right) - \frac{y}{4} = -\frac{11}{12}$
$-\frac{3}{4} - \frac{y}{4} = -\frac{11}{12}$
$-\frac{y}{4} = -\frac{1}{6}$

$y = \frac{2}{3}$

The solution is $(-\frac{1}{2}, \frac{2}{3})$.

39. (1) $4x - 5y = 3y + 4$
(2) $2x + 3y = 2x + 1$
Write the equations in the form $Ax + By = C$.
Solve the system of equations.
(1) $4x - 8y = 4$
(2) $3y = 1$

Solve equation (2) for y.
$3y = 1$
$y = \frac{1}{3}$

Replace y with $\frac{1}{3}$ in equation (1).
$4x - 5y = 3y + 4$
$4x - 5 \cdot \frac{1}{3} = 3 \cdot \frac{1}{3} + 4$
$4x - \frac{5}{3} = 1 + 4$
$4x - \frac{5}{3} = 5$
$4x = \frac{20}{3}$
$x = \frac{5}{3}$

The solution is $\left(\frac{5}{3}, \frac{1}{3}\right)$.

41. (1) $2x + 5y = 5x + 1$
(2) $3x - 2y = 3y + 3$
Write the equations in the form $Ax + By = C$.
Solve the system of equations.
(1) $-3x + 5y = 1$
(2) $3x - 5y = 3$

Add the equations.
$0 = 4$
This is not a true equation. The system of equations is inconsistent and therefore has no solution.

43. (1) $5x + 2y = 2x + 1$
(2) $2x - 3y = 3x + 2$
Write the equations in the form $Ax + By = C$.
Solve the system of equations.
(1) $3x + 2y = 1$
(2) $-x - 3y = 2$

Eliminate x.
$3x + 2y = 1$
$3(-x - 3y) = 3(2)$

$3x + 2y = 1$
$-3x - 9y = 6$
Add the equations.
$-7y = 7$
$y = -1$
Replace y with -1 in equation (1).
$5x + 2y = 2x + 1$
$5x + 2(-1) = 2x + 1$
$5x - 2 = 2x + 1$
$3x = 3$
$x = 1$
The solution is $(1, -1)$.

Objective B Exercises

45. (1) $x + 2y - z = 1$
(2) $2x - y + z = 6$
(3) $x + 3y - z = 2$
Eliminate z. Add equations (1) and (2).
$x + 2y - z = 1$
$2x - y + z = 6$

(4) $3x + y = 7$

Add equations (2) and (3).
$2x - y + z = 6$
$x + 3y - z = 2$

(5) $3x + 2y = 8$

Use equations (4) and (5) solve for x and y.
$3x + y = 7$
$3x + 2y = 8$
Eliminate x.
$-1(3x + y) = -1(7)$
$3x + 2y = 8$

$-3x - y = -7$
$3x + 2y = 8$
$y = 1$
Replace y with 1 in equation (4).
$3x + y = 7$
$3x + 1 = 7$

$3x = 6$
$x = 2$
Replace x with 2 and y with 1 in equation (1).
$x + 2y - z = 1$
$2 + 2(1) - z = 1$
$2 + 2 - z = 1$
$4 - z = 1$
$-z = -3$
$z = 3$
The solution is (2, 1, 3).

47. (1) $2x - y + 2z = 7$
(2) $x + y + z = 2$
(3) $3x - y + z = 6$
Eliminate y. Add equations (1) and (2).
$2x - y + 2z = 7$
$x + y + z = 2$

(4) $3x + 3z = 9$

Add equations (2) and (3).
$x + y + z = 2$
$3x - y + z = 6$

(5) $4x + 2z = 8$

Use equations (4) and (5) solve for x and z.
$3x + 3z = 9$
$4x + 2z = 8$
Eliminate z.
$-2(3x + 3z) = -2(9)$
$3(4x + 2z) = 3(8)$

$-6x - 6z = -18$
$12x + 6z = 24$
$6x = 6$
$x = 1$
Replace x with 1 in equation (4).
$3x + 3z = 9$
$3(1) + 3z = 9$
$3 + 3z = 9$
$3z = 6$
$z = 2$
Replace x with 1 and z with 2 in equation (1).
$2x - y + 2z = 7$
$2(1) - y + 2(2) = 7$
$2 - y + 4 = 7$
$6 - y = 7$
$-y = 1$
$y = -1$
The solution is (1, −1, 2).

49. (1) $3x + y = 5$
(2) $3y - z = 2$
(3) $x + z = 5$
Eliminate z. Add equations (2) and (3).
$3y - z = 2$
$x + z = 5$

(4) $x + 3y = 7$

Use equations (1) and (4). Solve for x and y.
$3x + y = 5$
$x + 3y = 7$

Eliminate y.
$-3(3x + y) = -3(5)$
$x + 3y = 7$

$-9x - 3y = -15$
$x + 3y = 7$
$-8x = -8$
$x = 1$
Replace x with 1 in equation (1).
$3x + y = 5$
$3(1) + y = 5$
$3 + y = 5$
$y = 2$
Replace y with 2 in equation (2).
$3y - z = 2$
$3(2) - z = 2$
$6 - z = 2$
$-z = -4$
$z = 4$
The solution is (1, 2, 4).

51. (1) $x - y + z = 1$
(2) $2x + 3y - z = 3$
(3) $-x + 2y - 4z = 4$
Eliminate z. Add equations (1) and (2).
$x - y + z = 1$
$2x + 3y - z = 3$

(4) $3x + 2y = 4$

Multiply equation (1) by 4 and add to equation (3).
$4(x - y + z) = 4(1)$
$-x + 2y - 4z = 4$

$4x - 4y + 4z = 4$
$-x + 2y - 4z = 4$

(5) $3x - 2y = 8$
Use equations (4) and (5) solve for x and y.
Add equation (4) and (5) to eliminate x.
$3x + 2y = 4$
$3x - 2y = 8$
$6x = 12$
$x = 2$
Replace x with 2 in equation (4).
$3x + 2y = 4$
$3(2) + 2y = 4$
$6 + 2y = 4$
$2y = -2$
$y = -1$
Replace x with 2 and y with -1 in equation (1).
$x - y + z = 1$
$2 - (-1) + z = 1$
$2 + 1 + z = 1$
$3 + z = 1$
$z = -2$
The solution is $(2, -1, -2)$.

53. (1) $2x + 3z = 5$
(2) $3y + 2z = 3$
(3) $3x + 4y = -10$
Eliminate z. Use equations (1) and (2).
$2x + 3z = 5$
$3y + 2z = 3$

$-2(2x + 3z) = -2(5)$
$3(3y + 2z) = 3(3)$

$-4x - 6z = -10$
$9y + 6z = 9$

(4) $-4x + 9y = -1$

Use equations (3) and (4), solve for x and y.
$3x + 4y = -10$
$-4x + 9y = -1$
Eliminate x.
$4(3x + 4y) = 4(-10)$
$3(-4x + 9y) = 3(-1)$

$12x + 16y = -40$
$-12x + 27y = -3$
$43y = -43$
$y = -1$

Replace y with -1 in equation (2).
$3y + 2z = 3$
$3(-1) + 2z = 3$
$-3 + 2z = 3$
$2z = 6$
$z = 3$
Replace z with 3 in equation (1).
$2x + 3z = 5$
$2x + 3(3) = 5$
$2x + 9 = 5$
$2x = -4$
$x = -2$
The solution is $(-2, -1, 3)$.

55. (1) $2x + 4y - 2z = 3$
(2) $x + 3y + 4z = 1$
(3) $x + 2y - z = 4$
Eliminate x. Use equations (1) and (2).
$2x + 4y - 2z = 3$
$x + 3y + 4z = 1$

$2x + 4y - 2z = 3$
$-2(x + 3y + 4z) = -2(1)$

$2x + 4y - 2z = 3$
$-2x - 6y - 8z = -2$

(4) $-2y - 10z = 1$

Use equations (2) and (3).
$x + 3y + 4z = 1$
$x + 2y - z = 4$

$x + 3y + 4z = 1$
$-1(x + 2y - z) = -1(4)$

$x + 3y + 4z = 1$
$-x - 2y + z = -4$

(5) $y + 5z = -3$

Use equations (4) and (5) solve for y and z.
$-2y - 10z = 1$
$y + 5z = -3$
Eliminate y.
$-2y - 10z = 1$
$2(y + 5z) = 2(-3)$

$-2y - 10z = 1$
$2y + 10z = -6$
$0 = -6$

This is not a true equation. The system of equations is inconsistent and therefore has no solution.

57. (1) $2x + y - z = 5$
(2) $x + 3y + z = 14$
(3) $3x - y + 2z = 1$
Eliminate z. Add equations (1) and (2).
$2x + y - z = 5$
$x + 3y + z = 14$

(4) $3x + 4y = 19$

Use equations (2) and (3).
$x + 3y + z = 14$
$3x - y + 2z = 1$

$-2(x + 3y + z) = -2(14)$
$3x - y + 2z = 1$

$-2x - 6y - 2z = -28$
$3x - y + 2z = 1$

(5) $x - 7y = -27$

Use equations (4) and (5) solve for x and y.
$3x + 4y = 19$
$x - 7y = -27$
Eliminate x.
$3x + 4y = 19$
$-3(x - 7y) = -3(-27)$

$3x + 4y = 19$
$-3x + 21y = 81$
$25y = 100$
$y = 4$
Replace y with 4 in equation (4).
$3x + 4y = 19$
$3x + 4(4) = 19$
$3x + 16 = 19$
$3x = 3$
$x = 1$
Replace x with 1 and y with 4 in equation (1).
$2x + y - z = 5$
$2(1) + 4 - z = 5$
$2 + 4 - z = 5$
$6 - z = 5$
$-z = -1$
$z = 1$

The solution is (1, 4, 1).

59. (1) $3x + y - 2z = 2$
(2) $x + 2y + 3z = 13$
(3) $2x - 2y + 5z = 6$
Eliminate x. Add equations (1) and (2).
$3x + y - 2z = 2$
$x + 2y + 3z = 13$

$3x + y - 2z = 2$
$-3(x + 2y + 3z) = -3(13)$

$3x + y - 2z = 2$
$-3x - 6y - 9z = -39$

(4) $-5y - 11z = -37$

Use equations (2) and (3).
$x + 2y + 3z = 13$
$2x - 2y + 5z = 6$

$-2(x + 2y + 3z) = -2(13)$
$2x - 2y + 5z = 6$

$-2x - 4y - 6z = -26$
$2x - 2y + 5z = 6$

(5) $-6y - z = -20$

Use equations (4) and (5) solve for y and z.
$-5y - 11z = -37$
$-6y - z = -20$
Eliminate z.
$-5y - 11z = -37$
$-11(-6y - z) = -11(-20)$

$-5y - 11z = -37$
$66y + 11z = 220$
$61y = 183$
$y = 3$
Replace y with 3 in equation (4).
$-5y - 11z = -37$
$-5(3) - 11z = -37$
$-15 - 11z = -37$
$-11z = -22$
$z = 2$
Replace y with 3 and z with 2 in equation (1).
$3x + y - 2z = 2$
$3x + 3 - 2(2) = 2$
$3x + 3 - 4 = 2$

$3x - 1 = 2$
$3x = 3$
$x = 1$
The solution is (1, 3, 2.)

61. (1) $2x - y + z = 6$
(2) $3x + 2y + z = 4$
(3) $x - 2y + 3z = 12$
Eliminate y. Use equations (1) and (2).
$2x - y + z = 6$
$3x + 2y + z = 4$

$2(2x - y + z) = 2(6)$
$3x + 2y + z = 4$

$4x - 2y + 2z = 12$
$3x + 2y + z = 4$

(4) $7x + 3z = 16$

Add equations (2) and (3).
$3x + 2y + z = 4$
$x - 2y + 3z = 12$

(5) $4x + 4z = 16$

Use equations (4) and (5) solve for x and z.
$7x + 3z = 16$
$4x + 4z = 16$
Eliminate z.
$4(7x + 3z) = 4(16)$
$-3(4x + 4z) = -3(16)$

$28x + 12z = 64$
$-12x - 12z = -48$
$16x = 16$
$x = 1$
Replace x with 1 in equation (4).
$7x + 3z = 16$
$7(1) + 3z = 16$
$7 + 3z = 16$
$3z = 9$
$z = 3$
Replace x with 1 and z with 3 in equation (1).
$2x - y + z = 6$
$2(1) - y + 3 = 6$
$-y + 5 = 6$
$-y = 1$
$y = -1$
The solution is (1, −1, 3).

63. (1) $3x - 2y + 3z = -4$
(2) $2x + y - 3z = 2$
(3) $3x + 4y + 5z = 8$
Eliminate y. Use equations (1) and (2).
$3x - 2y + 3z = -4$
$2x + y - 3z = 2$

$3x - 2y + 3z = -4$
$2(2x + y - 3z) = 2(2)$

$3x - 2y + 3z = -4$
$4x + 2y - 6z = 4$

(4) $7x - 3z = 0$

Use equations (2) and (3).
$2x + y - 3z = 2$
$3x + 4y + 5z = 8$

$-4(2x + y - 3z) = -4(2)$
$3x + 4y + 5z = 8$

$-8x - 4y + 12z = -8$
$3x + 4y + 5z = 8$

(5) $-5x + 17z = 0$

Use equations (4) and (5) solve for x and z.
$7x - 3z = 0$
$-5x + 17z = 0$
Eliminate x.
$5(7x - 3z) = 5(0)$
$7(-5x + 17z) = 7(0)$

$35x - 15z = 0$
$-35x + 119z = 0$
$104z = 0$
$z = 0$

Replace z with 0 in equation (4).
$7x - 3z = 0$
$7x - 3(0) = 0$
$7x = 0$
$x = 0$
Replace x with 0 and z with 0 in equation (1).
$3x - 2y + 3z = -4$
$3(0) - 2y + 3(0) = -4$
$0 - 2y + 0 = -4$

$-2y = -4$
$y = 2$
The solution is (0, 2, 0).

65. (1) $3x - y + 2z = 2$
(2) $4x + 2y - 7z = 0$
(3) $2x + 3y - 5z = 7$
Eliminate y. Use equations (1) and (2).
$3x - y + 2z = 2$
$4x + 2y - 7z = 0$

$2(3x - y + 2z) = 2(2)$
$4x + 2y - 7z = 0$

$6x - 2y + 4z = 4$
$4x + 2y - 7z = 0$

(4) $10x - 3z = 4$

Use equations (1) and (3).
$3x - y + 2z = 2$
$2x + 3y - 5z = 7$

$3(3x - y + 2z) = 3(2)$
$2x + 3y - 5z = 7$

$9x - 3y + 6z = 6$
$2x + 3y - 5z = 7$

(5) $11x + z = 13$

Use equations (4) and (5) solve for x and z.
$10x - 3z = 4$
$11x + z = 13$
Eliminate z.
$10x - 3z = 4$
$3(11x + z) = 3(13)$

$10x - 3z = 4$
$33x + 3z = 39$
$43x = 43$
$x = 1$

Replace x with 1 in equation (4).
$10x - 3z = 4$
$10(1) - 3z = 4$
$-3z = -6$
$z = 2$
Replace x with 1 and z with 2 in equation (1).
$3x - y + 2z = 2$
$3(1) - y + 2(2) = 2$
$3 - y + 4 = 2$
$7 - y = 2$
$-y = -5$
$y = 5$
The solution is (1, 5, 2).

67. (1) $2x - 3y + 7z = 0$
(2) $x + 4y - 4z = -2$
(3) $3x + 2y + 5z = 1$
Eliminate x. Use equations (1) and (2).
$2x - 3y + 7z = 0$
$x + 4y - 4z = -2$

$2x - 3y + 7z = 0$
$-2(x + 4y - 4z) = -2(-2)$

$2x - 3y + 7z = 0$
$-2x - 8y + 8z = 4$

(4) $-11y + 15z = 4$

Use equations (2) and (3).
$x + 4y - 4z = -2$
$3x + 2y + 5z = 1$

$-3(x + 4y - 4z) = -3(-2)$
$3x + 2y + 5z = 1$

$-3x - 12y + 12z = 6$
$3x + 2y + 5z = 1$

(5) $-10y + 17z = 7$

Use equations (4) and (5) solve for y and z.
$-11y + 15z = 4$
$-10y + 17z = 7$
Eliminate y.
$10(-11y + 15z) = 10(4)$
$-11(-10y + 17z) = -11(7)$

$-110y + 150z = 40$
$110y - 187z = -77$
$-37z = -37$
$z = 1$

Replace z with 1 in equation (4).
$-11y + 15z = 4$
$-11y + 15(1) = 4$
$-11y = -11$
$y = 1$

Replace y with 1 and z with 1 in equation (1).

$2x - 3y + 7z = 0$
$2x - 3(1) + 7(1) = 0$
$2x - 3 + 7 = 0$
$2x + 4 = 0$
$2x = -4$
$x = -2$

The solution is (−2, 1, 1).

69. a) (iii) no points
b) (ii) more than one point
c) (i) exactly one point

Applying the Concepts

71. (1) $\frac{1}{x} - \frac{2}{y} = 3$

(2) $\frac{2}{x} + \frac{3}{y} = -1$

Clear the fractions.

$xy\left(\frac{1}{x} - \frac{2}{y}\right) = xy(3)$

$xy\left(\frac{2}{x} + \frac{3}{y}\right) = xy(-1)$

$y - 2x = 3xy$
$2y + 3x = -xy$

Eliminate y.

$-2(y - 2x) = -2(3xy)$
$2y + 3x = -xy$

$-2y + 4x = -6xy$
$2y + 3x = -xy$
$7x = -7xy$
$y = -1$

Replace y with −1 in equation (1).

$\frac{1}{x} - \frac{2}{y} = 3$

$\frac{1}{x} - \frac{2}{-1} = 3$

$\frac{1}{x} + 2 = 3$

$\frac{1}{x} = 1$

$x = 1$

The solution is (1,−1).

73. (1) $\frac{3}{x} + \frac{2}{y} = 1$

(2) $\frac{2}{x} + \frac{4}{y} = -2$

Clear the fractions.

$xy\left(\frac{3}{x} + \frac{2}{y}\right) = xy(1)$

$xy\left(\frac{2}{x} + \frac{4}{y}\right) = xy(-2)$

$3y + 2x = xy$
$2y + 4x = -2xy$

Eliminate x.

$-2(3y + 2x) = -2(xy)$
$2y + 4x = -2xy$

$-6y - 4x = -2xy$
$2y + 4x = -2xy$
$-4y = -4xy$
$x = 1$

Replace x with 1 in equation (1).

$\frac{3}{x} + \frac{2}{y} = 1$

$\frac{3}{1} + \frac{2}{y} = 1$

$\frac{2}{y} = -2$

$y = -1$

The solution is (1,−1).

Section 5.3

Objective A Exercises

1. n is less than m.

3. **Strategy:** Let x represent the rate of the motorboat in calm water.
The rate of the current is y.

	Rate	Time	Distance
with current	$x+y$	2	$2(x+y)$
against current	$x-y$	3	$3(x-y)$

The distance traveled with the current is 36 miles. The distance traveled against the current is 36 miles.
$2(x+y)=36$
$3(x-y)=36$

Solution: $2(x+y)=36$
$3(x-y)=36$

$\frac{1}{2}\cdot 2(x+y)=\frac{1}{2}\cdot 36$

$\frac{1}{3}\cdot 3(x-y)=\frac{1}{3}\cdot 36$

$x+y=18$
$x-y=12$
$2x=30$
$x=15$

$x+y=18$
$15+y=18$
$y=3$

The rate of the motorboat in calm water is 15 mph. The rate of the current is 3 mph.

5. **Strategy:** Let p represent the rate of the plane in calm air.
The rate of the wind is w.

	Rate	Time	Distance
with wind	$p+w$	4	$4(p+w)$
against wind	$p-w$	4	$4(p-w)$

The distance traveled with the wind is 2200 miles. The distance traveled against the wind is 1820 miles.
$4(p+w)=2200$
$4(p-w)=1820$

Solution: $4(p+w)=2200$
$4(p-w)=1820$

$\frac{1}{4}\cdot 4(p+w)=\frac{1}{4}\cdot 2200$

$\frac{1}{4}\cdot 4(p-w)=\frac{1}{4}\cdot 1820$

$p+w=550$
$p-w=455$
$2p=1005$
$p=502.5$

$p+w=550$
$502.5+w=550$
$w=47.5$

The rate of the plane in calm air is 502.5 mph. The rate of the wind is 47.5 mph.

7. **Strategy:** Let x represent the rate of the team in calm water.
The rate of the current is y.

	Rate	Time	Distance
with current	$x+y$	2	$2(x+y)$
against current	$x-y$	2	$2(x-y)$

The distance traveled with the current is 20 km. The distance traveled against the current is 12 km.
$2(x+y)=20$
$2(x-y)=12$

Solution: $2(x+y)=20$
$2(x-y)=12$

$\frac{1}{2}\cdot 2(x+y)=\frac{1}{2}\cdot 20$

$\frac{1}{2}\cdot 2(x-y)=\frac{1}{2}\cdot 12$

$x+y=10$
$x-y=6$
$2x=16$
$x=8$

$x+y=10$
$8+y=10$
$y=2$
The rate of the team in calm water is 8 km/h.
The rate of the current is 2 km/h.

9. Strategy: Let x represent the rate of the plane in calm air.
The rate of the wind is y.

	Rate	Time	Distance
with wind	$x+y$	4	$4(x+y)$
against wind	$x-y$	5	$5(x-y)$

The distance traveled with the wind is 800 miles. The distance traveled against the wind is 800 miles.
$4(x+y)=800$
$5(x-y)=800$

Solution: $4(x+y)=800$
$5(x-y)=800$

$\frac{1}{4}\cdot 4(x+y)=\frac{1}{4}\cdot 800$
$\frac{1}{5}\cdot 5(x-y)=\frac{1}{5}\cdot 800$
$x+y=200$
$x-y=160$
$2x=360$
$x=180$

$x+y=200$
$180+y=200$
$y=20$

The rate of the plane in calm air is 180 mph.
The rate of the wind is 20 mph.

11. Strategy: Let x represent the rate of the plane in calm air.
The rate of the wind is y.

	Rate	Time	Distance
with wind	$x+y$	5	$5(x+y)$
against wind	$x-y$	6	$6(x-y)$

The distance traveled with the wind is 600 miles. The distance traveled against the wind is 600 miles.
$5(x+y)=600$
$6(x-y)=600$

Solution: $5(x+y)=600$
$6(x-y)=600$

$\frac{1}{5}\cdot 5(x+y)=\frac{1}{5}\cdot 600$
$\frac{1}{6}\cdot 6(x-y)=\frac{1}{6}\cdot 600$
$x+y=120$
$x-y=100$
$2x=220$
$x=110$

$x+y=120$
$110+y=120$
$y=10$

The rate of the plane in calm air is 110 mph.
The rate of the wind is 10 mph.

Objective B Exercises

13. The cost per pound of dark roast coffee is greater than the cost per pound of light roast coffee.

15. Strategy: Let n represent the number of nickels.
Number of dimes is d.

Coin	Number	Value	Total Value
Nickels	n	5	$5n$
Dimes	d	10	$10d$

Coins in the bank if the nickels were dimes and the dimes were nickels.
The value of the nickels and dimes in the bank is \$2.50.

Coin	Number	Value	Total Value
Nickels	d	5	$5d$
Dimes	n	10	$10n$

The value of the nickels and dimes in the bank would be \$3.50.

$5n+10d=250$
$10n+5d=350$

Solution: $5n + 10d = 250$
$10n + 5d = 350$

$-2(5n + 10d) = -2(250)$
$10n + 5d = 350$

$-10n - 20d = -500$
$10n + 5d = 350$

$-15d = -150$
$d = 10$

$5n + 10d = 250$
$5n + 10(10) = 250$
$5n = 150$
$n = 30$

There are 30 nickels in the bank.

17. Strategy: Let x represent the cost of redwood.
The cost of pine is y.
First purchase.

	Amount	Rate	Total Value
Redwood	60	x	$60x$
Pine	80	y	$80y$

Second purchase.

	Amount	Rate	Total Value
Redwood	100	x	$100x$
Pine	60	y	$60y$

The first purchase costs \$286. The second purchase costs \$396.

$60x + 80y = 286$
$100x + 60y = 396$

Solution: $60x + 80y = 286$
$100x + 60y = 396$

$3(60x + 80y) = 3(286)$
$-4(100x + 60y) = -4(396)$

$180x + 240y = 858$
$-400x - 240y = -1584$

$-220x = -726$
$x = 3.3$

$60x + 80y = 286$
$60(3.3) + 80y = 286$
$198 + 80y = 286$
$80y = 88$
$y = 1.1$
The cost of the pine is \$1.10/ft. The cost of the redwood is \$3.30/ft.

19. Strategy: Let x represent the cost of nylon carpet.
The cost of wool carpet is y.
First purchase.

	Amount	Rate	Total Cost
Nylon	16	x	$16x$
Wood	20	y	$20y$

Second purchase.

	Amount	Rate	Total Cost
Nylon	18	x	$18x$
Wool	25	y	$25y$

The first purchase costs \$1840. The second purchase costs \$2200.

$16x + 20y = 1840$
$18x + 25y = 2200$

Solution: $16x + 20y = 1840$
$18x + 25y = 2200$

$5(16x + 20y) = 5(1840)$
$-4(18x + 25y) = -4(2200)$

$80x + 100y = 9200$
$-72x - 100y = -8800$

$8x = 400$
$x = 50$

$16x + 20y = 1840$
$16(50) + 20y = 1840$
$800 + 20y = 1840$
$20y = 1040$
$y = 52$

The cost of the wool carpet is \$52/yd.

21. Strategy: Let m represent the number of mountain bikes to be manufactured.
The number of trail bikes to be manufactured is t.

Cost of materials.

Type	Number	Cost	Total Cost
Mountain	m	70	$70m$
Trail	t	50	$50t$

Cost of labor.

Type	Number	Cost	Total Cost
Mountain	m	80	$80m$
Trail	t	40	$40t$

The company has budgeted $2500 for materials and $2600 for labor.

$70m + 50t = 2500$
$80m + 40t = 2600$

Solution: $70m + 50t = 2500$
$80m + 40t = 2600$

$4(70m + 50t) = 4(2500)$
$-5(80m + 40t) = -5(2600)$

$280m + 200t = 10{,}000$
$-400m - 200t = -13{,}000$

$-120m = -3000$
$m = 25$

The company plans to manufacture 25 mountain bikes during the week.

23. Strategy: Let x represent the number of miles driven in the city.
The number of miles driven on the highway is $394 - x$.
Cost of hybrid driving.

	Number	Cost	Total Cost
City	x	0.09	$0.09x$
Highway	$394 - x$	0.08	$0.08(394 - x)$

The total amount spent on gasoline was $34.74.

Solution:
$0.09x + 0.08(394 - x) = 34.74$
$0.09x + 31.52 - 0.08x = 34.74$
$0.01x = 3.22$
$x = 322$

$394 - x = 394 - 322 = 72$

The owner drives 322 miles in the city and 72 on the highway.

25. Strategy: Let x represent the amount of the first alloy.
The amount of the second alloy is y.
Gold.

	Amount	Percent	Quantity
1st alloy	x	0.10	$0.10x$
2nd alloy	y	0.30	$0.30y$

Lead.

	Amount	Percent	Quantity
1st alloy	x	0.15	$0.15x$
2nd alloy	y	0.40	$0.40y$

The resulting alloy contains 60 g of gold and 88 g of lead.
$0.10x + 0.30y = 60$
$0.15x + 0.40y = 88$

Solution: $0.10x + 0.30y = 60$
$0.15x + 0.40y = 88$
$3(0.10x + 0.30y) = 3(60)$
$-2(0.15x + 0.40y) = -2(88)$

$0.30x + 0.90y = 180$
$-0.30x - 0.80y = -176$
$0.10y = 4$
$y = 40$

$0.10x + 0.30y = 60$
$0.10x + 0.30(40) = 60$
$0.10x + 12 = 60$
$0.10x = 48$
$x = 480$

The chemist should use 480 g of the first alloy and 40 g of the second alloy.

26. Strategy: Let x represent the cost of the Model II computer.
The cost of the Model IV computer is y.
The cost of the Model IX computer is z.

First shipment

	Number	Unit Cost	Value
Model II	4	x	$4x$
Model IV	6	y	$6y$
Model IX	10	z	$10z$

Second shipment

	Number	Unit Cost	Value
Model II	8	x	$8x$
Model IV	3	y	$3y$
Model IX	5	z	$5z$

Third shipment

	Number	Unit Cost	Value
Model II	2	x	$2x$
Model IV	9	y	$9y$
Model IX	5	z	$5z$

The value of the first shipment was $114,000. The value of the second shipment was $72,000. The value of the third shipment was $81,000.

Solution: (1) $4x + 6y + 10z = 114{,}000$
(2) $8x + 3y + 5z = 72{,}000$
(3) $2x + 9y + 5z = 81{,}000$

Multiply equation (2) by −2 and add to equation (1).
$4x + 6y + 10z = 114{,}000$
$-16x - 6y - 10z = -144{,}000$
$-12x = -30{,}000$
$x = 2500$

Multiply equation (3) by −1 and add to equation (2).
$8x + 3y + 5z = 72{,}000$
$-2x - 9y - 5z = -81{,}000$
$6x - 6y = -9000$

$6(2500) - 6y = -9000$
$-6y = -24{,}000$
$y = 4000$

The Model IV computer costs $4000.

29. Strategy: Let x represent the amount deposited at 8%.
The amount deposited at 6% is y.
The amount deposited at 4% is z.

	Principal	Rate	Interest
8%	x	0.08	$0.08x$
6%	y	0.06	$0.06y$
4%	z	0.04	$0.04z$

The amount deposited in the 8% account is twice the amount deposited in the 6% account. The total amount invested is $25,000. The total interest earned is $1520.

Solution: (1) $x = 2y$
(2) $x + y + z = 25{,}000$
(3) $0.08x + 0.06y + 0.04z = 1520$

Substitute $2y$ for x in equation (2) and equation (3).
$2y + y + z = 25000$
(4) $3y + z = 25000$
$0.08(2y) + 0.06y + 0.04z = 1520$
(5) $0.22y + 0.04z = 1520$

Solve equation (4) for z and substitute into equation (5).
$z = 25{,}000 - 3y$
$0.22y + 0.04(25{,}000 - 3y) = 1520$
$0.22y + 1000 - 0.12y = 1520$
$0.10y = 520$
$y = 5200$

$z = 25{,}000 - 3(5200)$
$z = 9400$

$x = 2(5200)$
$x = 10{,}400$

The investor placed $10,400 in the 8% account, $5200 in the 6% account and $9400 in the 4% account.

Applying the Concepts

31. Strategy: Let n represent the measure of the smaller angle.
The measure of the larger angle is m.
First relationship $m + n = 180$
Second relationship $m = 3n + 40$

Solution: Solve for n by substitution:
$(3n + 40) + n = 180$
$3n + 40 + n = 180$
$4n = 140$
$n = 35$
$m + n = 180$
$m + 35 = 180$
$m = 145$
The angles have measures of 35° and 145°.

33. Strategy: Let x represent the age of the oil painting.
The age of the watercolor is y.
First relationship $x - y = 35$
Second relationship
$x + 5 = 2(y - 5)$
$x + 5 = 2y - 10$
$x = 2y - 15$

$x - y = 35$
$x = 2y - 15$

Solution: Solve for y by substitution:
$(2y - 15) - y = 35$
$y - 15 = 35$
$y = 50$
$x - y = 35$
$x - 50 = 35$
$x = 85$
The age of the oil painting is 85 years and the age of the watercolor is 50 years.

Section 5.4

Objective A Exercises

1. Solve each inequality for y.
$x - y \geq 3$
$-y \geq -x + 3$
$y \leq x - 3$

$x + y \leq 5$
$y \leq 5 - x$

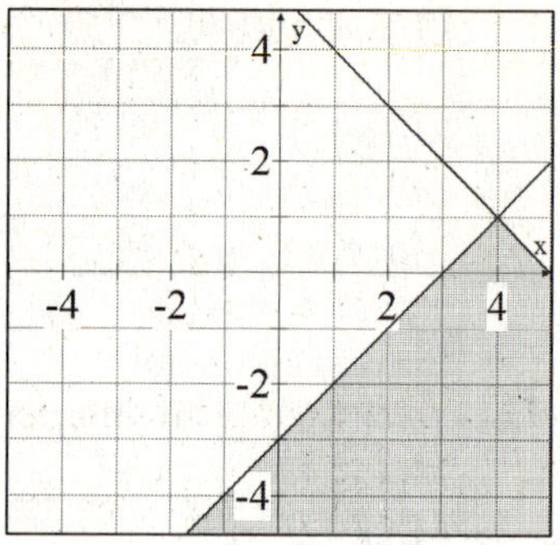

3. Solve each inequality for y.
$3x - y < 3$
$-y < -3x + 3$
$y > 3x - 3$

$2x + y \geq 2$
$y \geq 2 - 2x$

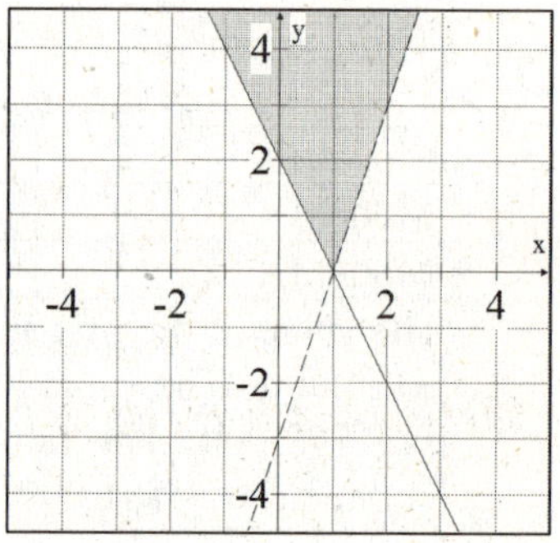

5. Solve each inequality for y.
$2x + y \geq -2$
$y \geq -2x - 2$

$6x + 3y \leq 6$
$3y \leq -6x + 6$
$y \leq -2x + 2$

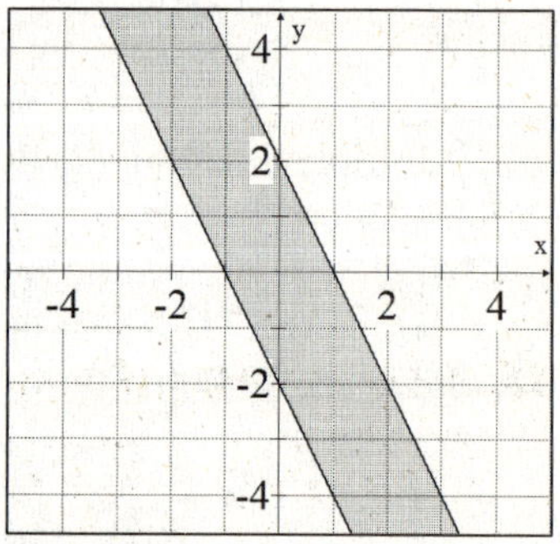

7. Solve the first inequality for y.
$3x - 2y < 6$
$-2y < -3x + 6$

$y > \frac{3}{2}x - 3$

$y \leq 3$

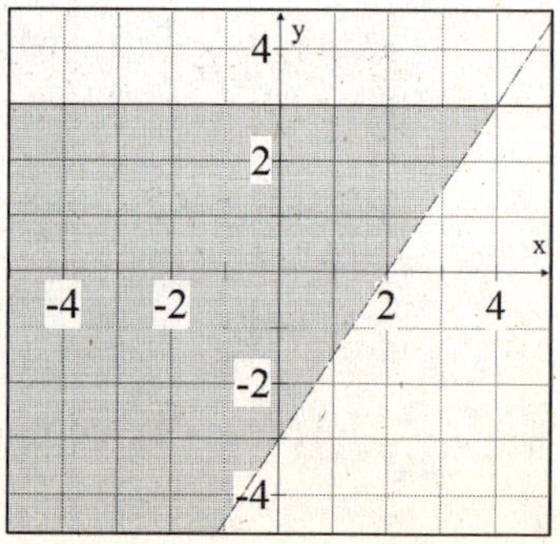

9. Solve the second inequality for y.

$y > 2x - 6$

$x + y < 0$

$y < -x$

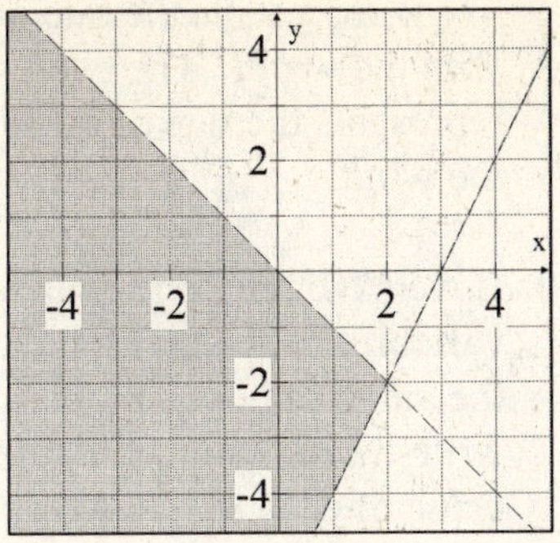

11. Solve each inequality for the variable.

$x + 1 \geq 0$

$x \geq -1$

$y - 3 \leq 0$

$y \leq 3$

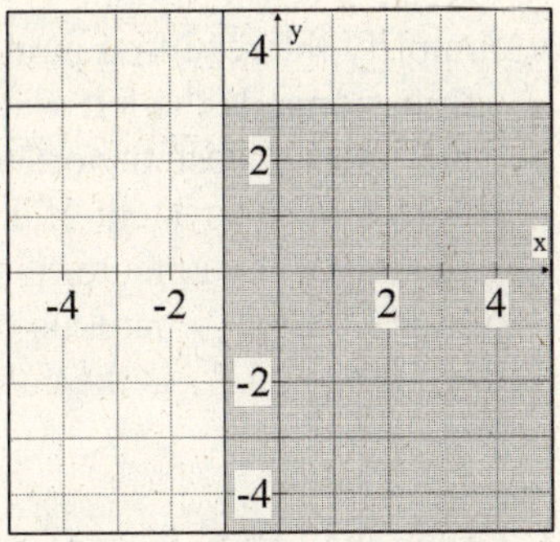

13. Solve each inequality for y.

$2x + y \geq 4$

$y \geq -2x + 4$

$3x - 2y < 6$

$-2y < -3x + 6$

$y > \frac{3}{2}x - 3$

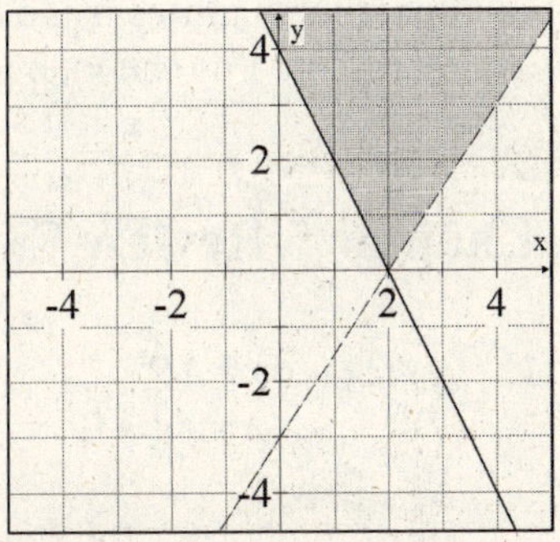

15. Solve each inequality for y.

$x - 2y \leq 6$

$-2y \leq -x + 6$

$y \geq \frac{1}{2}x - 3$

$2x + 3y \leq 6$

$3y \leq -2x + 6$

$y \geq -\frac{2}{3}x + 2$

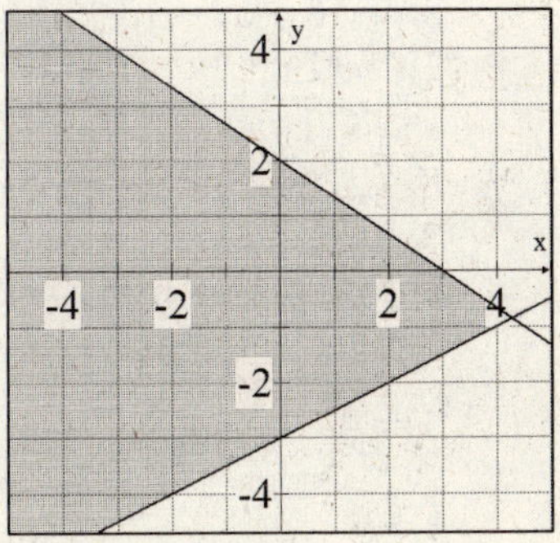

17. Solve each inequality for y.

$x - 2y \leq 4$

$-2y \leq -x + 4$

$y \geq \frac{1}{2}x - 2$

$3x + 2y \leq 8$

$2y \leq -3x + 8$

$y \leq -\frac{3}{2}x + 4$

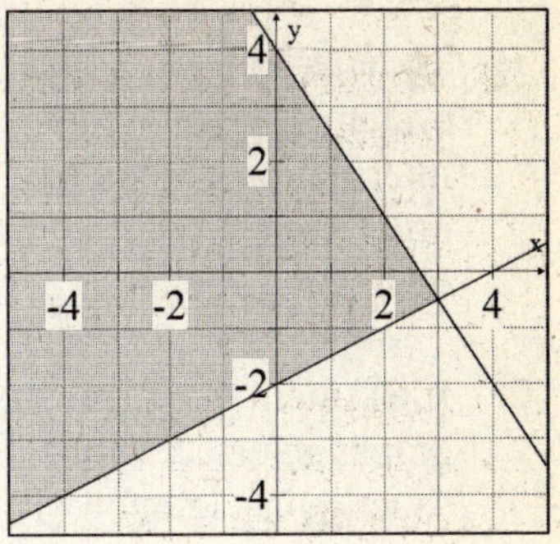

19. Points below the line $x + y = b$.

21. Region between the parallel lines $x + y = a$ and $x + y = b$.

Applying the Concepts

23. Solve each inequality for y.

$2x + 3y \leq 15$

$3y \leq -2x + 15$

$y \leq -\frac{2}{3}x + 5$

$3x - y \leq 6$

$-y \leq -3x + 6$

$y \geq 3x - 6$

$y \geq 0$

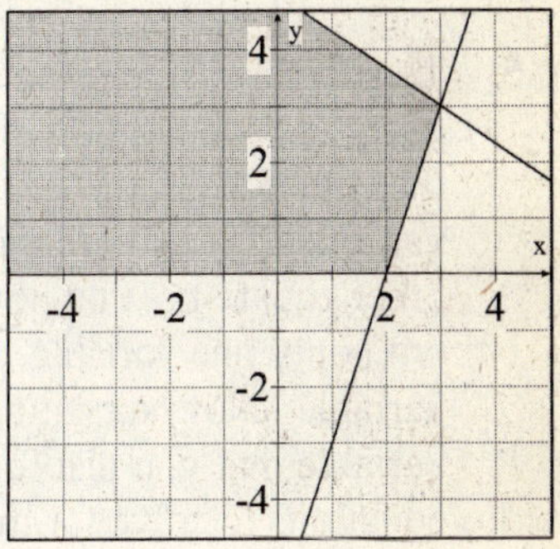

25. Solve each inequality for y.

$2x - y \leq 4$

$-y \leq -2x + 4$

$y \geq 2x - 4$

$3x + y < 1$

$y < -3x + 1$

$y \leq 0$

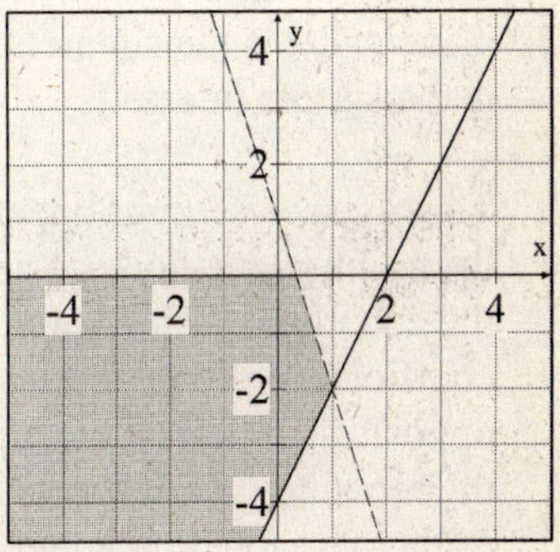

Concept Review

1. To determine the solution after graphing a system of linear equations find the point of intersection of the lines. This point is the ordered-pair solution for the system of linear equations.

2. If a system of equations is independent, the graphs intersect at one point and have exactly one solution. If a system of equations is dependent, the graphs are the same line and they intersect at infinity many points.

3. The graph of an inconsistent system of equations is the graph of two parallel lines.

4. The graph of a dependent system of equations looks like the graph of one line.

5. To solve a system of two linear equations using the substitution method solve one of the equations in the system for one of its variables. Substitute this expression into the other equation. You now have an equation with only one variable in it. Solve for this variable. Now substitute the value of the variable that you just found into one of the equations to solve for the other variable. The solution is the ordered pair that results from solving for both variables.

6. Use the formula $Pr = I$ where P is the principal, r is the simple interest rate, and I is the simple interest.

7. To solve a system of linear equations using the addition method rewrite one or both of the equations in the system so that the coefficients of one of the variables are opposites. Add the two equations. Solve for the remaining variable. Substitute the value of that variable into either of the original equations and solve for the remaining variable. The solution is the ordered pair that results from solving for both variables.

8. If the result is $0 = 0$ it means that the system is a dependent system of equations.

9. In a rate-of-wind problem the expression $p + w$ represents the rate of the plane flying with the wind. The expression $p - w$ represents the rate of the plane flying against the wind.

10. If we have only one equation in two variables we cannot solve that equation for the value of both variables. We need to write two equations in order to find the value of each of the two variables.

11. Substitute the values of the three variables into each of the equations in the system. Simplify the resulting numerical expressions. If the left and right sides are equal in each case the ordered triple is the solution of the system of equations.
The sign of the cofactor is $(-1)^{i+j}$ times the minor element, where i is the row number of the element and j is the column number of the element.

12. Graph the solution set of each linear inequality. The solution set of the system of linear inequalities is the region of the plane represented by the intersection of the shaded areas.

Chapter 5 Review Exercises

1. (1) $2x - 6y = 15$
(2) $x = 4y + 8$

Substitute $4y + 8$ for x in equation (1).
$2(4y + 8) - 6y = 15$
$8y + 16 - 6y = 15$
$2y = -1$
$y = -\frac{1}{2}$

Substitute $-\frac{1}{2}$ for y in equation (2).

$x = 4\left(-\frac{1}{2}\right) + 8$

$x = 6$

The solution is $\left(6, -\frac{1}{2}\right)$.

2. (1) $3x + 2y = 2$
(2) $x + y = 3$

Multiply equation (2) by −2 and add to equation (1).

$3x + 2y = 2$
$-2x - 2y = -6$
$x = -4$
Replace x with −4 in equation (2).
$-4 + y = 3$
$y = 7$
The solution is (−4, 7).

3. $x + y = 3$
$3x - 2y = -6$

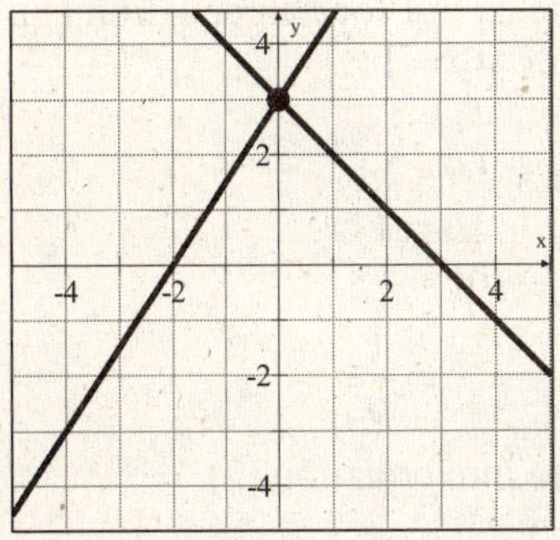

The solution is (0,3).

4. $2x - y = 4$
$y = 2x - 4$

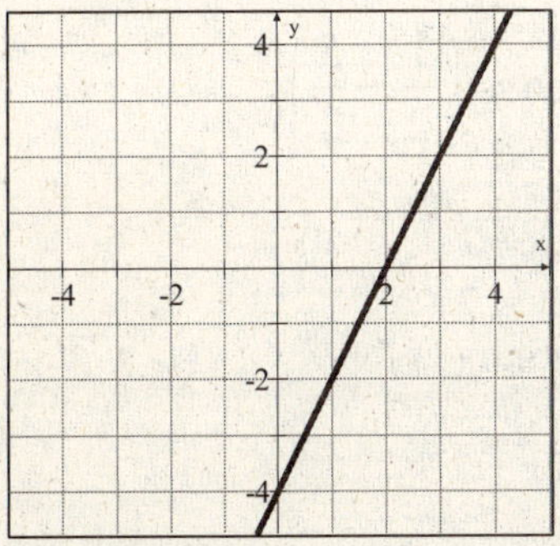

The two equations represent the same line. The system of equations is dependent. The solutions are the ordered pair $(x, 2x - 4)$.

5. (1) $3x + 12y = 18$
(2) $x + 4y = 6$

Solve equation (2) for x.
$x + 4y = 6$
$x = -4y + 6$
Substitute $-4y + 6$ for x in equation (1).
$3(-4y + 6) + 12y = 18$
$-12y + 18 + 12y = 18$
$18 = 18$
This is a true equation. The equations are dependent. The solutions are the ordered pairs $\left(x, -\frac{1}{4}x + \frac{3}{2}\right)$.

6. (1) $5x - 15y = 30$
(2) $x - 3y = 6$

Multiply equation (2) by −5 and add to equation (1).
$5x - 15y = 30$
$-5x + 15y = -30$
$0 = 0$
This is a true equation. The equations are dependent. The solutions are the ordered pairs $\left(x, \frac{1}{3}x - 2\right)$.

7. (1) $3x - 4y - 2z = 17$
(2) $4x - 3y + 5z = 5$
(3) $5x - 5y + 3z = 14$

Eliminate z. Multiply equation (1) by 3 and equation (3) by 2. Then add the equations.
$3(3x - 4y - 2z) = 3(17)$
$2(5x - 5y + 3z) = 2(14)$

$9x - 12y - 6z = 51$
$10x - 10y + 6z = 28$

(4) $19x - 22y = 79$

Multiply equation (1) by 5 and equation (2) by 2. Then add the equations
$5(3x - 4y - 2z) = 5(17)$
$2(4x - 3y + 5z) = 2(5)$

$15x - 20y - 10z = 85$
$8x - 6y + 10z = 10$
(5) $23x - 26y = 95$

Multiply equation (4) by 23 and equation (5) by −19. Then add the equations

$23(19x - 22y) = 23(79)$
$-19(23x - 26y) = -19(95)$

$437x - 506y = 1817$
$-437x + 494y = -1805$
$-12y = 12$
$y = -1$

Replace y with -1 in equation (4).
$19x - 22(-1) = 79$
$19x + 22 = 79$
$19x = 57$
$x = 3$
Replace x with 3 and y with -1 in equation (1).
$3(3) - 4(-1) - 2z = 17$
$9 + 4 - 2z = 17$
$-2z = 4$
$z = -2$
The solution is $(3, -1, -2)$.

8. (1) $3x + y = 13$
(2) $2y + 3z = 5$
(3) $x + 2z = 11$

Eliminate y. Multiply equation (1) by -2 then add to equation (2).
$-2(3x + y) = -2(13)$
$2y + 3z = 5$

$-6x - 2y = -26$
$2y + 3z = 5$

(4) $-6x + 3z = -21$

Multiply equation (3) by 6 then add to equation (4).
$6(x + 2z) = 6(11)$
$-6x + 3z = -21$

$6x + 12z = 66$
$-6x + 3z = -21$
$15z = 45$
$z = 3$

Replace z with 3 in equation (3).
$x + 2(3) = 11$
$x = 5$
Replace x with 5 in equation (1).
$3(5) + y = 13$
$y = -2$

The solution is $(5, -2, 3)$.

9.

$6x + y = 4$	
$6(1) + (-2)$	4
$6 - 2$	4
$4 = 4$	

$2x - 5y = 12$	
$2(1) - 5(-2)$	12
$2 + 10$	12
$12 = 12$	

Yes $(1, -2)$ is a solution of the system of equations.

10. (1) $2x - 4y = 11$
(2) $y = 3x - 4$
Substitute $3x - 4$ for y in equation (1).
$2x - 4y = 11$
$2x - 4(3x - 4) = 11$
$2x - 12x + 16 = 11$
$-10x + 16 = 11$
$-10x = -5$
$x = \frac{1}{2}$
Substitute into equation (2).
$y = 3x - 4$
$y = 3\left(\frac{1}{2}\right) - 4$
$y = -\frac{5}{2}$

The solution is $\left(\frac{1}{2}, -\frac{5}{2}\right)$.

11. (1) $2x - y = 7$
(2) $3x + 2y = 7$
Solve equation (1) for y.
$2x - y = 7$
$-y = -2x + 7$
$y = 2x - 7$
Substitute into equation (2).
$3x + 2y = 7$
$3x + 2(2x - 7) = 7$
$3x + 4x - 14 = 7$
$7x - 14 = 7$
$7x = 21$
$x = 3$
Substitute into equation (1).
$2x - y = 7$

$2(3) - y = 7$
$6 - y = 7$
$-y = 1$
$y = -1$
The solution is $(3,-1)$.

12. (1) $3x - 4y = 1$
(2) $2x + 5y = 16$
Eliminate y.
$5(3x - 4y) = 5(1)$
$4(2x + 5y) = 4(16)$
$15x - 20y = 5$
$8x + 20y = 64$
Add the equations.
$23x = 69$
$x = 3$
Substitute the value of x in equation (2).

$2x + 5y = 16$
$2(3) + 5y = 16$
$6 + 5y = 16$
$5y = 10$
$y = 2$
The solution is (3, 2).

13. (1) $x + y + z = 0$
(2) $x + 2y + 3z = 5$
(3) $2x + y + 2z = 3$

Eliminate x. Multiply equation (1) by −1 and add to equation (2).
$-1(x + y + z) = -1(0)$
$x + 2y + 3z = 5$

$-x - y - z = 0$
$x + 2y + 3z = 5$

(4) $y + 2z = 5$
Multiply equation (2) by −2 and add to equation (3).
$-2(x + 2y + 3z) = -2(5)$
$2x + y + 2z = 3$

$-2x - 4y - 6z = -10$
$2x + y + 2z = 3$

(5) $-3y - 4z = -7$
Multiply equation (4) by 2 and add to equation (5).
$2(y + 2z) = 2(5)$
$-3y - 4z = -7$

$2y + 4z = 10$
$-3y - 4z = -7$
$-y = 3$
$y = -3$
Replace y with −3 in equation (4).
$-3 + 2z = 5$
$2z = 8$
$z = 4$
Replace y with −3 and z with 4 in equation (1).
$x + y + z = 0$
$x - 3 + 4 = 0$
$x + 1 = 0$
$x = -1$
The solution is (−1, −3, 4).

14. (1) $x + 3y + z = 6$
(2) $2x + y - z = 12$
(3) $x + 2y - z = 13$

Eliminate z. Add equations (1) and (2).
$x + 3y + z = 6$
$2x + y - z = 12$

(4) $3x + 4y = 18$

Add equations (1) and (3).
$x + 3y + z = 6$
$x + 2y - z = 13$

(5) $2x + 5y = 19$
Multiply equation (4) by −2 and equation (5) by 3. Then add the equations.
$-2(3x + 4y) = -2(18)$
$3(2x + 5y) = 3(19)$

$-6x - 8y = -36$
$6x + 15y = 57$
$7y = 21$
$y = 3$
Replace y with 3 in equation (4).
$3x + 4(3) = 18$
$3x + 12 = 18$
$3x = 6$
$x = 2$
Replace x with 2 and y with 3 in equation (1).
$x + 3y + z = 6$
$2 + 3(3) + z = 6$

$2+9+z=6$
$11+z=6$
$z=-5$
The solution is (2, 3, −5).

15. Solve each inequality for y.
$x+3y\le 6$
$3y\le -x+6$
$y\le -\frac{1}{3}x+2$
$2x-y\ge 4$
$-y\ge -2x+4$
$y\le 2x-4$

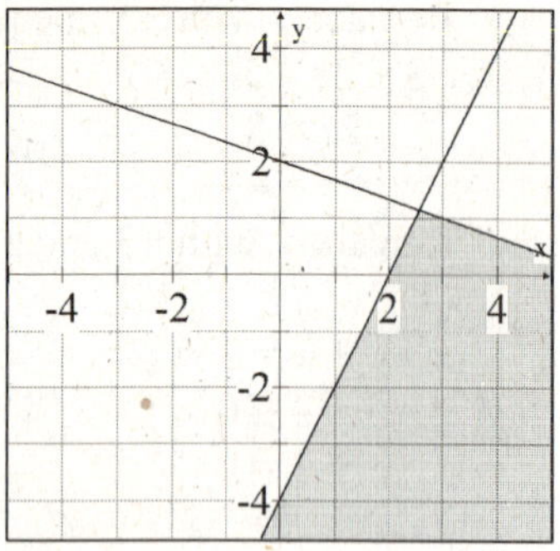

16. Solve each inequality for y.
$2x+4y\ge 8$
$4y\ge -2x+8$
$y\ge -\frac{1}{2}x+2$
$x+y\le 3$
$y\le -x+3$

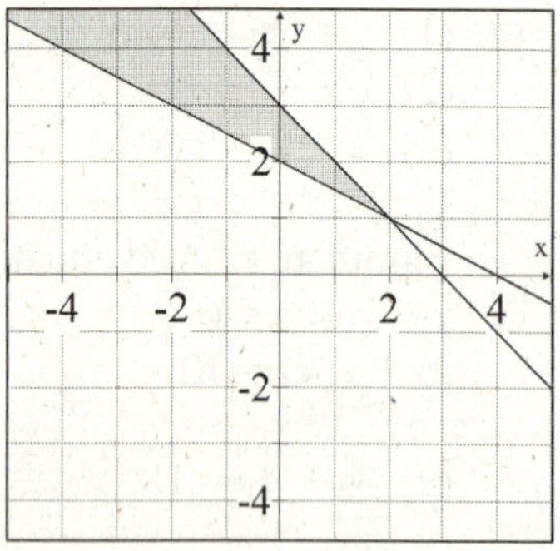

17. Strategy: Let x represent the rate of the cabin cruiser in calm water.
The rate of the current is y.

	Rate	Time	Distance
with current	$x+y$	3	$3(x+y)$
against current	$x-y$	5	$5(x-y)$

The distance traveled with the current is 60 mi. The distance traveled against the current is 60 mi.
$3(x+y)=60$
$5(x-y)=60$

Solution: $3(x+y)=60$
$5(x-y)=60$

$\frac{1}{3}\cdot 3(x+y)=\frac{1}{3}\cdot 60$
$\frac{1}{5}\cdot 5(x-y)=\frac{1}{5}\cdot 60$
$x+y=20$
$x-y=12$
$2x=32$
$x=16$

$x+y=20$
$16+y=20$
$y=4$
The rate of the boat in calm water is 16 mph.
The rate of the current is 4 mph.

18. Strategy: Let p represent the rate of the plane in calm air.
The rate of the wind is w.

	Rate	Time	Distance
with wind	$p+w$	3	$3(p+w)$
against wind	$p-w$	4	$4(p-w)$

The distance traveled with the wind is 600 mi. The distance traveled against the wind is 600 mi.
$3(p+w)=600$
$4(p-w)=600$

Solution: $3(p+w)=600$
$4(p-w)=600$

$\frac{1}{3}\cdot 3(p+w)=\frac{1}{3}\cdot 600$
$\frac{1}{4}\cdot 4(p-w)=\frac{1}{4}\cdot 600$
$p+w=200$
$p-w=150$
$2p=350$
$p=175$

$p+w=200$
$175+w=200$
$w=25$
The rate of the plane in calm air is 175 mph.
The rate of the wind is 25 mph.

19. Strategy: Let x represent the number of children's tickets sold.
The number of adult tickets sold is y.

Friday:

	Amount	Rate	Quantity
Children	x	5	$5x$
Adult	y	8	$8y$

Saturday:

	Amount	Rate	Quantity
Children	$3x$	5	$5(3x)$
Adult	$\frac{1}{2}y$	8	$8(\frac{1}{2})y$

The total receipts for Friday were \$2500.
The total receipts for Saturday were \$2500.
$5x + 8y = 2500$
$15x + 4y = 2500$

Solution: (1) $5x + 8y = 2500$
(2) $15x + 4y = 2500$
Multiply equation (2) by −2 then add to equation (1).
$5x + 8y = 2500$
$-2(15x + 4y) = -2(2500)$
$5x + 8y = 2500$
$-30x - 8y = -5000$
$-25x = -2500$
$x = 100$
On Friday, 100 children attended.

20. Strategy: Let x represent the amount invested at 3%.
The amount invested at 7% is y.

	Amount	Rate	Quantity
Amount at 3%	x	0.03	$0.03x$
Amount at 7%	y	0.07	$0.07y$

The total amount invested is \$20,000.
$x + y = 20{,}000$
The total annual interest earned is \$1200.
$0.03x + 0.07y = 1200$

Solution: (1) $x + y = 20{,}000$
(2) $0.03x + 0.07y = 1200$
Multiply equation (1) by −0.07 then add to equation (2).

$-0.07x - 0.07y = -1400$
$0.03x + 0.07y = 1200$
$-0.04x = -200$
$x = 5000$
Substitute 5000 for x in equation (1).
$x + y = 20{,}000$
$5000 + y = 20{,}000$
$y = 15{,}000$
The amount invested at 3% is \$5000.
The amount invested at 7% is \$15,000.

Chapter 5 Test

1. $2x - 3y = -6$
$2x - y = 2$

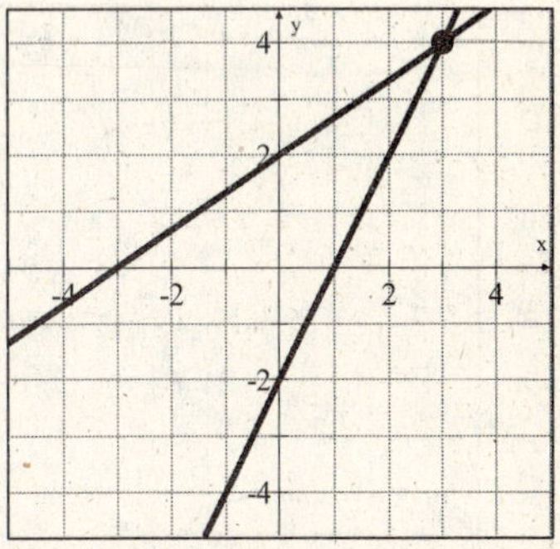

The solution is (3,4).

2. $x - 2y = -6$
$y = \frac{1}{2}x - 4$

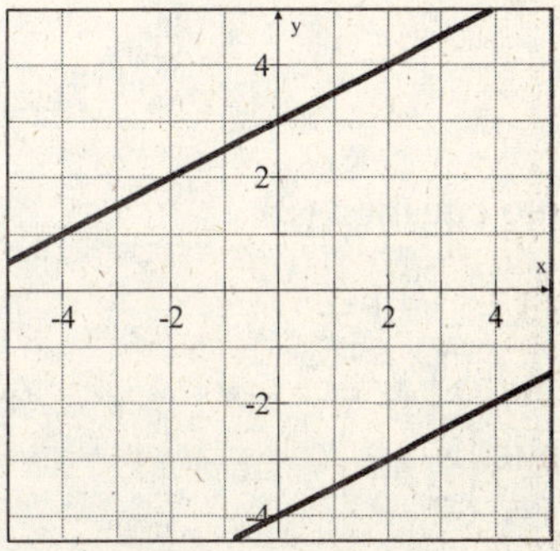

No solution.

3. Solve each inequality for y.
$2x - y < 3$
$-y < -2x + 3$
$y > 2x - 3$

$4x + 3y < 11$
$3y < -4x + 11$
$y < -\frac{4}{3}x + \frac{11}{3}$

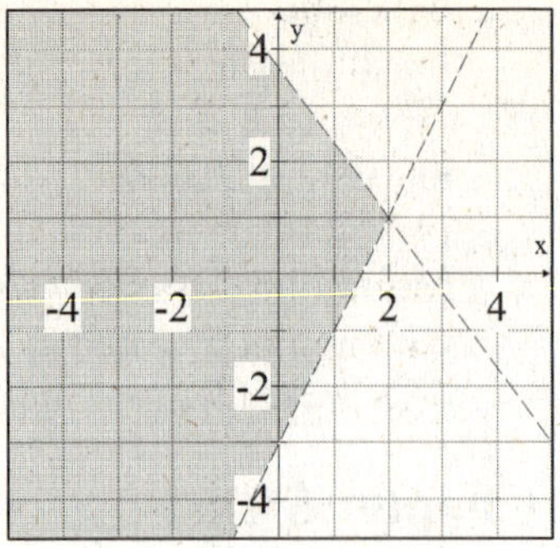

4. Solve each inequality for y.
$x + y > 2$
$y > -x + 2$

$2x - y < -1$
$-y < -2x - 1$
$y > 2x + 1$

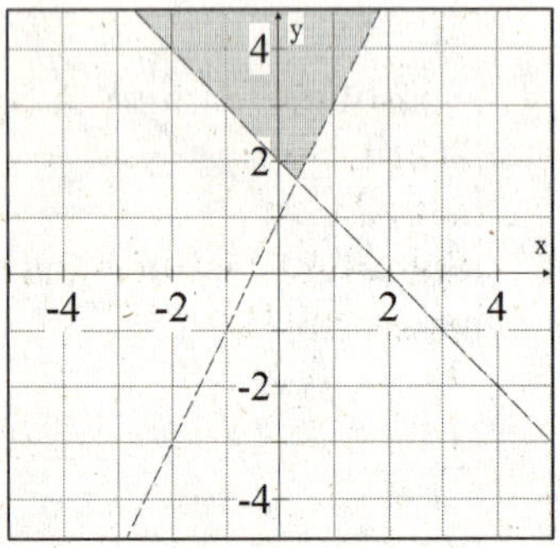

5. (1) $3x + 2y = 4$
(2) $x = 2y - 1$

Substitute $2y - 1$ for x in equation (1).
$3(2y - 1) + 2y = 4$
$6y - 3 + 2y = 4$
$8y = 7$
$y = \frac{7}{8}$
Substitute into equation (2).
$x = 2\left(\frac{7}{8}\right) - 1 = \frac{7}{4} - 1 = \frac{3}{4}$
The solution is $\left(\frac{3}{4}, \frac{7}{8}\right)$.

6. (1) $5x + 2y = -23$
(2) $2x + y = -10$

Solve equation (2) for y.
$y = -2x - 10$
Substitute $-2x - 10$ for y in equation (1).
$5x + 2(-2x - 10) = -23$
$5x - 4x - 20 = -23$
$x = -3$
Substitute in equation (2).
$2(-3) + y = -10$
$-6 + y = -10$
$y = -4$
The solution is $(-3, -4)$.

7. (1) $y = 3x - 7$
(2) $y = -2x + 3$

Substitute equation (2) into equation (1).
$-2x + 3 = 3x - 7$
$-5x + 3 = -7$
$-5x = -10$
$x = 2$
Substitute into equation (1).
$y = 3(2) - 7$
$y = -1$
The solution is $(2, -1)$.

8. (1) $3x + 4y = -2$
(2) $2x + 5y = 1$

Multiply equation (1) by 2 and equation (2) by -3 then add the new equations.
$2(3x + 4y) = 2(-2)$
$-3(2x + 5y) = -3(1)$

$6x + 8y = -4$
$-6x - 15y = -3$
$-7y = -7$
$y = 1$
Substitute into equation (1).
$3x + 4(1) = -2$
$3x = -6$
$x = -2$
The solution is $(-2, 1)$.

9. (1) $4x - 6y = 5$
(2) $6x - 9y = 4$

Multiply equation (1) by 3 and equation (2) by -2 then add the new equations.
$3(4x - 6y) = 3(5)$
$-2(6x - 9y) = -2(4)$

$12x - 18y = 15$
$-12x + 18y = -8$
$0 = 7$
This is not a true equation. The system of equations is inconsistent and therefore has no solution.

10. (1) $3x - y = 2x + y - 1$
(2) $5x + 2y = y + 6$
Write the equations in the form $Ax + By = C$
(3) $x - 2y = -1$
(4) $5x + y = 6$
$x - 2y = -1$
$10x + 2y = 12$
$11x = 11$
$x = 1$
Substitute in equation (4).
$5(1) + y = 6$
$y = 1$
The solution is (1,1).

11. (1) $2x + 4y - z = 3$
(2) $x + 2y + z = 5$
(3) $4x + 8y - 2z = 7$
Eliminate z. Add equations (1) and (2).
$2x + 4y - z = 3$
$x + 2y + z = 5$

(4) $3x + 6y = 8$
Multiply equation (2) by 2 and add to equation (3).
$2(x + 2y + z) = 2(5)$
$4x + 8y - 2z = 7$

$2x + 4y + 2z = 10$
$4x + 8y - 2z = 7$

(5) $6x + 12y = 17$
Multiply equation (4) by −2 then add to equation (5).
$-2(3x + 6y) = -2(8)$
$6x + 12y = 17$
$-6x - 12y = -16$
$6x + 12y = 17$
$0 = 1$
This is not a true equation. The system of equations is inconsistent and therefore has no solution.

12. (1) $x - y - z = 5$
(2) $2x + z = 2$
(3) $3y - 2z = 1$

Multiply equation (1) by 3 and then add to equation (3).
$3(x - y - z) = 3(5)$
$3x - 3y - 3z = 15$
$3y - 2z = 1$

Multiply equation (4) by 2 then add to equation (3).
$x - 2y = -1$
$2(5x + y) = 2(6)$

(4) $3x - 5z = 16$
Multiply equation (2) by 5 and add to equation (4).
$5(2x + z) = 5(2)$

$10x + 5z = 10$
$3x - 5z = 16$
$13x = 26$
$x = 2$

Substitute 2 in for x in equation (2).
$2(2) + z = 2$
$z = -2$
Substitute 2 in for x and −2 in for z in equation (1).
$2 - y - (-2) = 5$
$4 - y = 5$
$-y = 1$
$y = -1$

The solution is (2, −1, −2)

13. (1) $x - y = 3$
(2) $2x + y = -4$
Solve equation (2) for y.
$2x + y = -4$
$y = -2x - 4$

Substitute into equation (1).
$x - y = 3$
$x - (-2x - 4) = 3$
$x + 2x + 4 = 3$
$3x + 4 = 3$
$3x = -1$
$x = -\frac{1}{3}$

Substitute into equation (2).
$2x + y = -4$
$2\left(-\frac{1}{3}\right) + y = -4$

$-\frac{2}{3}+y=-4$

$y=-\frac{10}{3}$

The solution is $\left(-\frac{1}{3},-\frac{10}{3}\right)$.

14.
$$\begin{array}{c|c} \multicolumn{2}{c}{5x+2y=6} \\ \hline 5(2)+2(-2) & 6 \\ 10-4 & 6 \\ \multicolumn{2}{r}{6=6} \end{array}$$

$$\begin{array}{c|c} \multicolumn{2}{c}{3x+5y=-4} \\ \hline 3(2)+5(-2) & -4 \\ 6-10 & -4 \\ \multicolumn{2}{r}{-4=-4} \end{array}$$

Yes (2, −2) is a solution of the system of equations.

15. (1) $x-y+z=2$
(2) $2x-y-z=1$
(3) $x+2y-3z=-4$
Eliminate x. Multiply equation (1) by −2 and add to equation (2).
$-2(x-y+z)=-2(2)$
$2x-y-z=1$

$-2x+2y-2z=-4$
$2x-y-z=1$

(4) $y-3z=-3$
Multiply equation (1) by −1 and add to equation (3).
$-1(x-y+z)=-1(2)$
$x+2y-3z=-4$

$-x+y-z=-2$
$x+2y-3z=-4$

(5) $3y-4z=-6$
Multiply equation (4) by −3 and add to equation (5).
$-3(y-3z)=-3(-3)$
$3y-4z=-6$

$-3y+9z=9$
$3y-4z=-6$
$5z=3$

$z=\frac{3}{5}$

Substitute into equation (4).
$y-3z=-3$

$y-3z=-3$

$y-3\left(\frac{3}{5}\right)=-3$

$y-\frac{9}{5}=-3$

$y=-\frac{6}{5}$

Substitute into equation (1).
$x-y+z=2$

$x-\left(-\frac{6}{5}\right)+\left(\frac{3}{5}\right)=2$

$x+\frac{9}{5}=2$

$x=\frac{1}{5}$

The solution is $\left(\frac{1}{5},-\frac{6}{5},\frac{3}{5}\right)$.

16. Strategy: Let x represent the rate of the plane in calm air.
The rate of the wind is y.

	Rate	Time	Distance
with wind	$x+y$	2	$2(x+y)$
against wind	$x-y$	2.8	$2.8(x-y)$

The distance traveled with the wind is 350 mi. The distance traveled against the wind is 350 mi.
$2(x+y)=350$
$2.8(x-y)=350$

Solution: $2(x+y)=350$
$2.8(x-y)=350$

$\frac{1}{2}\cdot 2(x+y)=\frac{1}{2}\cdot 350$

$\frac{1}{2.8}\cdot 2.8(x-y)=\frac{1}{2.8}\cdot 350$

$x+y=175$
$x-y=125$
$2x=300$
$x=150$

$x+y=175$
$150+y=175$
$y=25$

The rate of the plane in calm air is 150 mph.
The rate of the wind is 25 mph.

17. Strategy: Let x represent the cost per yard of cotton.
The cost per yard of wool is y.
First purchase:

	Amount	Rate	Total Value
Cotton	60	x	$60x$
Wool	90	y	$90y$

Second purchase:

	Amount	Rate	Total Value
Cotton	80	x	$80x$
Wool	20	y	$20y$

The total cost of the first purchase was \$1800. The total cost of the second purchase was \$1000.

$60x+90y=1800$
$80x+20y=1000$

Solution: $-4(60x+90y)=-4(1800)$
$3(80x+20y)=3(1000)$

$-240x-360y=-7200$
$240x+60y=3000$
$-300y=-4200$
$y=14$

$60x+90(14)=1800$
$60x+1260=1800$
$60x=540$
$x=9$

The cost of cotton is \$9.00/yd.
The cost of wool is \$14.00/yd.

18. Strategy: Let x represent the amount invested at 2.7%.
The amount invested at 5.1% is y.

	Amount	Rate	Quantity
Amount at 2.7%	x	0.027	$0.027x$
Amount at 5.1%	y	0.051	$0.051y$

The total amount invested is \$15,000.
$x+y=15{,}000$
The total annual interest earned is \$549.
$0.027x+0.051y=549$

Solution: (1) $x+y=15{,}000$
(2) $0.027x+0.051y=549$
Multiply equation (1) by -0.051 then add to equation (2).

$-0.051x-0.051y=-765$
$0.027x+0.051y=549$
$-0.024x=-216$
$x=9000$
Substitute 9000 for x in equation (1).
$x+y=15{,}000$
$9000+y=15{,}000$
$y=6000$
The amount invested at 2.7% is \$9000.
The amount invested at 5.1% is \$6000.

Cumulative Review Exercises

1. $-2\sqrt{90}=-2\sqrt{2\cdot 3^2\cdot 5}=-2\sqrt{3^2}\sqrt{2\cdot 5}$
$=-2\cdot 3\sqrt{10}=-6\sqrt{10}$

2. $3(x-5)=2x+7$
$3x-15=2x+7$
$3x-2x-15=7$
$x-15+15=7+15$
$x=22$

3. $3[x-2(5-2x)-4x]+6$
$=3(x-10+4x-4x)+6$
$=3(x-10)+6$
$=3x-30+6$
$=3x-24$

4. $a+bc\div 2$
$4+8(-2)\div 2=4-16\div 2=4-8=-4$

5. $2x-3<9$ or $5x-1<4$
Solve each inequality.
$2x-3<9 \qquad 5x-1<4$
$2x<12 \qquad 5x<5$
$x<6$ or $x<1$

$x<6 \cup x<1$ which is $\{x \mid x<6\}$.

6. $|x-2|-4<2$
$|x-2|<6$
$-6<x-2<6$
$-6+2<x-2+2<6+2$
$-4<x<8$
$\{x \mid -4<x<8\}$

7. $|2x-3|>5$
Solve each inequality.
$2x-3<-5$ or $2x-3>5$
$2x<-2$ or $2x>8$
$x<-1$ or $x>4$

$\{x| x<-1\} \cup \{x| x>4\}$

$\{x| x<-1 \text{ or } x>4\}$

8. $F(x)=x^2-3$
$F(2)=(2)^2-3=1$

9. $\{x| x\le 2\} \cap \{x| x>-3\}$

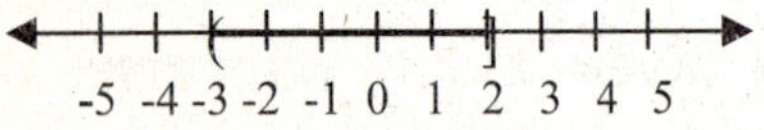

10. $(-2, 3)$, $m=-\frac{2}{3}$

$y-y_1=m(x-x_1)$

$y-3=-\frac{2}{3}(x-(-2))$

$y-3=-\frac{2}{3}x-\frac{4}{3}$

$y=-\frac{2}{3}x-\frac{4}{3}+3$

$y=-\frac{2}{3}x+\frac{5}{3}$

11. $m=\frac{y_2-y_1}{x_2-x_1}=\frac{4-(-1)}{3-2}=\frac{5}{1}=5$

$y-y_1=m(x-x_1)$

$y-(-1)=5(x-2)$

$y+1=5x-10$

$y=5x+11$

The equation of the line is $y=5x-11$.

12. The slope of the line $2x-3y=6$ is found by rearranging the equation as follows:
$-3y=-2x+6$

$y=\frac{2}{3}x-2$

The slope is $\frac{2}{3}$.

The perpendicular line has slope $-\frac{3}{2}$.

The line is found using

$y-2=-\frac{3}{2}(x-(-2))$

$y=-\frac{3}{2}(x+2)+2$

$y=-\frac{3}{2}x-3+2$

$y=-\frac{3}{2}x-1$

13. $2x-5y=10$
$-5y=-2x+10$

$y=\frac{2}{5}x-2$

slope is $\frac{2}{5}$; y-intercept is -2

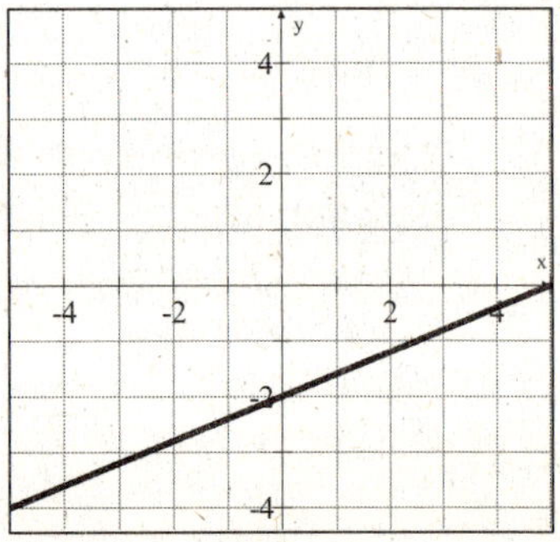

14. $3x - 4y \geq 8$
$-4y \geq -3x + 8$
$y \leq \frac{3}{4}x - 2$

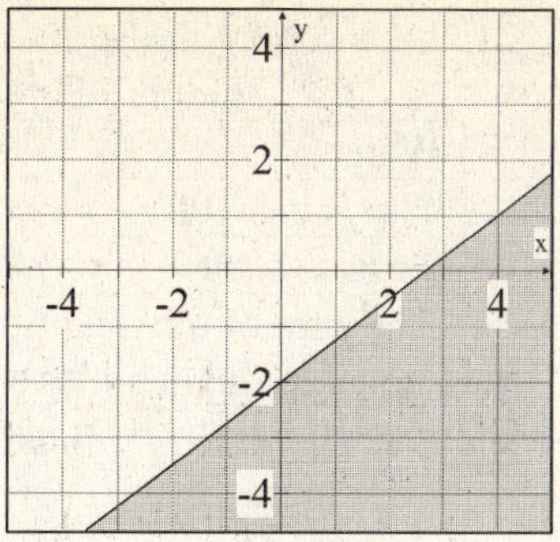

15. $5x - 2y = 10$
$3x + 2y = 6$

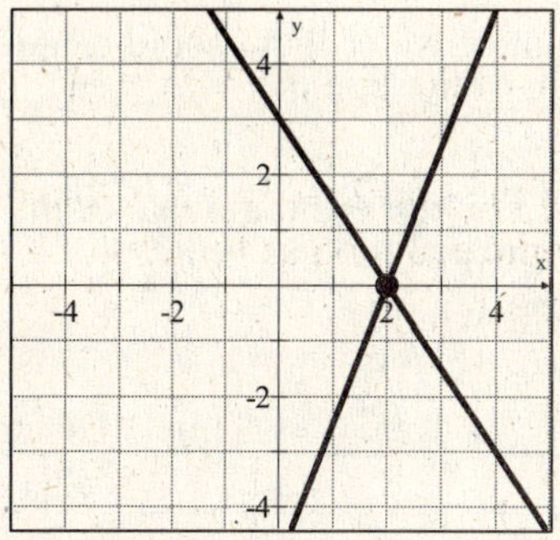

The solution is (2,0).

16. Solve each inequality for y.
$3x - 2y \geq 4$
$-2y \geq -3x + 4$
$y \leq \frac{3}{2}x - 2$

$x + y < 3$
$y < -x + 3$

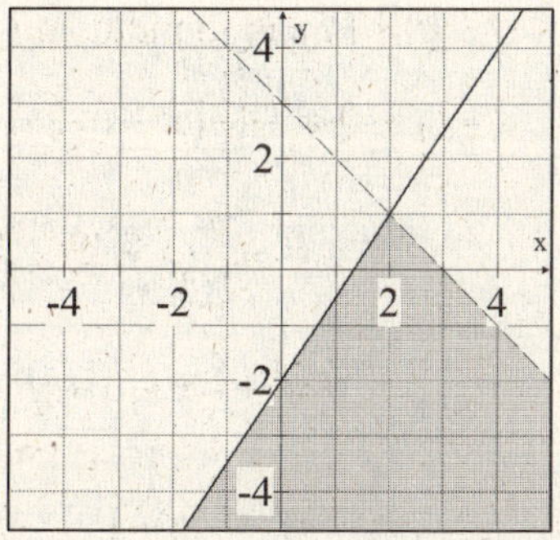

17. (1) $3x + 2z = 1$
(2) $2y - z = 1$
(3) $x + 2y = 1$

Multiply equation (2) by −1 and add to equation (3).
$-1(2y - z) = -1(1)$
$x + 2y = 1$
$-2y + z = -1$
$x + 2y = 1$

(4) $x + z = 0$

Multiply equation (4) by −2 and add to equation (1).
$-2(x + z) = -2(0)$
$3x + 2z = 1$
$-2x - 2z = 0$
$3x + 2z = 1$
$x = 1$
Substitute 1 for x in equation (3).
$1 + 2y = 1$
$2y = 0$
$y = 0$
Substitute 0 for y in equation (2).
$2(0) - z = 1$
$-z = 1$
$z = -1$
The solution is (1, 0, −1).

18. (1) $2x - y + z = 2$
(2) $3x + y + 2z = 5$
(3) $3x - y + 4z = 1$
Eliminate y. Add equation (1) to equation (2).
$2x - y + z = 2$
$3x + y + 2z = 5$

(4) $5x + 3z = 7$
Add equation (2) to equation (3).
$3x + y + 2z = 5$
$3x - y + 4z = 1$

(5) $6x + 6z = 6$
Multiply equation (4) by −2 and add to equation (5).
$-2(5x + 3z) = -2(7)$
$6x + 6z = 6$

$-10x - 6z = -14$
$6x + 6z = 6$
$-4x = -8$
$x = 2$
Substitute into equation (4).
$5x + 3z = 7$
$5(2) + 3z = 7$
$10 + 3z = 7$
$3z = -3$
$z = -1$

Substitute into equation (1).

$$2x - y + z = 2$$
$$2(2) - y + (-1) = 2$$
$$4 - y - 1 = 2$$
$$3 - y = 2$$
$$-y = -1$$
$$y = 1$$

The solution is (2, 1, −1).

19. (1) $4x - 3y = 17$
(2) $3x - 2y = 12$
Multiply equation (1) by −2 and equation (2) by 3. Then add the equations together.
$-2(4x - 3y) = -2(17)$
$3(3x - 2y) = 3(12)$

$-8x + 6y = -34$
$9x - 6y = 36$
$x = 2$

Substitute 2 for x in equation (1) and solve for y.
$4x - 3y = 17$
$4(2) - 3y = 17$
$8 - 3y = 17$
$-3y = 9$
$y = -3$
The solution is (2, −3).

20. (1) $3x - 2y = 7$
(2) $y = 2x - 1$

Solve by substitution.
$3x - 2(2x - 1) = 7$
$3x - 4x + 2 = 7$
$-x + 2 = 7$
$-x = 5$
$x = -5$

Substitute −5 for x in equation (2).
$y = 2(-5) - 1$
$y = -10 - 1$
$y = -11$
The solution is (−5,−11).

21. Let x represent the amount of pure water.

	Amount	Percent	Quantity
Water	x	0	$0x$
4%	100	0.04	0.04(100)
2.5%	$100 + x$	0.025	$0.025(100 + x)$

The sum of the quantities before mixing is equal to the quantity after mixing.
$0x + 0.04(100) = 0.025(100 + x)$

Solution: $0x + 0.04(100) = 0.025(100 + x)$
$0 + 4 = 2.5 + 0.025x$
$1.5 = 0.025x$
$x = 60$
The amount of water that should be added is 60ml.

22. Strategy: Let x represent the rate of the plane in calm air.
The rate of the wind is y.

	Rate	Time	Distance
with wind	$x + y$	2	$2(x + y)$
against wind	$x - y$	3	$3(x - y)$

The distance traveled with the wind is 150 mi. The distance traveled against the wind is 150 mi.
$2(x + y) = 150$
$3(x - y) = 150$

Solution: $2(x + y) = 150$
$3(x - y) = 150$

$$\frac{1}{2} \cdot 2(x + y) = \frac{1}{2} \cdot 150$$
$$\frac{1}{3} \cdot 3(x - y) = \frac{1}{3} \cdot 150$$

$x + y = 75$
$x - y = 50$
$2x = 125$
$x = 62.5$

$x + y = 75$
$62.5 + y = 75$
$y = 12.5$
The rate of the wind is 12.5 mph.

23. Let x represent the cost per pound of hamburger.
The cost per pound of steak is y.
First purchase:

	Amount	Cost	Quantity
Hamburger	100	x	$100x$
Steak	50	y	$50y$

Second purchase:

	Amount	Cost	Quantity
Hamburger	150	x	$150x$
Steak	100	y	$100y$

The total cost of the first purchase is \$541.
The total cost of the second purchase is \$960.
$100x + 50y = 540$
$150x + 100y = 960$

Solution: $100x + 50y = 540$
$150x + 100y = 960$

$-2(100x + 50y) = -2(540)$
$150x + 100y = 960$

$-200x - 100y = -1080$
$150x + 100y = 960$
$-50x = -120$
$x = 2.4$

$100(2.4) + 50y = 540$
$240 + 50y = 540$
$50y = 300$
$y = 6$
The cost of steak is \$6.00/lb.

24. Strategy: Let M represent the number of ohms, T the tolerance and r the value of the resistor. Find the tolerance and solve $|M - 12{,}000| \le T$ for M.

Solution: $T = 0.15 \cdot 12{,}000 = 1800$ ohms
$|M - 12{,}000| \le 1800$
$-1800 \le M - 12{,}000 \le 1800$
$-1800 + 12{,}000 \le M - 12{,}000 + 12{,}000 \le 1800 + 12{,}000$
$10{,}200 \le M \le 13{,}800$

The lower and upper limits of the resistor are 10,200 ohms and 13,800 ohms.

25. The slope of the line is
$$\frac{5000-1000}{100-0} = \frac{4000}{100} = 40.$$
The commission rate of the executive is \$40 for every \$1000 in sales.

Chapter 6: Polynomials

Prep Test

1. $-2-(-3)=-2+3=1$

2. $-3(6)=-18$

3. $-\dfrac{24}{-36}=\dfrac{2}{3}$

4. $3n^4=3(-2)^4=48$

5. $b\neq 0$

6. No

7. $3x^2-4x+1+2x^2-5x-7$
   $=5x^2-9x-6$

8. $-4y+4y=0$

9. $-3(2x-8)=-6x+24$

10. $3xy-4y-2(5xy-7y)$
    $=3xy-4y-10xy+14y$
    $=-7xy+10y$

Section 6.1

Objective A Exercises

1. $(ab^3)(a^3b)=a^4b^4$

3. $(9xy^2)(-2x^2y^2)=-18x^3y^4$

5. $(x^2y^4)^4=x^8y^{16}$

7. $(-3x^2y^3)^4=(-3)^4x^8y^{12}=81x^4y^{12}$

9. $(27a^5b^3)^2=(27)^2a^{10}b^6=729a^{10}b^6$

11. $[(2a^4b^3)^3]^2=(2a^4b^3)^6=(2)^6a^{24}b^{18}=64a^{24}b^{18}$

13. $(x^2y^2)(xy^3)^3=(x^2y^2)(x^3y^9)=x^5y^{11}$

15. $(-5ab)(3a^3b^2)^2=(-5ab)(3^2a^6b^4)$
    $=(-5ab)(9a^6b^4)$
    $=-45a^7b^5$

17. $(3x^5y)(-4x^3)^3=(3x^5y)((-4)^3x^9)$
    $=(3x^5y)(-64x^9)$
    $=-192x^{14}y$

19. $(-6a^4b^2)(-7a^2c^5)=42a^6b^2c^5$

21. $(-2ab^2)(-3a^4b^5)^3=(-2ab^2)((-3)^3a^{12}b^{15})$
    $=(-2ab^2)(-27a^{12}b^{15})$
    $=54a^{13}b^{17}$

23. $(-3ab^3)^3(-2^2a^2b)^2=((-3)^3a^3b^9)(2^4a^4b^2)$
    $=(-27a^3b^9)(16a^4b^2)$
    $=-432a^7b^{11}$

25. $(-2x^2y^3z)(3x^2yz^4)=-6x^4y^4z^5$

27. $(2xy)(-3x^2yz)(x^2y^3z^3)=-6x^5y^5z^4$

29. $(3b^5)(2ab^2)(-2ab^2c^2)=-12a^2b^9c^2$

31. $x^n\cdot x^{n+1}=x^{n+n+1}=x^{2n+1}$

33. $y^{3n}\cdot y^{3n-2}=y^{3n+3-2}=y^{6n-2}$

35. $(a^{n-3})^{2n}=a^{2n^2-6n}$

37. $(x^{3n+2})^5=x^{15n+10}$

39. The value of n is 33.

Objective B Exercises

41. $\dfrac{1}{3^{-5}}=3^5=243$

43. $\dfrac{1}{y^{-3}}=y^3$

45. $\dfrac{a^3}{4b^{-2}}=\dfrac{a^3b^2}{4}$

47. $xy^{-4}=\dfrac{x}{y^4}$

49. $\dfrac{1}{2x^0} = \dfrac{1}{2}$

51. $\dfrac{-3^{-2}}{(2y)^0} = \dfrac{-1}{3^2} = -\dfrac{1}{9}$

53. $(x^3y^5)^{-2} = x^{-6}y^{-10} = \dfrac{1}{x^6y^{10}}$

55. $(-3a^{-4}b^{-5})(-5a^{-2}b^4) = 15a^{-6}b^{-1} = \dfrac{15}{a^6b}$

57. $(4y^{-3}z^{-4})(-3y^3z^{-3})^{-2}$
$= (4y^{-3}z^{-4})((-3)^{-2}y^{-6}z^6)$
$= (4)(-3)^{-2}y^{-9}z^2 = \dfrac{4z^2}{(-3)^2y^9}$
$= \dfrac{4z^2}{9y^9}$

59. $(4x^{-3}y^2)^{-3}(2xy^{-3})^4$
$= (4^{-3}x^9y^{-6})(2^4x^4y^{-12})$
$= (4^{-3})(2)^4x^{13}y^{-18} = \dfrac{2^4x^{13}}{(4)^3y^{18}}$
$= \dfrac{16x^{13}}{64y^{18}} = \dfrac{x^{13}}{4y^{18}}$

61. $\dfrac{9x^5}{12x^8} = \dfrac{3}{4x^3}$

63. $\dfrac{-6x^2y}{12x^4y} = -\dfrac{1}{2x^2}$

65. $\dfrac{y^{-2}}{y^6} = y^{-2-(6)} = y^{-8} = \dfrac{1}{y^8}$

67. $\dfrac{x^4y^3}{x^{-1}y^{-2}} = x^5y^5$

69. $\dfrac{a^6b^{-4}}{a^{-2}b^5} = a^8b^{-9} = \dfrac{a^8}{b^9}$

71. $\dfrac{-3ab^2}{(9a^2b^4)^3} = \dfrac{-3ab^2}{9^3a^6b^{12}} = \dfrac{-3ab^2}{729a^6b^{12}}$
$= -\dfrac{1}{243a^5b^{10}}$

73. $\dfrac{(3a^2b)^3}{(-6ab^3)^2} = \dfrac{3^3a^6b^3}{(-6)^2a^2b^6} = \dfrac{27a^6b^3}{36a^2b^6} = \dfrac{3a^4}{4b^3}$

75. $\dfrac{(-8x^2y^2)^4}{(16x^3y^7)^2} = \dfrac{(-8)^4x^8y^8}{(16)^2x^6y^{14}} = \dfrac{4096x^8y^8}{256x^6y^{14}}$
$= \dfrac{16x^2}{y^6}$

77. $\dfrac{(3a^4b^{-2})^{-2}}{(2a^{-3}b)^3} = \dfrac{(3)^{-2}a^{-8}b^4}{(2)^3a^{-9}b^3} = \dfrac{ab}{(2)^3(3)^2} = \dfrac{ab}{72}$

79. $\dfrac{(-2x^{-5}y^2)^{-3}}{(4xy^{-2})^{-4}} = \dfrac{(-2)^{-3}x^{15}y^{-6}}{(4)^{-4}x^{-4}y^8} = \dfrac{(4)^4x^{19}}{(-2)^3y^{14}}$
$= -\dfrac{32x^{19}}{y^{14}}$

81. $\dfrac{b^{6n}}{b^{10n}} = b^{6n-10n} = b^{-4n} = \dfrac{1}{b^{4n}}$

83. $\dfrac{y^{2n}}{-y^{8n}} = -y^{2n-8n} = -y^{-6n} = -\dfrac{1}{y^{6n}}$

85. $\dfrac{y^{3n+2}}{y^{2n+4}} = y^{3n+2-(2n+4)} = y^{n-2}$

87. $\dfrac{x^{2n-1}y^{n-3}}{x^{n+4}y^{n+3}} = \dfrac{x^{2n-1-(n+4)}}{y^{n+3-(n-3)}} = \dfrac{x^{n-5}}{y^6}$

89. $\left(\dfrac{9ab^{-2}}{8a^{-2}b}\right)^{-2}\left(\dfrac{3a^{-2}b}{2a^2b^{-2}}\right)^3$
$= \left(\dfrac{9^{-2}a^{-2}b^4}{8^{-2}a^4b^{-2}}\right)\left(\dfrac{3^3a^{-6}b^3}{2^3a^6b^{-6}}\right)$

$$= \left(\frac{9^{-2}b^6}{8^{-2}a^6}\right)\left(\frac{3^3b^9}{2^3a^{12}}\right)$$

$$= \left(\frac{8^2b^6}{9^2a^6}\right)\left(\frac{3^3b^9}{2^3a^{12}}\right) = \frac{8b^{15}}{3a^{18}}$$

91. The value of $p - q$ is 0.

Objective C Exercises

93. 4.67×10^{-6}

95. 1.7×10^{-10}

97. 2×10^{11}

99. 0.000000123

101. 8,200,000,000,000,000

103. 0.039

105. $(3 \times 10^{-12})(5 \times 10^{16})$
$= (3)(5) \times 10^{-12+16}$
$= 15 \times 10^4$
$= 150,000$

107. $(0.0000065)(3,200,000,000,000)$
$= (6.5 \times 10^{-6})(3.2 \times 10^{12})$
$= (6.5)(3.2) \times 10^{-6+12}$
$= 20.8 \times 10^6$
$= 20,800,000$

109. $\dfrac{9\times10^{-3}}{6\times10^{5}} = 1.5\times10^{-3-5}$
$= 1.5\times10^{-8} = 0.000000015$

111. $\dfrac{0.0089}{500,000,000} = \dfrac{8.9\times10^{-3}}{5\times10^{8}}$
$= 1.78\times10^{-3-8} = 1.78\times10^{-11}$
$= 0.0000000000178$

113. $\dfrac{(3.3\times10^{-11})(2.7\times10^{15})}{8.1\times10^{-3}}$
$= \dfrac{(3.3)(2.7)\times10^{-11+15-(-3)}}{8.1}$
$= 1.1\times10^{7} = 11,000,000$

115. $\dfrac{(0.00000004)(84,000)}{(0.0003)(1,400,000)}$
$= \dfrac{4\times10^{-8}\times8.4\times10^{4}}{3\times10^{-4}\times1.4\times10^{6}}$
$= \dfrac{4(8.4)\times10^{-8+4-(-4)-6}}{3(1.4)}$
$= 8\times10^{-6} = 0.000008$

117. Greater than zero

Objective D Exercises

119. **Strategy:** To find the number of years needed to cross the galaxy, divide the width of the galaxy by the product of the rate of the space ship and the number of hours in a year.
Solution:
$$\frac{5.6\times10^{19}}{2.5\times10^{4}\times8.76\times10^{3}} \approx 2.6\times10^{11}$$
It would takes a space ship 2.6×10^{11} years to cross the galaxy.

121. **Strategy:** To find the number of times larger the mass of the proton is, divide the mass of the proton by the mass of an electron.

Solution: $\dfrac{1.673\times10^{-27}}{9.109\times10^{-31}} \approx 1.837\times10^{3}$
The mass of a proton is approximately 1.837×10^{3} times larger than an electron.

123. **Strategy:** To find the number of times larger the mass of the sun is, divide the mass of the sun by the mass of the earth.

Solution: $\dfrac{2\times10^{30}}{5.9\times10^{24}} \approx 3.39\times10^{5}$

The sun is approximately 3.39×10^{5} times larger than the earth.

125. **Strategy:** To find the rate of the signals divide the distance by the time.
Solution:

$$\frac{119{,}000{,}000}{11} = \frac{1.19\times10^{8}}{1.1\times10^{1}} = 1.0\overline{81}\times10^{7}$$

The signal travels $1.0\overline{81}\times10^{7}$ mi/min.

127. **Strategy:** To find the number of seeds produced, divide the number of seeds by the number of pine seedlings growing.
Solution:

$$\frac{2{,}000{,}000}{12{,}000} = \frac{2\times10^{6}}{1.2\times10^{4}} = 1.\overline{6}\times10^{2}$$

$1.\overline{6}\times10^{2}$ seeds are produced.

Applying the Concepts

129. a) $x^{0} = 0$ is incorrect
$x^{0} = 1$ is correct
The Definition of Zero as an Exponent was used incorrectly.

b) $(x^{4})^{5} = x^{9}$ is incorrect
$(x^{4})^{5} = x^{20}$ is correct
The Rule for Simplifying the Power of an Exponential Expression was used incorrectly.

c) $x^{2}\cdot x^{3} = x^{6}$ is incorrect
$x^{2}\cdot x^{3} = x^{5}$ is correct
The Rule for Multiplying Exponential Expressions was used incorrectly.

Section 6.2

Objective A Exercises

1. $P(x) = 3x^{2} - 2x - 8$
$P(3) = 3(3)^{2} - 2(3) - 8$
$P(3) = 13$

3. $R(x) = 2x^{3} - 3x^{2} + 4x - 2$
$R(2) = 2(2)^{3} - 3(2)^{2} + 4(2) - 2$
$R(2) = 10$

5. $f(x) = x^{4} - 2x^{2} - 10$
$f(-1) = (-1)^{4} - 2(-1)^{2} - 10$
$f(-1) = -11$

7. Polynomial: (a) −1 (b) 8 (c) 2

9. Not a polynomial. The terms are not monomials.

11. Not a polynomial. The terms are not monomials.

13. Polynomial: (a) 3 (b) π (c) 5

15. Polynomial: (a) −5 (b) 2 (c) 3

17. Polynomial: (a) 14 (b) 14 (c) 0

19. $P(x) = x^{2} - 1$

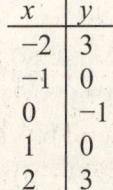

x	y
−2	3
−1	0
0	−1
1	0
2	3

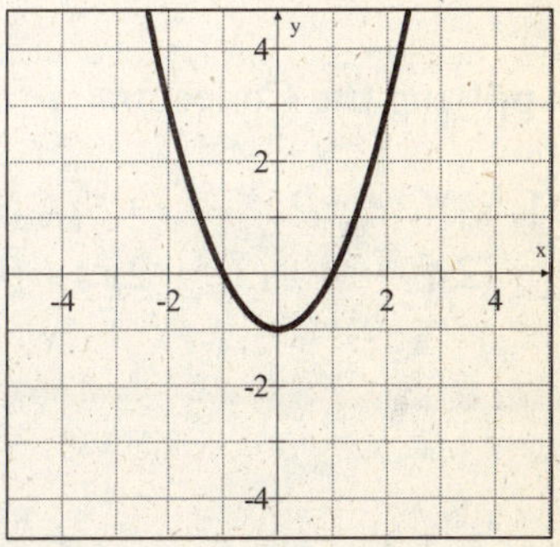

21. $R(x) = x^3 + 2$

x	y
−2	−6
−1	1
0	2
1	3
2	10

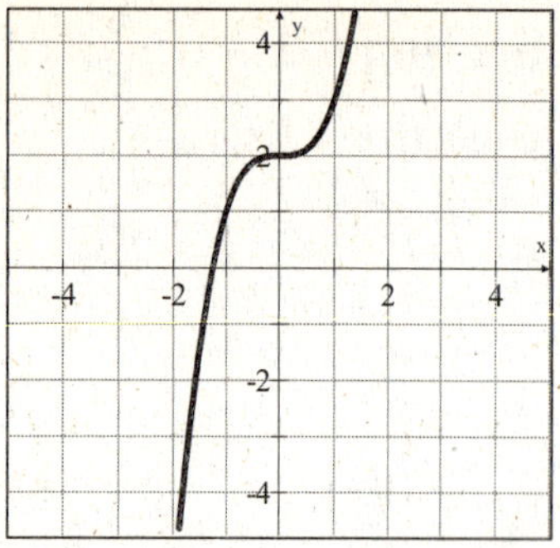

23. $f(x) = x^3 - 2x$

x	y
−2	−4
−1	1
0	0
1	−1
2	4

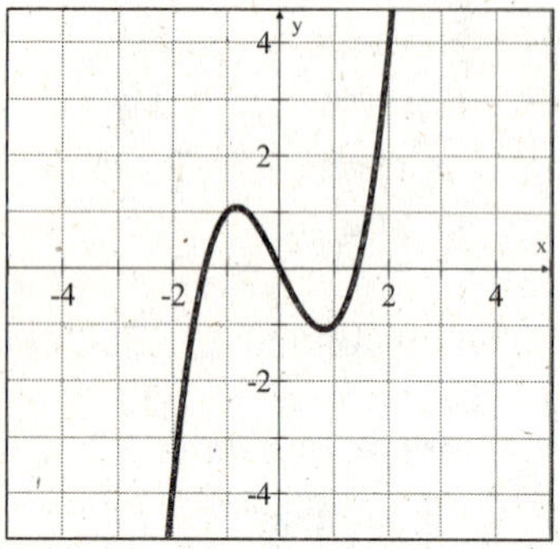

25. (a) $f(c) - g(c) > 0$

(b) $f(c) - g(c) < 0$

Objective B Exercises

27.
$$\begin{array}{r} 5x^2 + 2x - 7 \\ \underline{x^2 - 8x + 12} \\ 6x^2 - 6x + 5 \end{array}$$

29.
$$\begin{array}{r} x^2 - 3x + 8 \\ \underline{-2x^2 + 3x - 7} \\ -x^2 \qquad\quad + 1 \end{array}$$

31. $(3y^2 - 7y) + (2y^2 - 8y + 2)$
$= (3y^2 + 2y^2) + (-7y - 8y) + 2$
$= 5y^2 - 15y + 2$

33. $(2a^2 - 3a - 7) - (-5a^2 - 2a - 9)$
$= (2a^2 + 5a^2) + (-3a + 2a) + (-7 + 9)$
$= 7a^2 - a + 2$

35. $P(x) + R(x) = (x^2 - 3xy + y^2) + (2x^2 - 3y^2)$
$= (x^2 + 2x^2) - 3xy + (y^2 - 3y^2)$
$= 3x^2 - 3xy - 2y^2$

37. $P(x) - R(x)$
$= (3x^2 + 2y^2) - (-5x^2 + 2xy - 3y^2)$
$= (3x^2 + 5x^2) - 2xy + (2y^2 + 3y^2)$
$= 8x^2 - 2xy + 5y^2$

39. $S(x) = P(x) + R(x)$
$= (3x^4 - 3x^3 - x^2) + (3x^3 - 7x^2 + 2x)$
$= 3x^4 + (-3x^3 + 3x^3) + (-x^2 - 7x^2) + 2x$
$S(x) = 3x^4 - 8x^2 + 2x$
$S(2) = 3(2)^4 - 8(2)^2 + 2(2)$
$S(2) = 20$

Applying the Concepts

41. a)
$$\begin{aligned} (2x^3 + 3x^2 + kx + 5) - (x^3 + 2x^2 + 3x + 7) &= x^3 + x^2 + 5x - 2 \\ (2x^3 - x^3) + (3x^2 - 2x^2) + (kx - 3x) + (5 - 7) &= x^3 + x^2 + 5x - 2 \\ x^3 + x^2 + (k - 3)x - 2 &= x^3 + x^2 + 5x - 2 \\ (k - 3)x &= 5x \\ k - 3 &= 5 \\ k &= 8 \end{aligned}$$

b)
$$\begin{aligned} (6x^3 + kx^2 - 2x - 1) - (4x^3 - 3x^2 + 1) &= 2x^3 - x^2 - 2x - 2 \\ (6x^3 - 4x^3) + (kx^2 + 3x^2) - 2x + (-1 - 1) &= 2x^3 - x^2 - 2x - 2 \\ 2x^3 + (k + 3)x^2 - 2x - 2 &= 2x^3 - x^2 - 2x - 2 \\ (k + 3)x^2 &= -x^2 \\ k + 3 &= -1 \\ k &= -4 \end{aligned}$$

43. If $P(x)$ is a fifth degree polynomial and $Q(x)$ is a fourth degree polynomial, then $P(x) - Q(x)$ is a fifth degree polynomial.
Example: $P(x) = 3x^5 - 5x - 8$
$Q(x) = 3x^4 - 2x + 1$
$P(x) - Q(x) = 3x^5 - 3x^4 - 3x - 9$

Section 6.3

Objective A Exercises

1. The Distributive Property

3. $2x(x-3) = 2x^2 - 6x$

5. $3x^2(2x^2 - x) = 6x^4 - 3x^3$

7. $3xy(2x - 3y) = 6x^2y - 9xy^2$

9. $x^n(x+1) = x^{n+1} + x^n$

11. $x^n(x^n + y^n) = x^{2n} + x^ny^n$

13. $2b + 4b(2-b) = 2b + 8b - 4b^2 = -4b^2 + 10b$

15. $-2a^2(3a^2 - 2a + 3) = -6a^4 + 4a^3 - 6a^2$

17. $(-3y^2 - 4y + 2)(y^2) = -3y^4 - 4y^3 + 2y^2$

19. $-5x^2(4 - 3x + 3x^2 + 4x^3)$
$= -20x^2 + 15x^3 - 15x^4 - 20x^5$

21. $-2x^2y(x^2 - 3xy + 2y^2) = -2x^4y + 6x^3y^2 - 4x^2y^3$

23. $5x^3 - 4x(2x^2 + 3x - 7)$
$= 5x^3 - 8x^3 - 12x^2 + 28x$
$= -3x^3 - 12x^2 + 28x$

25. $2y^2 - y[3 - 2(y-4) - y]$
$= 2y^2 - y(3 - 2y + 8 - y)$
$= 2y^2 - y(11 - 3y)$
$= 2y^2 - 11y + 3y^2$
$= 5y^2 - 11y$

27. $2y - 3[y - 2y(y-3) + 4y]$
$= 2y - 3(y - 2y^2 + 6y + 4y)$
$= 2y - 3(-2y^2 + 11y)$
$= 2y + 6y^2 - 33y$
$= 6y^2 - 31y$

29. $P(b) = 3b$ and $Q(b) = 3b^4 - 3b^2 + 8$
$P(b)\cdot Q(b) = 3b(3b^4 - 3b^2 + 8)$
$= 9b^5 - 9b^3 + 24b$

Objective B Exercises

31. $(x-2)(x+7) = x^2 + 7x - 2x - 14$
$= x^2 + 5x - 14$

33. $(2y-3)(4y+7) = 8y^2 + 14y - 12y - 21$
$= 8y^2 + 2y - 21$

35. $(a+3c)(4a-5c) = 4a^2 - 5ac + 12ac - 15c^2$
$= 4a^2 + 7ac - 15c^2$

37. $(5x-7)(5x-7) = 25x^2 - 35x - 35x + 49$
$= 25x^2 - 70x + 49$

39. $2(2x-3y)(2x+5y)$
$= 2(4x^2 + 10xy - 6xy - 15y^2)$
$= 2(4x^2 + 4xy - 15y^2)$
$= 8x^2 + 8xy - 30y^2$

41. $(xy+4)(xy-3) = x^2y^2 - 3xy + 4xy - 12$
$= x^2y^2 + xy - 12$

43. $(2x^2-5)(x^2-5) = 2x^4 - 10x^2 - 5x^2 + 25$
$= 2x^4 - 15x^2 + 25$

45. $(5x^2 - 5y)(2x^2 - y)$
$= 10x^4 - 5x^2y - 10x^2y + 5y^2$
$= 10x^4 - 15x^2y + 5y^2$

47. $(x+5)(x^2 - 3x + 4)$
$= x(x^2 - 3x + 4) + 5(x^2 - 3x + 4)$
$= x^3 - 3x^2 + 4x + 5x^2 - 15x + 20$
$= x^3 + 2x^2 - 11x + 20$

49. $(2a-3b)(5a^2-6ab+4b^2)$
$=2a(5a^2-6ab+4b^2)-3b(5a^2-6ab+4b^2)$
$=10a^3-12a^2b+8ab^2-15a^2b+18ab^2-12b^3$
$=10a^3-27a^2b+26ab^2-12b^3$

51. $(2x^3+3x^2-2x+5)(2x-3)$
$=(2x^3+3x^2-2x+5)2x+(2x^3+3x^2-2x+5)(-3)$
$=4x^4+6x^3-4x^2+10x-6x^3-9x^2+6x-15$
$=4x^4-13x^2+16x-15$

53. $(2x-5)(2x^4-3x^3-2x+9)$
$=2x(2x^4-3x^3-2x+9)-5(2x^4-3x^3-2x+9)$
$=4x^5-6x^4-4x^2+18x-10x^4+15x^3+10x-45$
$=4x^5-16x^4+15x^3-4x^2+28x-45$

55. $(x^2+2x-3)(x^2-5x+7)$
$=x^2(x^2-5x+7)+2x(x^2-5x+7)-3(x^2-5x+7)$
$=x^4-5x^3+7x^2+2x^3-10x^2+14x-3x^2+15x-21$
$=x^4-3x^3-6x^2+29x-21$

57. $(a-2)(2a-3)(a+7)$
$=(2a^2-3a-4a+6)(a+7)$
$=(2a^2-7a+6)(a+7)$
$=(2a^2-7a+6)a+(2a^2-7a+6)7$
$=2a^3-7a^2+6a+14a^2-49a+42$
$=2a^3+7a^2-43a+42$

59. $(2x+3)(x-4)(3x+5)$
$=(2x^2-8x+3x-12)(3x+5)$
$=(2x^2-5x-12)(3x+5)$
$=(2x^2-5x-12)3x+(2x^2-5x-12)5$
$=6x^3-15x^2-36x+10x^2-25x-60$
$=6x^3-5x^2-61x-60$

61. $P(y)=2y^2-1$ and $Q(y)=y^3-5y^2-3$
$P(y)\cdot Q(y)=(2y^2-1)(y^3-5y^2-3)$
$=2y^2(y^3-5y^2-3)-1(y^3-5y^2-3)$
$=2y^5-10y^4-6y^2-y^3+5y^2+3$
$=2y^5-10y^4-y^3-y^2+3$

63. 5

Objective C Exercises

65. $(3x-2)(3x+2)=9x^2-4$

67. $(6-x)(6+x)=36-x^2$

69. $(2a-3b)(2a+3b)=4a^2-9b^2$

71. $(3ab+4)(3ab-4)=9a^2b^2-16$

73. $(x^2+1)(x^2-1)=x^4-1$

75. $(x-5)^2=x^2-10x+25$

77. $(3a+5b)^2=9a^2+30ab+25b^2$

79. $(x^2-3)^2=x^4-6x^2+9$

81. $(2x^2-3y^2)^2=4x^4-12x^2y^2+9y^4$

83. $(3mn-5)^2=9m^2n^2-30mn+25$

85. $y^2-(x-y)^2=y^2-(x^2-2xy+y^2)$
$=y^2-x^2+2xy-y^2$
$=-x^2+2xy$

87. $(x-y)^2-(x+y)^2$
$=x^2-2xy+y^2-(x^2+2xy+y^2)$
$=x^2-2xy+y^2-x^2-2xy-y^2$
$=-4xy$

89. False

Objective D Exercises

91. ft^2

93. Strategy: To find the area substitute the given values for L and W in the equation $A = L \cdot W$ and solve for A.

Solution: $A = L \cdot W$
$A = (3x - 2)(x + 4)$
$A = 3x^2 + 12x - 2x - 8$
$A = 3x^2 + 10x - 8$
The area is $(3x^2 + 10x - 8)$ ft^2.

95. Strategy: To find the area, add the area of the small rectangle to the area of the large rectangle.
Larger rectangle:
Length = $L_1 = x + 5$
Width = $W_1 = x - 2$
Smaller rectangle:
Length = $L_2 = 5$
Width = $W_2 = 2$

Solution: $A = L_1 \cdot W_1 + L_2 \cdot W_2$
$A = (x + 5)(x - 2) + (5)(2)$
$A = x^2 - 2x + 5x - 10 + 10$
$A = x^2 + 3x$
The area is $(x^2 + 3x)$ ft^2.

97. Strategy: To find the volume substitute the given value for s in the equation $V = s^3$ and solve for V.

Solution: $V = s^3$
$V = (x + 3)^3$
$V = (x + 3)(x + 3)(x + 3)$
$V = (x^2 + 6x + 9)(x + 3)$
$V = x^3 + 9x^2 + 27x + 27$
The volume is $(x^3 + 9x^2 + 27x + 27)$ cm^3.

99. Strategy: To find the volume subtract the volume of the small rectangular solid from the volume of the large rectangular solid.
Large rectangular solid:
Length = $L_1 = x + 2$
Width = $W_1 = 2x$
Height = $H_1 = x$
Small rectangular solid:
Length = $L_2 = x$
Width = $W_2 = 2x$
Height = $H_2 = 2$

Solution: $V = (L_1 \cdot W_1 \cdot H_1) - (L_2 \cdot W_2 \cdot H_2)$
$V = (x + 2)(2x)(x) - (x)(2x)(2)$
$V = (2x^2 + 4x)(x) - (2x^2)(2)$
$V = 2x^3 + 4x^2 - 4x^2$
$V = 2x^3$
The volume is $(2x^3)$ in^3.

101. Strategy: To find the area substitute the given value for r into the equation $A = \pi r^2$ and solve for A.

Solution: $A = \pi r^2$
$A = 3.14(5x + 4)^2$
$A = 3.14(25x^2 + 40x + 16)$
$A = 78.5x^2 + 125.6x + 50.24$
The area is $(78.5x^2 + 125.6x + 50.24)$ in^2.

Applying the Concepts

103. a) $(a - b)(a^2 + ab + b^2)$
$= a(a^2 + ab + b^2) - b(a^2 + ab + b^2)$
$= a^3 + a^2b + ab^2 - a^2b - ab^2 - b^3$
$= a^3 - b^3$

b) $(x + y)(x^2 - xy + y^2)$
$= x(x^2 - xy + y^2) + y(x^2 - xy + y^2)$
$= x^3 - x^2y + xy^2 + x^2y - xy^2 + y^3$
$= x^3 + y^3$

105. a) $(3x - k)(2x + k) = 6x^2 + 5x + k^2$
$6x^2 + xk - k^2 = 6x^2 + 5x + k^2$
$xk = 5x$
$k = 5$

b) $(4x + k)^2 = 16x^2 + 8x + k^2$
$16x^2 + 8xk + k^2 = 16x^2 + 8x + k^2$
$8xk = 8x$
$k = 1$

107. The product of $(4a + b)$ and $(2a - b)$
$(4a + b)(2a - b) = 8a^2 - 2ab - b^2$.
Subtract the product from $9a^2 - 2ab$.
$9a^2 - 2ab - (8a^2 - 2ab - b^2)$
$= 9a^2 - 2ab - 8a^2 + 2ab + b^2$
$= a^2 + b^2$

Section 6.4

Objective A Exercises

1. $\dfrac{3x^2-6x}{3x}=\dfrac{3x^2}{3x}-\dfrac{6x}{3x}=x-2$

3. $\dfrac{5x^2-10x}{-5x}=\dfrac{5x^2}{-5x}-\dfrac{10x}{-5x}=-x+2$

5. $\dfrac{5x^2y^2+10xy}{5xy}=\dfrac{5x^2y^2}{5xy}+\dfrac{10xy}{5xy}=xy+2$

7. $\dfrac{x^3+3x^2-5x}{x}=\dfrac{x^3}{x}+\dfrac{3x^2}{x}-\dfrac{5x}{x}=x^2+3x-5$

9. $\dfrac{9b^5+12b^4+6b^3}{3b^2}=\dfrac{9b^5}{3b^2}+\dfrac{12b^4}{3b^2}+\dfrac{6b^3}{3b^2}$
$=3b^3+4b^2+2b$

11. $\dfrac{a^5b-6a^3b+ab}{ab}=\dfrac{a^5b}{ab}-\dfrac{6a^3b}{ab}+\dfrac{ab}{ab}$
$=a^4-6a^2+1$

13. $P(x)=6x^3+21x^2-15x$

Objective B Exercises

15.
$$\begin{array}{r} x+8 \\ x-5\overline{)x^2+3x-40} \\ \underline{x^2-5x} \\ 8x-40 \\ \underline{8x-40} \\ 0 \end{array}$$

$(x^2+3x-40)\div(x-5)=x+8$

17.
$$\begin{array}{r} x^2+3x+6 \\ x-3\overline{)x^3+0x^2-3x+2} \\ \underline{x^3-3x^2} \\ 3x^2-3x \\ \underline{3x^2-9x} \\ 6x+2 \\ \underline{6x-18} \\ 20 \end{array}$$

$(x^3-3x+2)\div(x-3)=x^2+3x+6+\dfrac{20}{x-3}$

19.
$$\begin{array}{r} 3x+5 \\ 2x+1\overline{)6x^2+13x+8} \\ \underline{6x^2+3x} \\ 10x+8 \\ \underline{10x+5} \\ 3 \end{array}$$

$(6x^2+13x+8)\div(2x+1)=3x+5+\dfrac{3}{2x+1}$

21.
$$\begin{array}{r} 5x+7 \\ 2x-1\overline{)10x^2+9x-5} \\ \underline{10x^2-5x} \\ 14x-5 \\ \underline{14x-7} \\ 2 \end{array}$$

$(10x^2+9x-5)\div(2x-1)=5x+7+\dfrac{2}{2x-1}$

23.

$$\begin{array}{r}
4x^2+6x+9 \\
2x-3\overline{)\,8x^3+0x^2+0x-9} \\
\underline{8x^3-12x^2} \\
12x^2+0x \\
\underline{12x^2-18x} \\
18x-9 \\
\underline{18x-27} \\
18
\end{array}$$

$$(8x^3-9)\div(2x-3)=4x^2+6x+9+\frac{18}{2x-3}$$

25.

$$\begin{array}{r}
3x^2+1 \\
2x^2-5\overline{)\,6x^4+0x^3-13x^2+0x-4} \\
\underline{6x^4-15x^2} \\
2x^2+0x-4 \\
\underline{2x^2-5} \\
1
\end{array}$$

$$(6x^4-13x^2-4)\div(2x^2-5)$$

$$=3x^2+1+\frac{1}{2x^2-5}$$

27.

$$\begin{array}{r}
x^2-3x-10 \\
3x+1\overline{)\,3x^3-8x^2-33x-10} \\
\underline{3x^3+x^2} \\
-9x^2-33x \\
\underline{-9x^2-3x} \\
-30x-10 \\
\underline{-30x-10} \\
0
\end{array}$$

$$\frac{3x^3-8x^2-33x-10}{3x+1}=x^2-3x-10$$

29.

$$\begin{array}{r}
x^2-2x+1 \\
x-3\overline{)\,x^3-5x^2+7x-4} \\
\underline{x^3-3x^2} \\
-2x^2+7x \\
\underline{-2x^2+6x} \\
x-4 \\
\underline{x-3} \\
-1
\end{array}$$

$$\frac{x^3-5x^2+7x-4}{x-3}=x^2-2x+1-\frac{1}{x-3}$$

31.

$$\begin{array}{r}
2x^3-3x^2+x-4 \\
x-5\overline{)\,2x^4-13x^3+16x^2-9x+20} \\
\underline{2x^4-10x^3} \\
-3x^3+16x^2 \\
\underline{-3x^3+15x^2} \\
x^2-9x \\
\underline{x^2-5x} \\
-4x+20 \\
\underline{-4x+20} \\
0
\end{array}$$

$$\frac{2x^4-13x^3+16x^2-9x+20}{x-5}=2x^3-3x^2+x-4$$

33.

$$\begin{array}{r}
2x \\
x^2+2x-1\overline{)\,2x^3+4x^2-x+2} \\
\underline{2x^3+4x^2-2x} \\
x+2
\end{array}$$

$$\frac{2x^3+4x^2-x+2}{x^2+2x-1}=2x+\frac{x+2}{x^2+2x-1}$$

35.

$$\begin{array}{r} x^2+4x+6 \\ x^2-2x-1\overline{)x^4+2x^3-3x^2-6x+2} \\ \underline{x^4-2x^3-x^2} \\ 4x^3-2x^2-6x \\ \underline{4x^3-8x^2-4x} \\ 6x^2-2x+2 \\ \underline{6x^2-12x-6} \\ 10x+8 \end{array}$$

$$\frac{x^4+2x^3-3x^2-6x+2}{x^2-2x-1}=x^2+4x+6+\frac{10x+8}{x^2-2x-1}$$

37. $\dfrac{P(x)}{Q(x)}=\dfrac{2x^3+x^2+8x+7}{2x+1}$

$$\begin{array}{r} x^2+4 \\ 2x+1\overline{)2x^3+x^2+8x+7} \\ \underline{2x^3+x^2} \\ 8x+7 \\ \underline{8x+4} \\ 3 \end{array}$$

$$\frac{2x^3+x^2+8x+7}{2x+1}=x^2+4+\frac{3}{2x+1}$$

39. False

Objective C Exercise

41. Third

43.

$$\begin{array}{r|rrr} -1 & 2 & -6 & -8 \\ & & -2 & 8 \\ \hline & 2 & -8 & 0 \end{array}$$

$(2x^2-6x-8)\div(x+1)=2x-8$

45.

$$\begin{array}{r|rrr} 2 & 3 & -14 & 16 \\ & & 6 & -16 \\ \hline & 3 & -8 & 0 \end{array}$$

$(3x^2-14x+16)\div(x-2)=3x-8$

47.

$$\begin{array}{r|rrr} 1 & 3 & 0 & -4 \\ & & 3 & 3 \\ \hline & 3 & 3 & -1 \end{array}$$

$(3x^2-4)\div(x-1)=3x+3-\dfrac{1}{x-1}$

49.

$$\begin{array}{r|rrrr} -1 & 2 & -1 & 6 & 9 \\ & & -2 & 3 & -9 \\ \hline & 2 & -3 & 9 & 0 \end{array}$$

$(2x^3-x^2+6x+9)\div(x+1)=2x^2-3x+9$

51.

$$\begin{array}{r|rrrr} 2 & 4 & 0 & -1 & -18 \\ & & 8 & 16 & 30 \\ \hline & 4 & 8 & 15 & 12 \end{array}$$

$(4x^3-x-18)\div(x-2)=4x^2+8x+15+\dfrac{12}{x-2}$

53.

$$\begin{array}{r|rrrr} -4 & 2 & 5 & -5 & 20 \\ & & -8 & 12 & -28 \\ \hline & 2 & -3 & 7 & -8 \end{array}$$

$(2x^3+5x^2-5x+20)\div(x+4)=2x^2-3x+7-\dfrac{8}{x+4}$

55.

$$\begin{array}{r|rrrrr} 2 & 3 & -4 & 8 & -5 & -5 \\ & & 6 & 4 & 24 & 38 \\ \hline & 3 & 2 & 12 & 19 & 33 \end{array}$$

$(3x^4-4x^3+8x^2-5x-5)\div(x-2)$

$=3x^3+2x^2+12x+19+\dfrac{33}{x-2}$

57.

$$\begin{array}{r|rrrrr} -1 & 3 & 3 & -1 & 3 & 2 \\ & & -3 & 0 & 1 & -4 \\ \hline & 3 & 0 & -1 & 4 & -2 \end{array}$$

$$(3x^4 + 3x^3 - x^2 + 3x + 2) \div (x+1)$$
$$= 3x^3 - x + 4 - \frac{2}{x+1}$$

59. $\dfrac{P(x)}{Q(x)} = \dfrac{3x^2 - 5x + 6}{x - 2}$

$$\begin{array}{r|rrr} 2 & 3 & -5 & 6 \\ & & 6 & 2 \\ \hline & 3 & 1 & 8 \end{array}$$

$$(3x^2 - 5x + 6) \div (x-2) = 3x + 1 + \frac{8}{x-2}$$

Objective D Exercises

61. $x - 3$

63.

$$\begin{array}{r|rrr} 3 & 2 & -3 & -1 \\ & & 6 & 9 \\ \hline & 2 & 3 & 8 \end{array}$$

$P(3) = 8$

65.

$$\begin{array}{r|rrrr} 4 & 1 & -2 & 3 & -1 \\ & & 4 & 8 & 44 \\ \hline & 1 & 2 & 11 & 43 \end{array}$$

$R(4) = 43$

67.

$$\begin{array}{r|rrrr} -2 & 2 & -4 & 3 & -1 \\ & & -4 & 16 & -38 \\ \hline & 2 & -8 & 19 & -39 \end{array}$$

$P(-2) = -39$

69.

$$\begin{array}{r|rrrr} -3 & 2 & -1 & 0 & 3 \\ & & -6 & 21 & -63 \\ \hline & 2 & -7 & 21 & -60 \end{array}$$

$Z(-3) = -60$

71.

$$\begin{array}{r|rrrrrr} 2 & 1 & 0 & -4 & -2 & 5 & -2 \\ & & 2 & 4 & 0 & -4 & 2 \\ \hline & 1 & 2 & 0 & -2 & 1 & 0 \end{array}$$

$Q(2) = 0$

73.

$$\begin{array}{r|rrrrr} -3 & 2 & -1 & 0 & 2 & -5 \\ & & -6 & 21 & -63 & 183 \\ \hline & 2 & -7 & 21 & -61 & 178 \end{array}$$

$F(-3) = 178$

75.

$$\begin{array}{r|rrrr} 5 & 1 & 0 & 0 & -3 \\ & & 5 & 25 & 125 \\ \hline & 1 & 5 & 25 & 122 \end{array}$$

$P(5) = 122$

77.

$$\begin{array}{r|rrrrr} -3 & 4 & 0 & -3 & 0 & 5 \\ & & -12 & 36 & -99 & 297 \\ \hline & 4 & -12 & 33 & -99 & 302 \end{array}$$

$R(-3) = 302$

79.

$$\begin{array}{r|rrrrr} 3 & 1 & -2 & -3 & -1 & 7 \\ & & 3 & 3 & 0 & -3 \\ \hline & 1 & 1 & 0 & -1 & 4 \end{array}$$

$Q(2) = 0$

Applying the Concepts

81. a)

$$\begin{array}{r} a^2-ab+b^2 \\ a+b\overline{\big)a^3 \qquad\qquad +b^3} \\ \underline{a^3+a^2b} \\ -a^2b \\ \underline{-a^2b-ab^2} \\ ab^2+b^3 \\ \underline{ab^2+b^3} \\ 0 \end{array}$$

$$\frac{a^3+b^3}{a+b}=a^2-ab+b^2$$

b)

$$\begin{array}{r} x^4-x^3y+x^2y^2-xy^3+y^4 \\ x+y\overline{\big)x^5 \qquad\qquad\qquad\qquad +y^5} \\ \underline{x^5+x^4y} \\ -x^4y \\ \underline{-x^4y-x^3y^2} \\ x^3y^2 \\ \underline{x^3y^2+x^2y^3} \\ -x^2y^3 \\ \underline{-x^2y^3-xy^4} \\ xy^4+y^5 \\ \underline{xy^4+y^5} \\ 0 \end{array}$$

c)

$$\begin{array}{r} x^5-x^4y+x^3y^2-x^2y^3+xy^4-y^5 \\ x+y\overline{\big)x^6 \qquad\qquad\qquad\qquad\qquad -y^6} \\ \underline{x^6+x^5y} \\ -x^5y \\ \underline{-x^5y-x^4y^2} \\ x^4y^2 \\ \underline{x^4y^2+x^3y^3} \\ -x^3y^3 \\ \underline{-x^3y^3-x^2y^4} \\ x^2y^4 \\ \underline{x^2y^4+xy^5} \\ -xy^5-y^6 \\ \underline{-xy^5-y^6} \end{array}$$

83. Synthetic division can be modified so that the divisor may be of the form $ax+b$. Divide both the dividend and the divisor by a. The divisor is now in the form $x+\frac{b}{a}$ and the expression $\frac{-b}{a}$ can be used for a in the $x-a$ of synthetic division.

Concept Review

1. Writing the terms in descending order before adding helps us to identify and add the like terms of the polynomial.

2. We add the exponents of 3 and 6 to get $28p^9$.

3. To multiply two binomials use the FOIL method: Add the products of the first terms, the outer terms, the inner terms, and the last terms.

4. To square a binomial multiply it times itself.

5. To write a very large number in scientific notation, move the decimal point to the right of the first nonzero digit. This is the value of a. The exponent of 10 is positive and is

equal to the number of places the decimal point has been moved. To write a very small number in scientific notation, move the decimal point to the right of the first nonzero digit. This is the value of a. The exponent of 10 is negative and is equal to the number of places the decimal point has been moved.

6. After you divide a polynomial by a binomial check the answer by using the following: Dividend = (quotient X divisor) + remainder

7. A binomial is a factor of a polynomial if, when the polynomial is divided by the binomial, there is no remainder.

8. To use synthetic division for division of a polynomial the divisor must be in the form $x - a$.

9. When using synthetic division with a polynomial that has a missing term, insert a 0 for the missing term in the polynomial.

10. To evaluate a polynomial function at a given value for the variable, use synthetic division to divide the polynomial by the value at which the polynomial is being evaluated. The remainder is the value.

Chapter 6 Review Exercises

1. $(12y^2 + 17y - 4) + (9y^2 - 13y + 3)$
$= 21y^2 + 4y - 1$

2.
$$\begin{array}{r} 5x+4 \\ 3x-2\overline{\smash{\big)}15x^2+2x-2} \\ \underline{15x^2-10x} \\ 12x-2 \\ \underline{12x-8} \\ 6 \end{array}$$

$$\frac{15x^2+2x-2}{3x-2}=5x+4+\frac{6}{3x-2}$$

3. $(2x^{-1}y^2z^5)^4(-3x^3yz^{-3})^2$
$= (16x^{-4}y^8z^{20})(9x^6y^2z^{-6})$
$= 144x^2y^{10}z^{14}$

4. $(5y - 7)^2 = (5y - 7)(5y - 7)$
$= 25y^2 - 35y - 35y + 49$
$= 25y^2 - 70y + 49$

5. $\dfrac{a^{-1}b^3}{a^3b^{-3}} = \dfrac{b^6}{a^4}$

6.
$$\begin{array}{r|rrrr} 2 & 1 & -2 & 3 & -5 \\ & & 2 & 0 & 6 \\ \hline & 1 & 0 & 3 & 1 \end{array}$$

$P(2) = 1$

7. $(5x^2 - 8xy + 2y^2) - (x^2 - 3y^2)$
$= (5x^2 - x^2) - 8xy + (2y^2 + 3y^2)$
$= 4x^2 - 8xy + 5y^2$

8. $\dfrac{12b^7+36b^5-3b^3}{3b^3}=\dfrac{12b^7}{3b^3}+\dfrac{36b^5}{3b^3}-\dfrac{3b^3}{3b^3}$
$= 4b^4 + 12b^2 - 1$

9. $(\dfrac{3ab^4}{-6a^2b^4} = -\dfrac{1}{2a}$

10. $(-2a^2b^4)(3ab^2) = -6a^3b^6$

11. $\dfrac{8x^{12}}{12x^9} = \dfrac{2x^3}{3}$

12.
$$\begin{array}{r|rrrr} -6 & 4 & 27 & 10 & 2 \\ & & -24 & -18 & 48 \\ \hline & 4 & 3 & -8 & 50 \end{array}$$

$$\frac{4x^3+27x^2+10x+2}{x+6}=4x^2+3x-8+\frac{50}{x+6}$$

13. $P(x) = 2x^3 - x + 7$
$P(-2) = 2(-2)^3 - (-2) + 7$
$P(-2) = -16 + 2 + 7$
$P(-2) = -7$

14. $(13y^3 - 7y - 2) - (12y^2 - 2y - 1)$
$= (13y^3 - 7y - 2) + (-12y^2 + 2y + 1)$
$= 13y^3 - 12y^2 - 5y - 1$

15.

$$\begin{array}{r|rrrr} 7 & 1 & -2 & -33 & -7 \\ & & 7 & 35 & 14 \\ \hline & 1 & 5 & 2 & 7 \end{array}$$

$$\frac{b^3 - 2b^2 - 33b - 7}{b - 7} = b^2 + 5b + 2 + \frac{7}{b - 7}$$

16. $4x^2y(3x^3y^2 + 2xy - 7y^3)$
$= 12x^5y^3 + 8x^3y^2 - 28x^2y^4$

17. $(2a - b)(x - 2y) = 2ax - 4ay - bx + 2by$

18. $(2b - 3)(4b + 5) = 8b^2 + 10b - 12b - 15$
$= 8b^2 - 2b - 15$

19. $5x^2 - 4x[x - 3(3x + 2) + x]$
$= 5x^2 - 4x(x - 9x - 6 + x)$
$= 5x^2 - 4x(-7x - 6)$
$= 5x^2 + 28x^2 + 24x$
$= 33x^2 + 24x$

20. $(xy^5z^3)(x^3y^3z) = x^4y^8z^4$

21. $(4x - 3y)^2 = 16x^2 - 24xy + 9y^2$

22.

$$\begin{array}{r|rrrrr} 4 & 1 & 0 & 0 & 0 & -4 \\ & & 4 & 16 & 64 & 256 \\ \hline & 1 & 4 & 16 & 64 & 252 \end{array}$$

$$\frac{x^4 - 4}{x - 4} = x^3 + 4x^2 + 16x + 64 + \frac{252}{x - 4}$$

23. $(3x^2 - 2x - 6) + (-x^2 - 3x + 4)$
$= 2x^2 - 5x - 2$

24. $(5x^2yz^4)(2xy^3z^{-1})(7x^{-2}y^{-2}z^3)$
$= (10x^3y^4z^3)(7x^{-2}y^{-2}z^3)$
$= 70xy^2z^6$

25. $\dfrac{3x^4yz^{-1}}{-12xy^3z^2} = -\dfrac{x^3}{4y^2z^3}$

26. 9.48×10^8

27. $\dfrac{3\times10^{-3}}{15\times10^2} = 0.2\times10^{-5} = 2\times10^{-6}$

28.

$$\begin{array}{r|rrrr} -3 & -2 & 2 & 0 & -4 \\ & & 6 & -24 & 72 \\ \hline & -2 & 8 & -24 & 68 \end{array}$$

$P(-3) = 68$

29. $\dfrac{16x^5 - 8x^3 + 20x}{4x} = \dfrac{16x^5}{4x} - \dfrac{8x^3}{4x} + \dfrac{20x}{4x}$
$= 4x^4 - 2x^2 + 5$

30.

$$\begin{array}{r} 2x - 3 \\ 6x + 1 \overline{\big) 12x^2 - 16x - 7} \\ \underline{12x^2 + 2x} \\ -18x - 7 \\ \underline{-18x - 3} \\ -4 \end{array}$$

$$\frac{12x^2 - 16x - 7}{6x + 1} = 2x - 3 - \frac{4}{6x + 1}$$

31. $a^{2n+3}(a^n - 5a + 2) = a^{3n+3} - 5a^{2n+4} + 2a^{2n+3}$

32. $(x + 6)(x^3 - 3x^2 - 5x + 1)$
$= x(x^3 - 3x^2 - 5x + 1) + 6(x^3 - 3x^2 - 5x + 1)$
$= x^4 - 3x^3 - 5x^2 + x + 6x^3 - 18x^2 - 30x + 6$
$= x^4 + 3x^3 - 23x^2 - 29x + 6$

33. $-2x(4x^2 + 7x - 9) = -8x^3 - 14x^2 + 18x$

34.

$$\begin{array}{r} 3y^2 + 4y - 7 \\ \times \qquad\quad \underline{2y + 3} \\ 9y^2 + 12y - 21 \\ \underline{6y^3 + 8y^2 - 14y \qquad} \\ 6y^3 + 17y^2 - 2y - 21 \end{array}$$

35. $(-2u^3v^4)^4 = 16u^{12}v^{16}$

36.
$$\begin{array}{r} 2x^3 + 7x^2 + x \\ + 2x^2 - 4x - 12 \\ \hline 2x^3 + 9x^2 - 3x - 12 \end{array}$$

37. $(5x^2 - 2x - 1) - (3x^2 - 5x + 7)$
$= (5x^2 - 2x - 1) + (-3x^2 + 5x - 7)$
$= 2x^2 + 3x - 8$

38. $(a + 7)(a - 7) = a^2 - 7a + 7a - 49 = a^2 - 49$

39. $(5a^7b^6)^2(4ab) = (25a^{14}b^{12})(4ab) = 100a^{15}b^{13}$

40. $1.46 \times 10^7 = 14,600,000$

41. $(-2x^3)^2(-3x^4)^3 = (4x^6)(-27x^{12}) = -108x^{18}$

42.
$$\begin{array}{r} 2y-9 \\ 3y-4\overline{)6y^2-35y+36} \\ \underline{6y^2-8y} \\ -27y+36 \\ \underline{-27y+36} \\ 0 \end{array}$$

$$\frac{6y^2-35y+36}{3y-4} = 2y-9$$

43. $-4^{-2} = -\dfrac{1}{4^2} = -\dfrac{1}{16}$

44. $(5a - 7)(2a + 9) = 10a^2 + 45a - 14a - 63$
$= 10a^2 + 31a - 63$

45.
$$\begin{array}{r|rrr} -3 & -1 & -1 & 7 \\ & & 3 & -6 \\ \hline & -1 & 2 & 1 \end{array}$$

$$\frac{7-x-x^2}{x+3} = -x+2+\frac{1}{x+3}$$

46. $0.000000127 = 1.27 \times 10^{-7}$

47. $\dfrac{16y^2-32y}{-4y} = \dfrac{16y^2}{-4y} + \dfrac{32y}{4y} = -4y+8$

48. $\dfrac{(2a^4b^{-3}c^2)^3}{(2a^3b^2c^{-1})^4} = \dfrac{8a^{12}b^{-9}c^6}{16a^{12}b^8c^{-4}} = \dfrac{c^{10}}{2b^{17}}$

49. $(x - 4)(3x + 2)(2x - 3)$
$= (x - 4)(6x^2 - 5x - 6)$
$= x(6x^2 - 5x - 6) - 4(6x^2 - 5x - 6)$
$= 6x^3 - 5x^2 - 6x - 24x^2 + 20x + 24$
$= 6x^3 - 29x^2 + 14x + 24$

50. $(-3x^{-2}y^{-3})^{-2} = \dfrac{x^4y^6}{9}$

51. $(2a^{12}b^3)(-9b^2c^6)(3ac) = -54a^{13}b^5c^7$

52. $(5a + 2b)(5a - 2b) = 25a^2 - 4b^2$

53. 0.00254

54. $2ab^3(4a^2 - 2ab - 3b^2) = 8a^3b^3 - 4a^2b^4 + 6ab^5$

55. $y = x^2 + 1$

x	x
−2	5
−1	2
0	1
1	2
2	5

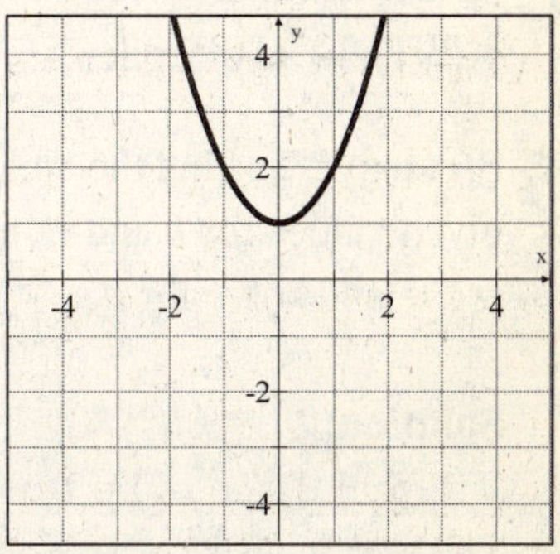

56. a) 3
b) 8
c) 5

57. Strategy: To find the mass of the moon multiply the mass of the sun by 3.7×10^{-8}.

Solution:
$(2.19\times10^{27})(3.7\times10^{-8}) = 8.103\times10^{19}$
The mass of the moon is 8.103×10^{19} tons.

58. Strategy: Let $3x - 2$ represent the side of a square. Find the area of the square.

Solution: $S^2 = A$
$(3x - 2)^2 = A$
$9x^2 - 6x - 6x + 4 = A$
$9x^2 - 12x + 4 = A$
The area is $(9x^2 - 12x + 4)$ in^2.

59. Strategy: To find how far Earth is from the Great Galaxy of Andromeda, use the equation $d = rt$ where $r = 6.7\times10^8$ mph and $t = 2.2\times10^6$ years.
$2.2\times10^6\times24\times365 = 1.9272\times10^{10}$ hours.

Solution: $d=rt$
$d = (6.7\times10^8)(1.9272\times10^{10})$
$= 1.291224\times10^{19}$
The distance from the Earth to the Great Galaxy of Andromeda is 1.291224×10^{19} miles.

60. Strategy: To find the area substitute the given values for L and W in the equation $A = LW$ and solve for A.

Solution: $A = LW$
$A = (5x + 3)(2x - 7) = 10x^2 - 29x - 21$
The area is $(10x^2 - 29x - 21)$ cm^2.

Chapter 6 Test

1. $2x(2x^2 - 3x) = 4x^3 - 6x^2$

2.
$$\begin{array}{r|rrrr} -2 & -1 & 0 & 4 & -8 \\ & & 2 & -4 & 0 \\ \hline & -1 & 2 & 0 & -8 \end{array}$$

$P(-2) = -8$

3. $\dfrac{12x^2}{-3x^8} = -\dfrac{4}{x^6}$

4. $(-2xy^2)(3x^2y^4) = -6x^3y^6$

5.
$$\begin{array}{r} x-1 \\ x+1\overline{)x^2+0+1} \\ \underline{x^2+x} \\ -x+1 \\ \underline{-x-1} \\ 2 \end{array}$$

$$\frac{x^2+1}{x+1} = x - 1 + \frac{2}{x+1}$$

6. $(x - 3)(x^2 - 4x + 5)$
$= x^3 - 4x^2 + 5x - 3x^2 + 12x - 15$
$= x^3 - 7x^2 + 17x - 15$

7. $(-2a^2b)^3 = -8a^6b^3$

8. $\dfrac{(3x^{-2}y^3)^3}{3x^4y^{-1}} = \dfrac{27x^{-6}y^9}{3x^4y^{-1}} = \dfrac{9y^{10}}{x^{10}}$

9. $(a - 2b)(a + 5b) = a^2 + 5ab - 2ab - 10b^2$
$= a^2 + 3ab - 10b^2$

10. $P(x) = 3x^2 - 8x + 1$
$P(2) = 3(2)^2 - 8(2) + 1$
$P(2) = 12 - 16 + 1$
$P(2) = -3$

11.
$$\begin{array}{r} x+7 \\ x-1\overline{)x^2+6x-7} \\ \underline{x^2-1x} \\ 7x-7 \\ \underline{7x-7} \end{array}$$

$$\frac{x^2+6x-7}{x-1} = x+7$$

12. $-3y^2(-2y^2+3y-6)=6y^4-9y^3+18y^2$

13. $(-2x^3+x^2-7)(2x-3)$
$=-4x^4+6x^3+2x^3-3x^2-14x+21$
$=-4x^4+8x^3-3x^2-14x+21$

14. $(4y-3)(4y+3)=16y^2+12y-12y-9$
$=16y^2-9$

15. $\dfrac{18x^5+9x^4-6x^3}{3x^2}=6x^3+3x^2-2x$

16. $\dfrac{2a^{-1}b}{2^{-2}a^{-2}b^{-3}}=2^3ab^4$

17. $\dfrac{(2a^{-4}b^2)^3}{4a^{-2}b^{-1}}=\dfrac{8a^{-12}b^6}{4a^{-2}b^{-1}}=\dfrac{2b^7}{a^{10}}$

18. $(3a^2-2a-7)-(5a^3+2a-10)$
$=-5a^3+3a^2-4a+3$

19. $(2x-5)^2=(2x-5)(2x-5)$
$=4x^2-10x-10x+25$
$=4x^2-20x+25$

20.
$$\begin{array}{r}
x^2-5x+10 \\
x+3\,\overline{)\,x^3-2x^2-5x+7} \\
\underline{x^3+3x^2} \\
-5x^2-5x \\
\underline{-5x^2-15x} \\
10x+7 \\
\underline{10x+30} \\
-23
\end{array}$$

$\dfrac{x^3-2x^2-5x+7}{x+3}=x^2-5x+10-\dfrac{23}{x+3}$

21. $(2x-7y)(5x-4y)$
$=10x^2-8xy-35xy+28y^2$
$=10x^2-43xy+28y^2$

22. $(3x^3-2x^2-4)+(8x^2-8x+7)$
$=3x^3+6x^2-8x+3$

23. $0.00000000302=3.02\times10^{-9}$

24. $10\ weeks\cdot\dfrac{7\ day}{1\ week}\cdot\dfrac{24\ h}{1\ day}\cdot\dfrac{60\text{ min}}{1\ h}\cdot\dfrac{60\ s}{1\text{ min}}$
$=6.048\times10^6$ s
There are 6.048×10^6 s in 10 weeks.

25. $r=(x-5)$
$A=\pi r^2=\pi(x-5)^2=\pi(x-5)(x-5)$
$=\pi(x^2-5x-5x+25)$
$=\pi(x^2-10x+25)$
The area of the circle is $\pi(x^2-10x+25)$ m^2.

Cumulative Review Exercises

1. $-8>-3$ False
$-3>-3$ True
$3>-3$ True
The inequality is true for −3 and 3.

2. The additive inverse of 83 is −83.

3. $8-2[-3-(-1)]^2\div4$
$=8-2(-3+1)^2\div4$
$=8-2(-2)^2\div4$
$=8-2(4)\div4$
$=8-2$
$=6$

4. $\dfrac{2a-b}{b-c}$

$\dfrac{2(4)-(-2)}{(-2)-6}=\dfrac{8+2}{-8}=\dfrac{10}{-8}=-\dfrac{5}{4}$

5. $-5\sqrt{300}=-5\sqrt{2^2\cdot5^2\cdot3}$
$=-5\cdot2\cdot5\sqrt{3}=-50\sqrt{3}$

6. Inverse Property of Addition

7. $2x-4[x-2(3-23x)+4]$
$=2x-4(x-6+46x+4)$
$=2x-4(47x-2)$
$=2x-188x+8$
$=-186x+8$

8. $\frac{2}{3} - y = \frac{5}{6}$

$\frac{2}{3} - y - \frac{2}{3} = \frac{5}{6} - \frac{2}{3}$

$-y = \frac{1}{6}$

$(-1)(-y) = -\frac{1}{6}(-1)$

$y = -\frac{1}{6}$

The solution is $-\frac{1}{6}$.

9. $8x - 3 - x = -6 + 3x - 8$

$7x - 3 = 3x - 14$

$4x - 3 = -14$

$4x = -11$

$x = -\frac{11}{4}$

The solution is $-\frac{11}{4}$.

10. $3 - |2 - 3x| = -2$

$-|2 - 3x| = -5$

$|2 - 3x| = 5$

$2 - 3x = 5 \quad 2 - 3x = -5$

$-3x = 3 \quad -3x = -7$

$x = -1 \quad x = \frac{7}{3}$

The solutions are -1 and $\frac{7}{3}$.

11. $P(x) = 3x^2 - 2x + 2$

$P(-2) = 3(-2)^2 - 2(-2) + 2$

$P(-2) = 3(4) + 4 + 2$

$P(-2) = 18$

12. The relation is a function. No two ordered pairs have the same first coordinate and different second coordinates.

13. $m = \frac{y_2 - y_1}{x_2 - x_1} = \frac{2-3}{4-(-2)} = -\frac{1}{6}$

14. Use the point-slope formula

$y - y_1 = m(x - x_1)$

$y - 2 = -\frac{3}{2}[x - (-1)]$

$y - 2 = -\frac{3}{2}x - \frac{3}{2}$

$y = -\frac{3}{2}x + \frac{1}{2}$

15. Solve the equation $3x + 2y = 4$ for y to find the slope of this line.

$3x + 2y = 4$

$2y = -3x + 4$

$y = -\frac{3}{2}x + 2$

$m = -\frac{3}{2}$

The perpendicular line will have a slope that is the negative reciprocal of $-\frac{3}{2}$.

$m = \frac{2}{3}$ and $(-2, 4)$

$y - y_1 = m(x - x_1)$

$y - 4 = \frac{2}{3}[x - (-2)]$

$y - 4 = \frac{2}{3}x + \frac{4}{3}$

$y = \frac{2}{3}x + \frac{16}{3}$

The equation of the perpendicular line is $y = \frac{2}{3}x + \frac{16}{3}$.

16. $2x - 3y = 4$
$x + y = -3$
$x = -y - 3$
$2(-y - 3) - 3y = 4$
$-2y - 6 - 3y = 4$
$-5y = 10$
$y = -2$
$x + (-2) = -3$
$x - 2 = -3$
$x = -1$
The solution is $(-1, -2)$.

17. (1) $x - y + z = 0$
(2) $2x + y - 3z = -7$
(3) $-x + 2y + 2z = 5$
Add equations (1) and (3) to eliminate x.
$x - y + z = 0$
$-x + 2y + 2z = 5$

(4) $y + 3z = 5$
Add −2 times equation (1) and equation (2) to eliminate x.
$-2x + 2y - 2z = 0$
$2x + y - 3z = -7$

(5) $3y - 5z = -7$
Add −3 times equation (4) to equation (5) to eliminate y.
$-3y - 9z = -15$
$3y - 5z = -7$
$-14z = -22$
$z = \frac{11}{7}$
Substitute $\frac{11}{7}$ for z in equation (4).
$y + 3(\frac{11}{7}) = 5$
$y = \frac{2}{7}$
Substitute in values for y and z.
$x - y + z = 0$
$x - \frac{2}{7} + \frac{11}{7} = 0$
$x = -\frac{9}{7}$
The solution is $\left(-\frac{9}{7}, \frac{2}{7}, \frac{11}{7}\right)$.

18. $-2x - (-xy) + 7x - 4xy = -2x + 7x + xy - 4xy$
$= 5x - 3xy$

19.
$$\begin{array}{r} 2x^2 - 3x + 1 \\ \times \quad 2x + 3 \\ \hline 6x^2 - 9x + 3 \\ 4x^3 - 6x^2 + 2x \quad \\ \hline 4x^3 \qquad - 7x + 3 \end{array}$$

20. $0.00000501 = 5.01 \times 10^{-6}$

21. $3x - 4y = 12$

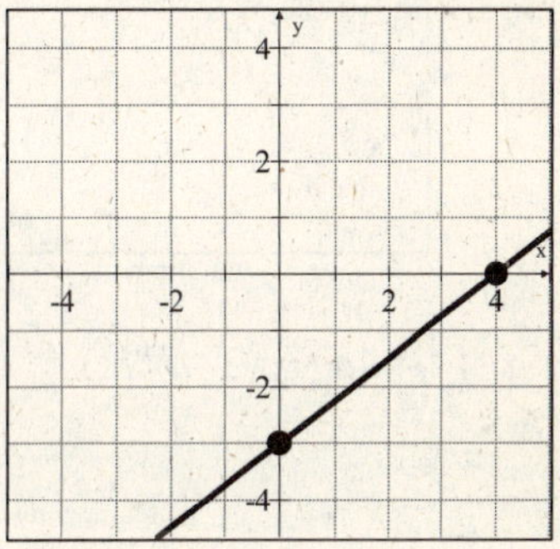

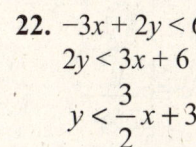

22. $-3x + 2y < 6$
$2y < 3x + 6$
$y < \frac{3}{2}x + 3$

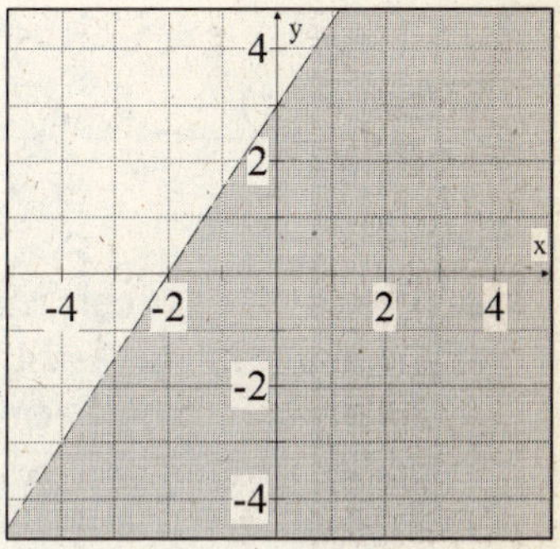

23. $x - 2y = 3$
$-2y = -x + 3$
$y = \frac{1}{2}x - \frac{3}{2}$
$-2x + y = -3$
$y = 2x - 3$

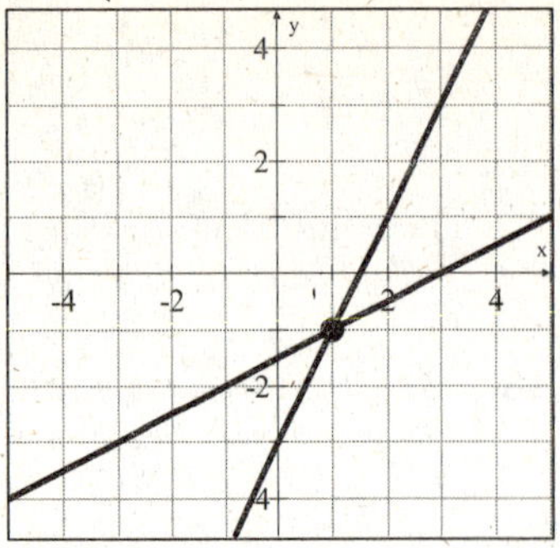

The solution is $(1, -1)$.

24. Solve each inequality for y.
$2x + y < 2$
$y < -2x + 2$

$-6x + 3y \geq 6$
$3y \geq 6x + 6$
$y \geq 2x + 2$

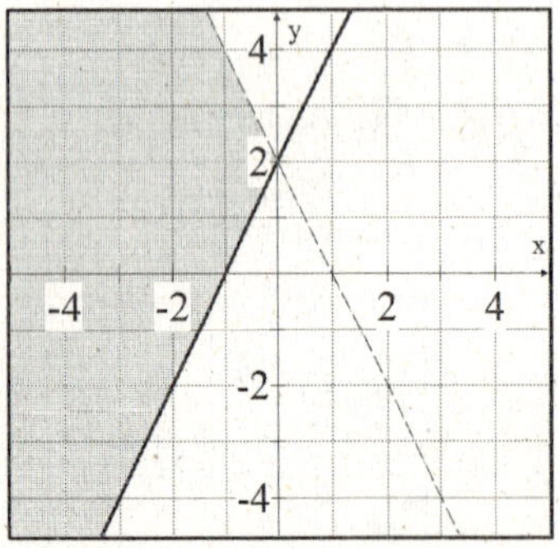

25. $(4a^{-2}b^3)(2a^3b^{-1})^{-2} = 4a^{-2}b^3(2^{-2}a^{-6}b^2)$
$= 4(2^{-2})a^{-8}b^5$
$= \frac{b^5}{a^8}$

26. $\frac{(5x^3y^{-3}z)^{-2}}{y^4z^{-2}} = \frac{5^{-2}x^{-6}y^6z^{-2}}{y^4z^{-2}} = \frac{y^2}{25x^6}$

27. Strategy Let x represent the smaller integer.
The larger integer is $24 - x$.
The difference between four times the smaller and nine is 3 less than twice the larger.
$4x - 9 = 2(24 - x) - 3$

Solution: $4x - 9 = 2(24 - x) - 3$
$4x - 9 = 48 - 2x - 3$
$4x - 9 = 45 - 2x$
$6x - 9 = 45$
$6x = 54$
$x = 9$
$24 - x = 24 - 9 = 15$
The integers are 9 and 15.

28. Strategy: Let x represent the number of ounces of pure gold.

	Amount	Cost	Value
Pure gold	x	360	$360x$
Alloy	80	120	80(120)
Mixture	$x + 80$	200	$200(x + 80)$

The sum of values before mixing is equal to the value after mixing.

Solution: $360x + 80(120) = 200(x + 80)$
$360x + 9600 = 200x + 16{,}000$
$160x + 9600 = 16{,}000$
$160x = 6400$
$x = 40$
40oz of pure gold must be mixed with the alloy.

29. Strategy: Let x represent the speed of the slower cyclist.
The speed of the faster cyclist is $1.5x$.

	Rate	Time	Distance
Faster cyclist	$1.5x$	2	$2(1.5x)$
Slower cyclist	x	2	$2x$

The sum of the distances is 25 mi.

Solution: $2x + 2(1.5x) = 25$
$2x + 3x = 25$
$5x = 25$
$x = 5$
$1.5x = 1.5(5) = 7.5$
The faster cyclist travels at 7.5 mph and the slower cyclist travels at 5 mph.

30. Strategy: Let x represent the amount invested at 4%.

	Principal	Rate	Interest
Amount at 4%	x	0.04	$0.04x$
Amount at 4.5%	$12{,}000 - x$	0.045	$0.045(12{,}000 - x)$

The total amount of interest earned is \$530.

Solution:
$$\begin{aligned} 0.045(12{,}000 - x) + 0.04x &= 530 \\ 540 - 0.045x + 0.04x &= 530 \\ -0.005x &= -10 \\ x &= 2000 \end{aligned}$$
The amount invested at 4% is \$2000.

31. $m = \dfrac{y_2 - y_1}{x_2 - x_1} = \dfrac{300 - 100}{6 - 2} = \dfrac{200}{4} = 50$

The average speed is 50 mph.

32. Strategy: Let x represent the length. The width is $0.40x$.
Use the formula for perimeter of a rectangle.

Solution:
$$\begin{aligned} 2L + 2W &= P \\ 2x + 2(0.40x) &= 42 \\ 2x + 0.80x &= 42 \\ 2.8x &= 42 \\ x &= 15 \end{aligned}$$
$0.40(x) = 0.40(15) = 6$
The length is 15 m and the width is 6 m.

33.
$$\begin{aligned} A &= s^2 \\ A &= (2x + 3)^2 \\ A &= (2x + 3)(2x + 3) \\ A &= 4x^2 + 6x + 6x + 9 \\ A &= 4x^2 + 12x + 9 \end{aligned}$$
The area is $(4x^2 + 12x + 9)$ m^2.

Chapter 7 Factoring

Prep Test

1. $30 = 2 \cdot 3 \cdot 5$

2. $-3(4y-5) = -12y+15$

3. $-(a-b) = -a+b$

4. $$\begin{aligned} 2(a-b)-5(a-b) &= 2a-2b-5a+5b \\ &= -3a+3b \end{aligned}$$

5. $$\begin{aligned} 4x &= 0 \\ \frac{4x}{4} &= \frac{0}{4} \\ x &= 0 \end{aligned}$$

The solution is 0.

6. $$\begin{aligned} 2x+1 &= 0 \\ 2x+1-1 &= 0-1 \\ 2x &= -1 \\ \frac{2x}{2} &= \frac{-1}{2} \\ x &= -\frac{1}{2} \end{aligned}$$

The solution is $-\frac{1}{2}$.

7. $$\begin{aligned} (x+4)(x-6) &= x^2-6x+4x-24 \\ &= x^2-2x-24 \end{aligned}$$

8. $$\begin{aligned} (2x-5)(3x+2) &= 6x^2+4x-15x-10 \\ &= 6x^2-11x-10 \end{aligned}$$

9. $\frac{x^5}{x^2} = x^{5-2} = x^3$

10. $\frac{6x^4y^3}{2xy^2} = 3x^{4-1}y^{3-2} = 3x^3y$

Section 7.1

Objective A Exercises

1. A common monomial factor is a monomial that is a factor of each term of a polynomial.

3. $5a+5 = 5(a)+5(1) = 5(a+1)$

5. $16-8a^2 = 8(2)+8(-a^2) = 8(2-a^2)$

7. $8x+12 = 4(2x)+4(3) = 4(2x+3)$

9. $30a-6 = 6(5a)+6(-1) = 6(5a-1)$

11. $7x^2-3x = x(7x)+x(-3) = x(7x-3)$

13. $3a^2+5a^5 = a^2(3)+a^2(5a^3) = a^2(3+5a^3)$

15. $14y^2+11y = y(14y)+y(11) = y(14y+11)$

17. $2x^4-4x = 2x(x^3)+2x(-2) = 2x(x^3-2)$

19. $10x^4-12x^2 = 2x^2(5x^2)+2x^2(-6) = 2x^2(5x^2-6)$

21. $8a^8-4a^5 = 4a^5(2a^3)+4a^5(-1) = 4a^5(2a^3-1)$

23. $x^2y^2-xy = xy(xy)+xy(-1) = xy(xy-1)$

25. $3x^2y^4-6xy = 3xy(xy^3)+3xy(-2) = 3xy(xy^3-2)$

27. $x^2y-xy^3 = xy(x)+xy(-y^2) = xy(x-y^2)$

29. $$\begin{aligned} 5y^3-20y^2+5y &= 5y(y^2)+5y(-4y)+5y(1) \\ &= 5y(y^2-4y+1) \end{aligned}$$

31. $$\begin{aligned} 3y^4-9y^3-6y^2 &= 3y^2(y^2)+3y^2(-3y)+3y^2(-2) \\ &= 3y^2(y^2-3y-2) \end{aligned}$$

33. $$\begin{aligned} 3y^3-9y^2+24y &= 3y(y^2)+3y(-3y)+3y(8) \\ &= 3y(y^2-3y+8) \end{aligned}$$

35. $$\begin{aligned} 6a^5-3a^3-2a^2 &= a^2(6a^3)+a^2(-3a)+a^2(-2) \\ &= a^2(6a^3-3a-2) \end{aligned}$$

37. $$\begin{aligned} &2a^2b-5a^2b^2+7ab^2 \\ &= ab(2a)+ab(-5ab)+ab(7b) \\ &= ab(2a-5ab+7b) \end{aligned}$$

39. $$\begin{aligned} &4b^5+6b^3-12b = 2b(2b^4)+2b(3b^2)+2b(-6) \\ &= 2b(2b^4+3b^2-6) \end{aligned}$$

41. $$\begin{aligned} 8x^2y^2-4x^2y+x^2 &= x^2(8y^2)+x^2(-4y)+x^2(1) \\ &= x^2(8y^2-4y+1) \end{aligned}$$

Objective B Exercises

43.a. (i) $x^2-15x+10x+6=x^2-5x+6$ Yes

(ii) $x^2-x-4x+6=x^2-5x+6$ Yes

(iii) $x^2-2x-3x+6=x^2-5x+6$ Yes

43. b. (i) $x^2-15x+10x+6=(x^2-15x)+(10x+6)$

$=x(x-15)+2(5x+3)$ No

(ii) $x^2-x-4x+6=(x^2-x)+(-4x+6)$

$=x(x-1)-2(x-3)$ No

(iii) $x^2-2x-3x+6=(x^2-2x)+(-3x+6)$

$=x(x-2)-3(x-2)$

$=(x-2)(x-3)$ Yes

45. $y(a+z)+7(a+z)=(a+z)(y+7)$

47. $3r(a-b)+s(a-b)=(a-b)(3r+s)$

49. $t(m-7)+7(7-m)=t(m-7)-7(m-7)$

$=(m-7)(t-7)$

51. $2y(4a+b)-(b+4a)$

$=2y(4a+b)-1(4a+b)$

$=(4a+b)(2y-1)$

53. $x^2+2x+2xy+4y=(x^2+2x)+(2xy+4y)$

$=x(x+2)+2y(x+2)$

$=(x+2)(x+2y)$

55. $p^2-2p-3rp+6r=(p^2-2p)+(-3rp+6r)$

$=p(p-2)-3r(p-2)$

$=(p-2)(p-3r)$

57. $ab+6b-4a-24=(ab+6b)+(-4a-24)$

$=b(a+6)-4(a+6)$

$=(a+6)(b-4)$

59. $2z^2-z+2yz-y=(2z^2-z)+(2yz-y)$

$=z(2z-1)+y(2z-1)$

$=(2z-1)(z+y)$

61. $8v^2-12vy+14v-21y=(8v^2-12vy)+(14v-21y)$

$=4v(2v-3y)+7(2v-3y)$

$=(2v-3y)(4v+7)$

63. $2x^2-5x-6xy+15y=(2x^2-5x)+(-6xy+15y)$

$=x(2x-5)-3y(2x-5)$

$=(2x-5)(x-3y)$

65. $3y^2-6y-ay+2a=(3y^2-6y)+(-ay+2a)$

$=3y(y-2)-a(y-2)$

$=(y-2)(3y-a)$

67. $3xy-y^2-y+3x=(3xy-y^2)+(-y+3x)$

$=y(3x-y)+1(-y+3x)$

$=y(3x-y)+1(3x-y)$

$=(3x-y)(y+1)$

69. $3st+t^2-2t-6s=(3st+t^2)+(-2t-6s)$

$=t(3s+t)-2(t+3s)$

$=t(3s+t)-2(3s+t)$

$=(3s+t)(t-2)$

Applying the Concepts

71. $P=2L+2W=2(L+W)$

When $L+W$ doubles, P also doubles.

Section 7.2

Objective A Exercises

1.

Factors	*Sum*
+1, +2	+3

$x^2+3x+2=(x+1)(x+2)$

3.

Factors	*Sum*
−1, +2	+1
+1, −2	−1

$x^2-x-2=(x+1)(x-2)$

5.

Factors	*Sum*
−1, +12	+11
+1, −12	−11
−2, +6	+4
+2, −6	−4
−3, +4	+1
+3, −4	−1

$a^2+a-12=(a+4)(a-3)$

7.

Factors	*Sum*
−1, −2	−3

$a^2 - 3a + 2 = (a-1)(a-2)$

9.

Factors	*Sum*
−1, +2	+1
+1, −2	−1

$a^2 + a - 2 = (a+2)(a-1)$

11.

Factors	*Sum*
−1, −9	−10
−3, −3	−6

$b^2 - 6b + 9 = (b-3)(b-3)$

13.

Factors	*Sum*
−1, +8	+7
+1, −8	−7
−2, +4	+2
+2, −4	−2

$b^2 + 7b - 8 = (b+8)(b-1)$

15.

Factors	*Sum*
−1, +55	+54
+1, −55	−54
−5, +11	+6
+5, −11	−6

$y^2 + 6y - 55 = (y+11)(y-5)$

17.

Factors	*Sum*
−1, −6	−7
−2, −3	−5

$y^2 - 5y + 6 = (y-2)(y-3)$

19.

Factors	*Sum*
−1, −45	−46
−3, −15	−18
−5, −9	−14

$z^2 - 14z + 45 = (z-5)(z-9)$

21.

Factors	*Sum*
−1, +160	+159
+1, −160	−159
−2, +80	+78
+2, −80	−78
−4, +40	+36
+4, −40	−36
−5, +32	+27
+5, −32	−27
−8, +20	+12
+8, −20	−12
−10, +16	+6
+10, −16	−6

$z^2 - 12z - 160 = (z+8)(z-20)$

23.

Factors	*Sum*
+1, +27	+28
+3, +9	+12

$p^2 + 12p + 27 = (p+3)(p+9)$

25.

Factors	*Sum*
+1, +100	+101
+2, +50	+52
+4, +25	+29
+5, +20	+25
+10, +10	+20

$x^2 + 20x + 100 = (x+10)(x+10)$

27.

Factors	*Sum*
+1, +20	+21
+2, +10	+12
+4, +5	+9

$b^2 + 9b + 20 = (b+4)(b+5)$

29.

Factors	*Sum*
−1, +42	+41
+1, −42	−41
−2, +21	+19
+2, −21	−19
−3, +14	+11
+3, −14	−11
−6, +7	+1
+6, −7	−1

$x^2 - 11x - 42 = (x+3)(x-14)$

31.

Factors	*Sum*
−1, +20	+19
+1, −20	−19
−2, +10	+8
+2, −10	−8
−4, +5	+1
+4, −5	−1

$b^2 - b - 20 = (b+4)(b-5)$

33.

Factors	*Sum*
−1, +51	+50
+1, −51	−50
−3, +17	+14
+3, −17	−14

$y^2 - 14y - 51 = (y+3)(y-17)$

35.

Factors	*Sum*
−1, +21	+20
+1, −21	−20
−3, +7	+4
+3, −7	−4

$p^2 - 4p - 21 = (p+3)(p-7)$

37.

Factors	*Sum*
−1, −32	−33
−2, −16	−18
−4, −8	−12

$y^2 - 8y + 32$ is nonfactorable over the integers.

39.

Factors	*Sum*
−1, −75	−76
−3, −25	−28
−5, −15	−20

$x^2 - 20x + 75 = (x - 5)(x - 15)$

41.

Factors	*Sum*
+1, +63	+64
+3, +21	+24

$p^2 + 24p + 63 = (p + 3)(p + 21)$

43.

Factors	*Sum*
+1, +38	+39
+2, +19	+21

$x^2 + 21x + 38 = (x + 2)(x + 19)$

45.

Factors	*Sum*
−1, +36	+35
+1, −36	−35
−2, +18	+16
+2, −18	−16
−3, +12	+9
+3, −12	−9
−4, +9	+5
+4, −9	−5
−6, +6	0

$x^2 + 5x - 36 = (x + 9)(x - 4)$

47.

Factors	*Sum*
−1, +44	+43
+1, −44	−43
−2, +22	+20
+2, −22	−20
−4, +11	+7
+4, −11	−7

$a^2 - 7a - 44 = (a + 4)(a - 11)$

49.

Factors	*Sum*
−1, −54	−55
−2, −27	−29
−3, −18	−21
−6, −9	−15

$a^2 - 21a + 54 = (a - 3)(a - 18)$

51.

Factors	*Sum*
−1, +147	+146
+1, −147	−146
−3, +49	+46
+3, −49	−46
−7, +21	+14
+7, −21	−14

$z^2 + 14z - 147 = (z + 21)(z - 7)$

53.

Factors	*Sum*
−1, +180	+179
+1, −180	−179
−2, +90	+88
+2, −90	−88
−3, +60	+57
+3, −60	−57
−4, +45	+41
+4, −45	−41
−5, +36	+31
+5, −36	−31
−6, +30	+24
+6, −30	−24
−9, +20	+11
+9, −20	−11
−10, +18	+8
+10, −18	−8
−12, +15	+3
+12, −15	−3

$c^2 - 3c - 180 = (c + 12)(c - 15)$

55.

Factors	*Sum*
+1, +135	+136
+3, +45	+48
+5, +27	+32
+9, +15	+24

$p^2 + 24p + 135 = (p + 9)(p + 15)$

57.

Factors	*Sum*
+1, +18	+19
+2, +9	+11
+3, +6	+9

$c^2 + 11c + 18 = (c + 2)(c + 9)$

59.

Factors	*Sum*
$-1, +75$	$+74$
$+1, -75$	-74
$-3, +25$	$+22$
$+3, -25$	-22
$-5, +15$	$+10$
$+5, -15$	-10

$x^2 + 10x - 75 = (x + 15)(x - 5)$

61.

Factors	*Sum*
$-1, +100$	$+99$
$+1, -100$	-99
$-2, +50$	$+48$
$+2, -50$	-48
$-4, +25$	$+21$
$+4, -25$	-21
$-5, +20$	$+15$
$+5, -20$	-15
$-10, +10$	0

$x^2 + 21x - 100 = (x + 25)(x - 4)$

63.

Factors	*Sum*
$-1, -72$	-73
$-2, -36$	-38
$-3, -24$	-27
$-4, -18$	-22
$-6, -12$	-18
$-8, -9$	-17

$b^2 - 22b + 72 = (b - 4)(b - 18)$

65.

Factors	*Sum*
$-1, +135$	$+134$
$+1, -135$	-134
$-3, +45$	$+42$
$+3, -45$	-42
$-5, +27$	$+22$
$+5, -27$	-22
$-9, +15$	$+6$
$+9, -15$	-6

$a^2 + 42a - 135 = (a + 45)(a - 3)$

67.

Factors	*Sum*
$-1, -126$	-127
$-2, -63$	-65
$-3, -42$	-45
$-6, -21$	-27
$-7, -18$	-25

$b^2 - 25b + 126 = (b - 7)(b - 18)$

69.

Factors	*Sum*
$+1, +144$	$+145$
$+2, +72$	$+74$
$+3, +48$	$+51$
$+4, +36$	$+40$
$+6, +24$	$+30$
$+8, +18$	$+26$
$+9, +16$	$+25$
$+12, +12$	$+24$

$z^2 + 24z + 144 = (z + 12)(z + 12)$

71.

Factors	*Sum*
$-1, -100$	-101
$-2, -50$	-52
$-4, -25$	-29
$-5, -20$	-25
$-10, -10$	-20

$x^2 - 29x + 100 = (x - 4)(x - 25)$

73.

Factors	*Sum*
$-1, +112$	$+111$
$+1, -112$	-111
$-2, +56$	$+54$
$+2, -56$	-54
$-4, +28$	$+24$
$+4, -28$	-24
$-7, +16$	$+9$
$+7, -16$	-9
$-8, +14$	$+6$
$+8, -14$	-6

$x^2 + 9x - 112 = (x + 16)(x - 7)$

75. Positive. The sum of two positive numbers is positive.

Objective B Exercises

77. The GCF is 3.

$3x^2+15x+18=3(x^2+5x+6)$

Factor the trinomial

x^2+5x+6.

Factors	*Sum*
+1, +6	+7
+2, +3	+5

$3x^2+15x+18=3(x+2)(x+3)$

79. The GCF is −1.

$-x^2-4x+12=-(x^2+4x-12)$

Factor the trinomial $x^2+4x-12$.

Factors	*Sum*
−1, +12	+11
+1, −12	−11
−2, +6	+4
+2, −6	−4

$12-4x-x^2=-(x-2)(x+6)$

81. The GCF is *a*.

$ab^2+7ab-8a=a(b^2+7b-8)$

Factor the trinomial b^2+7b-8.

Factors	*Sum*
−1, +8	+7
+1, −8	−7
−2, +4	+2
+2, −4	−2

$ab^2+7ab-8a=a(b+8)(b-1)$

83. The GCF is *x*.

$xy^2+8xy+15x=x(y^2+8y+15)$

Factor the trinomial

$y^2+8y+15$.

Factors	*Sum*
+1, +15	+16
+3, +5	+8

$xy^2+8xy+15x=x(y+3)(y+5)$

85. The GCF is $-2a$.

$-2a^3-6a^2-4a=-2a(a^2+3a+2)$

Factor the trinomial a^2+3a+2.

Factors	*Sum*
+1, +2	+3

$-2a^3-6a^2-4a=-2a(a+1)(a+2)$

87. The GCF is $4y$.

$4y^3+12y^2-72y=4y(y^2+3y-18)$

Factor the trinomial $y^2+3y-18$.

Factors	*Sum*
−1, +18	+17
+1, −18	−17
−2, +9	+7
+2, −9	−7
−3, +6	+3
+3, −6	−3

$4y^3+12y^2-72y=4y(y+6)(y-3)$

89. The GCF is $2x$.

$2x^3-2x^2+4x=2x(x^2-x+2)$

Factor the trinomial

x^2-x+2.

Factors	*Sum*
−1, −2	−3

x^2-x+2 is nonfactorable over the integers.

$2x^3-2x^2+4x=2x(x^2-x+2)$

91. The GCF is 6.

$6z^2+12z-90=6(z^2+2z-15)$

Factor the trinomial $z^2+2z-15$.

Factors	*Sum*
−1, +15	+14
+1, −15	−14
−3, +5	+2
+3, −5	−2

$6z^2+12z-90=6(z+5)(z-3)$

93. The GCF is $3a$.

$3a^3-9a^2-54a=3a(a^2-3a-18)$

Factor the trinomial

$a^2-3a-18$.

Factors	*Sum*
−1, +18	+17
+1, −18	−17
−2, +9	+7
+2, −9	−7
−3, +6	+3
+3, −6	−3

$3a^3-9a^2-54a=3a(a+3)(a-6)$

95. There is no common factor.
Factor the trinomial $x^2+4xy-21y^2$.

Factors	*Sum*
−1, +21	+20
+1, −21	−20
−3, +7	+4
+3, −7	−4

$x^2+4xy-21y^2=(x+7y)(x-3y)$

97. There is no common factor.
Factor the trinomial $a^2-15ab+50b^2$.

Factors	*Sum*
−1, −50	−51
−2, −25	−27
−5, −10	−15

$a^2-15ab+50b^2=(a-5b)(a-10b)$

99. There is no common factor.
Factor the trinomial $s^2+2st-48t^2$.

Factors	*Sum*
−1, +48	+47
+1, −48	−47
−2, +24	+22
+2, −24	−22
−3, +16	+13
+3, −16	−13
−4, +12	+8
+4, −12	−8
−6, +8	+2
+6, −8	−2

$s^2+2st-48t^2=(s+8t)(s-6t)$

101. There is no common factor.
Factor the trinomial $y^2+85yz+36z^2$.

Factors	*Sum*
+1, +36	+37
+2, +18	+20
+3, +12	+15
+4, +9	+13
+6, +6	+12

$y^2+85yz+36z^2$ is nonfactorable over the integers.

103. The GCF is z^2.
$z^4+2z^3-80z^2=z^2(z^2+2z-80)$

Factor the trinomial $z^2+2z-80$.

Factors	*Sum*
−1, +80	+79
+1, −80	−79
−2, +40	+38
+2, −40	−38
−4, +20	+16
+4, −20	−16
−5, +16	+11
+5, −16	−11
−8, +10	+2
+8, −10	−2

$z^4+2z^3-80z^2=z^2(z+10)(z-8)$

105. The GCF is b^2.
$b^4-3b^3-10b^2=b^2(b^2-3b-10)$

Factor the trinomial $b^2-3b-10$.

Factors	*Sum*
−1, +10	+9
+1, −10	−9
−2, +5	+3
+2, −5	−3

$b^4-3b^3-10b^2=b^2(b+2)(b-5)$

107. The GCF is $3y^2$.
$3y^4+54y^3+135y^2=3y^2(y^2+18y+45)$

Factor the trinomial $y^2+18y+45$.

Factors	*Sum*
+1, +45	+46
+3, +15	+18
+5, +9	+14

$3y^4+54y^3+135y^2=3y^2(y+3)(y+15)$

109. The GCF is $-x^2$.
$-x^4+11x^3+12x^2=-x^2(x^2-11x-12)$

Factor the trinomial $x^2-11x-12$.

Factors	*Sum*
−1, +12	+11
+1, −12	−11
−2, +6	+4
+2, −6	−4
−3, +4	+1
+3, −4	−1

$-x^4+11x^3+12x^2=-x^2(x+1)(x-12)$

111. The GCF is $3y$.

$3x^2y - 6xy - 45y = 3y(x^2 - 2x - 15)$

Factor the trinomial $x^2 - 2x - 15$.

Factors	*Sum*
−1, +15	+14
+1, −15	−14
−3, +5	+2
+3, −5	−2

$3x^2y - 6xy - 45y = 3y(x+3)(x-5)$

113. The GCF is $-3x$.

$-3x^3 + 36x^2 - 81x = -3x(x^2 - 12x + 27)$

Factor the trinomial $x^2 - 12x + 27$.

Factors	*Sum*
−1, −27	−28
−3, −9	−12

$-3x^3 + 36x^2 - 81x = -3x(x-3)(x-9)$

115. There is no common factor.

Factor the trinomial $x^2 - 8xy + 15y^2$.

Factors	*Sum*
−1, −15	−16
−3, −5	−8

$x^2 - 8xy + 15y^2 = (x - 3y)(x - 5y)$

117. There is no common factor.

Factor the trinomial $a^2 - 13ab + 42b^2$.

Factors	*Sum*
−1, −42	−43
−2, −21	−23
−3, −14	−17
−6, −7	−13

$a^2 - 13ab + 42b^2 = (a - 6b)(a - 7b)$

119. There is no common factor.

Factor the trinomial $y^2 + 8yz + 7z^2$.

Factors	*Sum*
+1, +7	+8

$y^2 + 8yz + 7z^2 = (y + z)(y + 7z)$

121. The GCF is $3y$.

$3x^2y + 60xy - 63y = 3y(x^2 + 20x - 21)$

Factor the trinomial $x^2 + 20x - 21$.

Factors	*Sum*
−1, +21	+20
+1, −21	−20
−3, +7	+4
+3, −7	−4

$3x^2y + 60xy - 63y = 3y(x + 21)(x - 1)$

123. The GCF is $3x$.

$3x^3 + 3x^2 - 36x = 3x(x^2 + x - 12)$

Factor the trinomial $x^2 + x - 12$.

Factors	*Sum*
−1, +12	+11
+1, −12	−11
−2, +6	+4
+2, −6	−4
−3, +4	+1
+3, −4	−1

$3x^3 + 3x^2 - 36x = 3x(x + 4)(x - 3)$

125. The GCF is 2.

$2t^2 - 24ts + 70s^2 = 2(t^2 - 12ts + 35s^2)$

Factor the trinomial $t^2 - 12ts + 35s^2$.

Factors	*Sum*
−1, −35	−36
−5, −7	−12

$2t^2 - 24ts + 70s^2 = 2(t - 5s)(t - 7s)$

127. The GCF is 3.

$3a^2 - 24ab - 99b^2 = 3(a^2 - 8ab - 33b^2)$

Factor the trinomial $a^2 - 8ab - 33b^2$.

Factors	*Sum*
−1, +33	32
+1, −33	−32
+3, −11	−8
−3, +11	8

$3a^2 - 24ab - 99b^2 = 3(a + 3b)(a - 11b)$

129. The GCF is $5x$.

$5x^3+30x^2y+40xy^2=5x(x^2+6xy+8y^2)$

Factor the trinomial $x^2+6xy+8y^2$.

Factors	*Sum*
+1, +8	+9
+2, +4	+6

$5x^3+30x^2y+40xy^2=5x(x+2y)(x+4y)$

131. a. The GCF is 2.

$2x^2-2xy-4y^2=2(x^2-xy-2y^2)$

Factor the trinomial $x^2-xy-2y^2$.

Factors	*Sum*
+1, −2	−1
+2, −1	1

$2x^2-2xy-4y^2=2(x+y)(x-2y)$

Yes, $x+y$ is a factor.

b. The GCF is $2y$.

$2x^2y-4xy-4y=2y(x^2-2x-2)$

Factor the trinomial x^2-2x-2.

Factors	*Sum*
+1, −2	−1
+2, −1	1

x^2-2x-2 is nonfactorable over the integers.

$2x^2y-4xy-4y=2y(x^2-2x-2)$

No, $x+y$ is not a factor.

Applying the Concepts

133. $x^2+kx+18$. The factors of 18 must sum to k.

Factors	*Sum*
−1, −18	−19
+1, +18	+19
−2, −9	−11
+2, +9	+11
−3, −6	−9
+3, +6	+9

k can be −19, 19, −11, 11, −9, or 9.

135. $y^2+4y+k,\ k>0$

Find two positive integers that sum to 4. Their product is k.

Integers	*Product*
+1, +3	3
+2, +2	4

k can be 3 or 4.

137. $a^2-6a+k,\ k>0$

Find two negative integers that sum to −6. Their product is k.

Integers	*Product*
−1, −5	5
−2, −4	8
−3, −3	9

k can be 5, 8, or 9.

139. $x^2-3x+k,\ k>0$

Find two negative integers that sum to −3. Their product is k.

Integers	*Product*
−2, −1	2

k can be 2.

141. If k is allowed to be any integer, then there will be an infinite number of values of k possible for each polynomial.

Section 7.3

Objective A Exercises

1.

Positive	Positive
Factors of 2: 1, 2	Factors of 1: +1, +1
Trial Factors	*Middle Term*
$(1x+1)(2x+1)$	$x+2x=3x$

$2x^2+3x+1=(x+1)(2x+1)$

3.

Positive	Positive
Factors of 2: 1, 2	Factors of 3: +1, +3
Trial Factors	*Middle Term*
$(1y+1)(2y+3)$	$3y+2y=5y$
$(1y+3)(2y+1)$	$y+6y=7y$

$2y^2+7y+3=(y+3)(2y+1)$

5.

Positive	Negative
Factors of 2: 1, 2	Factors of 1: −1, −1
Trial Factors	*Middle Term*
$(1a-1)(2a-1)$	$-a-2a=-3a$

$2a^2-3a+1=(a-1)(2a-1)$

7.

Positive	Negative
Factors of 2: 1, 2	Factors of 5: −1, −5
Trial Factors	*Middle Term*
$(1b-1)(2b-5)$	$-5b-2b=-7b$
$(1b-5)(2b-1)$	$-b-10b=-11b$

$2b^2-11b+5=(b-5)(2b-1)$

9.

Positive	
Factors of 2: 1, 2	Factors of −1: −1, +1
Trial Factors	*Middle Term*
$(1x-1)(2x+1)$	$x-2x=-x$
$(1x+1)(2x-1)$	$-x+2x=x$

$2x^2+x-1=(x+1)(2x-1)$

11.

Positive	
Factors of 2: 1, 2	Factors of −3: −1, +3
	+1, −3
Trial Factors	*Middle Term*
$(1x-1)(2x+3)$	$3x-2x=x$
$(1x+3)(2x-1)$	$-x+6x=5x$
$(1x+1)(2x-3)$	$-3x+2x=-x$
$(1x-3)(2x+1)$	$x-6x=-5x$

$2x^2-5x-3=(x-3)(2x+1)$

13.

Positive	
Factors of 2: 1, 2	Factors of −10: −1, +10
	+1, −10
	−2, +5
	+2, −5
Trial Factors	*Middle Term*
$(1t-1)(2t+10)$	Common factor
$(1t+10)(2t-1)$	$-t+20t=19t$
$(1t+1)(2t-10)$	Common factor
$(1t-10)(2t+1)$	$t-20t=-19t$
$(1t-2)(2t+5)$	$5t-4t=t$
$(1t+5)(2t-2)$	Common factor
$(1t+2)(2t-5)$	$-5t+4t=-t$
$(1t-5)(2t+2)$	Common factor

$2t^2-t-10=(t+2)(2t-5)$

15.

Positive	Negative
Factors of 3: 1, 3	Factors of 5: −1, −5
Trial Factors	*Middle Term*
$(1p-1)(3p-5)$	$-5p-3p=-8p$
$(1p-5)(3p-1)$	$-p-15p=-16p$

$3p^2-16p+5=(p-5)(3p-1)$

17.

Positive	Negative
Factors of 12: 1, 12	Factors of 1: −1, −1
2, 6	
3, 4	
Trial Factors	*Middle Term*
$(1y-1)(12y-1)$	$-y-12y=-13y$
$(2y-1)(6y-1)$	$-2y-6y=-8y$
$(3y-1)(4y-1)$	$-3y-4y=-7y$

$12y^2-7y+1=(3y-1)(4y-1)$

19.

Positive	Negative
Factors of 6: 1, 6	Factors of 3: −1, −3
2, 3	
Trial Factors	*Middle Term*
$(1z-1)(6z-3)$	Common factor
$(1z-3)(6z-1)$	$-z-18z=-19z$
$(2z-1)(3z-3)$	Common factor
$(2z-3)(3z-1)$	$-2z-9z=-11z$

$6z^2-7z+3$ is nonfactorable over the integers.

21.

Positive	Negative
Factors of 6: 1, 6	Factors of 4: −1, −4
2, 3	−2, −2
Trial Factors	*Middle Term*
$(1t-1)(6t-4)$	Common factor
$(1t-4)(6t-1)$	$-t-24t=-25t$
$(1t-2)(6t-2)$	Common factor
$(2t-1)(3t-4)$	$-8t-3t=-11t$
$(2t-4)(3t-1)$	Common factor
$(2t-2)(3t-2)$	Common factor

$6t^2-11t+4=(2t-1)(3t-4)$

23. Positive Positive

Factors of 8: 1, 8 Factors of 4: +1, +4

2, 4 +2, +2

Trial Factors	*Middle Term*
$(1x+1)(8x+4)$	Common factor
$(1x+4)(8x+1)$	$x+32x=33x$
$(1x+2)(8x+2)$	Common factor
$(2x+1)(4x+4)$	Common factor
$(2x+4)(4x+1)$	Common factor
$(2x+2)(4x+2)$	Common factor

$8x^2+33x+4=(x+4)(8x+1)$

25. Positive

Factors of 5: 1, 5 Factors of −7: −1, +7

+1, −7

Trial Factors	*Middle Term*
$(1x-1)(5x+7)$	$7x-5x=2x$
$(1x+7)(5x-1)$	$-x+35x=34x$
$(1x+1)(5x-7)$	$-7x+5x=-2x$
$(1x-7)(5x+1)$	$x-35x=-34x$

$5x^2-62x-7$ is nonfactorable over the integers.

27. Positive Positive

Factors of 12: 1, 12 Factors of 5: +1, +5

2, 6

3, 4

Trial Factors	*Middle Term*
$(1y+1)(12y+5)$	$5y+12y=17y$
$(1y+5)(12y+1)$	$y+60y=61y$
$(2y+1)(6y+5)$	$10y+6y=16y$
$(2y+5)(6y+1)$	$2y+30y=32y$
$(3y+1)(4y+5)$	$15y+4y=19y$
$(3y+5)(4y+1)$	$3y+20y=23y$

$12y^2+19y+5=(3y+1)(4y+5)$

29. Positive

Factors of 7: 1, 7 Factors of −14: −1, +14

+1, −14

−2, +7

+2, −7

Trial Factors	*Middle Term*
$(1a-1)(7a+14)$	Common factor
$(1a+14)(7a-1)$	$-a+98a=97a$
$(1a+1)(7a-14)$	Common factor
$(1a-14)(7a+1)$	$a-98a=-97a$
$(1a-2)(7a+7)$	Common factor
$(1a+7)(7a-2)$	$-2a+49a=47a$
$(1a+2)(7a-7)$	Common factor
$(1a-7)(7a+2)$	$2a-49a=-47a$

$7a^2+47a-14=(a+7)(7a-2)$

31. Positive Negative

Factors of 3: 1, 3 Factors of 16: −1, −16

−2, −8

−4, −4

Trial Factors	*Middle Term*
$(1b-1)(3b-16)$	$-16b-3b=-19b$
$(1b-16)(3b-1)$	$-b-48b=-49b$
$(1b-2)(3b-8)$	$-8b-6b=-14b$
$(1b-8)(3b-2)$	$-2b-24b=-26b$
$(1b-4)(3b-4)$	$-4b-12b=-16b$

$3b^2-16b+16=(b-4)(3b-4)$

33. Positive

Factors of 2: 1, 2 Factors of −14: −1, +14

+1, −14

−2, +7

+2, −7

Trial Factors	*Middle Term*
$(1z-1)(2z+14)$	Common factor
$(1z+14)(2z-1)$	$-z+28z=27z$
$(1z+1)(2z-14)$	Common factor
$(1z-14)(2z+1)$	$z-28z=-27z$
$(1z-2)(2z+7)$	$7z-4z=3z$
$(1z+7)(2z-2)$	Common factor
$(1z+2)(2z-7)$	$-7z+4z=-3z$
$(1z-7)(2z+2)$	Common factor

$2z^2-27z-14=(z-14)(2z+1)$

34. Positive

Factors of 4: 1, 4   Factors of −6: −1, +6
2, 2   +1, −6
−2, +3
+2, −3

Trial Factors	*Middle Term*
$(1z-1)(4z+6)$	Common factor
$(1z+6)(4z-1)$	$-z+24z=23z$
$(1z+1)(4z-6)$	Common factor
$(1z-6)(4z+1)$	$z-24z=-23z$
$(1z-2)(4z+3)$	$3z-8z=-5z$
$(1z+3)(4z-2)$	Common factor
$(1z+2)(4z-3)$	$-3z+8z=5z$
$(1z-3)(4z+2)$	Common factor
$(2z-1)(2z+6)$	Common factor
$(2z+1)(2z-6)$	Common factor
$(2z-2)(2z+3)$	Common factor
$(2z+2)(2z-3)$	Common factor

$4z^2+5z-6=(z+2)(4z-3)$

35. Positive

Factors of 3: 1, 3   Factors of −16: −1, +16
+1, −16
−2, +8
+2, −8
−4, +4

Trial Factors	*Middle Term*
$(1p-1)(3p+16)$	$16p-3p=13p$
$(1p+16)(3p-1)$	$-p+48p=47p$
$(1p+1)(3p-16)$	$-16p+3p=-13p$
$(1p-16)(3p+1)$	$p-48p=-47p$
$(1p-2)(3p+8)$	$8p-6p=2p$
$(1p+8)(3p-2)$	$-2p+24p=22p$
$(1p+2)(3p-8)$	$-8p+6p=-2p$
$(1p-8)(3p+2)$	$2p-24p=-22p$
$(1p-4)(3p+4)$	$4p-12p=-8p$
$(1p+4)(3p-4)$	$-4p+12p=8p$

$3p^2+22p-16=(p+8)(3p-2)$

37. The GCF is 2.

$4x^2+6x+2=2(2x^2+3x+1)$

Factor the trinomial.

Positive   Positive

Factors of 2: 1, 2   Factors of 1: +1, +1

Trial Factors	*Middle Term*
$(1x+1)(2x+1)$	$x+2x=3x$

$4x^2+6x+2=2(x+1)(2x+1)$

39. The GCF is 5.

$15y^2-50y+35=5(3y^2-10y+7)$

Factor the trinomial.

Positive   Negative

Factors of 3: 1, 3   Factors of 7: −1, −7

Trial Factors	*Middle Term*
$(1y-1)(3y-7)$	$-7y-3y=-10y$
$(1y-7)(3y-1)$	$-y-21y=-22y$

$15y^2-50y+35=5(y-1)(3y-7)$

41. The GCF is x.

$2x^3-11x^2+5x=x(2x^2-11x+5)$

Factor the trinomial $2x^2-11x+5$.

Positive   Negative

Factors of 2: 1, 2   Factors of 5: −1, −5

Trial Factors	*Middle Term*
$(1x-1)(2x-5)$	$-5x-2x=-7x$
$(1x-5)(2x-1)$	$-x-10x=-11x$

$2x^3-11x^2+5x=x(x-5)(2x-1)$

43. The GCF is b.

$3a^2b-16ab+16b=b(3a^2-16a+16)$

Factor the trinomial $3a^2-16a+16$.

Positive   Negative

Factors of 3: 1, 3   Factors of 16: −1, −16
−2, −8
−4, −4

Trial Factors	*Middle Term*
$(1a-1)(3a-16)$	$-16a-3a=-19a$
$(1a-16)(3a-1)$	$-a-48a=-49a$
$(1a-2)(3a-8)$	$-8a-6a=-14a$
$(1a-8)(3a-2)$	$-2a-24a=-26a$
$(1a-4)(3a-4)$	$-4a-12a=-16a$

$3a^2b-16ab+16b=b(a-4)(3a-4)$

45. There is no common factor.

Factor the trinomial.

Positive		Positive	
Factors of 3:	1, 3	Factors of 10:	+1, +10
			+2, +5

Trial Factors	*Middle Term*
$(1z+1)(3z+10)$	$10z+3z=13z$
$(1z+10)(3z+1)$	$z+30z=31z$
$(1z+2)(3z+5)$	$5z+6z=11z$
$(1z+5)(3z+2)$	$2z+15z=17z$

$3z^2+95z+10$ is nonfactorable over the integers.

47. The GCF is $-3x$.

$36x-3x^2-3x^3=-3x(x^2+x-12)$

Factor the trinomial.

Postive			
Factors of 1:	1, 1	Factors of -12:	$-1, +12$
			$+1, -12$
			$+2, -6$
			$-2, +6$
			$+3, -4$
			$-3, +4$

Trial Factors	*Middle Term*
$(1x-1)(1x+12)$	$12x-1x=11x$
$(1x+1)(1x-12)$	$-12x+1x=-11x$
$(1x+2)(1x-6)$	$-6x+2x=-4x$
$(1x-2)(1x+6)$	$6x-2x=4x$
$(1x+3)(1x-4)$	$-4x+3x=-x$
$(1x-3)(1x+4)$	$4x-3x=x$

$36x-3x^2-3x^3=-3x(x-3)(x+4)$

49. The GCF is 4.

$80y^2-36y+4=4(20y^2-9y+1)$

Factor the trinomial.

Positive		Negative	
Factors of 20:	1, 20	Factors of 1:	$-1, -1$
	2, 10		
	4, 5		

Trial Factors	*Middle Term*
$(1y-1)(20y-1)$	$-y-20y=-21y$
$(2y-1)(10y-1)$	$-2y-10y=-12y$
$(4y-1)(5y-1)$	$-4y-5y=-9y$

$80y^2-36y+4=4(4y-1)(5y-1)$

51. The GCF is z.

$8z^3+14z^2+3z=z(8z^2+14z+3)$

Factor the trinomial.

Positive		Positive	
Factors of 8:	1, 8	Factors of 3:	+1, +3
	2, 4		

Trial Factors	*Middle Term*
$(1z+1)(8z+3)$	$3z+8z=11z$
$(1z+3)(8z+1)$	$z+24z=25z$
$(2z+1)(4z+3)$	$6z+4z=10z$
$(2z+3)(4z+1)$	$2z+12z=14z$

$8z^3+14z^2+3z=z(2z+3)(4z+1)$

53. The GCF is y.

$6x^2y-11xy-10y=y(6x^2-11x-10)$

Factor the trinomial.

Positive			
Factors of 6:	1, 6	Factors of -10:	$-1, +10$
	2, 3		$+1, -10$
			$-2, +5$
			$+2, -5$

Trial Factors	*Middle Term*
$(1x-1)(6x+10)$	Common factor
$(1x+10)(6x-1)$	$-x+60x=59x$
$(1x+1)(6x-10)$	Common factor
$(1x-10)(6x+1)$	$x-60x=-59x$
$(1x-2)(6x+5)$	$5x-12x=-7x$
$(1x+5)(6x-2)$	Common factor
$(1x+2)(6x-5)$	$-5x+12x=7x$
$(1x-5)(6x+2)$	Common factor
$(2x-1)(3x+10)$	$20x-3x=17x$
$(2x+10)(3x-1)$	Common factor
$(2x+1)(3x-10)$	$-20x+3x=-17x$
$(2x-10)(3x+1)$	Common factor
$(2x-2)(3x+5)$	Common factor
$(2x+5)(3x-2)$	$-4x+15x=11x$
$(2x+2)(3x-5)$	Common factor
$(2x-5)(3x+2)$	$4x-15x=-11x$

$6x^2y-11xy-10y=y(2x-5)(3x+2)$

55. The GCF is 5. $10t^2-5t-50=5(2t^2-t-10)$

Factor the trinomial.

Positive Factors of 2:	1, 2	Factors of -10:	$-1, +10$
			$+1, -10$
			$-2, +5$
			$+2, -5$

Trial Factors	*Middle Term*
$(1t-1)(2t+10)$	Common factor
$(1t+10)(2t-1)$	$-t+20t=19t$
$(1t+1)(2t-10)$	Common factor
$(1t-10)(2t+1)$	$t-20t=-19t$
$(1t-2)(2t+5)$	$5t-4t=t$
$(1t+5)(2t-2)$	Common factor
$(1t+2)(2t-5)$	$-5t+4t=-t$
$(1t-5)(2t+2)$	Common factor

$10t^2-5t-50=5(t+2)(2t-5)$

57. The GCF is p. $3p^3-16p^2+5p=p(3p^2-16p+5)$

Factor the trinomial.

Positive	Negative
Factors of 3: 1, 3	Factors of 5: $-1, -5$

Trial Factors	*Middle Term*
$(1p-1)(3p-5)$	$-5p-3p=-8p$
$(1p-5)(3p-1)$	$-p-15p=-16p$

$3p^3-16p^2+5p=p(p-5)(3p-1)$

59. The GCF is 2. $26z^2+98z-24=2(13z^2+49z-12)$

Factor the trinomial.

Positive Factors of 13:	1, 13	Factors of -12:	$-1, +12$
			$+1, -12$
			$-2, +6$
			$+2, -6$
			$-3, +4$
			$+3, -4$

Trial Factors	*Middle Term*
$(1z-1)(13z+12)$	$2z-13z=-z$
$(1z+12)(13z-1)$	$-z+156z=155z$
$(1z+1)(13z-12)$	$-12z+13z=z$
$(1z-12)(13z+1)$	$z-156z=-155z$
$(1z-2)(13z+6)$	$6z-26z=-20z$
$(1z+6)(13z-2)$	$-2z+78z=76z$
$(1z+2)(13z-6)$	$-6z+26z=20z$
$(1z-6)(13z+2)$	$2z-78z=-76z$
$(1z-3)(13z+4)$	$4z-39z=-35z$
$(1z+4)(13z-3)$	$-3z+52z=49z$
$(1z+3)(13z-4)$	$-4z+39z=35z$
$(1z-4)(13z+3)$	$3z-52z=-49z$

$26z^2+98z-24=2(z+4)(13z-3)$

61. The GCF is $2y$.

$10y^3-44y^2+16y=2y(5y^2-22y+8)$

Factor the trinomial.

Positive	Negative
Factors of 5: 1, 5	Factors of 8: $-1, -8$
	$-2, -4$

Trial Factors	*Middle Term*
$(1y-1)(5y-8)$	$-8y-5y=-13y$
$(1y-8)(5y-1)$	$-y-40y=-41y$
$(1y-2)(5y-4)$	$-4y-10y=-14y$
$(1y-4)(5y-2)$	$-2y-20y=-22y$

$10y^3-44y^2+16y=2y(y-4)(5y-2)$

63. The GCF is yz.

$4yz^3 + 5yz^2 - 6yz = yz(4z^2 + 5z - 6)$

Factor the trinomial.

Positive Factors of 4:		Factors of -6:	
	1, 4		-1, $+6$
	2, 2		$+1$, -6
			-2, $+3$
			$+2$, -3

Trial Factors	*Middle Term*
$(1z-1)(4z+6)$	Common factor
$(1z+6)(4z-1)$	$-z+24z=23z$
$(1z+1)(4z-6)$	Common factor
$(1z-6)(4z+1)$	$z-24z=-23z$
$(1z-2)(4z+3)$	$3z-8z=-5z$
$(1z+3)(4z-2)$	Common factor
$(1z+2)(4z-3)$	$-3z+8z=5z$
$(1z-3)(4z+2)$	Common factor
$(2z-1)(2z+6)$	Common factor
$(2z+1)(2z-6)$	Common factor
$(2z-2)(2z+3)$	Common factor
$(2z+2)(2z-3)$	Common factor

$4yz^3 + 5yz^2 - 6yz = yz(z+2)(4z-3)$

65. The GCF is $3a$.

$42a^3 + 45a^2 - 27a = 3a(14a^2 + 15a - 9)$

Factor the trinomial.

Positive Factors of 14:		Factors of -9:	
	1, 14		-1, $+9$
	2, 7		$+1$, -9
			-3, $+3$

Trial Factors	*Middle Term*
$(1a-1)(14a+9)$	$9a-14a=-5a$
$(1a+9)(14a-1)$	$-a+126a=125a$
$(1a+1)(14a-9)$	$-9a+14a=5a$
$(1a-9)(14a+1)$	$a-126a=-125a$
$(1a-3)(14a+3)$	$3a-42a=-39a$
$(1a+3)(14a-3)$	$-3a+42a=39a$
$(2a-1)(7a+9)$	$18a-7a=11a$
$(2a+9)(7a-1)$	$-2a+63a=61a$
$(2a+1)(7a-9)$	$-18a+7a=-11a$
$(2a-9)(7a+1)$	$2a-63a=-61a$
$(2a-3)(7a+3)$	$6a-21a=-15a$
$(2a+3)(7a-3)$	$-6a+21a=15a$

$42a^3 + 45a^2 - 27a = 3a(2a+3)(7a-3)$

67. The GCF is y.

$9x^2y - 30xy^2 + 25y^3 = y(9x^2 - 30xy + 25y^2)$

Factor the trinomial.

Positive Factors of 9:		Negative Factors of 25:	
	1, 9		-1, -25
	3, 3		-5, -5

Trial Factors	*Middle Term*
$(1x-1y)(9x-25y)$	$-25xy-9xy=-34xy$
$(1x-25y)(9x-1y)$	$-xy-225xy=-226xy$
$(1x-5y)(9x-5y)$	$-5xy-45xy=-50xy$
$(3x-1y)(3x-25y)$	$-75xy-3xy=-78xy$
$(3x-5y)(3x-5y)$	$-15xy-15xy=-30xy$

$9x^2y - 30xy^2 + 25y^3 = y(3x-5y)(3x-5y)$

69. The GCF is xy.

$9x^3y - 24x^2y^2 + 16xy^3 = xy(9x^2 - 24xy + 16y^2)$

Factor the trinomial.

Positive Factors of 9:		Negative Factors of 16:	
	1, 9		-1, -16
	3, 3		-2, -8
			-4, -4

Trial Factors	*Middle Term*
$(1x-1y)(9x-16y)$	$-16xy-9xy=-25xy$
$(1x-16y)(9x-1y)$	$-xy-144xy=-145xy$
$(1x-2y)(9x-8y)$	$-8xy-18xy=-26xy$
$(1x-8y)(9x-2y)$	$-2xy-72xy=-74xy$
$(1x-4y)(9x-4y)$	$-4xy-36xy=-40xy$
$(3x-1y)(3x-16y)$	$-48xy-3xy=-51xy$
$(3x-2y)(3x-8y)$	$-24xy-6xy=-30xy$
$(3x-4y)(3x-4y)$	$-12xy-12xy=-24xy$

$9x^3y - 24x^2y^2 + 16xy^3 = xy(3x-4y)(3x-4y)$

71. p must be odd. If p were even, $(nx+p)$ would have a common factor of 2 since n is even.

Objective B Exercises

73. $6x^2 - 17x + 12 \qquad 6 \cdot 12 = 72$

Factors of 72 whose sum is -17: -9 and -8

$$\begin{aligned} 6x^2 - 17x + 12 &= 6x^2 - 9x - 8x + 12 \\ &= (6x^2 - 9x) + (-8x + 12) \\ &= 3x(2x-3) - 4(2x-3) \\ &= (2x-3)(3x-4) \end{aligned}$$

75. $5b^2+33b-14 \quad 5(-14)=-70$

Factors of -70 whose sum is 33: 35 and –2

$$\begin{aligned}5b^2+33b-14&=5b^2+35b-2b-14\\&=(5b^2+35b)+(-2b-14)\\&=5b(b+7)-2(b+7)\\&=(b+7)(5b-2)\end{aligned}$$

77. $6a^2+7a-24 \quad 6(-24)=-144$

Factors of -144 whose sum is 7: 16 and –9

$$\begin{aligned}6a^2+7a-24&=6a^2+16a-9a-24\\&=(6a^2+16a)+(-9a-24)\\&=2a(3a+8)-3(3a+8)\\&=(3a+8)(2a-3)\end{aligned}$$

79. $4z^2+11z+6 \quad 4\cdot 6=24$

Factors of 24 whose sum is 11: 8 and 3

$$\begin{aligned}4z^2+11z+6&=4z^2+8z+3z+6\\&=(4z^2+8z)+(3z+6)\\&=4z(z+2)+3(z+2)\\&=(z+2)(4z+3)\end{aligned}$$

81. $22p^2+51p-10 \quad 22(-10)=-220$

Factors of -220 whose sum is 51: 55 and –4

$$\begin{aligned}22p^2+51p-10&=22p^2+55p-4p-10\\&=(22p^2+55p)+(-4p-10)\\&=11p(2p+5)-2(2p+5)\\&=(2p+5)(11p-2)\end{aligned}$$

83. $8y^2+17y+9 \quad 8\cdot 9=72$

Factors of 72 whose sum is 17: 9 and 8

$$\begin{aligned}8y^2+17y+9&=8y^2+8y+9y+9\\&=(8y^2+8y)+(9y+9)\\&=8y(y+1)+9(y+1)\\&=(y+1)(8y+9)\end{aligned}$$

85. $18t^2-9t-5 \quad 18(-5)=-90$

Factors of -90 whose sum is –9: –15 and 6

$$\begin{aligned}18t^2-9t-5&=18t^2-15t+6t-5\\&=(18t^2-15t)+(6t-5)\\&=3t(6t-5)+1(6t-5)\\&=(6t-5)(3t+1)\end{aligned}$$

87. $6b^2+71b-12 \quad 6(-12)=-72$

Factors of -72 whose sum is 71: 72 and –1

$$\begin{aligned}6b^2+71b-12&=6b^2+72b-b-12\\&=(6b^2+72b)+(-b-12)\\&=6b(b+12)-1(b+12)\\&=(b+12)(6b-1)\end{aligned}$$

89. $9x^2+12x+4 \quad 9\cdot 4=36$

Factors of 36 whose sum is 12: 6 and 6

$$\begin{aligned}9x^2+12x+4&=9x^2+6x+6x+4\\&=(9x^2+6x)+(6x+4)\\&=3x(3x+2)+2(3x+2)\\&=(3x+2)(3x+2)\end{aligned}$$

91. $6b^2-13b+6 \quad 6\cdot 6=36$

Factors of 36 whose sum is –13: –9 and –4

$$\begin{aligned}6b^2-13b+6&=6b^2-9b-4b+6\\&=(6b^2-9b)+(-4b+6)\\&=3b(2b-3)-2(2b-3)\\&=(2b-3)(3b-2)\end{aligned}$$

93. $33b^2+34b-35 \quad 33(-35)=-1155$

Factors of -1155 whose sum is 34: 55 and –21

$$\begin{aligned}33b^2+34b-35&=33b^2+55b-21b-35\\&=(33b^2+55b)+(-21b-35)\\&=11b(3b+5)-7(3b+5)\\&=(3b+5)(11b-7)\end{aligned}$$

95. $18y^2-39y+20 \quad 18\cdot 20=360$

Factors of 360 whose sum is -39: –24 and –15

$$\begin{aligned}18y^2-39y+20&=18y^2-24y-15y+20\\&=(18y^2-24y)+(-15y+20)\\&=6y(3y-4)-5(3y-4)\\&=(3y-4)(6y-5)\end{aligned}$$

97. $15a^2+26a-21 \qquad 15(-21)=-315$

Factors of -315 whose sum is 26: 35 and –9

$$\begin{aligned}15a^2+26a-21&=15a^2+35a-9a-21\\&=(15a^2+35a)+(-9a-21)\\&=5a(3a+7)-3(3a+7)\\&=(3a+7)(5a-3)\end{aligned}$$

99. $8y^2-26y+15 \qquad 8\cdot 15=120$

Factors of 120 whose sum is –26: –20 and –6

$$\begin{aligned}8y^2-26y+15&=8y^2-20y-6y+15\\&=(8y^2-20y)+(-6y+15)\\&=4y(2y-5)-3(2y-5)\\&=(2y-5)(4y-3)\end{aligned}$$

101. $8z^2+2z-15 \qquad 8(-15)=-120$

Factors of -120 whose sum is 2: 12 and -10

$$\begin{aligned}8z^2+2z-15&=8z^2+12z-10z-15\\&=(8z^2+12z)+(-10z-15)\\&=4z(2z+3)-5(2z+3)\\&=(2z+3)(4z-5)\end{aligned}$$

103. $15x^2-82x+24 \qquad 15\cdot 24=360$

Factors of 360 whose sum is –82: none

$15x^2-82x+24$ is nonfactorable over the integers.

105. $10z^2-29z+10 \qquad 10\cdot 10=100$

Factors of 100 whose sum is –29: –25 and –4

$$\begin{aligned}10z^2-29z+10&=10z^2-25z-4z+10\\&=(10z^2-25z)+(-4z+10)\\&=5z(2z-5)-2(2z-5)\\&=(2z-5)(5z-2)\end{aligned}$$

107. $36z^2+72z+35 \qquad 36\cdot 35=1260$

Factors of 1260 whose sum is 72: 30 and 42

$$\begin{aligned}36z^2+72z+35&=36z^2+30z+42z+35\\&=(36z^2+30z)+(42z+35)\\&=6z(6z+5)+7(6z+5)\\&=(6z+5)(6z+7)\end{aligned}$$

109. $3x^2+xy-2y^2 \qquad 3(-2)=-6$

Factors of -6 whose sum is 1: 3 and -2

$$\begin{aligned}3x^2+xy-2y^2&=3x^2+3xy-2xy-2y^2\\&=(3x^2+3xy)+(-2xy-2y^2)\\&=3x(x+y)-2y(x+y)\\&=(x+y)(3x-2y)\end{aligned}$$

111. $3a^2+5ab-2b^2 \qquad 3(-2)=-6$

Factors of -6 whose sum is 5: 6 and -1

$$\begin{aligned}3a^2+5ab-2b^2&=3a^2+6ab-ab-2b^2\\&=(3a^2+6ab)+(-ab-2b^2)\\&=3a(a+2b)-b(a+2b)\\&=(a+2b)(3a-b)\end{aligned}$$

113. $4y^2-11yz+6z^2 \qquad 4\cdot 6=24$

Factors of 24 whose sum is –11: –8 and –3

$$\begin{aligned}4y^2-11yz+6z^2&=4y^2-8yz-3yz+6z^2\\&=(4y^2-8yz)+(-3yz+6z^2)\\&=4y(y-2z)-3z(y-2z)\\&=(y-2z)(4y-3z)\end{aligned}$$

115. $28+3z-z^2 \qquad 28(-1)=-28$

Factors of -28 whose sum is : 7 and -4

$$\begin{aligned}28+3z-z^2&=-z^2+3z+28=-z^2-4z+7z+28\\&=(-z^2-4z)+(7z+28)\\&=-z(z+4)+7(z+4)\\&=-(z-7)(z+4)\end{aligned}$$

117. $8-7x-x^2 \qquad 8(-1)=-8$

Factors of -8 whose sum is –7: –8 and 1

$$\begin{aligned}&8-7x-x^2=-x^2-7x+8=-x^2+x-8x+8\\&=(-x^2+x)+(-8x+8)\\&=-x(x-1)-8(x-1)\\&=-(x-1)(x+8)\end{aligned}$$

119. $9x^2+33x-60$

Common factor 3: $3(3x^2+11x-20)$

$3(-20)=-60$

Factors of -60 whose sum is 11: 15 and -4

$$\begin{aligned}3(3x^2+11x-20)&=3(3x^2+15x-4x-20)\\&=3\left[(3x^2+15x)+(-4x-20)\right]\\&=3\left[3x(x+5)-4(x+5)\right]\\&=3(x+5)(3x-4)\end{aligned}$$

121. $24x^2-52x+24$

Common factor 4: $4(6x^2-13x+6)$

$6\cdot 6=36$

Factors of 36 whose sum is -13: -9 and -4

$$\begin{aligned}4(6x^2-13x+6)&=4(6x^2-9x-4x+6)\\&=4\left[(6x^2-9x)+(-4x+6)\right]\\&=4\left[3x(2x-3)-2(2x-3)\right]\\&=4(2x-3)(3x-2)\end{aligned}$$

123. $35a^4+9a^3-2a^2$

Common factor a^2: $a^2(35a^2+9a-2)$

$35(-2)=-70$

Factors of -70 whose sum is 9: 14 and -5

$$\begin{aligned}a^2(35a^2+9a-2)&=a^2(35a^2+14a-5a-2)\\&=a^2\left[(35a^2+14a)+(-5a-2)\right]\\&=a^2\left[7a(5a+2)-1(5a+2)\right]\\&=a^2(5a+2)(7a-1)\end{aligned}$$

125. $15b^2-115b+70$

Common factor 5: $5(3b^2-23b+14)$

$3\cdot 14=42$

Factors of 42 whose sum is -23: -21 and -2

$$\begin{aligned}5(3b^2-23b+14)&=5(3b^2-21b-2b+14)\\&=5\left[(3b^2-21b)+(-2b+14)\right]\\&=5\left[3b(b-7)-2(b-7)\right]\\&=5(b-7)(3b-2)\end{aligned}$$

127. $3x^2-26xy+35y^2$ $\quad 3\cdot 35=105$

Factors of 105 whose sum is -26: -21 and -5

$$\begin{aligned}3x^2-25xy+36y^2&=3x^2-21xy-5xy+35y^2\\&=(3x^2-21xy)+(-5xy+35y^2)\\&=3x(x-7y)-5y(x-7y)\\&=(x-7y)(3x-5y)\end{aligned}$$

129. $216y^2-3y-3$

Common factor 3: $3(72y^2-y-1)$

$72(-1)=-72$

Factors of -72 whose sum is -1: -9 and 8

$$\begin{aligned}3(72y^2-y-1)&=3(72y^2-9y+8y-1)\\&=3\left[(72y^2-9y)+(8y-1)\right]\\&=3\left[9y(8y-1)+1(8y-1)\right]\\&=3(8y-1)(9y+1)\end{aligned}$$

131. $21-20x-x^2$ $\quad 21(-1)=-21$

Factors of -21 whose sum is -20: -21 and 1

$$\begin{aligned}21-20x-x^2&=-x^2-20x+21=-x^2-21x+x+21\\&=(-x^2-21x)+(x+21)\\&=-x(x+21)+(x+21)\\&=-(x-1)(x+21)\end{aligned}$$

133. If c is positive, both the signs are the same as b, so if b is positive, both signs are positive.

135. If c is positive, both the signs are the same as b, so if b is negative, both signs are negative.

Applying the Concepts

137. Students should explain that the sign of the product of the last terms of the two binomial factors must be the same as the sign of the last term of the trinomial. Thus if the last term of the trinomial is positive, the last terms of the two binomial factors are either both positive or both negative, depending on the middle term of the trinomial. If the last term of the trinomial is negative, the last terms of the two binomial factors will have different signs.

139. $(x-2)^2+3(x-2)+2$ Let $a=x-2$
$=a^2+3a+2$
$=(a+1)(a+2)$
$=(x-2+1)(x-2+2)$
$=(x-1)x$ or $x(x-1)$

141. $2(y+2)^2-(y+2)-3$ Let $a=y+2$
$=2a^2-a-3$
$=(2a-3)(a+1)$
$=[2(y+2)-3][y+2+1]$
$=(2y+4-3)(y+3)$
$=(2y+1)(y+3)$

143. $4(y-1)^2-7(y-1)-2$ Let $a=y-1$
$=4a^2-7a-2$
$=(4a+1)(a-2)$
$=[4(y-1)+1][y-1-2]$
$=(4y-4+1)(y-3)$
$=(4y-3)(y-3)$

Section 7.4

Objective A Exercises

1. perfect squares: 4; $25x^6$; $100x^4y^4$

3. $4z^4$

5. $9a^2b^3$

7. (iv) m^4-n^2

9. $x^2-16=x^2-4^2=(x+4)(x-4)$

11. $4x^2-1=(2x)^2-1^2=(2x+1)(2x-1)$

13. $16x^2-121=(4x)^2-11^2=(4x+11)(4x-11)$

15. $1-9a^2=1^2-(3a)^2=(1+3a)(1-3a)$

17. $x^2y^2-100=(xy)^2-10^2=(xy+10)(xy-10)$

19. Not factorable

21. $25-a^2b^2=5^2-(ab)^2=(5+ab)(5-ab)$

23. $a^{2n}-1=(a^n)^2-1^2=(a^n+1)(a^n-1)$

25. $x^2-12x+36=(x-6)^2$

27. $b^2-2b+1=(b-1)^2$

29. $16x^2-40x+25=(4x-5)^2$

31. Not factorable

33. Not factorable

35. $x^2+6xy+9y^2=(x+3y)^2$

37. $25a^2-40ab+16b^2=(5a-4b)^2$

39. $x^{2n}+6x^n+9=(x^n+3)^2$

41. $(x-4)^2-9=[(x-4)-3][(x-4)+3]$
$=(x-7)(x-1)$

43. $(x-y)^2-(a+b)^2$
$=[(x-y)-(a+b)][(x-y)+(a+b)]$
$=(x-y-a-b)(x-y+a+b)$

Objective B Exercises

45. 8; x^9; $27c^{15}d^{18}$

47. $2x^3$

49. $4a^2b^6$

51. Yes

53. Yes

55. $x^3-27=x^3-3^3$
$=(x-3)(x^2+3x+9)$

57. $8x^3-1=(2x)^3-1^3$
$=(2x-1)(4x^2+2x+1)$

59. $x^3-y^3=(x-y)(x^2+xy+y^2)$

61. $m^3+n^3=(m+n)(m^2-mn+n^2)$

63. $64x^3+1=(4x)^3+1^3$
$=(4x+1)(16x^2-4x+1)$

65. $27x^3-8y^3=(3x)^3-(2y)^3$
$=(3x-2y)(9x^2+6xy+4y^2)$

67. $x^3y^3+64=(xy)^3+(4)^3$
$=(xy+4)(x^2y^2-4xy+16)$

69. Not factorable

71. Not factorable

73. $(a-b)^3-b^3$
$= [(a-b)-b][(a-b)^2+b(a-b)+b^2]$
$= (a-2b)(a^2-2ab+b^2+ab-b^2+b^2)$
$= (a-2b)(a^2-ab+b^2)$

75. $x^{6n}+y^{3n}=(x^{2n})^3+(y^n)^3$
$= (x^{2n}+y^n)(x^{4n}-x^{2n}y^n+y^{2n})$

77. $x^{3n}+8=(x^n)^3+(2)^3$
$= (x^n+2)(x^{2n}-2x^n+4)$

Objective C Exercises

79. No. Polynomials cannot have square roots as variable terms.

81. Let $u=xy$
$x^2y^2-8xy+15=u^2-8u+15$
$= (u-3)(u-5)$
$= (xy-3)(xy-5)$

83. Let $u=xy$
$x^2y^2-17xy+60=u^2-17u+60$
$= (u-12)(u-5)$
$= (xy-12)(xy-5)$

85. Let $u=x^2$
$x^4-9x^2+18=u^2-9u+18$
$= (u-3)(u-6)$
$= (x^2-3)(x^2-6)$

87. Let $u=b^2$
$b^4-13b^2-90=u^2-13u-90$
$= (u+5)(u-18)$
$= (b^2+5)(b^2-18)$

89. Let $u=x^2y^2$
$x^4y^4-8x^2y^2+12=u^2-8u+12$
$= (u-2)(u-6)$
$= (x^2y^2-2)(x^2y^2-6)$

91. Let $u=x^n$
$x^{2n}+3x^n+2=u^2+3u+2$
$= (u+1)(u+2)$
$= (x^n+1)(x^n+2)$

93. Let $u=xy$
$3x^2y^2-14xy+15=3u^2-14u+15$
$= (3u-5)(u-3)$
$= (3xy-5)(xy-3)$

95. Let $u=ab$
$6a^2b^2-23ab+21=6u^2-23u+21$
$= (2u-3)(3u-7)$
$= (2ab-3)(3ab-7)$

97. Let $u=x^2$
$2x^4-13x^2-15=2u^2-13u-15$
$= (2u-15)(u+1)$
$= (2x^2-15)(x^2+1)$

99. Let $u=x^n$
$2x^{2n}-7x^n+3=2u^2-7u+3$
$= (2u-1)(u-3)$
$= (2x^n-1)(x^n-3)$

101. Let $u=a^n$
$6a^{2n}+19a^n+10=6u^2+19u+10$
$= (2u+5)(3u+2)$
$= (2a^n+5)(3a^n+2)$

Objective D Exercises

103. $12x^2-36x+27=3(4x^2-12x+9)$
$= 3(2x-3)^2$

105. $27a^4-a=a(27a^3-1)$
$= a(3a-1)(9a^2+3a+1)$

107. $20x^2-5=5(4x^2-1)$
$= 5(2x+1)(2x-1)$

109. $y^5+6y^4-55y^3=y^3(y^2+6y-55)$
$= y^3(y+11)(y-5)$

111. $16x^4-81=(4x^2+9)(4x^2-9)$
$= (4x^2+9)(2x+3)(2x-3)$

113. $16a-2a^4=2a(8-a^3)$
$= 2a(2-a)(4+2a+a^2)$

115. $a^3b^6-b^3=b^3(a^3b^3-1)$
$= b^3(ab-1)(a^2b^2+ab+1)$

117. $8x^4 - 40x^3 + 50x^2 = 2x^2(4x^2 - 20x + 25)$
$= 2x^2(2x - 5)^2$

119. $x^4 - y^4 = (x^2 + y^2)(x^2 - y^2)$
$= (x^2 + y^2)(x + y)(x - y)$

121. $x^6 + y^6 = (x^2 + y^2)(x^4 - x^2y^2 + y^4)$

123. Not factorable

125. $16a^4 - 2a = 2a(8a^3 - 1)$
$= 2a(2a - 1)(4a^2 + 2a + 1)$

127. $a^4b^2 - 8a^3b^3 - 48a^2b^4$
$= a^2b^2(a^2 - 8ab - 48b^2)$
$= a^2b^2(a + 4b)(a - 12b)$

129. $x^3 - 2x^2 - 4x + 8 = x^2(x - 2) - 4(x - 2)$
$= (x - 2)(x^2 - 4)$
$= (x - 2)(x + 2)(x - 2)$
$= (x - 2)^2(x + 2)$

131. $2x^3 + x^2 - 32x - 16$
$= x^2(2x + 1) - 16(2x + 1)$
$= (2x + 1)(x^2 - 16)$
$= (2x + 1)(x + 4)(x - 4)$

133. $4x^4 - x^2 - 4x^2y^2 + y^2$
$= x^2(4x^2 - 1) - y^2(4x^2 - 1)$
$= (4x^2 - 1)(x^2 - y^2)$
$= (2x + 1)(2x - 1)(x + y)(x - y)$

135. $3b^{n+2} + 4b^{n+1} - 4b^n$
$= b^n(3b^2 + 4b - 4)$
$= b^n(b + 2)(3b - 2)$

137. The coefficient is 8.

Applying the Concepts

139. If $x - 3$ and $x + 4$ are factors of $x^3 + 6x^2 - 7x - 60$ then $x^3 + 6x^2 - 7x - 60$ is divisible by $x - 3$ and $x + 4$. Divide $x^3 + 6x^2 - 7x - 60$ by $x - 3$. The quotient is $x^2 + 9x + 20$. Divide this quotient by $x + 4$. The quotient is $x + 5$. Therefore $x + 5$ is a third first-degree factor of $x^3 + 6x^2 - 7x - 60$.

Section 7.5

Objective A Exercises

1. The Principle of Zero Products states that if the product of two numbers equals zero, then one or both of the numbers is (are) zero.

3. $(y+3)(y+2)=0$
$y+3=0 \quad y+2=0$
$y=-3 \quad y=-2$

The solutions are –2 and –3.

5. $(z-7)(z-3)=0$
$z-7=0 \quad z-3=0$
$z=7 \quad z=3$

The solutions are 3 and 7.

7. $x(x-5)=0$
$x=0 \quad x-5=0$
$x=5$

The solutions are 0 and 5.

9. $a(a-9)=0$
$a=0 \quad a-9=0$
$a=9$

The solutions are 0 and 9.

11. $y(2y+3)=0$
$y=0 \quad 2y+3=0$
$2y=-3$
$y=-\frac{3}{2}$

The solutions are 0 and $-\frac{3}{2}$.

13. $2a(3a-2)=0$
$2a=0 \quad 3a-2=0$
$a=0 \quad 3a=2$
$a=\frac{2}{3}$

The solutions are 0 and $\frac{2}{3}$.

15. $(b+2)(b-5)=0$
$b+2=0 \quad b-5=0$
$b=-2 \quad b=5$

The solutions are −2 and 5.

17. $x^2 - 81 = 0$

$(x+9)(x-9) = 0$

$x+9=0 \quad x-9=0$

$x=-9 \quad x=9$

The solutions are −9 and 9.

19. $4x^2 - 49 = 0$

$(2x+7)(2x-7)=0$

$2x+7=0 \quad 2x-7=0$

$2x=-7 \quad 2x=7$

$x=-\frac{7}{2} \quad x=\frac{7}{2}$

The solutions are $-\frac{7}{2}$ and $\frac{7}{2}$.

21. $9x^2 - 1 = 0$

$(3x+1)(3x-1)=0$

$3x+1=0 \quad 3x-1=0$

$3x=-1 \quad 3x=1$

$x=-\frac{1}{3} \quad x=\frac{1}{3}$

The solutions are $-\frac{1}{3}$ and $\frac{1}{3}$.

23. $x^2+6x+8=0$

$(x+2)(x+4)=0$

$x+2=0 \quad x+4=0$

$x=-2 \quad x=-4$

The solutions are −2 and −4.

25. $z^2+5z-14=0$

$(z+7)(z-2)=0$

$z+7=0 \quad z-2=0$

$z=-7 \quad z=2$

The solutions are −7 and 2.

27. $2a^2-9a-5=0$

$(2a+1)(a-5)=0$

$2a+1=0 \quad a-5=0$

$2a=-1 \quad a=5$

$a=-\frac{1}{2}$

The solutions are $-\frac{1}{2}$ and 5.

29. $6z^2+5z+1=0$

$(3z+1)(2z+1)=0$

$3z+1=0 \quad 2z+1=0$

$3z=-1 \quad 2z=-1$

$z=-\frac{1}{3} \quad z=-\frac{1}{2}$

The solutions are $-\frac{1}{3}$ and $-\frac{1}{2}$.

31. $x^2-3x=0$

$x(x-3)=0$

$x=0 \quad x-3=0$

$x=3$

The solutions are 0 and 3.

33. $x^2-7x=0$

$x(x-7)=0$

$x=0 \quad x-7=0$

$x=7$

The solutions are 0 and 7.

35. $a^2+5a=-4$

$a^2+5a+4=0$

$(a+1)(a+4)=0$

$a+1=0 \quad a+4=0$

$a=-1 \quad a=-4$

The solutions are −1 and −4.

37. $y^2-5y=-6$

$y^2-5y+6=0$

$(y-2)(y-3)=0$

$y-2=0 \quad y-3=0$

$y=2 \quad y=3$

The solutions are 2 and 3.

39. $2t^2+7t=4$

$2t^2+7t-4=0$

$(2t-1)(t+4)=0$

$2t-1=0 \quad t+4=0$

$2t=1 \quad t=-4$

$t=\frac{1}{2}$

The solutions are $\frac{1}{2}$ and −4.

41. $3t^2 - 13t = -4$

$3t^2 - 13t + 4 = 0$

$(3t - 1)(t - 4) = 0$

$3t - 1 = 0 \quad t - 4 = 0$

$3t = 1 \quad t = 4$

$t = \frac{1}{3}$

The solutions are $\frac{1}{3}$ and 4.

43. $x(x - 12) = -27$

$x^2 - 12x = -27$

$x^2 - 12x + 27 = 0$

$(x - 3)(x - 9) = 0$

$x - 3 = 0 \quad x - 9 = 0$

$x = 3 \quad x = 9$

The solutions are 3 and 9.

45. $y(y - 7) = 18$

$y^2 - 7y = 18$

$y^2 - 7y - 18 = 0$

$(y + 2)(y - 9) = 0$

$y + 2 = 0 \quad y - 9 = 0$

$y = -2 \quad y = 9$

The solutions are −2 and 9.

47. $p(p + 3) = -2$

$p^2 + 3p = -2$

$p^2 + 3p + 2 = 0$

$(p + 1)(p + 2) = 0$

$p + 1 = 0 \quad p + 2 = 0$

$p = -1 \quad p = -2$

The solutions are −1 and −2.

49. $y(y + 4) = 45$

$y^2 + 4y = -45$

$y^2 + 4y - 45 = 0$

$(y + 9)(y - 5) = 0$

$y + 9 = 0 \quad y - 5 = 0$

$y = -9 \quad y = 5$

The solutions are −9 and 5.

51. $x(x+3) = 28$

$x^2 + 3x = 28$

$x^2 + 3x - 28 = 0$

$(x+7)(x-4) = 0$

$x + 7 = 0 \quad x - 4 = 0$

$x = -7 \quad x = 4$

The solutions are −7 and 4.

53. $(x+8)(x-3) = -30$

$x^2 + 5x - 24 = -30$

$x^2 + 5x + 6 = 0$

$(x+2)(x+3) = 0$

$x + 2 = 0 \quad x + 3 = 0$

$x = -2 \quad x = -3$

The solutions are −2 and −3.

55. $(z - 5)(z + 4) = 52$

$z^2 - z - 20 = 52$

$z^2 - z - 72 = 0$

$(z + 8)(z - 9) = 0$

$z + 8 = 0 \quad z - 9 = 0$

$z = -8 \quad z = 9$

The solutions are −8 and 9.

57. $(z - 6)(z + 1) = -10$

$z^2 - 5z - 6 = -10$

$z^2 - 5z + 4 = 0$

$(z - 1)(z - 4) = 0$

$z - 1 = 0 \quad z - 4 = 0$

$z = 1 \quad z = 4$

The solutions are 1 and 4.

59. $(a - 4)(a + 7) = -18$

$a^2 + 3a - 28 = -18$

$a^2 + 3a - 10 = 0$

$(a + 5)(a - 2) = 0$

$a + 5 = 0 \quad a - 2 = 0$

$a = -5 \quad a = 2$

The solutions are −5 and 2.

61. To have one positive solution and one negative solution, the factors must have one negative sign and one positive sign, so c must be less than zero

Objective B Exercises

63. **Strategy** The positive number: x
The square of the positive number is six more than five times the positive number.

Solution

$$x^2 = 5x + 6$$
$$x^2 - 5x - 6 = 0$$
$$(x - 6)(x + 1) = 0$$
$$x - 6 = 0 \quad x + 1 = 0$$
$$x = 6 \quad x = -1$$

Because -1 is not a positive number, it is not a solution. The number is 6.

65. **Strategy** One of the numbers: x
The other number: $6 - x$
The sum of the squares of the numbers is twenty.

Solution

$$x^2 + (-x+6)^2 = 20$$
$$2x^2 - 12x + 36 = 20$$
$$2x^2 - 12x + 16 = 0$$
$$(2x-4)(x-4) = 0$$
$$2x - 4 = 0 \quad x - 4 = 0$$
$$x = 2 \quad x = 4$$

The numbers are 2 and 4.

67. **Strategy** First positive integer: x
Next positive integer: $x + 1$
The sum of the squares of two consecutive positive integers is 113.

Solution

$x^2 + (x+1)^2 = 113$ is (ii).

69. **Strategy** The first positive integer: x
The next positive integer: $x + 1$
The sum of the squares of two consecutive positive integers is forty-one.

Solution

$$x^2 + (x+1)^2 = 41$$
$$2x^2 + 2x + 1 = 41$$
$$2x^2 + 2x - 40 = 0$$
$$(2x+10)(x-4) = 0$$
$$2x + 10 = 0 \quad x - 4 = 0$$
$$x = -5 \quad x = 4$$

Because -5 is not a positive number, it is not a solution. The numbers are 4 and 5.

71. **Strategy** One of the numbers: x
The other number: $10 - x$
The product of two numbers is twenty-one.

Solution

$$x(10-x) = 21$$
$$10x - x^2 = 21$$
$$x^2 - 10x + 21 = 0$$
$$(x-3)(x-7) = 0$$
$$x - 3 = 0 \quad x - 7 = 0$$
$$x = 3 \quad x = 7$$

The numbers are 3 and 7.

73. **Strategy** Known: $S = 78$
Unknown: n

Solution

$$S = \frac{n^2 + n}{2}$$
$$78 = \frac{n^2 + n}{2}$$
$$156 = n^2 + n$$
$$n^2 + n - 156 = 0$$
$$(n+13)(n-12) = 0$$
$$n + 13 = 0 \quad n - 12 = 0$$
$$n = -13 \quad n = 12$$

There cannot be a negative number of numbers, so -13 is not a solution.
Twelve consecutive numbers beginning with 1 and giving a sum of 78.

75. **Strategy** Known: $N = 15$
Unknown: t

Solution

$$N = \frac{t^2 - t}{2}$$
$$15 = \frac{t^2 - t}{2}$$
$$30 = t^2 - t$$
$$t^2 - t - 30 = 0$$
$$(t+5)(t-6) = 0$$

$t + 5 = 0 \qquad t - 6 = 0$

$t = -5 \qquad t = 6$

There cannot be a negative number of teams, so −5 is not a solution. There are 6 teams in the league.

77. **Strategy** Known: $S = 192$; $v = 16$
Unknown: t

Solution

$$S = vt + 16t^2$$
$$192 = 16t + 16t^2$$
$$16t^2 + 16t - 192 = 0$$
$$16(t^2 + t - 12) = 0$$
$$(t+4)(t-3) = 0$$

$t + 4 = 0 \qquad t - 3 = 0$

$t = -4 \qquad t = 3$

Time cannot be negative, so −4 cannot be a solution. In 3 s, the object will hit the ground.

79. **Strategy** Known: $h = 0$; $v = 60$
Unknown: t

Solution

$$h = vt - 16t^2$$
$$0 = 60t - 16t^2$$
$$16t^2 - 60t = 0$$
$$t(16t - 60) = 0$$
$$16t - 60 = 0$$
$$t = 3.75$$

In 3.75 s the ball will hit the ground.

81. **Strategy** Width of the rectangle: x
Length of the rectangle: $2x + 5$
The area of the rectangle is 75 in^2.
The equation for the area of a rectangle is $A = L \times W$. Substitute in the equation and solve for x.

Solution

$$A = L \times W$$
$$75 = (2x+5)(x)$$
$$75 = 2x^2 + 5x$$
$$0 = 2x^2 + 5x - 75$$
$$0 = (2x+15)(x-5)$$

$2x - 15 = 0 \qquad x - 5 = 0$

$x = -\frac{15}{2} \qquad x = 5$

Because $-\frac{15}{2}$ is not positive, it can not be a solution.

$2x + 5 = 2(5) + 5 = 10 + 5 = 15$

The length is 15 in.
The width is 5 in.

83. **Strategy** Base of the triangle: x
Height of the triangle: $2x + 4$
The area of the rectangle is 35 m^2.
The equation for the area of a triangle is $A = \frac{1}{2}bh$. Substitute in the equation and solve for x.

Solution

$$35 = \frac{1}{2}x(2x+4)$$
$$35 = x^2 + 2x$$
$$0 = x^2 + 2x - 35$$
$$0 = (x+7)(x-5)$$

$x + 7 = 0 \qquad x - 5 = 0$

$x = -7 \qquad x = 5$

Because −7 is not positive, it cannot be a solution.

$2x + 4 = 2(5) + 4 = 14$

The base is 5 m. The height is 14 m.

85. **Strategy** Base of the trapezoid: x

Height of the trapezoid: $x + 3$

The area of the trapezoid is 304 ft^2.

The equation for the area of a trapezoid is $A = bh$. Substitute in the equation and solve for x.

Solution

$$A = bh$$
$$304 = x(x+3)$$
$$304 = x^2 + 3x$$
$$0 = x^2 + 3x - 304$$
$$0 = (x+19)(x-16)$$
$$x+19=0 \qquad x-16=0$$
$$x=-19 \qquad x=16$$

Because the base of a trapezoid cannot be a negative number, −19 is not a solution. The base of the trapezoid is 16 ft.

87. **Strategy** • The width of the uniform border: $\frac{x}{2}$

New width: $6-2\left(\frac{x}{2}\right)=6-x$

New length: $9-2\left(\frac{x}{2}\right)=9-x$

• The required area is 28 in^2. The equation for the area of a rectangle is $A = LW$.

Solution

$$(6-x)(9-x)=28$$
$$54-15x+x^2=28$$
$$x^2-15x+26=0$$
$$(x-13)(x-2)=0$$
$$x-13=0 \qquad x=2$$
$$x=13 \qquad x=2$$

Since 6 – 13 would give a width of −7, 13 is not a solution.

$$6-x=6-2=4$$
$$9-x=9-2=7$$

The dimensions of the type area are 4 in. by 7 in.

89. **Strategy** Area: 27π

The equation for the area of a square is $A = 12\pi x - \pi x^2$. Substitute in the equation and solve for x.

Solution

$$A = 12\pi x - \pi x^2$$
$$27\pi = 12\pi x - \pi x^2$$
$$\pi x^2 - 12\pi x + 27\pi = 0$$
$$\pi(x^2 - 12x + 27) = 0$$
$$\pi(x-9)(x-3) = 0$$
$$x-9=0 \qquad x-3=0$$
$$x=9 \qquad x=3$$

The width of the iris is 3 mm.

Applying the Concepts

91.

$$n(n+3)=4$$
$$n^2+3n-4=0$$
$$(n+4)(n-1)=0$$
$$n+4=0 \qquad n-1=0$$
$$n=-4 \qquad n=1$$
$$2n^2=2(-4)^2 \qquad 2n^2=2(1)^2$$
$$=2(16) \qquad =2(1)$$
$$=32 \qquad =2$$

$2n^2$ equals 32 or 2.

93.

$$(b+5)^2=16$$
$$b^2+10b+25=16$$
$$b^2+10b+9=0$$
$$(b+1)(b+9)=0$$
$$b+1=0 \qquad b+9=0$$
$$b=-1 \qquad b=-9$$

The solutions are −1 and −9.

95.

$$(x+3)(2x-1)=(3-x)(5-3x)$$
$$2x^2+5x-3=15-14x+3x^2$$
$$-x^2+19x-18=0$$
$$x^2-19x+18=0$$
$$(x-1)(x-18)=0$$
$$x-1=0 \qquad x-18=0$$
$$x=1 \qquad x=18$$

The solutions are 1 or 18.

97. The error of accidentally dividing by zero takes place in the solution

$$x^2 = x$$
$$\frac{x^2}{x} = \frac{x}{x}$$
$$x = 1$$

Since one of the solutions of $x^2 = x$ is $x = 0$, an error occurs when the quotients in the equation $\frac{x^2}{x} = \frac{x}{x}$ are introduced. Here is the correct solution.

$$x^2 = x$$
$$x^2 - x = 0$$
$$x(x-1) = 0$$
$$x = 0 \quad x - 1 = 0$$
$$x = 1.$$

The solutions are 0 and 1.

Chapter 7 Concept Review

1. In factoring a polynomial, we always first check to see if the terms of the polynomial have a common factor. If they do, we factor out the GCF of the terms. [7.4A]

2. When factoring, the terms of a polynomial do not have to be like terms. If they were like terms, we would combine them, and the result would be a monomial. [7.1A]

3. To check the answer to a factorization, multiply the factors. The product must be the original polynomial. [7.2A]

4. When factoring by grouping, after we group the first two terms and group the last two terms, we factor the GCF from each group. [7.1B]

5. A polynomial of the form $x^2 + bx + c$ or $ax^2 + bx + c$ is nonfactorable over the integers when it does not factor into the product of two binomials that have integer coefficients and constants. [7.2A]

6. To factor a polynomial completely means to write the polynomial as a product of factors that are nonfactorable over the integers. [7.2B]

7. When factoring a polynomial of the form $x^2 + bx + c$, we begin by finding the possible factors of *c* because we are looking for two numbers whose product is *c* and whose sum is *b*. [7.2A]

8. Trial factors can be used when factoring a trinomial of the form $ax^2 + bx + c$. We use the factors of *a* and the factors of *c* to write all possible binomial pairs that, when multiplied, have ax^2 and *c* in their product. We test each pair of trial factors to find which one has *bx* as the middle term of the product when the factors are multiplied. [7.3A]

9. The middle term of a trinomial of the form $x^2 + bc + c$ or $ax^2 + bx + c$ is *bx*. [7.3A]

10. The binomial factors of the difference of two squares $a^2 - b^2$ are $a + b$ and $a - b$. [7.4A]

11. The square of a binomial is perfect-square trinomial. For example, $(2x-5)^2$ is the square of a binomial. $(2x-5)(2x-5) = 4x^2 - 20x + 25$, so $4x^2 - 20x + 25$ is a perfect-square trinomial. [7.4A]

12. To solve an equation by factoring, the equation must be set equal to zero in order to use the Principle of Zero Products, which state that if the product of two factors is zero, then at least one of the factors must be zero. [7.5A]

Chapter 7 Review Exercises

1.
$$\begin{aligned} b^2 - 13b + 30 &= b - 10b - 3b + 30 \\ &= (b - 10b) + (-3b + 30) \\ &= b(b-10) - 3(b-10) \\ &= (b-10)(b-3) \end{aligned}$$

2.
$$\begin{aligned} 4x(x-3) - 5(3-x) &= 4x(x-3) + 5(x-3) \\ &= (x-3)(4x+5) \end{aligned}$$

3. $2x^2 - 5x + 6$

Factors of 2	*Negative Factors of* 6
1 and 2	−1 and −6
	−2 and −3

Trial Factors	*Middle Term*
$(x-1)(2x-6)$	Common factor
$(x-2)(2x-3)$	$-3x-4x=-7x$
$(2x-1)(x-6)$	$-12x-x=-13x$
$(2x-2)(x-3)$	Common factor

$2x^2 - 5x + 6$ is nonfactorable over the integers.

4. $21x^4y^4 + 23x^2y^2 + 6$

Let $u = x^2y^2$

$21u^2 + 23u + 6$

$(3u+2)(7u+3)$

$(3x^2y^2+2)(7x^2y^2+3)$

5. $14y^9 - 49y^6 + 7y^3$

The GCF is $7y^3$: $7y^3(2y^6 - 7y^3 + 1)$

6. $y^2 + 5y - 36$

Factors of −36 whose sum is 5: 9 and −4

$$y^2 + 9y - 4y - 36 = y(y+9) - 4(y+9)$$
$$= (y+9)(y-4)$$

7. $6x^2 - 29x + 28$

Factors of 6	*Negative Factors of* 28
1 and 6	−1 and −28
2 and 3	−2 and −14
	−4 and −7

Trial Factors	*Middle Term*
$(x-1)(6x-28)$	Common factor
$(x-4)(6x-7)$	$-7x-24x=-31x$
$(2x-1)(3x-28)$	$-56x-6x=-62x$
$(2x-4)(3x-7)$	Common factor
$(2x-7)(3x-4)$	$-8x-21x=-29x$

$6x^2 - 29x + 28 = (2x-7)(3x-4)$

8. $12a^2b + 3ab^2$

The GCF is $3ab$: $3ab(4a+b)$

9. $a^6 - 100 = (a^3)^2 - 10^2 = (a^3+10)(a^3-10)$

10. $n^4 - 2n^3 - 3n^2$

The GCF is n^2: $n^2(n^2 - 2n - 3)$

Factors of −3 whose sum is −2: −3 and 1

$$n^2(n^2 - 3n + n - 3) = n^2[n(n-3) + 1(n-3)]$$
$$= n^2(n-3)(n+1)$$

11. $12y^2 + 16y - 3$

Factors of 12	*Factors of* −3
1 and 12	1 and −3
2 and 6	−1 and 3
3 and 4	

Trial Factors	*Middle Term*
$(3y+1)(4y-3)$	$-9y+4y=-5y$
$(3y-1)(4y+3)$	$9y-4y=5y$
$(2y+1)(6y-3)$	$-6y+6y=0$
$(6y-1)(2y+3)$	$18y-2y=16y$

$12y^2 + 16y - 3 = (6y-1)(2y+3)$

12. $12b^3 - 58b^2 + 56b$

The GCF is $2b$: $2b(6b^2 - 29b + 28)$

$6 \times 28 = 168$

Factors of 168 whose sum is −29: −21 and −8

$2b(6b^2 - 21b - 8b + 28)$
$2b[3b(2b-7) - 4(2b-7)]$
$2b(2b-7)(3b-4)$

$12b^3 - 58b^2 + 56b = 2b(2b-7)(3b-4)$

13.
$$9y^4 - 25z^2 = (3y^2)^2 - (5z)^2$$
$$= (3y^2 + 5z)(3y^2 - 5z)$$

14. $c^2 + 8c + 12$

Factors of 12 whose sum is 8: 6 and 2

$$c^2 + 6c + 2c + 12 = (c^2 + 6c) + (2c + 12)$$
$$= c(c+6) + 2(c+6)$$
$$= (c+6)(c+2)$$

15. $18a^2 - 3a - 10$

$18(-10) = -180$

Factors of −180 whose sum is −3: −15 and 12

$$18a^2 - 15a + 12a - 10 = 3a(6a-5) + 2(6a-5)$$
$$= (6a-5)(3a+2)$$
$$18a^2 - 3a - 10 = (6a-5)(3a+2)$$

16.
$$4x^2+27x=7$$
$$4x^2+27x-7=0$$
$$(4x-1)(x+7)=0$$

$4x-1=0$ $x+7=0$

$4x=1$ $x=-7$

$x=\frac{1}{4}$

The solutions are $\frac{1}{4}$ and –7.

17. $4x^3-20x^2-24x$

The GCF is $4x$: $4x(x^2-5x-6)$

Factors of –6 whose sum is –5: –6 and 1

$$4x(x^2-5x-6)=4x(x^2-6x+x-6)$$
$$=4x[(x^2-6x)+(x-6)]$$
$$=4x[x(x-6)+1(x-6)]$$
$$=4x(x-6)(x+1)$$

18.
$$64a^3-27b^3=(4a)^3-(3b)^3$$
$$=(4a-3b)(16a^2+12ab+9b^2)$$

19.
$$2a^2-19a-60$$
$$2(-60)=-120$$

Factors of –120 whose sum is –19: –24 and 5

$$2a^2-24a+5a-60$$
$$=(2a^2-24a)+(5a-60)$$
$$=2a(a-12)+5(a-12)$$
$$=(a-12)(2a+5)$$
$$2a^2-19a-60=(a-12)(2a+5)$$

20.
$$(x+1)(x-5)=16$$
$$x^2-4x-5=16$$
$$x^2-4x-21=0$$
$$(x-7)(x+3)=0$$

$x-7=0$ $x+3=0$

$x=7$ $x=-3$

The solutions are 7 and –3.

21.
$$21ax-35bx-10by+6ay$$
$$=(21ax-35bx)+(-10by+6ay)$$
$$=7x(3a-5b)+2y(-5b+3a)$$
$$=7x(3a-5b)+2y(3a-5b)$$
$$=(3a-5b)(7x+2y)$$

22.
$$36x^8-36x^4+5$$

Let $u=x^4$

$$36u^2-36u+5$$
$$(6u-5)(6u-1)$$
$$(6x^4-5)(6x^4-1)$$

23.
$$10x^2+25x+4xy+10y$$
$$=(10x^2+25x)+(4xy+10y)$$
$$=5x(2x+5)+2y(2x+5)$$
$$=(2x+5)(5x+2y)$$

24. $5x^2-5x-30$

The GCF is 5: $5(x^2-x-6)$

$1\times(-6)=-6$ Factors of –6 whose sum is –1: –3 and 2

$$5(x^2-3x+2x-6)$$
$$=5[(x^2-3x)+(2x-6)]$$
$$=5[x(x-3)+2(x-3)]=5(x-3)(x+2)$$
$$5x^2-5x-30=5(x-3)(x+2)$$

25. $3x^2+36x+108$

The GCF is 3: $3(x^2+12x+36)$

$\sqrt{x^2}=x$ $2(x\cdot 6)=12x$

$\sqrt{36}=6$ The trinomial is a perfect square.

$$3x^2+36x+108=3(x+6)^2$$

26.
$$3x^2-17x+10$$
$$3\cdot 10=30$$

Factors of 30 whose sum is –17: –15 and –2

$$3x^2-15x-2x+10$$
$$=(3x^2-15x)+(-2x+10)$$
$$=3x(x-5)-2(x-5)=(x-5)(3x-2)$$
$$3x^2-17x+10=(x-5)(3x-2)$$

27. **Strategy** • Width: x

Length: $2x-20$

• Use the equation for the area of a rectangle:

$A = LW$

Solution

$$LW = A$$
$$x(2x-20)=6000$$
$$2x^2-20x-6000=0$$
$$2(x^2-10x-3000)=0$$
$$2(x-60)(x+50)=0$$
$$x-60=0 \quad x+50=0$$
$$x=60 \quad x=-50$$

Because the width of a rectangle cannot be a negative number, –50 is not a solution.

$2x-20=120-20=100$

The width is 60 yd and the length is 100 yd.

28. **Strategy** Known: $S = 400$

Unknown: d

Solution

$$S=d^2$$
$$400=d^2$$
$$0=d^2-400$$
$$0=(d+20)(d-20)$$
$$d+20=0 \quad d-20=0$$
$$d=-20 \quad d=20$$

Since the distance cannot be negative, −20 is not a solution.

The distance is 20 ft.

29. **Strategy** •Width of the picture frame: x

New width: $12+2x$

New length: $15+2x$

•Use the equation for the area of a rectangle.

Solution

$$LW = A$$
$$(12+2x)(15+2x)=270$$
$$180+24x+30x+4x^2=270$$
$$4x^2+54x-90=0$$
$$2(2x^2+27x-45)=0$$
$$2(2x-3)(x+15)=0$$
$$2x-3=0 \quad x+15=0$$
$$2x=3 \quad x=-15$$
$$x=1.5$$

Because the width of a picture frame cannot be a negative number, –15 is not a solution. The width of the frame is 1.5 in.

30. **Strategy** • Side of original square: x

Side of new square: $x+4$

• Use the equation for the area of a square.

Solution

$$s^2=A$$
$$(x+4)^2=576$$
$$x^2+8x+16=576$$
$$x^2+8x-560=0$$
$$(x+28)(x-20)=0$$
$$x+28=0 \quad x-20=0$$
$$x=-28 \quad x=20$$

Since the side cannot be negative, −28 is not a solution. The side of the original square is 20 ft.

Chapter 7 Test

1.
$$ab+6a-3b-18$$
$$=(ab+6a)+(-3b-18)$$
$$=a(b+6)-3(b+6)=(b+6)(a-3)$$

2.
$$2y^4-14y^3-16y^2$$
$$=2y^2(y^2-7y-8)$$
$$=2y^2[y^2-8y+y-8]$$
$$=2y^2[(y^2-8y)+(y-8)]$$
$$=2y^2[y(y-8)+(y-8)]$$
$$=2y^2(y+1)(y-8)$$

3. $8x^2 + 20x - 48$

$8(-48) = -384$

Factors of –384 whose sum is 20: 32 and –12

$8x^2 + 20x - 48$

$= 8x^2 - 12x + 32x - 48$

$= (8x^2 - 12x) + (32x - 48)$

$= 4x(2x - 3) + 16(2x - 3)$

$= (4x + 16)(2x - 3) = 4(x + 4)(2x - 3)$

4. There is no common factor.
Factor the trinomial.
Positive

Factors of 6	*Positive Factors of* 8
1 and 6	1 and 8
2 and 3	2 and 4
Trial Factors	*Middle Term*
$(x+1)(6x+8)$	Common factor
$(x+8)(6x+1)$	$x + 48x = 49x$
$(2x+1)(3x+8)$	$16x + 3x = 19x$
$(2x+8)(3x+1)$	Common factor

$6x^2 + 19x + 8 = (2x + 1)(3x + 8)$

5.

Factors	*Sum*
−1, −48	−49
−2, −24	−26
−3, −16	−19
−4, −12	−16
−6, −8	−14

$a^2 - 19a + 48 = (a - 3)(a - 16)$

6. $6x^3 - 8x^2 + 10x = 2x(3x^2 - 4x + 5)$

7.

Factors	*Sum*
−1, +15	+14
+1, −15	−14
−3, +5	+2
+3, −5	−2

$x^2 + 2x - 15 = (x + 5)(x - 3)$

8. $4x^2 - 1 = 0$

$(2x - 1)(2x + 1) = 0$

$2x - 1 = 0 \quad 2x + 1 = 0$

$x = \frac{1}{2} \quad x = -\frac{1}{2}$

The solutions are $-\frac{1}{2}$ and $\frac{1}{2}$.

9. $5x^2 - 45x - 15 = 5(x^2 - 9x - 3)$

10. $p^2 + 12p + 36 = (p + 6)^2$

11. $x(x - 8) = -15$

$x^2 - 8x = -15$

$x^2 - 8x + 15 = 0$

$(x - 3)(x - 5) = 0$

$x - 3 = 0 \quad x - 5 = 0$

$x = 3 \quad x = 5$

The solutions are 3 and 5.

12. $3x^2 + 12xy + 12y^2 = 3(x^2 + 4xy + 4y^2)$

$3(x + 2y)(x + 2y) = 3(x + 2y)^2$

13. $b^2 - 16 = (b + 4)(b - 4)$

14. $6x^2y^2 + 9xy^2 + 3y^2$

Common Factor $3y^2$:

$3y^2(2x^2 + 3x + 1)$

$2 \cdot 2 = 2$

Factors of 2 whose sum is 3: 2, 1

$3y^2(2x^2 + 3x + 1) = 3y^2(2x^2 + 2x + x + 1)$

$= 3y^2[(2x^2 + 2x) + (x + 1)]$

$= 3y^2[2x(x + 1) + (x + 1)]$

$= 3y^2(2x + 1)(x + 1)$

15. $27x^3 - 8 = (3x)^2 - (2)^2$

$= (3x - 2)(9x^2 + 6x + 4)$

16. $6a^4 - 13a^2 - 5$

Let $u = a^2$

$6u^2 - 13u - 5$

$(2u - 5)(3u + 1)$

$(2a^2 - 5)(3a^2 + 1)$

17. $x(p + 1) - (p + 1) = (p + 1)(x - 1)$

18. $3a^2 - 75 = 3(a^2 - 25)$

$= 3(a + 5)(a - 5)$

19. There is no common factor.
Factor the trinomial.

Factors of 2	*Factors of* −5
1 and 2	−1 and 5
	1 and −5

Trial Factors	*Middle Term*
$(x-1)(2x+5)$	$5x-2x=3x$
$(x+1)(2x-5)$	$-5x+2x=-3x$

$2x^2+4x-5$ is nonfactorable over the integers.

20.

Factors	*Sum*
−1, +36	+35
+1, −36	−35
−2, +18	+16
+2, −18	−16
−3, +12	+9
+3, −12	−9
−4, +9	+5
+4, −9	−5
−6, +6	0

$x^2-9x-36=(x+3)(x-12)$

21. $4a^2-12ab+9b^2=(2a-3b)^2$

22. $4x^2-49y^2=(2x+7y)(2x-7y)$

23. $(2a-3)(a+7)=0$

$2a-3=0 \quad a+7=0$

$a=\frac{3}{2} \quad a=-7$

The solutions are $\frac{3}{2}$ and −7.

24. **Strategy** The first number: x
The second number: $10-x$
The sum of the squares of the two numbers is fifty-eight.

Solution

$x^2+(-x+10)^2=58$

$2x^2-20x+100=58$

$2x^2-20x+42=0$

$(2x-6)(x-7)=0$

$2x-6=0 \quad x-7=0$

$x=3 \quad x=7$

The numbers are 3 and 7.

25. **Strategy** The width of the rectangle: w
The length of a rectangle is 3 cm more than twice its width.
Length of the rectangle: $2w+3$
The area of the rectangle is 90 cm^2.

Solution

$(2w+3)w=90$

$2w^2+3w=90$

$2w^2+3w-90=0$

$(2w+15)(w-6)=0$

$2w+15=0 \quad w-6=0$

$w=-7.5 \quad w=6$

Because −7.5 is not positive, it cannot be a solution. Therefore the width is 6 cm and the length is 15 cm.

Cumulative Review Exercises

1. $-2-(-3)-5-(-11)=-2+3-5+11=1-5+11$
$=-4+11=7$

2. $(3-7)^2\div(-2)-3(-4)=(-4)^2\div(-2)-3(-4)$
$=16\div(-2)-(-12)$
$=-8+12=4$

3. $-2a^2\div(2b)-c$

$-2(-4)^2\div[2(2)]-(-1)$

$=-2(16)\div4+1$

$=-32\div4+1=-8+1=-7$

4. $-\frac{3}{4}(-20x^2)=\frac{3\cdot20x^2}{4}=15x^2$

5. $-2[4x-2(3-2x)-8x]=-2[4x-6+4x-8x]$
$=-2[-6]=12$

6. $-\frac{5}{7}x=-\frac{10}{21}$

$-\frac{7}{5}\left(-\frac{5}{7}x\right)=-\frac{7}{5}\left(-\frac{10}{21}\right)$

$x=\frac{2}{3}$

The solution is $\frac{2}{3}$.

7.

$$\begin{aligned} 3x-2&=12-5x \\ 3x+5x&=12+2 \\ 8x&=14 \\ x&=\frac{14}{8} \\ x&=\frac{7}{4} \end{aligned}$$

The solution is $\frac{7}{4}$.

8.

$$\begin{aligned} -2+4[3x-2(4-x)-3]&=4x+2 \\ -2+4[3x-8+2x-3]&=4x+2 \\ -2+4[5x-11]&=4x+2 \\ -2+20x-44&=4x+2 \\ 20x-46&=4x+2 \\ 16x&=48 \\ x&=3 \end{aligned}$$

The solution is 3.

9.

$$\begin{aligned} P\cdot B&=A \\ 120\%\cdot B&=54 \\ 1.2B&=54 \\ B&=\frac{54}{1.2} \\ B&=45 \end{aligned}$$

The number is 45.

10.

$$\begin{aligned} f(x)&=-x^2+3x-1 \\ f(2)&=-(2)^2+3(2)-1 \\ f(2)&=-4+6-1 \\ f(2)&=1 \end{aligned}$$

11. $y=\frac{1}{4}x+3$

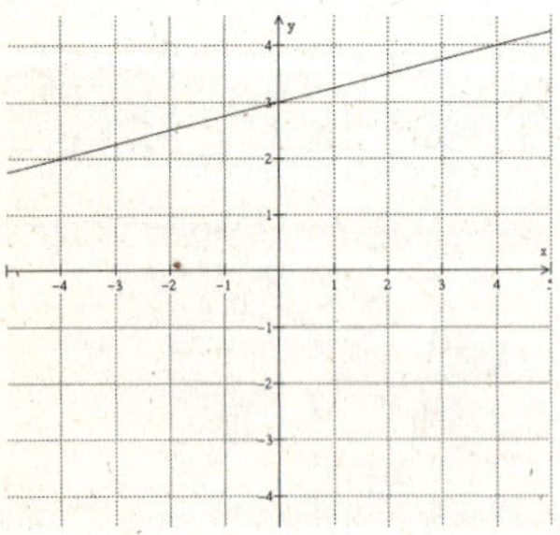

12. $5x+3y=15$

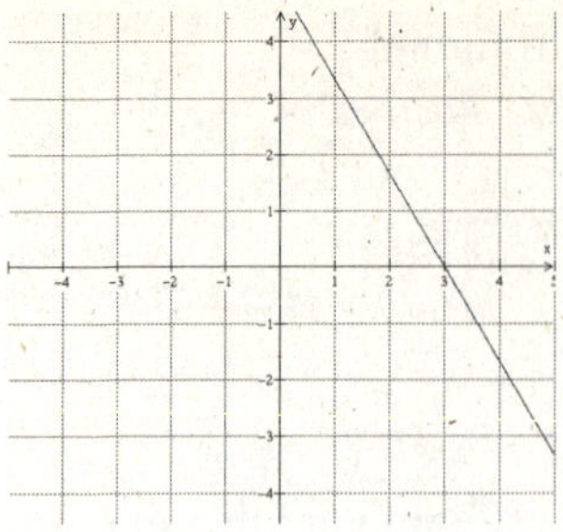

13.

$$\begin{aligned} y-4&=\frac{2}{3}(x-(-3)) \\ y-4&=\frac{2}{3}(x+3) \\ y-4&=\frac{2}{3}x+2 \\ y&=\frac{2}{3}x+6 \end{aligned}$$

14.

$$\begin{aligned} 8x-y&=2 \\ y&=5x+1 \end{aligned}$$

$$\begin{aligned} 8x-(5x+1)&=2 \\ 8x-5x-1&=2 \\ 3x&=3 \\ x&=1 \end{aligned}$$

$$\begin{aligned} y&=5x+1 \\ y&=5(1)+1 \\ y&=6 \end{aligned}$$

The solution is (1, 6).

15.

$$\begin{aligned} 5x+2y&=-9 \qquad -12[5x+2y=-9] \\ 12x-7y&=2 \qquad 5[12x-7y=2] \end{aligned}$$

$$\begin{aligned} -60x-24y&=108 \\ 60x-35y&=10 \\ \hline -59y&=118 \\ y&=-2 \end{aligned}$$

$$\begin{aligned} 5x+2y&=-9 \\ 5x+2(-2)&=-9 \\ 5x-4&=-9 \\ 5x&=-5 \\ x&=-1 \end{aligned}$$

The solution is (–1, –2)

16. $\left(-3a^3b^2\right)^2=9a^6b^4$

17.
$$\begin{array}{r} x^2 - 5x + 4 \\ x + 2 \\ \hline 2x^2 - 10x + 8 \\ x^3 - 5x^2 + 4x \quad \\ \hline x^3 - 3x^2 - 6x + 8 \end{array}$$

18.
$$\begin{array}{r} 4x + 8 \\ 2x-3\overline{)8x^2 + 4x - 3} \\ \underline{8x^2 - 12x} \quad \\ 16x - 3 \\ \underline{16x - 24} \\ 21 \end{array}$$

$$(8x^2+4x-3)\div(2x-3) = 4x+8+\frac{21}{2x-3}$$

19. $(x^{-4}y^3)^2 = x^{-8}y^6 = \dfrac{y^6}{x^8}$

20.
$$\begin{aligned} 3a-3b-ax+bx &= (3a-3b)+(-ax+bx) \\ &= 3(a-b)-x(a-b) \\ &= (a-b)(3-x) \end{aligned}$$

21. $15xy^2 - 20xy^4$

The GCF is $5xy^2$: $5xy^2(3-4y^2)$

22. $x^2 - 5xy - 14y^2$

Factors of –14 whose sum is –5: –7 and 2

$$\begin{aligned} x^2-7xy+2xy-14y^2 &= (x^2-7xy)+(2xy-14y^2) \\ &= x(x-7y)+2y(x-7y) \\ &= (x-7y)(x+2y) \end{aligned}$$

23.
$$3x^2 + 19x - 14 = 0$$
$$(3x-2)(x+7) = 0$$
$$3x - 2 = 0 \qquad x + 7 = 0$$
$$3x = 2 \qquad x = -7$$
$$x = \frac{2}{3}$$

The solutions are $\frac{2}{3}$ and –7.

24.
$$6x^2 + 60 = 39x$$
$$6x^2 - 39x + 60 = 0$$
$$3(2x^2 - 13x + 20) = 0$$
$$3(2x-5)(x-4) = 0$$
$$2x - 5 = 0 \qquad x - 4 = 0$$
$$2x = 5 \qquad x = 4$$
$$x = \frac{5}{2}$$

The solutions are $\frac{5}{2}$ and 4.

25. **Strategy** The sum of the measure of the angles of a triangle is 180^0.

Solution

$$31° + 90° + x = 180°$$
$$121° + x = 180°$$
$$x = 59°$$

The measure of the third angle is 59°.

26. **Strategy** Use the formula for the perimeter of a rectangle where $P = 86$ and $L = 28$.

Solution

$$P = 2L + 2W$$
$$86 = 2(28) + 2W$$
$$86 = 56 + 2W$$
$$30 = 2W$$
$$15 = W$$

The width is 15 ft.

27. **Strategy** • Shorter piece: x

Longer piece: $10 - x$

• Four times the length of the shorter piece is 2 ft less than three times the length of the longer piece.

Solution

$$4x = 3(10-x)-2$$
$$4x = 30 - 3x - 2$$
$$4x = 28 - 3x$$
$$7x = 28$$
$$x = 4$$
$$10 - x = 10 - 4 = 6$$

The shorter piece is 4 ft long and the longer piece is 6 ft long.

28. **Strategy** Amount invested at 4.5%: x

Amount invested at 4%: $15000 - x$

	Amount	Rate	Interest
4.5%	x	0.045	$0.045x$
4%	$15000 - x$	0.04	$0.04(15000 - x)$

The total interest is $635.

Solution

$$0.045x + 0.04(15000 - x) = 635$$
$$0.045x + 600 - 0.04 = 635$$
$$0.005x = 35$$
$$x = 7000$$
$$15000 - x = 15000 - 7000 = 8000$$

$8000 was invested at 4%.

29. **Strategy** • Time driving to resort: x

	Rate	Time	Distance
To resort	42	x	$42x$
From resort	56	$7 - x$	$56(7 - x)$

• The distances are the same.

Solution

$$42x = 56(7 - x)$$
$$42x = 392 - 56x$$
$$98x = 392$$
$$x = 4$$
$$42x = 42(4) = 168$$

The distance to the resort is 168 mi.

30. **Strategy** • Height: x

Base: $3x$

• Use the equation for the area of a triangle.

Solution

$$\frac{1}{2}bh = A$$
$$\frac{1}{2}(3x)(x) = 24$$
$$\frac{3}{2}x^2 = 24$$
$$\frac{2}{3}\left(\frac{3}{2}x^2\right) = 24 \cdot \frac{2}{3}$$
$$x^2 = 16$$
$$x = 4$$
$$3x = 3(4) = 12$$

The base of the triangle is 12 in.

Chapter 8 Rational Expressions

Prep Test

1. $10 = 2 \cdot 5$
   $25 = 5 \cdot 5$
   $\text{LCM} = 2 \cdot 5 \cdot 5 = 50$

2. $-\frac{3}{8} \cdot \frac{4}{9} = -\frac{\overset{1}{\cancel{3}} \cdot \overset{1}{\cancel{2}} \cdot \overset{1}{\cancel{2}}}{\underset{1}{\cancel{2}} \cdot \underset{1}{\cancel{2}} \cdot 2 \cdot \underset{1}{\cancel{3}} \cdot 3} = -\frac{1}{6}$

3. $-\frac{4}{5} \div \frac{8}{15} = -\frac{4}{5} \cdot \frac{15}{8} = -\frac{\overset{1}{\cancel{2}} \cdot \overset{1}{\cancel{2}} \cdot 3 \cdot \overset{1}{\cancel{5}}}{\underset{1}{\cancel{5}} \cdot \underset{1}{\cancel{2}} \cdot \underset{1}{\cancel{2}} \cdot 2} = -\frac{3}{2}$

4. $-\frac{5}{6} + \frac{7}{8} = -\frac{20}{24} + \frac{21}{24} = \frac{1}{24}$

5. $-\frac{3}{8} - \left(\frac{7}{12}\right) = -\frac{9}{24} + \frac{14}{24} = \frac{5}{24}$

6. $\frac{2x-3}{x^2-x+1}$
   $\frac{2(2)-3}{(2)^2-2+1} = \frac{4-3}{4-2+1} = \frac{1}{3}$

7. $4(2x+1) = 3(x-2)$
   $8x+4 = 3x-6$
   $5x = -10$
   $x = -2$

8. $10\left(\frac{t}{2} + \frac{t}{5}\right) = 10(1)$
   $5t + 2t = 10$
   $7t = 10$
   $t = \frac{10}{7}$

9. **Strategy** • rate of second plane: r
   • rate of first plane: $r - 20$

	Rate	Time	Distance
1st Plane	$r-20$	2	$2(r-20)$
2nd Plane	r	2	$2r$

•The sum of the two distances is 480 mi.

Solution

$2(r-20) + 2r = 480$
$2r - 40 + 2r = 480$
$4r = 520$
$r = 130$
$r - 20 = 130 - 20 = 110$

The first plane is flyng 110 mph, the second is flying 130 mph.

Section 8.1

Objective A Exercises

1. To write a rational expression in simplest form, factor the numerator and denominator. Then divide by the common factors.

3. $\frac{9x^3}{12x^4} = \frac{9}{12x^{4-3}} = \frac{3}{4x}$

5. $\frac{(x+3)^2}{(x+3)^3} = \frac{1}{(x+3)^{3-2}} = \frac{1}{x+3}$

7. $\frac{3n-4}{4-3n} = \frac{-\overset{1}{\cancel{(4-3n)}}}{\underset{1}{\cancel{4-3n}}} = -1$

9. $\frac{6y(y+2)}{9y^2(y+2)} = \frac{6\overset{1}{\cancel{(y+2)}}}{9y^{2-1}\underset{1}{\cancel{(y+2)}}} = \frac{2}{3y}$

11. $\frac{6x(x-5)}{8x^2(5-x)} = \frac{6\overset{1}{\cancel{(x-5)}}}{8x^{2-1}(-1)\underset{1}{\cancel{(x-5)}}} = -\frac{3}{4x}$

13. $\frac{a^2+4a}{ab+4b} = \frac{a\overset{1}{\cancel{(a+4)}}}{b\underset{1}{\cancel{(a+4)}}} = \frac{a}{b}$

15. $\frac{4-6x}{3x^2-2x} = \frac{-1(2)\overset{1}{\cancel{(3x-2)}}}{x\underset{1}{\cancel{(3x-2)}}} = -\frac{2}{x}$

17. $\frac{y^2-3y+2}{y^2-4y+3} = \frac{\overset{1}{\cancel{(y-1)}}(y-2)}{\underset{1}{\cancel{(y-1)}}(y-3)} = \frac{y-2}{y-3}$

19. $\dfrac{x^2+3x-10}{x^2+2x-8}=\dfrac{(x+5)\cancel{(x-2)}}{(x+4)\cancel{(x-2)}}=\dfrac{x+5}{x+4}$

21. $\dfrac{x^2+x-12}{x^2-6x+9}=\dfrac{(x+4)\cancel{(x-3)}}{(x-3)\cancel{(x-3)}}=\dfrac{x+4}{x-3}$

23. $\dfrac{x^2-3x-10}{25-x^2}=\dfrac{(x+2)\cancel{(x-5)}}{-1(x+5)\cancel{(x-5)}}=-\dfrac{x+2}{x+5}$

25. $\dfrac{2x^3+2x^2-4x}{x^3+2x^2-3x}$

$=\dfrac{2x(x^2+x-2)}{x(x^2+2x-3)}=\dfrac{2x(x+2)\cancel{(x-1)}}{x(x+3)\cancel{(x-1)}}=\dfrac{2(x+2)}{x+3}$

27. $\dfrac{6x^2-7x+2}{6x^2+5x-6}=\dfrac{\cancel{(3x-2)}(2x-1)}{\cancel{(3x-2)}(2x+3)}=\dfrac{2x-1}{2x+3}$

29. $\dfrac{x^2+3x-28}{24-2x-x^2}=\dfrac{(x+7)\cancel{(x-4)}}{-1(x+6)\cancel{(x-4)}}=-\dfrac{x+7}{x+6}$

Objective B Exercises

31. $\dfrac{8x^2}{9y^3}\cdot\dfrac{3y^2}{4x^3}=\dfrac{8\cdot 3}{9y^{3-2}4x^{3-2}}=\dfrac{2}{3xy}$

33. $\dfrac{12x^3y^4}{7a^2b^3}\cdot\dfrac{14a^3b^4}{9x^2y^2}=\dfrac{12x^{3-2}y^{4-2}\cdot 14a^{3-2}b^{4-3}}{7\cdot 9}=\dfrac{8abxy^2}{3}$

35. $\dfrac{3x-6}{5x-20}\cdot\dfrac{10x-40}{27x-54}=\dfrac{3(x-2)}{5(x-4)}\cdot\dfrac{10(x-4)}{27(x-2)}$

$=\dfrac{3\cancel{(x-2)}10\cancel{(x-4)}}{5\cancel{(x-4)}\cdot 27\cancel{(x-2)}}=\dfrac{2}{9}$

37. $\dfrac{3x^2+2x}{2xy-3y}\cdot\dfrac{2xy^3-3y^3}{3x^3+2x^2}$

$=\dfrac{x(3x+2)}{y(2x-3)}\cdot\dfrac{y^3(2x-3)}{x^2(3x+2)}$

$=\dfrac{\cancel{(3x+2)}}{\cancel{(2x-3)}}\cdot\dfrac{y^{3-2}\cancel{(2x-3)}}{x^{2-1}\cancel{(3x+2)}}=\dfrac{y^2}{x}$

39. $\dfrac{x^2+5x+4}{x^3y^2}\cdot\dfrac{x^2y^3}{x^2+2x+1}$

$=\dfrac{(x+4)\cancel{(x+1)}}{x^{3-2}}\cdot\dfrac{y^{3-2}}{\cancel{(x+1)}(x+1)}=\dfrac{y(x+4)}{x(x+1)}$

41. $\dfrac{x^4y^2}{x^2+3x-28}\cdot\dfrac{x^2-49}{xy^4}$

$=\dfrac{x^{4-1}}{\cancel{(x+7)}(x-4)}\cdot\dfrac{(x-7)\cancel{(x+7)}}{y^{4-2}}=\dfrac{x^3(x-7)}{y^2(x-4)}$

43. $\dfrac{2x^2-5x}{2xy+y}\cdot\dfrac{2xy^2+y^2}{5x^2-2x^3}$

$=\dfrac{x(2x-5)}{y(2x+1)}\cdot\dfrac{y^2(2x+1)}{-x^2(2x-5)}$

$=\dfrac{\cancel{(2x-5)}}{\cancel{(2x+1)}}\cdot\dfrac{y^{2-1}\cancel{(2x+1)}}{-1x^{2-1}\cancel{(2x-5)}}=-\dfrac{y}{x}$

45. $\dfrac{x^2-2x-24}{x^2-5x-6}\cdot\dfrac{x^2+5x+6}{x^2+6x+8}$

$=\dfrac{\cancel{(x+4)}\cancel{(x-6)}}{(x+1)\cancel{(x-6)}}\cdot\dfrac{\cancel{(x+2)}(x+3)}{\cancel{(x+2)}\cancel{(x+4)}}=\dfrac{x+3}{x+1}$

47. $\dfrac{x^2+2x-35}{x^2+4x-21}\cdot\dfrac{x^2+3x-18}{x^2+9x+18}$

$=\dfrac{\cancel{(x+7)}(x-5)}{\cancel{(x+7)}\cancel{(x-3)}}\cdot\dfrac{\cancel{(x+6)}\cancel{(x-3)}}{(x+3)\cancel{(x+6)}}=\dfrac{x-5}{x+3}$

49. $\dfrac{x^2-3x-4}{x^2+6x+5}\cdot\dfrac{x^2+5x+6}{8+2x-x^2}$

$=\dfrac{\cancel{(x-4)}\cancel{(x+1)}}{(x+5)\cancel{(x+1)}}\cdot\dfrac{\cancel{(x+2)}(x+3)}{\cancel{(4-x)}\cancel{(2+x)}}=-\dfrac{x+3}{x+5}$

51. $\dfrac{16+6x-x^2}{x^2-10x-24}\cdot\dfrac{x^2-6x-27}{x^2-17x+72}$

$=\dfrac{\cancel{(8-x)}\cancel{(2+x)}}{\cancel{(x+2)}(x-12)}\cdot\dfrac{(x+3)\cancel{(x-9)}}{\cancel{(x-8)}\cancel{(x-9)}}=-\dfrac{x+3}{x-12}$

53. $\dfrac{2x^2+5x+2}{2x^2+7x+3}\cdot\dfrac{x^2-7x-30}{x^2-6x-40}$

$=\dfrac{\cancel{(2x+1)}(x+2)}{\cancel{(2x+1)}\cancel{(x+3)}}\cdot\dfrac{\cancel{(x+3)}\cancel{(x-10)}}{(x+4)\cancel{(x-10)}}=\dfrac{x+2}{x+4}$

55. Since $a>d$, the x will be in the numerator, and since $c>b$, the y will also be in the numerator, the denominator will be 1.

57. Since $a<d$, the x will be in the denominator, and since $c=b$, there will be no y variable. The numerator will be 1.

Objective C Exercises

59. To divide rational expressions, multiply by the reciprocal of the divisor.

61. $\dfrac{9x^3y^4}{16a^4b^2}\div\dfrac{45x^4y^2}{14a^7b}$

$=\dfrac{9x^3y^4}{16a^4b^2}\cdot\dfrac{14a^7b}{45x^4y^2}=\dfrac{9y^{4-2}\cdot14a^{7-4}}{16b^{2-1}\cdot45x^{4-3}}=\dfrac{7a^3y^2}{40bx}$

63. $\dfrac{28x+14}{45x-30}\div\dfrac{14x+7}{30x-20}$

$=\dfrac{14(2x+1)}{15(3x-2)}\cdot\dfrac{10(3x-2)}{7(2x+1)}$

$=\dfrac{14\cancel{(2x+1)}\cdot10\cancel{(3x-2)}}{15\cancel{(3x-2)}\cdot7\cancel{(2x+1)}}=\dfrac{4}{3}$

65. $\dfrac{5a^2y+3a^2}{2x^3+5x^2}\div\dfrac{10ay+6a}{6x^3+15x^2}$

$=\dfrac{a^2(5y+3)}{x^2(2x+5)}\cdot\dfrac{3x^2(2x+5)}{2a(5y+3)}$

$=\dfrac{a^{2-1}\cancel{(5y+3)}}{x^2\cancel{(2x+5)}}\cdot\dfrac{3x^2\cancel{(2x+5)}}{2\cancel{(5y+3)}}=\dfrac{3a}{2}$

67. $\dfrac{x^3y^2}{x^2-3x-10}\div\dfrac{xy^4}{x^2-x-20}$

$=\dfrac{x^3y^2}{(x-5)(x+2)}\cdot\dfrac{(x-5)(x+4)}{xy^4}$

$=\dfrac{x^{3-1}}{\cancel{(x-5)}(x+2)}\cdot\dfrac{\cancel{(x-5)}(x+4)}{y^{4-2}}=\dfrac{x^2(x+4)}{y^2(x+2)}$

69. $\dfrac{x^2y^5}{x^2-11x+30}\div\dfrac{xy^6}{x^2-7x+10}$

$=\dfrac{x^{2-1}}{\cancel{(x-5)}(x-6)}\cdot\dfrac{(x-2)\cancel{(x-5)}}{y^{6-5}}=\dfrac{x(x-2)}{y(x-6)}$

71. $\dfrac{3x^2y-9xy}{a^2b}\div\dfrac{3x^2-x^3}{ab^2}$

$=\dfrac{3xy(x-3)}{a^2b}\cdot\dfrac{ab^2}{x^2(3-x)}$

$=\dfrac{3xy\cancel{(x-3)}}{a^{2-1}}\cdot\dfrac{b^{2-1}}{x^{2-1}\cancel{(3-x)}}=-\dfrac{3by}{ax}$

73. $\dfrac{x^2+3x-40}{x^2+2x-35}\div\dfrac{x^2+2x-48}{x^2+3x-18}$

$=\dfrac{\cancel{(x+8)}\cancel{(x-5)}}{(x+7)\cancel{(x-5)}}\cdot\dfrac{(x+6)(x-3)}{\cancel{(x+8)}(x-6)}=\dfrac{(x+6)(x-3)}{(x+7)(x-6)}$

75. $\dfrac{y^2-y-56}{y^2+8y+7}\div\dfrac{y^2-13y+40}{y^2-4y-5}$

$=\dfrac{\cancel{(y+7)}\cancel{(y-8)}}{\cancel{(y+1)}\cancel{(y+7)}}\cdot\dfrac{\cancel{(y+1)}\cancel{(y-5)}}{\cancel{(y-5)}\cancel{(y-8)}}=1$

77. $\frac{x^2-x-2}{x^2-7x+10} \div \frac{x^2-3x-4}{40-3x-x^2}$

$= \frac{\overset{1}{\cancel{(x+1)}}\,\overset{1}{\cancel{(x-2)}}}{\underset{1}{\cancel{(x-2)}}\,\underset{1}{\cancel{(x-5)}}} \cdot \frac{(8+x)\,\overset{-1}{\cancel{(5-x)}}}{\underset{1}{\cancel{(x+1)}}\,(x-4)} = -\frac{x+8}{x-4}$

79. $\frac{6n^2+13n+6}{4n^2-9} \div \frac{6n^2+n-2}{4n^2-1}$

$= \frac{\overset{1}{\cancel{(2n+3)}}\,\overset{1}{\cancel{(3n+2)}}}{\underset{1}{\cancel{(2n+3)}}\,(2n-3)} \cdot \frac{(2n+1)\,\overset{1}{\cancel{(2n-1)}}}{\underset{1}{\cancel{(3n+2)}}\,\underset{1}{\cancel{(2n-1)}}} = \frac{2n+1}{2n-3}$

81. $\frac{x+1}{x+6} \div \frac{x-1}{x-4} = \frac{x+1}{x+6} \cdot \frac{x-4}{x-1}$

$= \frac{(x+1)(x-4)}{(x+6)(x-1)} = \frac{x^2-3x-4}{x^2+5x-6}$ Yes.

83. $\frac{x-1}{x+1} \div \frac{x-4}{x+6} = \frac{x-1}{x+1} \cdot \frac{x+6}{x-4}$

$= \frac{(x-1)(x+6)}{(x+1)(x-4)} = \frac{x^2+5x-6}{x^2-3x-4}$ No.

85. $\frac{\text{shaded area}}{\text{total area}} = \frac{\pi(2x)^2}{\pi(5x)^2} = \frac{4x^2\pi}{25x^2\pi} = \frac{4}{25}$

Section 8.2

Objective A Exercises

1. a) six
b) four

3. The LCM is $12x^2y^4$.

$\frac{3}{4x^2y} = \frac{3}{4x^2y} \cdot \frac{3y^3}{3y^3} = \frac{9y^3}{12x^2y^4}$

$\frac{17}{12xy^4} = \frac{17}{12xy^4} \cdot \frac{x}{x} = \frac{17x}{12x^2y^4}$

5. The LCM is $6x^2(x-2)$.

$\frac{x-2}{3x(x-2)} = \frac{x-2}{3x(x-2)} \cdot \frac{2x}{2x} = \frac{2x^2-4x}{6x^2(x-2)}$

$\frac{3}{6x^2} = \frac{3}{6x^2} \cdot \frac{x-2}{x-2} = \frac{3x-6}{6x^2(x-2)}$

7. The LCM is $2x(x-5)$.

$\frac{3x-1}{2x(x-5)}$

$-3x = \frac{-3x}{1} = \frac{-3x}{1} \cdot \frac{2x(x-5)}{2x(x-5)} = -\frac{6x^3-30x^2}{2x(x-5)}$

9. The LCM is $(2x-3)(2x+3)$.

$\frac{3x}{2x-3} = \frac{3x}{2x-3} \cdot \frac{2x+3}{2x+3} = \frac{6x^2+9x}{(2x-3)(2x+3)}$

$\frac{5x}{2x+3} = \frac{5x}{2x+3} \cdot \frac{2x-3}{2x-3} = \frac{10x^2-15x}{(2x-3)(2x+3)}$

11. The LCM is $(x-3)(x+3)$.

$\frac{2x}{x^2-9} = \frac{2x}{(x-3)(x+3)}$

$\frac{x+1}{x-3} = \frac{x+1}{x-3} \cdot \frac{x+3}{x+3} = \frac{x^2+4x+3}{(x-3)(x+3)}$

13. $3x^2-12y^2 = 3(x+2y)(x-2y)$
$6x-12y = 6(x-2y)$
The LCM is $6(x+2y)(x-2y)$.

$\frac{3}{3x^2-12y^2} = \frac{3}{3(x+2y)(x-2y)} \cdot \frac{2}{2}$

$= \frac{6}{6(x+2y)(x-2y)}$

$\frac{5}{6x-12y} = \frac{5}{6(x-2y)} \cdot \frac{x+2y}{x+2y}$

$= \frac{5x+10y}{6(x-2y)(x+2y)}$

15. $x^2-1=(x+1)(x-1)$

$x^2-2x+1=(x-1)(x-1)$

The LCM is $(x+1)(x-1)(x-1)=(x+1)(x-1)^2$.

$$\frac{3x}{x^2-1}=\frac{3x}{(x+1)(x-1)}\cdot\frac{x-1}{x-1}$$

$$=\frac{3x^2-3x}{(x+1)(x-1)^2}$$

$$\frac{5x}{x^2-2x+1}=\frac{5x}{(x-1)(x-1)}\cdot\frac{x+1}{x+1}$$

$$=\frac{5x^2+5x}{(x+1)(x-1)^2}$$

17. $8-x^3=-(x^3-8)=-(x-2)(x^2+2x+4)$

The LCM is $-(x-2)(x^2+2x+4)$.

$$\frac{x-3}{8-x^3}=-\frac{x-3}{(x-2)(x^2+2x+4)}$$

$$\frac{2}{4+2x+x^2}=\frac{2}{x^2+2x+4}\cdot\frac{x-2}{x-2}$$

$$=\frac{2x-4}{(x-2)(x^2+2x+4)}$$

19. $x^2+2x-3=(x+3)(x-1)$

$x^2+6x+9=(x+3)(x+3)$

The LCM is

$(x-1)(x+3)(x+3)=(x-1)(x+3)^2$.

$$\frac{2x}{x^2+2x-3}=\frac{2x}{(x+3)(x-1)}\cdot\frac{x+3}{x+3}$$

$$=\frac{2x^2+6x}{(x-1)(x+3)^2}$$

$$\frac{-x}{x^2+6x+9}=\frac{-x}{(x+3)(x+3)}\cdot\frac{x-1}{x-1}$$

$$=-\frac{x^2-x}{(x-1)(x+3)^3}$$

21. $4x^2-16x+15=(2x-3)(2x-5)$

$6x^2-19x+10=(2x-5)(3x-2)$

The LCM is $(2x-3)(2x-5)(3x-2)$.

$$\frac{-4x}{4x^2-16x+15}=\frac{-4x}{(2x-3)(2x-5)}\cdot\frac{3x-2}{3x-2}$$

$$=-\frac{12x^2-8x}{(2x-3)(2x-5)(3x-2)}$$

$$\frac{3x}{6x^2-19x+10}=\frac{3x}{(2x-5)(3x-2)}\cdot\frac{2x-3}{2x-3}$$

$$=\frac{6x^2-9x}{(2x-3)(2x-5)(3x-2)}$$

23. $6x^2-17x+12=(3x-4)(2x-3)$

$4-3x=-(3x-4)$

The LCM is $(3x-4)(2x-3)$.

$$\frac{5}{6x^2-17x+12}=\frac{5}{(3x-4)(2x-3)}$$

$$\frac{2x}{4-3x}=-\frac{2x}{3x-4}\cdot\frac{2x-3}{2x-3}$$

$$=-\frac{4x^2-6x}{(3x-4)(2x-3)}$$

$$\frac{x+1}{2x-3}=\frac{x+1}{2x-3}\cdot\frac{3x-4}{3x-4}=\frac{3x^2-x-4}{(3x-4)(2x-3)}$$

25. $15-2x-x^2=-(x^2+2x-15)$

$=-(x+5)(x-3)$

The LCM is $(x+5)(x-3)$.

$$\frac{2x}{x-3}=\frac{2x}{x-3}\cdot\frac{x+5}{x+5}=\frac{2x^2+10x}{(x-3)(x+5)}$$

$$\frac{-2}{x+5}=\frac{-2}{x+5}\cdot\frac{x-3}{x-3}=-\frac{2x-6}{(x-3)(x+5)}$$

$$\frac{x-1}{20-x-x^2}=-\frac{x-1}{(x-3)(x+5)}$$

Objective B Exercises

27. True

29. The LCM is $4x^2$.

$$-\frac{3}{4x^2}+\frac{8}{4x^2}-\frac{3}{4x^2}=\frac{-3+8-3}{4x^2}$$
$$=\frac{2}{4x^2}=\frac{1}{2x^2}$$

31. The LCM is $3x^2+x-10$.

$$\frac{3x}{3x^2+x-10}-\frac{5}{3x^2+x-10}=\frac{3x-5}{3x^2+x-10}$$
$$=\frac{3x-5}{(3x-5)(x+2)}=\frac{1}{x+2}$$

33. The LCM is $30a^2b^2$.

$$\frac{2}{5ab}-\frac{3}{10a^2b}+\frac{4}{15ab^2}$$
$$=\frac{2}{5ab}\cdot\frac{6ab}{6ab}-\frac{3}{10a^2b}\cdot\frac{3b}{3b}+\frac{4}{15ab^2}\cdot\frac{2a}{2a}$$
$$=\frac{12ab-9b+8a}{30a^2b^2}$$

35. The LCM is $40ab$.

$$\frac{3}{4ab}-\frac{2}{5a}+\frac{3}{10b}-\frac{5}{8ab}$$
$$=\frac{3}{4ab}\cdot\frac{10}{10}-\frac{2}{5a}\cdot\frac{8b}{8b}+\frac{3}{10b}\cdot\frac{4a}{4a}-\frac{5}{8ab}\cdot\frac{5}{5}$$
$$=\frac{30-16b+12a-25}{30ab}$$
$$=\frac{5-16b+12a}{40ab}$$

37. The LCM is $12x$.

$$\frac{3x-4}{6x}-\frac{2x-5}{4x}=\frac{3x-4}{6x}\cdot\frac{2}{2}-\frac{2x-5}{4x}\cdot\frac{3}{3}$$
$$=\frac{2(3x-4)-3(2x-5)}{12x}=\frac{6x-8-6x+15}{12x}$$
$$=\frac{7}{12x}$$

39. The LCM is $10x^2y^2$.

$$\frac{2y-4}{5xy^2}+\frac{3-2x}{10x^2y}=\frac{2y-4}{5xy^2}\cdot\frac{2x}{2x}+\frac{3-2x}{10x^2y}\cdot\frac{y}{y}$$
$$=\frac{2x(2y-4)+y(3-2x)}{10x^2y^2}=\frac{4xy-8x+3y-2xy}{10x^2y^2}$$
$$=\frac{2xy-8x+3y}{10x^2y^2}$$

41. The LCM is $(a-2)(a+1)$.

$$\frac{3a}{a-2}-\frac{5a}{a+1}=\frac{3a}{a-2}\cdot\frac{a+1}{a+1}-\frac{5a}{a+1}\cdot\frac{a-2}{a-2}$$
$$=\frac{3a(a+1)-5a(a-2)}{(a-2)(a+1)}=\frac{3a^2+3a-5a^2+10a}{(a-2)(a+1)}$$
$$=\frac{-2a^2+13a}{(a-2)(a+1)}=-\frac{a(2a-13)}{(a-2)(a+1)}$$

43. The LCM is $(2x-5)(5x-2)$.

$$\frac{x}{2x-5}-\frac{2}{5x-2}=\frac{x}{2x-5}\cdot\frac{5x-2}{5x-2}-\frac{2}{5x-2}\cdot\frac{2x-5}{2x-5}$$
$$=\frac{x(5x-2)-2(2x-5)}{(2x-5)(5x-2)}=\frac{5x^2-2x-4x+10}{(2x-5)(5x-2)}$$
$$=\frac{5x^2-6x+10}{(2x-5)(5x-2)}$$

45. The LCM is $b(a-b)$.

$$\frac{1}{a-b}+\frac{1}{b}=\frac{1}{a-b}\cdot\frac{b}{b}+\frac{1}{b}\cdot\frac{a-b}{a-b}$$
$$=\frac{b+a-b}{b(a-b)}=\frac{a}{b(a-b)}$$

47. The LCM is $a(a-3)$.

$$\frac{6a}{a-3}-5+\frac{3}{a}=\frac{6a}{a-3}\cdot\frac{a}{a}-\frac{5}{1}\cdot\frac{a(a-3)}{a(a-3)}+\frac{3}{a}\cdot\frac{a-3}{a-3}$$
$$=\frac{6a^2-5a^2+15a+3a-9}{a(a-3)}$$
$$=\frac{a^2+18a-9}{a(a-3)}$$

49. The LCM is $x(6x-5)$.

$$\frac{5}{x}-\frac{5x}{5-6x}+2$$

$$=\frac{5}{x}\cdot\frac{6x-5}{6x-5}-\frac{-5x}{6x-5}\cdot\frac{x}{x}+\frac{2}{1}\cdot\frac{x(6x-5)}{x(6x-5)}$$

$$=\frac{5(6x-5)+5x(x)+2(x)(6x-5)}{x(6x-5)}$$

$$=\frac{30x-25+5x^2+12x^2-10x}{x(6x-5)}$$

$$=\frac{17x^2+20x-25}{x(6x-5)}$$

51. $x^2-6x+9=(x-3)(x-3)$

$x^2-9=(x+3)(x-3)$

The LCM is

$(x+3)(x-3)(x-3)=(x+3)(x-3)^2$.

$$\frac{1}{x^2-6x+9}-\frac{1}{x^2-9}$$

$$=\frac{1}{(x-3)(x-3)}\cdot\frac{x+3}{x+3}-\frac{1}{(x+3)(x-3)}\cdot\frac{x-3}{x-3}$$

$$=\frac{x+3-x+3}{(x+3)(x-3)^2}$$

$$=\frac{6}{(x+3)(x-3)^2}$$

53. $x^2+4x+4=(x+2)(x+2)$

The LCM is $(x+2)(x+2)=(x+2)^2$.

$$\frac{1}{x+2}-\frac{3x}{x^2+4x+4}$$

$$=\frac{1}{(x+2)}\cdot\frac{x+2}{x+2}-\frac{3x}{(x+2)(x+2)}$$

$$=\frac{x+2-3x}{(x+2)(x+2)}=\frac{-2x+2}{(x+2)^2}$$

$$=-\frac{2(x-1)}{(x+2)^2}$$

55. $x^2+2x-8=(x+4)(x-2)$

The LCM is $(x+4)(x-2)$.

$$\frac{-3x^2+8x+2}{x^2+2x-8}-\frac{2x-5}{x+4}$$

$$=\frac{-3x^2+8x+2}{(x+4)(x-2)}-\frac{2x-5}{x+4}\cdot\frac{x-2}{x-2}$$

$$=\frac{-3x^2+8x+2-2x^2+9x-10}{(x+4)(x-2)}$$

$$=\frac{-5x^2+17x-8}{(x+4)(x-2)}$$

$$=-\frac{5x^2-17x+8}{(x+4)(x-2)}$$

57. $x^{2n}-1=(x^n+1)(x^n-1)$

The LCM is $(x^n+1)(x^n-1)$.

$$\frac{2}{x^n-1}+\frac{x^n}{x^{2n}-1}$$

$$=\frac{2}{x^n-1}\cdot\frac{x^n+1}{x^n+1}+\frac{x^n}{(x^n+1)(x^n-1)}$$

$$=\frac{2(x^n+1)+x^n}{(x^n+1)(x^n-1)}=\frac{2x^n+2+x^n}{(x^n+1)(x^n-1)}$$

$$=\frac{3x^n+2}{(x^n+1)(x^n-1)}$$

59. $x^{2n}-x^n-6=(x^n-3)(x^n+2)$

The LCM is

$(x^n-3)(x^n+2)$.

$$\frac{2x^n-6}{x^{2n}-x^n-6}+\frac{x^n}{x^n+2}$$
$$=\frac{2x^n-6}{(x^n-3)(x^n+2)}+\frac{x^n}{(x^n+2)}\cdot\frac{x^n-3}{x^n-3}$$
$$=\frac{2x^n-6+x^{2n}-3x^n}{(x^n-3)(x^n+2)}$$
$$=\frac{x^{2n}-x^n-6}{(x^n-3)(x^n+2)}$$
$$=-\frac{(x^n-3)(x^n+2)}{(x^n-3)(x^n+2)}$$
$$=1$$

61. $4x^2-36=4(x^2-9)=4(x-3)(x+3)$

The LCM is $4(x-3)(x+3)$.

$$\frac{x^2+4}{4x^2-36}-\frac{13}{x+3}$$
$$=\frac{x^2+4}{4(x-3)(x+3)}-\frac{13}{x+3}\cdot\frac{4(x-3)}{4(x-3)}$$
$$=\frac{x^2+4-13(4)(x-3)}{4(x-3)(x+3)}=\frac{x^2+4-52x+156}{4(x-3)(x+3)}$$
$$=\frac{x^2-52x+160}{4(x-3)(x+3)}$$

63. $4x^2+9x+2=(4x+1)(x+2)$

The LCM is $(4x+1)(x+2)$.

$$\frac{3x-4}{4x+1}+\frac{3x+6}{4x^2+9x+2}$$
$$=\frac{3x-4}{4x+1}\cdot\frac{x+2}{x+2}+\frac{3x+6}{(4x+1)(x+2)}$$
$$=\frac{3x^2+2x-8+3x+6}{(4x+1)(x+2)}=\frac{3x^2+5x-2}{(4x+1)(x+2)}$$
$$=\frac{(3x-1)(x+2)}{(4x+1)(x+2)}$$
$$=\frac{3x-1}{4x+1}$$

65. $x^2+x-12=(x+4)(x-3)$

$x^2+7x+12=(x+4)(x+3)$

The LCM is $(x+4)(x-3)(x+3)$.

$$\frac{x+1}{x^2+x-12}-\frac{x-3}{x^2+7x+12}$$
$$=\frac{x+1}{(x+4)(x-3)}\cdot\frac{x+3}{x+3}-\frac{x-3}{(x+4)(x+3)}\cdot\frac{x-3}{x-3}$$
$$=\frac{x^2+4x+3-x^2+6x-9}{(x+4)(x-3)(x+3)}$$
$$=\frac{10x-6}{(x+4)(x-3)(x+3)}=\frac{2(5x-3)}{(x+4)(x-3)(x+3)}$$

67. $x^2-2x-15=(x-5)(x+3)$

$5-x=-(x-5)$

The LCM is $(x-5)(x+3)$.

$$\frac{2x^2-2x}{x^2-2x-15}-\frac{2}{x+3}+\frac{x}{5-x}$$
$$=\frac{2x^2-2x}{(x-5)(x+3)}-\frac{2}{x+3}\cdot\frac{x-5}{x-5}+\frac{-(x)}{x-5}\cdot\frac{x+3}{x+3}$$
$$=\frac{2x^2-2x-2x+10-x^2-3x}{(x-5)(x+3)}$$
$$=\frac{x^2-7x+10}{(x-5)(x+3)}=\frac{(x-5)(x-2)}{(x-5)(x+3)}$$
$$=\frac{x-2}{x+3}$$

69. $3x^2-11x-20=(3x+4)(x-5)$

The LCM is $(3x+4)(x-5)$.

$$\frac{x}{3x+4}+\frac{3x+2}{x-5}-\frac{7x^2+24x+28}{3x^2-11x-20}$$

$$=\frac{x}{3x+4}\cdot\frac{x-5}{x-5}+\frac{3x+2}{x-5}\cdot\frac{3x+4}{3x+4}-\frac{7x^2+24x+28}{(3x+4)(x-5)}$$

$$=\frac{x^2-5x+9x^2+18x+8-7x^2-24x-28}{(3x+4)(x-5)}$$

$$=\frac{3x^2-11x-20}{(2x-5)(x-2)}=\frac{(3x+4)(x-5)}{(3x+4)(x-5)}$$

$$=1$$

71. $8x^2-10x+3=(4x-3)(2x-1)$

$1-2x=-(2x-1)$

The LCM is $(4x-3)(2x-1)$.

$$\frac{x+1}{1-2x}-\frac{x+3}{4x-3}+\frac{10x^2+7x-9}{8x^2-10x+3}$$

$$=\frac{-(x+1)}{2x-1}\cdot\frac{4x-3}{4x-3}-\frac{x+3}{4x-3}\cdot\frac{2x-1}{2x-1}+\frac{10x^2+7x-9}{(4x-3)(2x-1)}$$

$$=\frac{-4x^2-x+3-2x^2-5x+3+10x^2+7x-9}{(4x-3)(2x-1)}$$

$$=\frac{4x^2+x-3}{(4x-3)(2x-1)}=\frac{(4x-3)(x+1)}{(4x-3)(2x-1)}$$

$$=\frac{x+1}{2x-1}$$

73. $8x^3-1=(2x-1)(4x^2+2x+1)$

The LCM is $(2x-1)(4x^2+2x+1)$.

$$\frac{2x}{4x^2+2x+1}+\frac{4x+1}{8x^3-1}$$

$$=\frac{2x}{4x^2+2x+1}\cdot\frac{2x-1}{2x-1}+\frac{4x+1}{(2x-1)(4x^2+2x+1)}$$

$$=\frac{4x^2-2x+4x+1}{(2x-1)(4x^2+2x+1)}=\frac{4x^2+2x+1}{(2x-1)(4x^2+2x+1)}$$

$$=\frac{1}{2x-1}$$

75. $x^4-16=(x^2+4)(x^2-4)$

The LCM is $(x^2+4)(x^2-4)$.

$$\frac{x^2-12}{x^4-16}+\frac{1}{x^2-4}-\frac{1}{x^2+4}$$

$$=\frac{x^2-12}{(x^2+4)(x^2-4)}+\frac{1}{x^2-4}\cdot\frac{x^2+4}{x^2+4}-\frac{1}{x^2+4}\cdot\frac{x^2-4}{x^2-4}$$

$$=\frac{x^2-12+x^2+4-x^2+4}{(x^2+4)(x^2-4)}$$

$$=\frac{x^2-4}{(x^2+4)(x^2-4)}$$

$$=\frac{1}{x^2+4}$$

77. The LCM is $9a^2$.

$$\left(\frac{a-3}{a^2}-\frac{a-3}{9}\right)\div\frac{a^2-9}{3a}$$

$$=\left(\frac{a-3}{a^2}\cdot\frac{9}{9}-\frac{a-3}{9}\cdot\frac{a^2}{a^2}\right)\div\frac{a^2-9}{3a}$$

$$=\left(\frac{9(a-3)-a^2(a-3)}{9a^2}\right)\div\frac{a^2-9}{3a}$$

$$=\frac{(9-a^2)(a-3)}{9a^2}\cdot\frac{3a}{a^2-9}$$

$$=\frac{(3-a)(3+a)(a-3)}{9a^2}\cdot\frac{3a}{(a+3)(a-3)}$$

$$=\frac{3-a}{3a}$$

79. $\dfrac{x^2-4x+4}{2x+1}\cdot\dfrac{2x^2+x}{x^3-4x}-\dfrac{3x-2}{x+1}$

$=\dfrac{(x-2)(x-2)}{2x+1}\cdot\dfrac{x(2x+1)}{x(x+2)(x-2)}-\dfrac{3x-2}{x+1}$

$=\dfrac{x-2}{x+2}-\dfrac{3x-2}{x+1}$

The LCM is $(x+2)(x+1)$.

$\dfrac{x-2}{x+2}-\dfrac{3x-2}{x+1}$

$=\dfrac{x-2}{x+2}\cdot\dfrac{x+1}{x+1}-\dfrac{3x-2}{x+1}\cdot\dfrac{x+2}{x+2}$

$=\dfrac{x^2-x-2-3x^2-4x+4}{x(x-2)9x+2)}=\dfrac{-2x^2-5x+2}{(x+2)(x+1)}$

$=-\dfrac{2x^2+5x-2}{(x+2)(x+1)}$

81. The LCM is ab.

$\left(\dfrac{a-2b}{b}+\dfrac{b}{a}\right)\div\left(\dfrac{b+a}{a}-\dfrac{2a}{b}\right)$

$=\left(\dfrac{a-2b}{b}\cdot\dfrac{a}{a}+\dfrac{b}{a}\cdot\dfrac{b}{b}\right)\div\left(\dfrac{b+a}{a}\cdot\dfrac{b}{b}-\dfrac{2a}{b}\cdot\dfrac{a}{a}\right)$

$=\left(\dfrac{a^2-2ab+b^2}{ab}\right)\div\left(\dfrac{b^2+ab-2a^2}{ab}\right)$

$=\dfrac{(a-b)(a-b)}{ab}\cdot\dfrac{ab}{(b+2a)(b-a)}$

$=\dfrac{a-b}{-(b+2a)}$

$=-\dfrac{a-b}{b+2a}=\dfrac{b-a}{b+2a}$

83. $\dfrac{2x}{x^2-x-6}-\dfrac{6x-6}{2x^2-9x+9}\div\dfrac{x^2+x-2}{2x-3}$

$=\dfrac{2x}{x^2-x-6}-\dfrac{6(x-1)}{(2x-3)(x-3)}\div\dfrac{(x+2)(x-1)}{2x-3}$

$=\dfrac{2x}{(x-3)(x+2)}-\dfrac{6(x-1)}{(2x-3)(x-3)}\cdot\dfrac{2x-3}{(x+2)(x-1)}$

$=\dfrac{2x}{(x-3)(x+2)}-\dfrac{6}{(x-3)(x+2)}$

$=\dfrac{2x-6}{(x-3)(x+2)}=\dfrac{2(x-3)}{(x-3)(x+2)}$

$=\dfrac{2}{x+2}$

Applying the Concepts

85. a) $f(x)=\dfrac{x}{x+2},\quad f(4)=\dfrac{2}{3}$

$g(x)=\dfrac{4}{x-3},\quad g(4)=4$

$S(x)=\dfrac{x^2+x+8}{x^2-x-6},\quad S(4)=4\dfrac{2}{3}$

b) $f(4)+g(4)=\dfrac{2}{3}+4=4\dfrac{2}{3}=S(4)$ Yes

Section 8.3

Objective A Exercises

1. A complex fraction is a fraction whose numerator or denominator contains one or more fractions.

3. The LCM is 3.

$\dfrac{2-\frac{1}{3}}{4+\frac{11}{3}}=\dfrac{2-\frac{1}{3}}{4+\frac{11}{3}}\cdot\dfrac{3}{3}=\dfrac{2\cdot 3-\frac{1}{3}\cdot 3}{4\cdot 3+\frac{11}{3}\cdot 3}$

$=\dfrac{6-1}{12+11}=\dfrac{5}{23}$

5. The LCM is 6.

$$\frac{3-\frac{2}{3}}{5+\frac{5}{6}}=\frac{3-\frac{2}{3}}{5+\frac{5}{6}}\cdot\frac{6}{6}=\frac{3\cdot 6-\frac{2}{3}\cdot 6}{5\cdot 6+\frac{5}{6}\cdot 6}$$

$$=\frac{18-4}{30+5}=\frac{14}{35}=\frac{2}{5}$$

7. The LCM is x^2.

$$\frac{1+\frac{1}{x}}{1-\frac{1}{x^2}}=\frac{1+\frac{1}{x}}{1-\frac{1}{x^2}}\cdot\frac{x^2}{x^2}=\frac{1\cdot x^2+\frac{1}{x}\cdot x^2}{1\cdot x^2-\frac{1}{x^2}\cdot x^2}$$

$$=\frac{x^2+x}{x^2-1}=\frac{x(x+1)}{(x-1)(x+1)}$$

$$=\frac{x}{x-1}$$

9. The LCM is a.

$$\frac{a-2}{\frac{4}{a}-a}=\frac{a-2}{\frac{4}{a}-a}\cdot\frac{a}{a}=\frac{a\cdot a-2\cdot a}{\frac{4}{a}\cdot a-a\cdot a}$$

$$=\frac{a^2-2a}{4-a^2}=\frac{a(a-2)}{(2-a)(2+a)}$$

$$=-\frac{a}{a+2}$$

11. The LCM is $x(x+h)$.

$$\frac{\frac{1}{x+h}-\frac{1}{x}}{h}=\frac{\frac{1}{x+h}-\frac{1}{x}}{h}\cdot\frac{x(x+h)}{x(x+h)}$$

$$=\frac{x-(x+h)}{hx(x+h)}=\frac{-h}{hx(x+h)}$$

$$=-\frac{1}{x(x+h)}$$

13. The LCM is x.

$$\frac{x-\frac{1}{x}}{x+\frac{1}{x}}=\frac{x-\frac{1}{x}}{x+\frac{1}{x}}\cdot\frac{x}{x}$$

$$=\frac{x^2-1}{x^2+1}=\frac{(x-1)(x+1)}{x^2+1}$$

15. The LCM is a^2.

$$\frac{\frac{1}{a^2}-\frac{1}{a}}{\frac{1}{a^2}+\frac{1}{a}}=\frac{\frac{1}{a^2}-\frac{1}{a}}{\frac{1}{a^2}+\frac{1}{a}}\cdot\frac{a^2}{a^2}$$

$$=\frac{1-a}{1+a}=-\frac{a-1}{a+1}$$

17. The LCM is $x+2$.

$$\frac{2-\frac{4}{x+2}}{5-\frac{10}{x+2}}=\frac{2-\frac{4}{x+2}}{5-\frac{10}{x+2}}\cdot\frac{x+2}{x+2}$$

$$=\frac{2(x+2)-4}{5(x+2)-10}=\frac{2x+4-4}{5x+10-10}$$

$$=\frac{2x}{5x}=\frac{2}{5}$$

19. The LCM is $2a-3$.

$$\frac{\frac{3}{2a-3}+2}{\frac{-6}{2a-3}-4}=\frac{\frac{3}{2a-3}+2}{\frac{-6}{2a-3}-4}\cdot\frac{2a-3}{2a-3}$$

$$=\frac{3+2(2a-3)}{-6-4(2a-3)}=\frac{3+4a-6}{-6-8a+12}$$

$$=\frac{4a-3}{-8a+6}=\frac{4a-3}{-2(4a-3)}$$

$$=-\frac{1}{2}$$

21. The LCM is $(x-4)(x+1)$.

$$\frac{1-\dfrac{1}{x-4}}{1-\dfrac{6}{x+1}}=\frac{1-\dfrac{1}{x-4}}{1-\dfrac{6}{x+1}}\cdot\frac{(x-4)(x+1)}{(x-4)(x+1)}$$

$$=\frac{(x-4)(x+1)-(x+1)}{(x-4)(x+1)-6(x-4)}$$

$$=\frac{x^2-3x-4-x-1}{x^2-3x-4-6x+24}$$

$$=\frac{x^2-4x-5}{x^2-9x+20}=\frac{(x-5)(x+1)}{(x-5)(x-4)}$$

$$=\frac{x+1}{x-4}$$

23. The LCM is $(x-3)(2-x)$.

$$\frac{1-\dfrac{2}{x-3}}{1+\dfrac{3}{2-x}}=\frac{1-\dfrac{2}{x-3}}{1+\dfrac{3}{2-x}}\cdot\frac{(x-3)(2-x)}{(x-3)(2-x)}$$

$$=\frac{(x-3)(2-x)-2(2-x)}{(x-3)(2-x)+3(x-3)}$$

$$=\frac{2x-x^2-6+3x-4+2x}{2x-x^2-6+3x+3x-9}$$

$$=\frac{-x^2+7x-10}{-x^2+8x-15}=\frac{-(x^2-7x+10)}{-(x^2-8x+15)}$$

$$=\frac{(x-5)(x-2)}{(x-5)(x-3)}$$

$$=\frac{x-2}{x-3}$$

25. The LCM is $(2x+3)$.

$$\frac{x-4+\dfrac{9}{2x+3}}{x+3-\dfrac{5}{2x+3}}=\frac{x-4+\dfrac{9}{2x+3}}{x+3-\dfrac{5}{2x+3}}\cdot\frac{2x+3}{2x+3}$$

$$=\frac{x(2x+3)-4(2x+3)+9}{x(2x+3)+3(2x+3)-5}$$

$$=\frac{2x^2+3x-8x-12+9}{2x^2+3x+6x+9-5}$$

$$=\frac{2x^2-5x-3}{2x^2+9x+4}=\frac{(2x+1)(x-3)}{(2x+1)(x+4)}$$

$$=\frac{x-3}{x+4}$$

27. The LCM is $(x+4)(x-3)$.

$$\frac{x-3+\dfrac{10}{x+4}}{x+7+\dfrac{16}{x-3}}=\frac{x-3+\dfrac{10}{x+4}}{x+7+\dfrac{16}{x-3}}\cdot\frac{(x+4)(x-3)}{(x+4)(x-3)}$$

$$=\frac{x(x+4)(x-3)-3(x+4)(x-3)+10(x-3)}{x(x+4)(x-3)+7(x+4)(x-3)+16(x+4)}$$

$$=\frac{(x-3)(x^2+4x-3x-12+10)}{(x+4)(x^2-3x+7x-21+16)}$$

$$=\frac{(x-3)(x^2+x-2)}{(x+4)(x^2+4x-5)}=\frac{(x-3)(x+2)(x-1)}{(x+4)(x+5)(x-1)}$$

$$=\frac{(x-3)(x+2)}{(x+4)(x+5)}$$

29. The LCM is x^2.

$$\frac{1-\dfrac{1}{x}-\dfrac{6}{x^2}}{1-\dfrac{4}{x}+\dfrac{3}{x^2}}=\frac{1-\dfrac{1}{x}-\dfrac{6}{x^2}}{1-\dfrac{4}{x}+\dfrac{3}{x^2}}\cdot\frac{x^2}{x^2}$$

$$=\frac{x^2-x-6}{x^2-4x+3}=\frac{(x-3)(x+2)}{(x-3)(x-1)}$$

$$=\frac{x+2}{x-1}$$

31. The LCM is x^2.

$$\frac{1+\frac{1}{x}-\frac{12}{x^2}}{\frac{9}{x^2}+\frac{3}{x}-2}=\frac{1+\frac{1}{x}-\frac{12}{x^2}}{\frac{9}{x^2}+\frac{3}{x}-2}\cdot\frac{x^2}{x^2}$$

$$=\frac{x^2+x-12}{9+3x-2x^2}=\frac{(x-3)(x+4)}{(3-x)(3+2x)}$$

$$=-\frac{x+4}{2x+3}$$

33. The LCM is x^2y^2.

$$\frac{\frac{1}{y^2}-\frac{1}{xy}-\frac{2}{x^2}}{\frac{1}{y^2}-\frac{3}{xy}+\frac{2}{x^2}}=\frac{\frac{1}{y^2}-\frac{1}{xy}-\frac{2}{x^2}}{\frac{1}{y^2}-\frac{3}{xy}+\frac{2}{x^2}}\cdot\frac{x^2y^2}{x^2y^2}$$

$$=\frac{x^2-xy-2y^2}{x^2-3xy+2y^2}=\frac{(x+y)(x-2y)}{(x-y)(x-2y)}$$

$$=\frac{x+y}{x-y}$$

35. The LCM is $x(x+1)$.

$$\frac{\frac{x}{x+1}-\frac{1}{x}}{\frac{x}{x+1}+\frac{1}{x}}=\frac{\frac{x}{x+1}-\frac{1}{x}}{\frac{x}{x+1}+\frac{1}{x}}\cdot\frac{x(x+1)}{x(x+1)}$$

$$=\frac{x^2-(x+1)}{x^2+(x+1)}=\frac{x^2-x-1}{x^2+x+1}$$

37. The LCM is $a(a-2)$.

$$\frac{\frac{1}{a}-\frac{3}{a-2}}{\frac{2}{a}+\frac{5}{a-2}}=\frac{\frac{1}{a}-\frac{3}{a-2}}{\frac{2}{a}+\frac{5}{a-2}}\cdot\frac{a(a-2)}{a(a-2)}$$

$$=\frac{a-2-3a}{2a-4+5a}=\frac{-2a-2}{7a-4}$$

$$=-\frac{2a+2}{7a-4}=-\frac{2(a+1)}{7a-4}$$

39. The LCM is $(x-1)(x+1)$.

$$\frac{\frac{x-1}{x+1}-\frac{x+1}{x-1}}{\frac{x-1}{x+1}+\frac{x+1}{x-1}}=\frac{\frac{x-1}{x+1}-\frac{x+1}{x-1}}{\frac{x-1}{x+1}+\frac{x+1}{x-1}}\cdot\frac{(x-1)(x+1)}{(x-1)(x+1)}$$

$$=\frac{(x-1)(x-1)-(x+1)(x+1)}{(x-1)(x-1)+(x+1)(x+1)}$$

$$=\frac{x^2-2x+1-x^2-2x-1}{x^2-2x+1+x^2+2x+1}=\frac{-4x}{2x^2+2}$$

$$=\frac{-4x}{2(x^2+1)}=-\frac{2x}{x^2+1}$$

41. The LCM is a.

$$a+\frac{a}{a+\frac{1}{a}}=a+\frac{a}{a+\frac{1}{a}}\cdot\frac{a}{a}$$

$$=a+\frac{a^2}{a^2+1}$$

The LCM is a^2+1.

$$a+\frac{a^2}{a^2+1}=a\cdot\frac{a^2+1}{a^2+1}+\frac{a^2}{a^2+1}$$

$$=\frac{a^3+a+a^2}{a^2+1}$$

$$=\frac{a^3+a^2+a}{a^2+1}=\frac{a\left(a^2+a+1\right)}{a^2+1}$$

43. The LCM is $1-a$.

$$a-\frac{a}{1-\frac{a}{1-a}}=a-\frac{a}{1-\frac{a}{1-a}}\cdot\frac{1-a}{1-a}$$

$$=a-\frac{a-a^2}{1-a-a}=a-\frac{a-a^2}{1-2a}$$

The LCM is $1-2a$.

$$a-\frac{a-a^2}{1-2a}=a\cdot\frac{1-2a}{1-2a}-\frac{a-a^2}{1-2a}$$

$$=\frac{a-2a^2-a+a^2}{1-2a}$$

$$=\frac{-a^2}{1-2a}=-\frac{a^2}{1-2a}$$

45. The LCM is x.

$$3-\frac{2}{1-\frac{2}{3-\frac{2}{x}}}=3-\frac{2}{1-\frac{2}{3-\frac{2}{x}}\cdot\frac{x}{x}}$$

$$=3-\frac{2}{1-\frac{2x}{3x-2}}$$

The LCM is $3x-2$.

$$3-\frac{2}{1-\frac{2x}{3x-2}}=3-\frac{2}{1-\frac{2x}{3x-2}}\cdot\frac{3x-2}{3x-2}$$

$$=3-\frac{6x-4}{3x-2-2x}=3-\frac{6x-4}{x-2}$$

The LCM is $x-2$.

$$3-\frac{6x-4}{x-2}=3\cdot\frac{x-2}{x-2}-\frac{6x-4}{x-2}$$

$$=\frac{3x-6-6x+4}{x-2}=\frac{-3x-2}{x-2}$$

$$=-\frac{3x+2}{x-2}$$

47. $\dfrac{1}{1-\frac{1}{a}}=\dfrac{1}{1-\frac{1}{a}}\cdot\dfrac{a}{a}=\dfrac{a}{a-1}$

The reciprocal is $\dfrac{a-1}{a}$.

Applying the Concepts

49. $(3+3^{-1})^{-1}=\left(3+\dfrac{1}{3}\right)^{-1}=\left(\dfrac{10}{3}\right)^{-1}=\dfrac{3}{10}$

51. $\dfrac{x^{-1}}{x^{-1}+2^{-1}}=\dfrac{\frac{1}{x}}{\frac{1}{x}+\frac{1}{2}}=\dfrac{\frac{1}{x}}{\frac{2+x}{2x}}=\dfrac{1}{x}\cdot\dfrac{2x}{2+x}=\dfrac{2}{2+x}$

Section 8.4

Objective A Exercises

1. The following values would make a denominator 0.

$x+1=0$ and $x-2=0$

$x=-1$ $\quad x=2$

3. The following values would make a denominator 0.

$x^2-9x=x(x-9)$

$x=0$ and $x-9=0$

$x=9$

5. The LCM is 12.

$$\frac{x}{3}-\frac{1}{4}=\frac{1}{12}$$

$$\frac{12}{1}\left(\frac{x}{3}-\frac{1}{4}\right)=\frac{12}{1}\left(\frac{1}{12}\right)$$

$$\frac{\overset{4}{\cancel{12}}}{1}\cdot\frac{x}{\underset{1}{\cancel{3}}}-\frac{\overset{3}{\cancel{12}}}{1}\cdot\frac{1}{\underset{1}{\cancel{4}}}=1$$

$$4x-3=1$$
$$4x=4$$
$$x=1$$

1 checks as a solution. The solution is 1.

7. The LCM is 18.

$$\frac{2y}{9}-\frac{1}{6}=\frac{y}{9}+\frac{1}{6}$$

$$\frac{18}{1}\left(\frac{2y}{9}-\frac{1}{6}\right)=\frac{18}{1}\left(\frac{y}{9}+\frac{1}{6}\right)$$

$$\frac{\overset{2}{\cancel{18}}}{1}\cdot\frac{2y}{\underset{1}{\cancel{9}}}-\frac{\overset{3}{\cancel{18}}}{1}\cdot\frac{1}{\underset{1}{\cancel{6}}}=\frac{\overset{2}{\cancel{18}}}{1}\cdot\frac{y}{\underset{1}{\cancel{9}}}+\frac{\overset{3}{\cancel{18}}}{1}\cdot\frac{1}{\underset{1}{\cancel{6}}}$$

$$4y-3=2y+3$$
$$2y-3=3$$
$$2y=6$$
$$y=3$$

3 checks as a solution. The solution is 3.

9. The LCM is 12.

$$\frac{3x+4}{12}-\frac{1}{3}=\frac{5x+2}{12}-\frac{1}{2}$$

$$\frac{12}{1}\left(\frac{3x+4}{12}-\frac{1}{3}\right)=\frac{12}{1}\left(\frac{5x+2}{12}-\frac{1}{2}\right)$$

$$\frac{\overset{1}{\cancel{12}}}{1}\cdot\frac{3x+4}{\underset{1}{\cancel{12}}}-\frac{\overset{4}{\cancel{12}}}{1}\cdot\frac{1}{\underset{1}{\cancel{3}}}=\frac{\overset{1}{\cancel{12}}}{1}\cdot\frac{5x+2}{\underset{1}{\cancel{12}}}-\frac{\overset{6}{\cancel{12}}}{1}\cdot\frac{1}{\underset{1}{\cancel{2}}}$$

$$3x+4-4=5x+2-6$$

$$3x=5x-4$$

$$-2x=-4$$

$$x=2$$

2 checks as a solution. The solution is 2.

11. The LCM is $3x-2$.

$$\frac{12}{3x-2}=3$$

$$\frac{\overset{1}{\cancel{3x-2}}}{1}\cdot\frac{12}{\underset{1}{\cancel{3x-2}}}=\frac{3x-2}{1}\cdot\frac{3}{1}$$

$$12=9x-6$$

$$18=9x$$

$$2=x$$

2 checks as a solution. The solution is 2.

13. The LCM is $4-3x$.

$$\frac{6}{4-3x}=3$$

$$\frac{\overset{1}{\cancel{4-3x}}}{1}\cdot\frac{6}{\underset{1}{\cancel{4-3x}}}=\frac{4-3x}{1}\cdot\frac{3}{1}$$

$$6=12-9x$$

$$-6=-9x$$

$$\frac{-6}{-9}=x$$

$$\frac{2}{3}=x$$

$\frac{2}{3}$ checks as a solution. The solution is $\frac{2}{3}$.

15. The LCM is n.

$$3+\frac{8}{n}=5$$

$$\frac{n}{1}\left(3+\frac{8}{n}\right)=\frac{n}{1}\cdot 5$$

$$\frac{n}{1}\cdot 3+\frac{n}{1}\cdot\frac{8}{n}=5n$$

$$3n+8=5n$$

$$8=2n$$

$$4=n$$

4 checks as a solution. The solution is 4.

17. The LCM is x.

$$3-\frac{12}{x}=7$$

$$\frac{x}{1}\left(3-\frac{12}{x}\right)=\frac{x}{1}\cdot 7$$

$$\frac{x}{1}.3-\frac{x}{1}\cdot\frac{12}{x}=7x$$

$$3x-12=7x$$

$$-12=4x$$

$$-3=x$$

−3 checks as a solution. The solution is −3.

19. The LCM is x.

$$\frac{6}{x}+3=11$$

$$\frac{x}{1}\left(\frac{6}{x}+3\right)=\frac{x}{1}\cdot 11$$

$$\frac{x}{1}\cdot\frac{6}{x}+\frac{x}{1}\cdot 3=11x$$

$$6+3x=11x$$

$$6=8x$$

$$\frac{6}{8}=x$$

$$\frac{3}{4}=x$$

$\frac{3}{4}$ checks as a solution. The solution is $\frac{3}{4}$.

21. The LCM is $(x+3)(x-1)$.

$$\frac{5}{x+3}=\frac{3}{x-1}$$

$$\frac{(x+3)(x-1)}{1}\cdot\frac{5}{x+3}=\frac{(x+3)(x-1)}{1}\cdot\frac{3}{x-1}$$

$$5x-5=3x+9$$
$$2x=14$$
$$x=7$$

7 checks as a solution. The solution is 7.

23. The LCM is $(3x-4)(1-2x)$.

$$\frac{5}{3x-4}=\frac{-3}{1-2x}$$

$$\frac{(3x-4)(1-2x)}{1}\cdot\frac{5}{3x-4}=\frac{(3x-4)(1-2x)}{1}\cdot\frac{-3}{1-2x}$$

$$5-10x=-9x+12$$
$$-x=7$$
$$x=-7$$

−7 checks as a solution. The solution is −7.

25. The LCM is $(5y-1)(2y-1)$.

$$\frac{4}{5y-1}=\frac{2}{2y-1}$$

$$\frac{(5y-1)(2y-1)}{1}\cdot\frac{4}{5y-1}=\frac{(5y-1)(2y-1)}{1}\cdot\frac{2}{2y-1}$$

$$8y-4=10y-2$$
$$-2y=2$$
$$y=-1$$

−1 checks as a solution. The solution is −1.

27. The LCM is $x+2$.

$$\frac{2x}{x+2}-5=\frac{7x}{x+2}$$

$$\frac{x+2}{1}\left(\frac{2x}{x+2}-5\right)=\frac{x+2}{1}\cdot\frac{7x}{x+2}$$

$$\frac{x+2}{1}\cdot\frac{2x}{x+2}-\frac{x+2}{1}\cdot5=7x$$

$$2x-5x-10=7x$$
$$-3x-10=7x$$
$$-10x=10$$
$$x=-1$$

−1 checks as a solution. The solution is −1.

29. The LCM is $x+4$.

$$\frac{x}{x+4}=3-\frac{4}{x+4}$$

$$\frac{x+4}{1}\cdot\frac{x}{x+4}=\frac{x+4}{1}\left(3-\frac{4}{x+4}\right)$$

$$x=\frac{x+4}{1}\cdot3-\frac{x+4}{1}\cdot\frac{4}{x+4}$$

$$x=3x+12-4$$
$$x=3x+8$$
$$-2x=8$$
$$x=-4$$

−4 does not check as a solution. The equation has no solution.

31. The LCM is $(x+12)(x+5)$.

$$\frac{x}{x+12}=\frac{1}{x+5}$$

$$\frac{(x+12)(x+5)}{1}\cdot\frac{x}{x+12}=\frac{(x+12)(x+5)}{1}\cdot\frac{1}{x+5}$$

$$x^2+5x=x+12$$
$$x^2+4x-12=0$$
$$(x+6)(x-2)=0$$

$x+6=0$ $\quad$ $x-2=0$

$x=-6$ $\quad$ $x=2$

Both −6 and 2 check as solutions. The solutions are −6 and 2.

33. The LCM is $(3n-8)(n+2)$.

$$\frac{5}{3n-8}=\frac{n}{n+2}$$

$$\frac{(3n-8)(n+2)}{1}\cdot\frac{5}{3n-8}=\frac{(3n-8)(n+2)}{1}\cdot\frac{n}{n+2}$$

$$5n+10=3n^2-8n$$
$$0=3n^2-13n-10$$
$$0=(3n+2)(n-5)$$

$3n+2=0$ $\quad$ $n-5=0$

$3n=-2$ $\quad$ $n=5$

$n=-\frac{2}{3}$

Both $-\frac{2}{3}$ and 5 check as solutions. The solutions are $-\frac{2}{3}$ and 5.

35. The LCM is $x-3$.

$$x-\frac{6}{x-3}=\frac{2x}{x-3}$$

$$\frac{x-3}{1}\cdot\left(x-\frac{6}{x-3}\right)=\frac{\cancel{x-3}}{1}\cdot\frac{2x}{\cancel{x-3}}$$

$$\frac{x-3}{1}\cdot x-\frac{\cancel{x-3}}{1}\cdot\frac{6}{\cancel{x-3}}=2x$$

$$x^2-3x-6=2x$$

$$x^2-5x-6=0$$

$$(x-6)(x+1)=0$$

$$x-6=0 \qquad x+1=0$$

$$x=6 \qquad x=-1$$

Both 6 and -1 check as solutions. The solutions are 6 and -1.

Applying the Concepts

37. Students' explanations should include each of the following steps:

(1) Find the LCM of the denominators of the fractions in the equation.

(2) Multiply each side of the equation by the LCM of the denominators.

(3) Simplify each side of the equation.

(4) Solve for the variable.

(5) Check the solution in the original equation.

39. The LCM is 4.

$$\frac{3}{4}a=\frac{1}{2}(3-a)+\frac{a-2}{4}$$

$$\frac{\cancel{4}}{1}\cdot\frac{3}{\cancel{4}}a=\frac{\cancel{4}}{1}\cdot\frac{1}{\cancel{2}}(3-a)+\frac{\cancel{4}}{1}\cdot\frac{a-2}{\cancel{4}}$$

$$3a=6-2a+a-2$$

$$3a=-a+4$$

$$4a=4$$

$$a=1$$

1 checks as a solution. The solution is 1.

41. The LCM is $(2x+1)(x+1)(x-1)$.

$$\frac{x}{2x^2-x-1}=\frac{3}{x^2-1}+\frac{3}{2x+1}$$

$$\frac{x}{(2x+1)(x-1)}=\frac{3}{(x-1)(x+1)}+\frac{3}{2x+1}$$

$$\frac{\cancel{(2x+1)}\cancel{(x-1)}(x+1)}{1}\cdot\frac{x}{\cancel{(2x+1)}\cancel{(x-1)}}$$

$$=\frac{(2x+1)\cancel{(x-1)}\cancel{(x+1)}}{1}\cdot\frac{3}{\cancel{(x+1)}\cancel{(x-1)}}$$

$$+\frac{\cancel{(2x+1)}(x-1)(x+1)}{1}\cdot\frac{3}{\cancel{(2x+1)}}$$

$$(x+1)x=(2x+1)3+(x-1)(x+1)3$$

$$x^2+x=6x+3+\left(x^2-1\right)3$$

$$x^2+x=6x+3+3x^2-3$$

$$0=2x^2+5x$$

$$0=x(2x+5)$$

$$x=0 \qquad 2x+5=0$$

$$2x=-5$$

$$x=-\frac{5}{2}$$

Both 0 and $-\frac{5}{2}$ check as solutions. The solutions are 0 and $-\frac{5}{2}$.

43. The LCM is $(y-2)(y+2)(y+1)$.

$$\frac{y+2}{y^2-y-2}+\frac{y+1}{y^2-4}=\frac{1}{y+1}$$

$$\frac{y+2}{(y-2)(y+1)}+\frac{y+1}{(y-2)(y+2)}=\frac{1}{y+1}$$

$$\frac{\overset{1}{\cancel{(y-2)}}\,\overset{1}{\cancel{(y+1)}}(y+2)}{1}\cdot\frac{y+2}{\underset{1}{\cancel{(y-2)}}\,\underset{1}{\cancel{(y+1)}}}$$

$$+\frac{\overset{1}{\cancel{(y-2)}}(y+1)\overset{1}{\cancel{(y+2)}}}{1}\cdot\frac{y+1}{\underset{1}{\cancel{(y-2)}}\,\underset{1}{\cancel{(y+2)}}}$$

$$=\frac{(y-2)\overset{1}{\cancel{(y+1)}}(y+2)}{1}\cdot\frac{1}{\underset{1}{\cancel{y+1}}}$$

$$(y+2)(y+2)+(y+1)(y+1)=(y-2)(y+2)$$
$$y^2+4y+4+y^2+2y+1=y^2-4$$
$$2y^2+6y+5=y^2-4$$
$$y^2+6y+9=0$$
$$(y+3)(y+3)=0$$

$$y+3=0 \qquad y+3=0$$
$$y=-3 \qquad y=-3$$

-3 checks as a solution. The solution is -3.

Section 8.5

Objective A Exercises

1.
$$\frac{x}{12}=\frac{3}{4}$$
$$\overset{1}{\cancel{12}}\cdot\frac{x}{\underset{1}{\cancel{12}}}=\overset{3}{\cancel{12}}\cdot\frac{3}{\underset{1}{\cancel{4}}}$$
$$x=9$$

The solution is 9.

3.
$$\frac{4}{9}=\frac{x}{27}$$
$$\overset{3}{\cancel{27}}\cdot\frac{4}{\underset{1}{\cancel{9}}}=\overset{1}{\cancel{27}}\cdot\frac{x}{\underset{1}{\cancel{27}}}$$
$$12=x$$

The solution is 12.

5.
$$\frac{x+3}{12}=\frac{5}{6}$$
$$\overset{1}{\cancel{12}}\cdot\frac{x+3}{\underset{1}{\cancel{12}}}=\overset{2}{\cancel{12}}\cdot\frac{5}{\underset{1}{\cancel{6}}}$$
$$x+3=10$$
$$x=7$$

The solution is 7.

7.
$$\frac{18}{x+4}=\frac{9}{5}$$
$$5\overset{1}{\cancel{(x+4)}}\cdot\frac{18}{\underset{1}{\cancel{x+4}}}=\overset{1}{\cancel{5}}(x+4)\cdot\frac{9}{\underset{1}{\cancel{5}}}$$
$$90=9x+36$$
$$54=9x$$
$$6=x$$

The solution is 6.

9.
$$\frac{2}{x}=\frac{4}{x+1}$$
$$\cancel{x}(x+1)\cdot\frac{2}{\cancel{x}}=x\overset{1}{\cancel{(x+1)}}\cdot\frac{4}{\underset{1}{\cancel{x+1}}}$$
$$2x+2=4x$$
$$2=2x$$
$$1=x$$

The solution is 1.

11.
$$\frac{x+3}{4}=\frac{x}{8}$$
$$\overset{2}{\cancel{8}}\cdot\frac{x+3}{\underset{1}{\cancel{4}}}=\overset{1}{\cancel{8}}\cdot\frac{x}{\underset{1}{\cancel{8}}}$$
$$2x+6=x$$
$$6=-x$$
$$-6=x$$

The solution is -6.

13.
$$\frac{2}{x-1}=\frac{6}{2x+1}$$
$$\overset{1}{\cancel{(x-1)}}(2x+1)\cdot\frac{2}{\underset{1}{\cancel{x-1}}}=(x-1)\overset{1}{\cancel{(2x+1)}}\cdot\frac{6}{\underset{1}{\cancel{2x+1}}}$$
$$4x+2=6x-6$$
$$8=2x$$
$$4=x$$

The solution is 4.

15. $\frac{2x}{7} = \frac{x-2}{14}$

$$\overset{2}{\cancel{14}} \cdot \frac{2x}{\underset{1}{\cancel{7}}} = \overset{1}{\cancel{14}} \cdot \frac{x-2}{\underset{1}{\cancel{14}}}$$

$$4x = x - 2$$

$$3x = -2$$

$$x = -\frac{2}{3}$$

The solution is $-\frac{2}{3}$.

Objective B

17. **Strategy** To solve for the number of Americans with no health insurance, write and solve a proportion using x to represent the number of millions of Americans with no health insurance.

Solution

$$\frac{1}{6} = \frac{x}{300}$$

$$\overset{50}{\cancel{300}} \cdot \frac{1}{\underset{1}{\cancel{6}}} = \overset{1}{\cancel{300}} \cdot \frac{x}{\underset{1}{\cancel{300}}}$$

$$50 = x$$

There are approximately 50 million Americans with no health insurance.

19. **Strategy** To solve for the number of voters who favor the amendment, write and solve a proportion using x to represent the number of voters who favor the amendment.

Solution

$$\frac{4}{7} = \frac{x}{35,000}$$

$$\overset{5,000}{\cancel{35,000}} \cdot \frac{4}{\underset{1}{\cancel{7}}} = \overset{1}{\cancel{35,000}} \cdot \frac{x}{\underset{1}{\cancel{35,000}}}$$

$$20,000 = x$$

There are approximately 20,000 voters who favor the amendment.

21. **Strategy** To find the amount of sugar needed, write and solve a proportion using x to represent the required amount of sugar.

Solution

$$\frac{2}{\frac{2}{3}} = \frac{x}{2}$$

$$\frac{2}{\frac{2}{\underset{1}{\cancel{3}}}} \cdot \frac{3}{\underset{1}{\cancel{3}}} = \frac{x}{2}$$

$$\frac{6}{2} = \frac{x}{2}$$

$$\overset{1}{\cancel{2}} \cdot \frac{6}{\underset{1}{\cancel{2}}} = \overset{1}{\cancel{2}} \cdot \frac{x}{\underset{1}{\cancel{2}}}$$

$$6 = x$$

6 cups of sugar will be required for 2 cups of water.

23. **Strategy** To find the number of fish in the lake, write and solve a proportion using x to represent the number of fish in the lake.

Solution

$$\frac{4}{80} = \frac{40}{x}$$

$$\overset{1}{\cancel{80}}x \cdot \frac{4}{\underset{1}{\cancel{80}}} = 80x \cdot \frac{40}{x}$$

$$4x = 3200$$

$$x = 800$$

There are approximately 800 fish in the lake.

25. **Strategy** To find the height, write and solve a proportion using x to find the standing height of the person.

Solution

$$\frac{3}{4} = \frac{48}{x}$$

$$\overset{1}{\cancel{4}}x \cdot \frac{3}{\underset{1}{\cancel{4}}} = 4x \cdot \frac{48}{x}$$

$$3x = 192$$

$$x = 64$$

The standing height of the person is 64 in.

27. Strategy To find the deduction, write and solve a proportion using x to find the deduction for 2008.

Solution

$$\frac{505}{1000}=\frac{x}{2200}$$

$$\overset{22}{\cancel{22{,}000}}\cdot\frac{505}{\underset{1}{\cancel{1000}}}=\overset{10}{\cancel{22{,}000}}\cdot\frac{x}{\underset{1}{\cancel{2200}}}$$

$$11110=10x$$

$$1111=x$$

The deduction for 2200 miles is \$1111.

29. Strategy To find the length of the claw, write and solve a proportion using x to find the length of a claw of a 7 in. scorpion.
•Change 8.2 ft to in. $8.2\times12=98.4$ in

Solution

$$\frac{18}{98.4}=\frac{x}{7}$$

$$\overset{1}{\cancel{98.4}}\cdot7\cdot\frac{18}{\underset{1}{\cancel{98.4}}}=98.4\cdot\overset{1}{\cancel{7}}\cdot\frac{x}{\underset{1}{\cancel{7}}}$$

$$126=98.4x$$

$$1.28=x$$

The claw of a 7-inch scorpion is approximately 1.28 in.

Objective C

31. Strategy To find AC, write a proportion using the fact that in similar triangles, the ratios of corresponding sides are equal. Solve the proportion for AC.

Solution

$$\frac{AC}{AB}=\frac{DF}{DE}$$

$$\frac{AC}{4}=\frac{15}{9}$$

$$\overset{9}{\cancel{36}}\cdot\frac{AC}{\underset{1}{\cancel{4}}}=\frac{15}{\underset{1}{\cancel{9}}}\cdot\overset{4}{\cancel{36}}$$

$$9AC=60$$

$$AC\approx6.7$$

The length of AC is 6.7 cm.

33. Strategy To find the height of triangle ABC, write a proportion using the fact that in similar triangles, the ratio of corresponding sides equals the ratio of corresponding heights. Solve the proportion of the height.

Solution

$$\frac{h_{ABC}}{BC}=\frac{h_{DFE}}{FE}$$

$$\frac{h}{5}=\frac{7}{12}$$

$$\overset{12}{\cancel{60}}\cdot\frac{h}{\underset{1}{\cancel{5}}}=\frac{7}{\underset{1}{\cancel{12}}}\cdot\overset{5}{\cancel{60}}$$

$$12h=35$$

$$h\approx2.9$$

The height of triangle ABC is 2.9 m.

35. Strategy To find the perimeter:

- Find side DF by writing a proportion using the fact that the ratios of corresponding sides of similar triangles are equal.
- Use the formula for the perimeter of a triangle.

Solution

$$\frac{AC}{BC}=\frac{DF}{EF}$$

$$\frac{5}{6}=\frac{DF}{9}$$

$$\overset{3}{\cancel{18}}\cdot\frac{5}{\underset{1}{\cancel{6}}}=\frac{DF}{\underset{1}{\cancel{9}}}\cdot\overset{2}{\cancel{18}}$$

$$15=2DF$$

$$7.5=DF$$

$$P=a+b+c=7.5+9+6=22.5$$

The perimeter of triangle DEF is 22.5 ft.

37. Strategy To find the area:

- Find the height of triangle ABC by writing a proportion using the fact that in similar triangles, the ratio of corresponding sides equals the ratio of corresponding heights. Solve the proportion for the height (h).
- Use the formula for the area of a triangle.

Solution

$$\frac{h_{ABC}}{AB}=\frac{h_{DEF}}{DE}$$
$$\frac{h}{12}=\frac{12}{18}$$
$$\overset{3}{\cancel{36}}\cdot\frac{h}{\underset{1}{\cancel{12}}}=\frac{12}{\underset{1}{\cancel{18}}}\cdot\overset{2}{\cancel{36}}$$
$$3h=24$$
$$h=8$$
$$A=\frac{1}{2}bh=\frac{1}{2}(12)(8)=48$$

The area of triangle ABC is 48 m^2.

39. True

41. Strategy To find BC, write a proportion using the fact that in similar triangles, the ratios of corresponding sides are equal. Solve the proportion for BC.

Solution

$$\frac{BD}{BC}=\frac{AE}{AC}$$
$$\frac{5}{BC}=\frac{8}{10}$$
$$10\cancel{(BC)}\frac{5}{\cancel{BC}}=\frac{8}{\underset{1}{\cancel{10}}}(BC)\overset{1}{\cancel{10}}$$
$$50=8BC$$
$$BC=6.25$$

The length of BC is 6.25 cm.

43. Strategy To find QO, write a proportion using the fact that in similar triangles, the ratios of corresponding sides are equal. Solve the proportion for QO.

Solution

$$\frac{PO}{QO}=\frac{MO}{NO}$$
$$\frac{8}{QO}=\frac{20}{25}$$
$$25\cancel{QO}\frac{8}{\cancel{QO}}=\frac{20}{\underset{1}{\cancel{25}}}(QO)\overset{1}{\cancel{25}}$$
$$200=20QO$$
$$10=QO$$

The length QO is 10 ft.

45. Strategy To find the width of the river, write a proportion using the fact that in similar triangles, the ratios of corresponding sides are equal. Solve the proportion for the width CD.

Solution

$$\frac{BO}{AB}=\frac{OC}{CD}$$
$$\frac{8}{14}=\frac{20}{CD}$$
$$\overset{1}{\cancel{14}}(CD)\frac{8}{\underset{1}{\cancel{14}}}=\frac{20}{CD}CD14$$
$$8CD=280$$
$$CD=35$$

The width of the river is 35 m.

Applying the Concepts

47. Strategy Write and solve a proportion using x to represent the first person's share of the winnings.

- Total number of shares = 25 + 30 + 35
  = 90

Solution

$$\frac{25}{90}=\frac{x}{4.5}$$
$$\overset{1}{\cancel{90}}\cdot\frac{25}{\underset{1}{\cancel{90}}}=\frac{x}{\underset{1}{\cancel{4.5}}}\cdot\overset{20}{\cancel{90}}$$
$$25=20x$$
$$1.25=x$$

The first person won \$1.25 million.

Section 8.6

Objective A Exercises

1.
$$\begin{aligned} 3x + y &= 10 \\ 3x - 3x + y &= -3x + 10 \\ y &= -3x + 10 \end{aligned}$$

3.
$$\begin{aligned} 4x - y &= 3 \\ 4x - 4x - y &= -4x + 3 \\ -y &= -4x + 3 \\ -1(-y) &= -1(-4x + 3) \\ y &= 4x - 3 \end{aligned}$$

5.
$$\begin{aligned} 3x + 2y &= 6 \\ 3x - 3x + 2y &= -3x + 6 \\ 2y &= -3x + 6 \\ \frac{2y}{2} &= \frac{-3x + 6}{2} \\ y &= -\frac{3}{2}x + 3 \end{aligned}$$

7.
$$\begin{aligned} 2x - 5y &= 10 \\ 2x - 2x - 5y &= -2x + 10 \\ -5y &= -2x + 10 \\ \frac{-5y}{-5} &= \frac{-2x + 10}{-5} \\ y &= \frac{2}{5}x - 2 \end{aligned}$$

9.
$$\begin{aligned} 2x + 7y &= 14 \\ 2x - 2x + 7y &= -2x + 14 \\ 7y &= -2x + 14 \\ \frac{7y}{7} &= \frac{-2x + 14}{7} \\ y &= -\frac{2}{7}x + 2 \end{aligned}$$

11.
$$\begin{aligned} x + 3y &= 6 \\ x - x + 3y &= -x + 6 \\ 3y &= -x + 6 \\ \frac{3y}{3} &= \frac{-x + 6}{3} \\ y &= -\frac{1}{3}x + 2 \end{aligned}$$

13.
$$\begin{aligned} y - 2 &= 3(x + 2) \\ y - 2 &= 3x + 6 \\ y - 2 + 2 &= 3x + 6 + 2 \\ y &= 3x + 8 \end{aligned}$$

15.
$$\begin{aligned} y - 1 &= -\frac{2}{3}(x + 6) \\ y - 1 &= -\frac{2}{3}x - 4 \\ y - 1 + 1 &= -\frac{2}{3}x - 4 + 1 \\ y &= -\frac{2}{3}x - 3 \end{aligned}$$

17.
$$\begin{aligned} x + 6y &= 10 \\ x + 6y - 6y &= -6y + 10 \\ x &= -6y + 10 \end{aligned}$$

19.
$$\begin{aligned} 2x - y &= 6 \\ 2x - y + y &= y + 6 \\ 2x &= y + 6 \\ \frac{2x}{2} &= \frac{y + 6}{2} \\ x &= \frac{1}{2}x + 3 \end{aligned}$$

21.
$$\begin{aligned} 4x + 3y &= 12 \\ 4x + 3y - 3y &= -3y + 12 \\ 4x &= -3y + 12 \\ \frac{4x}{4} &= \frac{-3y + 12}{4} \\ x &= -\frac{3}{4}y + 3 \end{aligned}$$

23.
$$\begin{aligned} x - 4y - 3 &= 0 \\ x - 4y - 3 + 3 &= 0 + 3 \\ x - 4y &= 3 \\ x - 4y + 4y &= 4y + 3 \\ x &= 4y + 3 \end{aligned}$$

25.
$$\begin{aligned} d &= rt \\ \frac{d}{r} &= \frac{rt}{r} \\ \frac{d}{r} &= t \end{aligned}$$

27.
$$\begin{aligned} PV &= nRT \\ \frac{PV}{nR} &= \frac{nRT}{nR} \\ \frac{PV}{nR} &= T \end{aligned}$$

29.
$$\begin{aligned} P &= 2l + 2w \\ P - 2w &= 2l + 2w - 2w \\ P - 2w &= 2l \\ \frac{P - 2w}{2} &= \frac{2l}{2} \\ \frac{P - 2w}{2} &= l \end{aligned}$$

31.
$$\begin{aligned} A &= \frac{1}{2}h(b_1 + b_2) \\ 2 \cdot A &= 2 \cdot \frac{1}{2}h(b_1 + b_2) \\ 2A &= h(b_1 + b_2) \\ 2A &= hb_1 + hb_2 \\ 2A - hb_2 &= hb_1 + hb_2 - hb_2 \\ 2A - hb_2 &= hb_1 \\ \frac{2A - hb_2}{h} &= \frac{hb_1}{h} \\ \frac{2A - hb_2}{h} &= b_1 \end{aligned}$$

33.
$$\begin{aligned} V &= \frac{1}{3}Ah \\ 3 \cdot V &= 3 \cdot \frac{1}{3}Ah \\ 3V &= Ah \\ \frac{3V}{A} &= \frac{Ah}{A} \\ \frac{3V}{A} &= h \end{aligned}$$

35.
$$\begin{aligned} R &= \frac{C - S}{t} \\ t \cdot R &= t \cdot \frac{C - S}{t} \\ Rt &= C - S \\ Rt - C &= C - C - S \\ Rt - C &= -S \\ -1(Rt - C) &= -1(-S) \\ C - Rt &= S \end{aligned}$$

37.
$$\begin{aligned} A &= P + Prt \\ A &= P(1 + rt) \\ \frac{A}{1 + rt} &= \frac{P(1 + rt)}{1 + rt} \\ \frac{A}{1 + rt} &= P \end{aligned}$$

39.
$$\begin{aligned} A &= Sw + w \\ A &= w(S + 1) \\ \frac{A}{S + 1} &= \frac{w(S + 1)}{S + 1} \\ \frac{A}{S + 1} &= w \end{aligned}$$

Applying the Concepts

41. a.
$$\begin{aligned} B &= \frac{F}{S - V} \\ (S - V) \cdot B &= (S - V) \cdot \frac{F}{S - V} \\ BS - BV &= F \\ BS - BV + BV &= F + BV \\ BS &= F + BV \\ \frac{BS}{B} &= \frac{F + BV}{B} \\ S &= \frac{F + BV}{B} \end{aligned}$$

b. $S = \dfrac{20{,}000 + 200(80)}{200} = 180$

The required selling price is \$180.

c. $S = \dfrac{15{,}000 + 600(50)}{600} = 75$

The required selling price is \$75.

Section 8.7

Objective A Exercises

1. The rate of work is the amount of a task that is competed per unit of time.

3. Together it will take them less time than the fastest one alone, so it would take less than *k*.

5. **Strategy** •Time to remove the earth with both skiploaders working together: *t*

	Rate	Time	Part
First	$\frac{1}{12}$	t	$\frac{t}{12}$
Second	$\frac{1}{4}$	t	$\frac{t}{4}$

• The sum of the parts of the task completed by each must equal 1.

Solution

$$\frac{t}{12}+\frac{t}{4}=1$$

$$12\left(\frac{t}{12}+\frac{t}{4}\right)=12\cdot 1$$

$$t+3t=12$$

$$4t=12$$

$$t=3$$

With both skiploaders, it would take 3 hr to remove the earth.

7. Strategy • Time for the new machine: t

•Time for the old machine: $3t$

	Rate	Time	Part
New machine	$\frac{1}{t}$	9	$\frac{9}{t}$
Old machine	$\frac{1}{3t}$	9	$\frac{9}{3t}$

• The sum of the parts of the task completed by both machines must equal 1.

Solution

$$\frac{9}{t}+\frac{9}{3t}=1$$

$$3t\left(\frac{9}{t}+\frac{9}{3t}\right)=3t\cdot 1$$

$$27+9=3t$$

$$36=3t$$

$$12=t$$

It would take the new machine 12 h to complete the task.

9. Strategy •Time to print the first edition with both presses operating: t

	Rate	Time	Part
First press	$\frac{1}{55}$	t	$\frac{t}{55}$
Second press	$\frac{1}{66}$	t	$\frac{t}{66}$

• The sum of the parts printed by each press must equal 1.

Solution

$$\frac{t}{55}+\frac{t}{66}=1$$

$$330\left(\frac{t}{55}+\frac{t}{66}\right)=330\cdot 1$$

$$6t+5t=330$$

$$11t=330$$

$$t=30$$

It would take 30 minutes to print the first edition with both presses operating.

11. Strategy •Time for the apprentice working alone to complete the wall: t

	Rate	Time	Part
Mason	$\frac{1}{10}$	6	$\frac{6}{10}$
Apprentice	$\frac{1}{t}$	6	$\frac{6}{t}$

• The sum of the parts completed by both must equal 1.

Solution

$$\frac{6}{10}+\frac{6}{t}=1$$

$$10t\cdot\left(\frac{6}{10}+\frac{6}{t}\right)=10t\cdot 1$$

$$6t+60=10t$$

$$60=4t$$

$$15=t$$

It would take the apprentice 15 h to construct the wall.

13. Strategy •Time for the second technician to complete the task: t

	Rate	Time	Part
First tech	$\frac{1}{4}$	2	$\frac{2}{4}$
Second tech	$\frac{1}{6}$	t	$\frac{t}{6}$

• The sum completed by each must equal 1.

Solution

$$\frac{2}{4}+\frac{t}{6}=1$$
$$12\left(\frac{2}{4}+\frac{t}{6}\right)=12\cdot 1$$
$$6+2t=12$$
$$2t=6$$
$$t=3$$

It would take the second technician 3 h to complete the wiring.

15. Strategy •Time for one of the welders to complete the job: t

	Rate	Time	Part
First Welder	$\frac{1}{t}$	10	$\frac{10}{t}$
Second Welder	$\frac{1}{t}$	30	$\frac{30}{t}$

• The sum of the parts of the task completed by each welder must equal 1.

Solution

$$\frac{10}{t}+\frac{30}{t}=1$$
$$t\left(\frac{10}{t}+\frac{30}{t}\right)=t\cdot 1$$
$$10+30=t$$
$$40=t$$

It would take one of the welders 40 h to complete the welds.

17. Strategy •Time for one machine to complete the task: t

	Rate	Time	Part
First machine	$\frac{1}{t}$	7	$\frac{7}{t}$
Second machine	$\frac{1}{t}$	21	$\frac{21}{t}$

• The sum of the parts of the task completed by each machine must equal 1.

Solution

$$\frac{7}{t}+\frac{21}{t}=1$$
$$t\left(\frac{7}{t}+\frac{21}{t}\right)=t\cdot 1$$
$$7+21=t$$
$$28=t$$

It would take one machine 28 h to fill the boxes.

19. Strategy •Time both to pick the peas is m and time for Zachary is n.

	Rate	Time	Part
Zachary	$\frac{1}{n}$	m	$\frac{m}{n}$
Eli			x

• The sum of the parts of the task completed by each must equal 1.

Solution

$$\frac{m}{n}+x=1$$
$$n\left(\frac{m}{n}+x\right)=n(1)$$
$$m+nx=n$$
$$nx=n-m$$
$$x=\frac{n-m}{x}$$

The part completed by Zachary is $\frac{m}{n}$. The part completed by Eli is $\frac{n-m}{n}$.

Objective B

21. If distance = rate · time, then

$$\text{Time}=\frac{\text{distance}}{\text{rate}} \text{ and Rate}=\frac{\text{distance}}{\text{time}}.$$

23. Strategy • Rate of the prop plane: r

• Rate of the jet plane: $4r$

	Distance	Rate	Time
Prop plane	300	r	$\frac{300}{r}$
Jet plane	1800	$4r$	$\frac{1800}{4r}$

• The total time of the trip was 5 h.

Solution

$$\begin{aligned} \frac{300}{r}+\frac{1800}{4r}&=5 \\ 4r\left(\frac{300}{r}+\frac{1800}{4r}\right)&=4r\cdot 5 \\ 1200+1800&=20r \\ 3000&=20r \\ 150&=r \\ 4r=4\cdot 150&=600 \end{aligned}$$

The rate of the jet plane was 600 mph.

25. Strategy • Rate of the freight train: r

• Rate of the express train: $r+20$

	Distance	Rate	Time
Freight train	360	r	$\frac{360}{r}$
Express train	600	$r+20$	$\frac{600}{r+20}$

• The time of the freight train equals the time of the express train.

Solution

$$\begin{aligned} \frac{360}{r}&=\frac{600}{r+20} \\ r(r+20)\cdot\frac{360}{r}&=r(r+20)\cdot\frac{600}{r+20} \\ 360(r+20)&=600r \\ 360r+7200&=600r \\ 7200&=240r \\ 30&=r \\ 50&=r+20 \end{aligned}$$

The rate of the freight train is 30 mph.
The rate of the express train is 50 mph.

27. Strategy • Walking rate: r

• Running rate: $r+3$

Note: convert minutes to hours. $\frac{40}{60}=\frac{2}{3}$ hr

	Distance	Rate	Time
Walking	$\frac{2}{3}r$	r	$\frac{2}{3}$
Running	$\frac{1}{3}(r+3)$	$r+3$	$\frac{1}{3}$

• The total distance walked was 5 mi.

Solution

$$\begin{aligned} \frac{2}{3}r+\frac{1}{3}(r+3)&=5 \\ \frac{2}{3}r+\frac{1}{3}r+1&=5 \\ r&=4 \end{aligned}$$

Camille's walking rate is 4 mph.

29. Strategy •Rate of the cyclist: r

• Rate of the car: $r+36$

	Distance	Rate	Time
Cyclist	96	r	$\frac{96}{r}$
Car	384	$r+36$	$\frac{384}{r+36}$

• The time of the cyclist equals the time of the car.

Solution

$$\begin{aligned} \frac{96}{r}&=\frac{384}{r+36} \\ r(r+36)\cdot\frac{96}{r}&=r(r+36)\cdot\frac{384}{r+36} \\ 96(r+36)&=384r \\ 96r+3456&=384r \\ 3456&=288r \\ 12&=r \\ 48&=r+36 \end{aligned}$$

The rate of the car is 48 mph.

31. Strategy •original rate: r

• Reduced rate: $r-1$

	Distance	Rate	Time
Original rate	9	r	$\frac{9}{r}$
Reduced rate	4	$r-1$	$\frac{4}{r-1}$

• The time at the reduced rate is one hour less than the time at the original rate.

Solution

$$\frac{9}{r}-1=\frac{4}{r-1}$$
$$r(r-1)\left(\frac{9}{r}-1\right)=r(r-1)\cdot\frac{4}{r-1}$$
$$9(r-1)-r(r-1)=4r$$
$$9r-9-r^2+r=4r$$
$$-r^2+10r-9=4r$$
$$0=r^2-6r+9$$
$$0=(r-3)(r-3)$$
$$0=r-3$$
$$3=r$$

The original rate is 3 mph.

33. Strategy •Rate of the jet stream: r

	Distance	Rate	Time
With jet stream	2400	$550+r$	$\frac{2400}{550+r}$
Against jet stream	2000	$550-r$	$\frac{2000}{550-r}$

• The time traveled with the jet stream equals the time traveled against the jet stream.

Solution

$$\frac{2400}{550+r}=\frac{2000}{550-r}$$
$$(550+r)(550-r)\frac{2400}{550+r}=(550+r)(550-r)\frac{2000}{550-r}$$
$$(550-r)2400=2000(550+r)$$
$$1{,}320{,}000-2400r=1{,}100{,}000+2000r$$
$$220{,}000=4400r$$
$$50=r$$

The rate of the jet stream is 50 mph.

35. Strategy • Rate of the current: r

	Distance	Rate	Time
With current	25	$20+r$	$\frac{25}{20+r}$
Against current	15	$20-r$	$\frac{15}{20-r}$

• The time traveled with the current equals the time traveled against the current.

Solution

$$\frac{25}{20+r}=\frac{15}{20-r}$$
$$(20+r)(20-r)\frac{25}{20+r}=(20+r)(20-r)\frac{15}{20-r}$$
$$500-25r=300+15r$$
$$200=40r$$
$$5=r$$

T

he rate of the current is 5 mph.

37. $\frac{1440}{380-r}$ is the time it takes the airplane to fly 1440 mi against the wind.

$\frac{1600}{380+r}$ is the time it takes the airplane to fly 1600 with the wind.

Applying the Concepts

39. Strategy •Usual speed: r

• Speed in bad weather: $r-10$

	Distance	Rate	Time
Usual	150	r	$\frac{150}{r}$
Bad weather	150	$r-10$	$\frac{150}{r-10}$

• The time during bad weather is $\frac{1}{2}$ h more than the usual time.

Solution

$$\frac{150}{r-10}=\frac{150}{r}+\frac{1}{2}$$

$$2r(r-10)\cdot\frac{150}{r-10}=2r(r-10)\left(\frac{150}{r}+\frac{1}{2}\right)$$

$$300r=300(r-10)+r(r-10)$$

$$300r=300r-3000+r^2-10r$$

$$0=r^2-10r-3000$$

$$0=(r-60)(r+50)$$

$r-60=0$ $\quad$ $r+50=0$

$r=60$ $\quad$ $r=-50$

The solution −50 is not possible because the rate cannot be negative. The bus usually travels 60 mph.

Section 8.8

Objective A Exercises

1. **Strategy:** Write the basic direct variation equation replacing the variable with the given values. Solve for *k*.
Write the direct variation equation replacing *k* with its value. Substitute 5000 for *s* and solve for *P*.

Solution:

$P=ks$

$4000=250k$

$16=k$

$P=16s=16(5000)=80{,}000$

When the company sells 5000 products its profit will be $80,000.

3. **Strategy:** Write the basic direct variation equation replacing the variable with the given values. Solve for *k*.
Write the direct variation equation replacing *k* with its value. Substitute 15 for *d* and solve for *p*.

Solution:

$p=kd$

$4.5=10k$

$0.45=k$

$p=0.45d=0.45(15)=6.75$

The pressure is 6.75 lb/in^2.

5. **Strategy:** Write the basic direct variation equation replacing the variable with the given values. Solve for *k*.
Write the direct variation equation replacing *k* with its value. Substitute 10 for *t* and solve for *d*.

Solution:

$d=kt^2$

$144=k(3)^2$

$144=9k$

$16=k$

$d=16t^2=16(10^2)=16(100)=1600$

In 10 s the object will fall 1600 ft.

7. **Strategy:** Write the basic inverse variation equation replacing the variable with the given values. Solve for *k*.
Write the inverse variation equation replacing *k* with its value. Substitute 5 for *n* and solve for *T*.

Solution:

$T=\frac{k}{n}$

$500=\frac{k}{1}$

$500=k$

$T=\frac{500}{n}=\frac{500}{5}=100$

It will take 5 computers 100 s to solve the same problem.

9. Strategy: Write the basic direct variation equation replacing the variable with the given values. Solve for k.
Write the direct variation equation replacing k with its value. Substitute 15 for v and solve for L.

Solution:

$$L = kv^2$$
$$640 = k(20)^2$$
$$640 = 400k$$
$$1.6 = k$$
$$L = 1.6v^2 = 1.6(15)^2 = 360$$

The load on the sail will be 360 lbs.

11. Strategy: Write the basic direct variation equation replacing the variable with the given values. Solve for k.
Write the direct variation equation replacing k with its value. Substitute 230000 for p and solve for c.

Solution:

$$c = kp$$
$$15600 = 260{,}000k$$
$$0.06 = k$$
$$c = 0.06p = 0.06(230{,}000) = 13{,}800$$

The commission on a \$230,000 home is \$13,800.

13. Strategy: Write the basic combined variation equation replacing the variable with the given values. Solve for k.
Write the combined variation equation replacing k with its value.
Substitute 24 for r, 180 for v and solve for I.

Solution:

$$I = \frac{kv}{r}$$
$$10 = \frac{110k}{11}$$
$$110 = 110k$$
$$1 = k$$
$$I = \frac{1v}{r} = \frac{180}{24} = 7.5$$

The current is 7.5 amps.

15. Strategy: Write the basic inverse variation equation replacing the variable with the given values. Solve for k.
Write the inverse variation equation replacing k with its value. Substitute 5 for d and solve for I.

Solution:

$$I = \frac{k}{d^2}$$
$$12 = \frac{k}{10^2}$$
$$12 = \frac{k}{100}$$
$$1200 = k$$
$$I = \frac{1200}{d^2} = \frac{1200}{5^2} = \frac{1200}{25} = 48$$

The intensity is 48 foot-candles when the distance is 5 ft.

17. If x is doubled than y is doubled.

Applying the Concepts

19. inversely

21. inversely

Chapter 8 Concept Review

1. A rational expression is in simplest form when the numerator and denominator have no common factors. [8.1A]

2. To divide two rational expressions, change the division to a multiplication and change the divisor to its reciprocal. Then multiply. [8.1C]

3. To find the LCM of two polynomials, first factor each polynomial completely. The LCM is the product of each factor the greatest number of times it occurs in any one factorization. [8.2A]

4. When subtracting two rational expressions, both expressions must have the same denominator before the subtraction can take place. [8.2A]

5. To add rational expressions:
1. Find the LCM of the denominators.
2. Write each fraction as an equivalent fraction using the LCM as the denominator.
3. Add the numerators and place the result over the common denominator.
4. Write the answer in simplest form. [8.2B]

6. To simplify a complex fraction by Method 1:
1. Determine the LCM of the denominators of the fractions in the numerator and denominator of the complex fraction.
2. Multiply the numerator and denominator of the complex fraction by the LCM.
3. Simplify

To simplify a complex fraction by Method 2:
1. Simplify the numerator to a single fraction and simplify the denominator to a single fraction.
2. Using the definition for dividing fractions, multiply the numerator by the reciprocal of the denominator.
3. Simplify. [8.3A]

7. When solving an equation that contains fractions, first we clear the denominators in order to rewrite the equation without any fractions. [8.3A]

8. If the units in a comparison are different, the comparison is a rate. [8.5A]

9. The ratios of corresponding sides of similar triangles are equal. Therefore, we can write a proportion by setting one ratio of corresponding sides equal to the ratio of another pair of corresponding sides. If there is one unknown side in the proportion, we can solve the proportion for the unknown side. [8.5B]

10. When solving a literal equation for a particular variable, the goal is to rewrite the equation so that the variable being solved for is on one side of the equation and all numbers and other variables are on the other side. [8.6A]

11. If a job is completed in x hours, the rate of work is $\frac{1}{x}$ of the job each hour. [8.7A]

Chapter 8 Review Exercises

1. $\frac{6a^2b^7}{25x^3y} \div \frac{12a^3b^4}{5x^2y^2} = \frac{6a^2b^7}{25x^3y} \cdot \frac{5x^2y^2}{12a^3b^4}$

$= \frac{\overset{1}{\cancel{6}} \cdot \overset{1}{\cancel{5}} a^2b^7x^2y^2}{\underset{5}{\cancel{25}} \cdot \underset{2}{\cancel{12}} a^3b^4x^3y} = \frac{b^3y}{10ax}$

2. $\frac{x+7}{15x} + \frac{x-2}{20x} = \frac{4}{4}\left(\frac{x+7}{15x}\right) + \frac{3}{3}\left(\frac{x-2}{20x}\right)$

$= \frac{4x+28}{60x} + \frac{3x-6}{60x} = \frac{7x+22}{60x}$

3.
$$\frac{x-\frac{16}{5x-2}}{3x-4-\frac{88}{5x-2}}=\frac{x-\frac{16}{5x-2}}{3x-4-\frac{88}{5x-2}}\cdot\frac{5x-2}{5x-2}$$
$$=\frac{x(5x-2)-\frac{16}{5x-2}\cdot(5x-2)}{3x(5x-2)-4(5x-2)-\frac{88}{5x-2}\cdot(5x-2)}$$
$$=\frac{5x^2-2x-16}{15x^2-6x-20x+8-88}=\frac{5x^2-2x-16}{15x^2-26x-80}$$
$$=\frac{(5x+8)(x-2)}{(5x+8)(3x-10)}=\frac{x-2}{3x-10}$$

4.
$$\frac{x^2+x-30}{15+2x-x^2}=\frac{(x+6)(x-5)}{(3+x)(5-x)}=-\frac{x+6}{x+3}$$

5.
$$\frac{16x^5y^3}{24xy^{10}}=\frac{2\cdot 2\cdot 2\cdot 2x^{5-1}}{2\cdot 2\cdot 2\cdot 3y^{10-3}}=\frac{2x^4}{3y^7}$$

6.
$$\frac{20}{x+2}=\frac{5}{16}$$
$$16(x+2)\cdot\frac{20}{x+2}=16(x+2)\cdot\frac{5}{16}$$
$$320=5x+10$$
$$310=5x$$
$$62=x$$

62 checks as a solution. 62 is the solution.

7.
$$\frac{10-23x+12y^2}{6y^2-y-5}\div\frac{4y^2-13y+10}{18y^2+3y-10}$$
$$=\frac{12y^2-23y+10}{6y^2-y-5}\cdot\frac{18y^2+3y-10}{4y^2-13y+10}$$
$$=\frac{(3y-2)(4y-5)}{(6y+5)(y-1)}\cdot\frac{(3y-2)(6y+5)}{(4y-5)(y-2)}$$
$$=\frac{(3y-2)^2}{(y-1)(y-2)}$$

8.
$$\frac{8ab^2}{15x^3y}\cdot\frac{5xy^4}{16a^2b}$$
$$\frac{8\cdot 5ab^2xy^4}{15\cdot 16a^2bx^3y}=\frac{by^3}{6ax^2}$$

9.
$$\frac{1-\frac{1}{x}}{1-\frac{8x-7}{x^2}}=\frac{1-\frac{1}{x}}{1-\frac{8x-7}{x^2}}\cdot\frac{x^2}{x^2}$$
$$=\frac{1\cdot x^2-\frac{1}{x}\cdot x^2}{1\cdot x^2-\frac{8x-7}{x^2}\cdot x^2}=\frac{x^2-x}{x^2-8x+7}$$
$$=\frac{x(x-1)}{(x-1)(x-7)}=\frac{x}{x-7}$$

10.
$$\frac{x}{12x^2+16x-3},\ \frac{4x^2}{6x^2+7x-3}$$
$$\frac{x}{(6x-1)(2x+3)},\ \frac{4x^2}{(2x+3)(3x-1)}$$

The LCM is $(6x-1)(2x+3)(3x-1)$.

$$\frac{x}{(6x-1)(2x+3)}\cdot\frac{3x-1}{3x-1}=\frac{3x^2-x}{(6x-1)(2x+3)(3x-1)}$$
$$\frac{4x^2}{(2x+3)(3x-1)}\cdot\frac{6x-1}{6x-1}=\frac{24x^3-4x^2}{(6x-1)(2x+3)(3x-1)}$$

11.
$$T=2(ab+bc+ca)$$
$$T=2ab+2bc+2ca$$
$$T-2bc=2a(b+c)$$
$$\frac{T-2bc}{2(b+c)}=a$$
$$\frac{T-2bc}{2b+2C}=a$$

12.
$$\frac{5}{7}+\frac{x}{2}=2-\frac{x}{7}$$
$$14\left(\frac{5}{7}+\frac{x}{2}\right)=14\left(2-\frac{x}{7}\right)$$
$$14\cdot\frac{5}{7}+14\cdot\frac{x}{2}=14\cdot 2-14\cdot\frac{x}{7}$$
$$10+7x=28-2x$$
$$9x=18$$
$$x=2$$

2 checks as a solution. The solution is 2.

13.
$$i=\frac{100m}{c}$$
$$c(i)=\frac{100m}{c}(c)$$
$$ci=100m$$
$$c=\frac{100m}{i}$$

14.
$$\frac{x+8}{x+4}=1+\frac{5}{x+4}$$
$$(x+4)\left(\frac{x+8}{x+4}\right)=(x+4)\cdot 1+(x+4)\left(\frac{5}{x+4}\right)$$
$$x+8=x+4+5$$
$$0=1$$

There is no solution.

15.
$$\frac{20x^2-45x}{6x^3+4x^2}\div\frac{40x^3-90x^2}{12x^2+8x}$$
$$=\frac{20x^2-45x}{6x^3+4x^2}\cdot\frac{12x^2+8x}{40x^3-90x^2}$$
$$=\frac{5x(4x-9)}{2x^2(3x+2)}\cdot\frac{4x(3x+2)}{10x^2(4x-9)}$$
$$=\frac{5x(4x-9)}{2x^2(3x+2)}\cdot\frac{2\cdot 2\cdot x(3x+2)}{5\cdot 2x^2(4x-9)}=\frac{1}{x^2}$$

16.
$$\frac{2y}{5y-7}+\frac{3}{7-5y}=\frac{2y}{5y-7}+\frac{3}{-1(5y-7)}$$
$$=\frac{2y}{5y-7}-\frac{3}{5y-7}=\frac{2y-3}{5y-7}$$

17.
$$\frac{5x+3}{2x^2+5x-3}-\frac{3x+4}{2x^2+5x-3}$$
$$=\frac{5x+3-(3x+4)}{2x^2+5x-3}=\frac{5x+3-3x-4}{2x^2+5x-3}$$
$$=\frac{2x-1}{(2x-1)(x+3)}=\frac{1}{x+3}$$

18. $10x^2-11x+3=(5x-3)(2x-1)$

$20x^2-17x+3=(5x-3)(4x-1)$

LCM = $(5x-3)(2x-1)(4x-1)$

19.
$$4x+9y=18$$
$$4x-4x+9y=-4x+18$$
$$9y=-4x+18$$
$$\frac{9y}{9}=\frac{-4x+18}{9}$$
$$y=-\frac{4}{9}x+2$$

20.
$$\frac{24x^2-94x+15}{12x^2-49x+15}\cdot\frac{24x^2+7x-5}{4-27x+18x^2}$$
$$=\frac{(6x-1)(4x-15)}{(3x-1)(4x-15)}\cdot\frac{(3x-1)(8x+5)}{(3x-4)(6x-1)}=\frac{8x+5}{3x-4}$$

21.
$$\frac{20}{2x+3}=\frac{17x}{2x+3}-5$$
$$(2x+3)\cdot\frac{20}{2x+3}=(2x+3)\cdot\frac{17x}{2x+3}-5(2x+3)$$
$$20=17x-10x-15$$
$$20=7x-15$$
$$35=7x$$
$$5=x$$

5 checks as a solution. The solution is 5.

22.
$$\frac{x-1}{x+2}+\frac{3x-2}{5-x}+\frac{5x^2+15x-11}{x^2-3x-10}$$
LCM $=(x-5)(x+2)$
$$\frac{x-1}{x+2}+\frac{3x-2}{5-x}+\frac{5x^2+15x-11}{x^2-3x-10}$$
$$=\frac{x-1}{x+2}-\frac{3x-2}{x-5}+\frac{5x^2+15x-11}{(x+2)(x-5)}$$
$$=\frac{x-1}{x+2}\cdot\frac{x-5}{x-5}-\frac{3x-2}{x-5}\cdot\frac{x+2}{x+2}+\frac{5x^2+15x-11}{(x+2)(x-5)}$$
$$=\frac{x^2-6x+5}{(x+2)(x-5)}-\frac{3x^2+4x-4}{(x+2)(x-5)}+\frac{5x^2+15x-11}{(x+2)(x-5)}$$
$$=\frac{x^2-6x+5-(3x^2+4x-4)+5x^2+15x-11}{(x+2)(x-5)}$$
$$=\frac{x^2-6x+5-3x^2-4x+4+5x^2+15x-11}{(x+2)(x-5)}$$
$$=\frac{3x^2+5x-2}{(x+2)(x-5)}=\frac{(3x-1)(x+2)}{(x+2)(x-5)}=\frac{3x-1}{x-5}$$

23.
$$\frac{6}{x-7}=\frac{8}{x-6}$$
$$(x-6)(x-7)\frac{6}{x-7}=(x-6)(x-7)\frac{8}{x-6}$$
$$6x-36=8x-56$$
$$20=2x$$
$$10=x$$

10 checks as a solution. The solution is 10.

24.
$$\frac{3}{20}=\frac{x}{80}$$
$$80\cdot\frac{3}{20}=\frac{x}{80}\cdot 80$$
$$12=x$$

12 checks as a solution. The solution is 12.

25. Strategy •Triangle ABC is similar to triangle DEF. Solve a proportion to find the length of AC. Let x represent the length of AC.

• Find the perimeter of the triangle by adding the lengths of the three sides.

Solution
$$\frac{BC}{AC}=\frac{EF}{DF}$$
$$\frac{6}{x}=\frac{9}{12}$$
$$12x\left(\frac{6}{x}\right)=12x\left(\frac{9}{12}\right)$$
$$72=9x$$
$$8=x$$
$$P=10+6+8$$
$$P=24$$

The perimeter of triangle ABC is 24 in.

26. Strategy •Time to fill the pool using both hoses: t

	Rate	Time	Part
First hose	$\frac{1}{15}$	t	$\frac{t}{15}$
Second hose	$\frac{1}{10}$	t	$\frac{t}{10}$

• The sum of the parts of the task completed by each hose must equal 1.

Solution
$$\frac{t}{15}+\frac{t}{10}=1$$
$$30\cdot\frac{t}{15}+30\cdot\frac{t}{10}=30\cdot 1$$
$$2t+3t=30$$
$$5t=30$$
$$t=6$$

It would take 6 hours to fill the pool.

27. Strategy •Rate of the bus: r

•Rate of the car: $r+10$

	Distance	Rate	Time
Bus	245	r	$\frac{245}{r}$
Car	315	$r+10$	$\frac{315}{r+10}$

• The time of the bus equals the time of the car.

Solution
$$\frac{315}{r+10}=\frac{245}{r}$$
$$r(r+10)\frac{315}{r+10}=r(r+10)\frac{245}{r}$$
$$315r=245r+2450$$
$$70r=2450$$
$$r=35$$
$$r+10=45$$

The rate of the car is 45 mph.

28. Strategy •Rate of the wind: r

	Distance	Rate	Time
With the wind	2100	$400+r$	$\frac{2100}{400+r}$
Against the wind	1900	$400-r$	$\frac{1900}{400-r}$

• The time traveled with the wind equals the time traveled against the wind.

Solution

$$\frac{2100}{400+r}=\frac{1900}{400-r}$$

$$(400+r)(400-r)\frac{2100}{400+r}=(400+r)(400-r)\frac{1900}{400-r}$$

$$840{,}000-2100r=760{,}000+1900r$$

$$80{,}000=4000r$$

$$20=r$$

The rate of the wind is 20 mph.

29. Strategy • Unknown runs: x

Write and solve a proportion.

Solution

$$\frac{\text{ERA}}{9}=\frac{15}{100}$$

$$900\cdot\frac{\text{ERA}}{9}=900\cdot\frac{15}{100}$$

$$100\text{ERA}=135$$

$$\text{ERA}=1.35$$

The pitcher's ERA is 1.35.

30. Strategy: • Write the basic inverse variation equation replacing the variable with the given values. Solve for k.
• Write the inverse variation equation replacing k with its value. Substitute 100 for R and solve for I.

Solution:

$$I=\frac{k}{R}$$

$$4=\frac{k}{50}$$

$$200=k$$

$$I=\frac{200}{R}$$

$$I=\frac{200}{100}$$

$$I=2$$

The current is 2 amps.

Chapter 8 Test

1. $\dfrac{16x^5y}{24x^2y^4}=\dfrac{16x^5y}{24x^2y^4}=\dfrac{2x^3}{3y^3}$

2. $\dfrac{x^2+4x-5}{1-x^2}=\dfrac{(x+5)(x-1)}{(1+x)(1-x)}=-\dfrac{x+5}{x+1}$

3. $\dfrac{x^3y^4}{x^2-4x+4}\cdot\dfrac{x^2-x-2}{x^6y^4}$

$=\dfrac{x^3y^4}{(x-2)(x-2)}\cdot\dfrac{(x-2)(x+1)}{x^6y^4}=\dfrac{x+1}{x^3(x-2)}$

4. $\dfrac{x^2+2x-3}{x^2+6x+9}\cdot\dfrac{2x^2-11x+5}{2x^2+3x-5}$

$\dfrac{(x+3)(x-1)}{(x+3)(x+3)}\cdot\dfrac{(2x-1)(x-5)}{(2x+5)(x-1)}=\dfrac{(2x-1)(x-5)}{(x+3)(2x+5)}$

5. $\dfrac{x^2+3x+2}{x^2+5x+4}\div\dfrac{x^2-x-6}{x^2+2x-15}$

$=\dfrac{(x+2)(x+1)}{(x+1)(x+4)}\cdot\dfrac{(x+5)(x-3)}{(x-3)(x+2)}=\dfrac{x+5}{x+4}$

6. $6x-3=3(2x-1)$

$2x^2+x-1=(2x-1)(x+1)$

$\text{LCM}=3(2x-1)(x+1)$

7. $\dfrac{3}{x^2-2x}=\dfrac{3}{x(x-2)}\qquad \dfrac{x}{x^2-4}=\dfrac{x}{(x-2)(x+2)}$

$\text{LCM}=x(x-2)(x+2)$

$\dfrac{3}{x(x-2)}\cdot\dfrac{x+2}{x+2}=\dfrac{3x+6}{x(x-2)(x+2)}$

$\dfrac{x}{(x-2)(x+2)}\cdot\dfrac{x}{x}=\dfrac{x^2}{x(x-2)(x+2)}$

8. $\dfrac{2x}{x^2+3x-10}-\dfrac{4}{x^2+3x-10}=\dfrac{2x-4}{x^2+3x-10}$

$=\dfrac{2(x-2)}{(x+5)(x-2)}=\dfrac{2}{x+5}$

9. $\frac{2}{2x-1}-\frac{3}{3x+1}=\frac{2}{2x-1}\cdot\frac{3x+1}{3x+1}-\frac{3}{3x+1}\cdot\frac{2x-1}{2x-1}$

$=\frac{6x+2}{(2x-1)(3x+1)}-\frac{6x-3}{(2x-1)(3x+1)}$

$=\frac{6x+2-(6x-3)}{(2x-1)(3x+1)}$

$=\frac{6x+2-6x+3}{(2x-1)(3x+1)}=\frac{5}{(2x-1)(3x+1)}$

10. $\frac{x}{x+3}-\frac{2x-5}{x^2+x-6}$

$\frac{x}{x+3}-\frac{2x-5}{(x+3)(x-2)}$

LCM $=(x+3)(x-2)$

$\frac{x}{x+3}\cdot\frac{x-2}{x-2}-\frac{2x-5}{(x+3)(x-2)}$

$=\frac{x^2-2x}{(x+3)(x-2)}-\frac{2x-5}{(x+3)(x-2)}$

$=\frac{x^2-2x-(2x-5)}{(x+3)(x-2)}=\frac{x^2-2x-2x+5}{(x+3)(x-2)}$

$=\frac{x^2-4x+5}{(x+3)(x-2)}$

11. $\frac{1+\frac{1}{x}-\frac{12}{x^2}}{1+\frac{2}{x}-\frac{8}{x^2}}=\frac{1+\frac{1}{x}-\frac{12}{x^2}}{1+\frac{2}{x}-\frac{8}{x^2}}\cdot\frac{x^2}{x^2}$

$=\frac{x^2\cdot1+x^2\cdot\frac{1}{x}-x^2\cdot\frac{12}{x^2}}{x^2\cdot1+x^2\cdot\frac{2}{x}-x^2\cdot\frac{8}{x^2}}=\frac{x^2+x-12}{x^2+2x-8}$

$=\frac{\overset{1}{\cancel{(x+4)}}(x-3)}{\underset{1}{\cancel{(x+4)}}(x-2)}=\frac{x-3}{x-2}$

12. $\frac{6}{x}-2=1$

$\overset{1}{\cancel{x}}\cdot\frac{6}{\underset{1}{\cancel{x}}}-2\cdot x=1\cdot x$

$6-2x=x$

$6=3x$

$2=x$

2 checks as a solution. The solution is 2.

13. $\frac{2x}{x+1}-3=\frac{-2}{x+1}$

$\overset{1}{\cancel{(x+1)}}\left(\frac{2x}{\underset{1}{\cancel{x+1}}}\right)-3(x+1)=\overset{1}{\cancel{(x+1)}}\left(\frac{-2}{\underset{1}{\cancel{x+1}}}\right)$

$2x-3x-3=-2$

$-x=1$

$x=-1$

−1 does not check as a solution. There is no solution.

14. $\frac{3}{x+4}=\frac{5}{x+6}$

$(x+6)\overset{1}{\cancel{(x+4)}}\cdot\frac{3}{\underset{1}{\cancel{x+4}}}=\overset{1}{\cancel{(x+6)}}(x+4)\frac{5}{\underset{1}{\cancel{x+6}}}$

$3x+18=5x+20$

$-2=2x$

$-1=x$

−1 checks as a solution. The solution is −1.

15. Strategy •Triangle *ABC* is similar to triangle *DEF*. Solve a proportion to find the length of *DF*. Let *x* represent the length of *DF* .

• Find the area of the triangle by using the formula $A=\frac{1}{2}bh.$

Solution

$\frac{h}{AC}=\frac{h}{DF}$

$\frac{5}{8}=\frac{9}{x}$

$8x\left(\frac{5}{8}\right)=8x\left(\frac{9}{x}\right)$

$5x=72$

$x=14.4$

$A=\frac{1}{2}\cdot14.4\cdot9$

$A=64.8$

The area of triangle *DEF* is 64.8 m^2.

16. $d=s+rt$

$d-s=rt$

$\frac{d-s}{r}=\frac{rt}{r}$

$\frac{d-s}{r}=t$

17. Strategy Write and solve a proportion using x to represent the number of rolls of wallpaper for 315 ft^2 of wall space.

Solution

$$\frac{2}{45}=\frac{x}{315}$$
$$315\cdot 45\left(\frac{2}{45}\right)=315\cdot 45\left(\frac{x}{315}\right)$$
$$630=45x$$
$$14=x$$

14 rolls of wallpaper are needed to cover 315 ft^2 of wall space.

18. Strategy•Time for both landscapers t

	Rate	Time	Part
First	$\frac{1}{30}$	t	$\frac{t}{30}$
Second	$\frac{1}{15}$	t	$\frac{t}{15}$

- The sum of the parts completed by each pipe must equal 1.

Solution

$$\frac{t}{30}+\frac{t}{15}=1$$
$$30\cdot\frac{t}{30}+30\cdot\frac{t}{15}=1\cdot 30$$
$$t+2t=30$$
$$3t=30$$
$$t=10$$

It would take both landscapers 10 min.

19. Strategy • Rate of the hiker: r
- Rate of the cyclist: $r+7$

	Distance	Rate	Time
Cyclist	20	$r+7$	$\frac{20}{r+7}$
Hiker	6	r	$\frac{6}{r}$

- The time traveled by the cyclist equals the time traveled by the hiker.

Solution

$$\frac{20}{r+7}=\frac{6}{r}$$
$$r(r+7)\left(\frac{20}{r+7}\right)=r(r+7)\frac{6}{r}$$
$$20r=6r+42$$
$$14r=42$$
$$r=3$$
$$r+7=3+7=10$$

The rate of the cyclist is 10 mph.

20. Strategy: Write the combined variation equation replacing the variable with the given values. Solve for k.
Write the inverse variation equation replacing k with its value. Substitute 8000 for l and $\frac{1}{2}$ for d. Then solve for r.

Solution:

$$r=\frac{kl}{d^2}$$
$$3.2=\frac{k16{,}000}{\left(\frac{1}{4}\right)^2}$$
$$3.2=256{,}000k$$
$$\frac{3.2}{256000}=k$$
$$r=\frac{\frac{3.2}{256000}l}{d^2}$$
$$r=\frac{\frac{3.2}{256000}\cdot 8000}{\left(\frac{1}{2}\right)^2}$$
$$r=0.4$$

The resistance is 0.4 ohms.

Cumulative Review Exercises

1. $\left(\frac{2}{3}\right)^2\div\left(\frac{3}{2}-\frac{2}{3}\right)+\frac{1}{2}=\left(\frac{2}{3}\right)^2\div\left(\frac{9}{6}-\frac{4}{6}\right)+\frac{1}{2}$

$$= \frac{4}{9} \div \left(\frac{5}{6}\right) + \frac{1}{2} = \frac{4}{9} \cdot \frac{6}{5} + \frac{1}{2}$$

$$= \frac{2 \cdot 2 \cdot 2 \cdot \overset{1}{\cancel{3}}}{\underset{1}{\cancel{3}} \cdot 3 \cdot 5} + \frac{1}{2} = \frac{8}{15} + \frac{1}{2}$$

$$= \frac{16}{30} + \frac{15}{30} = \frac{31}{30}$$

2. $-a^2 + (a-b)^2$

$$-(-2)^2 + (-2-3)^2 = -4 + (-5)^2 = -4 + 25 = 21$$

3. $-2x - (-3y) + 7x - 5y = -2x + 3y + 7x - 5y$

$$= -2x + 7x + 3y - 5y$$
$$= 5x - 2y$$

4. $2[3x - 7(x-3) - 8] = 2[3x - 7x + 21 - 8]$

$$= 2[-4x + 13] = -8x + 26$$

5. $4 - \frac{2}{3}x = 7$

$$-\frac{2}{3}x = 3$$
$$-\frac{3}{2}\left(-\frac{2}{3}x\right) = 3\left(-\frac{3}{2}\right)$$
$$x = -\frac{9}{2}$$

The solution is $-\frac{9}{2}$.

6. $3[x - 2(x-3)] = 2(3-2x)$

$$3[x - 2x + 6] = 6 - 4x$$
$$3[-x + 6] = 6 - 4x$$
$$-3x + 18 = 6 - 4x$$
$$x = -12$$

The solution is -12.

7. $P \cdot B = A$

$$16\frac{2}{3}\% \cdot 60 = A$$
$$\frac{1}{6} \cdot 60 = A$$
$$10 = A$$

8. $x - 3(1-2x) \geq 1 - 4(3-2x)$

$$x - 3 + 6x \geq 1 - 12 + 8x$$
$$7x - 3 \geq -11 + 8x$$
$$-x - 3 \geq -11$$
$$-x \geq -8$$
$$x \leq 8$$

$\{x \mid x \leq 8\}$

9. Strategy Use the formula for the volume of a rectangular solid.

Solution

$V = LWH$

$V = 10 \cdot 5 \cdot 4$

$V = 200$

The volume is 200 ft^3.

10. $x - 2y = 2$

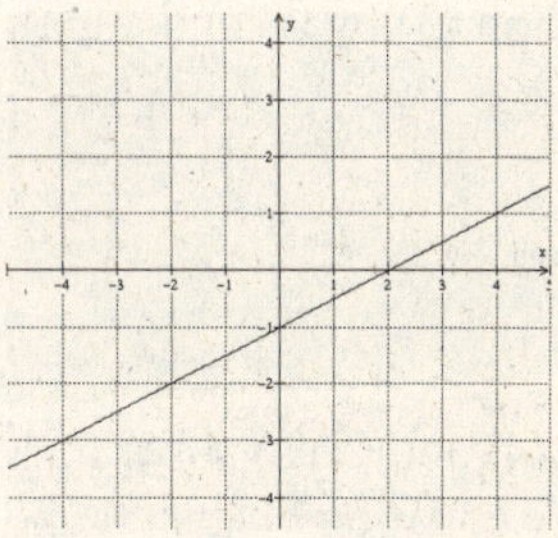

11. $P(x) = \frac{x-1}{2x-3}$

$$P(-2) = \frac{-2-1}{2(-2)-3}$$
$$P(-2) = \frac{-3}{-4-3}$$
$$P(-2) = \frac{-3}{-7}$$
$$P(-2) = \frac{3}{7}$$

12. $3x - 2y = 6$

$$-2y = -3x + 6$$
$$\frac{-2y}{-2} = \frac{-3x}{-2} + \frac{6}{-2}$$
$$y = \frac{3}{2}x - 3$$

$m = \frac{3}{2}$

$$y - y_1 = m(x - x_1)$$
$$y - (-1) = \frac{3}{2}(x - (-2))$$
$$y + 1 = \frac{3}{2}(x + 2)$$
$$y + 1 = \frac{3}{2}x + 3$$
$$y = \frac{3}{2}x + 2$$

13. (1) $2x - y + z = 2$
(2) $3x + y - 2z = 9$
(3) $x - y + z = 0$

Multiply equation (3) by –2 and add to equation (1).

$$\begin{aligned} 2x - y + z &= 2 \\ -2x + 2y - 2z &= 0 \\ \hline (4) \quad y - z &= 2 \end{aligned}$$

Multiply equation (3) by –3 and add to equation (2).

$$\begin{aligned} 3x + y - 2z &= 9 \\ -3x + 3y - 3z &= 0 \\ \hline (5) \quad 4y - 5z &= 9 \end{aligned}$$

Use equations (4) and (5) to solve for y and z. Multiply equation (4) by –4 and add to equation (5).

$$\begin{aligned} -4y + 4z &= -8 \\ 4y - 5z &= 9 \\ \hline -z &= 1 \\ z &= -1 \end{aligned}$$

Replace z with –1 in equation (4).

$$\begin{aligned} y - z &= 2 \\ y - (-1) &= 2 \\ y + 1 &= 2 \\ y &= 1 \end{aligned}$$

Replace place y with 1 and z with –1 in equation (3).

$$\begin{aligned} x - 1 - 1 &= 0 \\ x - 2 &= 0 \\ x &= 2 \end{aligned}$$

The solution is (2, 1, –1).

14. $(a^2b^5)(ab^2) = a^{2+1}b^{5+2} = a^3b^7$

15. $\frac{(2a^{-2}b^3)^{-2}}{(4a)^{-1}} = \frac{2^{-2}a^4b^{-6}}{4^{-1}a^{-1}} = \frac{4^1a^{4+1}}{2^2b^6} = \frac{4a^5}{4b^6} = \frac{a^4}{b^6}$

16. $0.000000035 = 3.5 \times 10^{-8}$

17. $(2a^2 - 3a + 1)(-2a^2)$
$= 2a^2(-2a^2) - 3a(-2a^2) + 1(-2a^2)$
$= -4a^4 + 6a^3 - 2a^2$

18. $(a - 3b)(a + 4b) = a^2 + 4ab - 3ab - 12b^2$
$= a^2 + ab - 12b^2$

19.
$$\begin{array}{r} x^2 + 2x + 4 \\ x - 2 \overline{)\, x^3 + 0x^2 + 0x - 8} \\ \underline{x^3 - 2x^2} \qquad\qquad \\ 2x^2 + 0x \qquad \\ \underline{2x^2 - 4x} \qquad \\ 4x - 8 \\ \underline{4x - 8} \\ 0 \end{array}$$

$(x^3 - 8) \div (x - 2) = x^2 + 2x + 4$

20. $y^2 - 7y + 6$
$1(6) = 6$
Factors of 6 whose sum is –7: –6 and –1
$y^2 - 6y - y + 6$
$y(y - 6) - 1(y - 6) = (y - 6)(y - 1)$

21. $12x^2 - x - 1$
$12 \cdot (-1) = -12$
Factors of –12 whose sum is –1: –4 and 3
$12x^2 - 4x + 3x - 1$
$4x(3x - 1) + 1(3x - 1) = (3x - 1)(4x + 1)$

22. $2a^3 + 7a^2 - 15a$
The GCF is a: $a(2a^2 + 7a - 15)$
$2 \cdot (-15) = -30$
Factors of –30 whose sum is 7: 10 and –3
$a(2a^2 + 10a - 3a - 15)$
$a[2a(a + 5) - 3(a + 5)] = a(a + 5)(2a - 3)$

23. $4b^2 - 100$

The GCF is 4: $4(b^2-25)$

$4[b^2-(5)^2] = 4(b+5)(b-5)$

24. $(x+3)(2x-5)=0$

$x+3=0$

$x=-3$

$x+3=0 \qquad 2x-5=0$

$x=-3 \qquad 2x=5$

$x=\frac{5}{2}$

The solutions are −3 and $\frac{5}{2}$.

25. $\frac{12x^4y^2}{18xy^7} = \frac{2\cdot 2\cdot 3x^{4-1}}{2\cdot 3\cdot 3y^{7-2}} = \frac{2x^3}{3y^5}$

26. $\frac{x^2-7x+10}{25-x^2} = \frac{(x-5)(x-2)}{(5+x)(5-x)} = -\frac{x-2}{x+5}$

27. $\frac{x^2-x-56}{x^2+8x+7} \div \frac{x^2-13x+40}{x^2-4x-5}$

$= \frac{(x-8)(x+7)}{(x+7)(x+1)} \cdot \frac{(x-5)(x+1)}{(x-8)(x-5)} = 1$

28. The GCF is $(2x-1)(x+1)$.

$\frac{2}{2x-1} - \frac{1}{x+1} = \frac{2}{2x-1}\cdot\frac{x+1}{x+1} - \frac{1}{x+1}\cdot\frac{2x-1}{2x-1}$

$= \frac{2x+2}{(2x-1)(x+1)} - \frac{2x-1}{(2x-1)(x+1)}$

$= \frac{2x+2-(2x-1)}{(2x-1)(x+1)}$

$= \frac{2x+2-2x+1}{(2x-1)(x+1)} = \frac{3}{(2x-1)(x+1)}$

29. $\frac{1-\frac{2}{x}-\frac{15}{x^2}}{1-\frac{25}{x^2}} = \frac{1-\frac{2}{x}-\frac{15}{x^2}}{1-\frac{25}{x^2}}\cdot\frac{x^2}{x^2}$

$\frac{1\cdot x^2 - \frac{2}{x}\cdot x^2 - \frac{15}{x^2}\cdot x^2}{1\cdot x^2 - \frac{25}{x^2}\cdot x^2} = \frac{x^2-2x-15}{x^2-25}$

$= \frac{(x-5)(x+3)}{(x-5)(x+5)} = \frac{x+3}{x+5}$

30. $\frac{3x}{x-3} - 2 = \frac{10}{x-3}$

$(x-3)\left(\frac{3x}{x-3}\right) - 2(x-3) = (x-3)\left(\frac{10}{x-3}\right)$

$3x-2(x-3)=10$

$3x-2x+6=10$

$x+6=10$

$x=4$

4 checks as a solution. The solution is 4.

31. Strategy • Percent of silver in the alloy: x

	Amount	Percent	Quantity
40% silver	60	0.40	0.40(60)
Silver alloy	120	x	$120x$
Mixture	180	0.60	0.60(180)

• The sum of the quantities before mixing is equal to the quantity after mixing.

Solution

$0.40(60)+120x=0.60(180)$

$24+120x=108$

$120x=84$

$x=0.70$

The silver alloy is 70% silver.

32. Strategy To find the cost of a $5000 policy, write and solve a proportion using x to represent the cost for a $5000 policy.

Solution

$\frac{32}{1000} = \frac{x}{5000}$

$5000\cdot\frac{32}{1000} = 5000\cdot\frac{x}{5000}$

$160 = x$

The cost of a $5000 policy is $160.

33. Strategy •Time to fill the pool with both pipes: t

	Rate	Time	Part
First pipe	$\frac{1}{9}$	t	$\frac{t}{9}$
Second pipe	$\frac{1}{18}$	t	$\frac{t}{18}$

• The sum of the parts completed by each pipe must equal 1.

Solution

$$\frac{t}{9}+\frac{t}{18}=1$$

$$\overset{2}{\cancel{18}}\cdot\frac{t}{\underset{1}{\cancel{9}}}+\overset{1}{\cancel{18}}\cdot\frac{t}{\underset{1}{\cancel{18}}}=18\cdot 1$$

$$2t+t=18$$

$$3t=18$$

$$t=6$$

It would take both pipes 6 min to fill the tank.

Chapter 9: Exponents and Radicals

Prep Test

1. $48 = ? \cdot 3$

$\left(\frac{1}{3}\right)48 = ? \cdot 3\left(\frac{1}{3}\right)$

$16 = ?$

2. $2^5 = 2 \cdot 2 \cdot 2 \cdot 2 \cdot 2 = 32$

3. $6\left(\frac{3}{2}\right) = \frac{6}{1}\left(\frac{3}{2}\right) = \frac{3 \cdot 2}{1}\left(\frac{3}{2}\right) = \frac{3 \cdot 3}{1 \cdot 1} = 9$

4. $\frac{1}{2} - \frac{2}{3} + \frac{1}{4} = \frac{6}{12} - \frac{8}{12} + \frac{3}{12}$

$= \frac{6 - 8 + 3}{12}$

$= \frac{1}{12}$

5. $(3 - 7x) - (4 - 2x)$

$= 3 - 7x - 4 + 2x$

$= -5x - 1$

6. $\frac{3x^5y^6}{12x^4y} = \frac{xy^5}{4}$

7. $(3x - 2)^2 = (3x - 2)(3x - 2)$

$= 9x^2 - 6x - 6x + 4$

$= 9x^2 - 12x + 4$

8. $(2 + 4x)(5 - 3x)$

$= 10 - 6x + 20x - 12x^2$

$= -12x^2 + 14x + 10$

9. $(6x - 1)(6x + 1)$

$= 36x^2 + 6x - 6x - 1$

$= 36x^2 - 1$

10. $x^2 - 14x - 5 = 10$

$x^2 - 14x - 15 = 0$

$(x - 15)(x + 1) = 0$

$x - 15 = 0 \quad x + 1 = 0$

$x = 15 \qquad x = -1$

The solutions are −1 and 15.

Section 9.1

Objective A Exercises

1. (i) and (iii) are not real numbers.

3. $8^{1/3} = (2^3)^{1/3} = 2$

5. $9^{3/2} = (3^2)^{3/2} = 3^3 = 27$

7. $27^{-2/3} = (3^3)^{-2/3} = 3^{-2} = \frac{1}{3^2} = \frac{1}{9}$

9. $32^{2/5} = (2^5)^{2/5} = 2^2 = 4$

11. $(-25)^{5/2}$

Not a real number.

The base of the exponential expression is a negative number and the denominator of the exponent is a positive even number.

13. $\left(\frac{25}{49}\right)^{-3/2}$

$= \left(\frac{5^2}{7^2}\right)^{-3/2} = \left(\left(\frac{5}{7}\right)^2\right)^{-3/2}$

$= \left(\frac{5}{7}\right)^{-3} = \frac{5^{-3}}{7^{-3}} = \frac{7^3}{5^3}$

$= \frac{343}{125}$

15. $x^{1/2}x^{1/2} = x$

17. $y^{-1/4}y^{3/4} = y^{1/2}$

19. $x^{-2/3} \cdot x^{3/4} = x^{1/12}$

21. $a^{1/3} \cdot a^{3/4} \cdot a^{-1/2} = a^{7/12}$

23. $\dfrac{a^{1/2}}{a^{3/2}} = a^{-1} = \dfrac{1}{a}$

25. $\dfrac{y^{-3/4}}{y^{1/4}} = y^{-1} = \dfrac{1}{y}$

27. $\dfrac{y^{2/3}}{y^{-5/6}} = y^{9/6} = y^{3/2}$

29. $(x^2)^{-1/2} = x^{-1} = \dfrac{1}{x}$

31. $(x^{-2/3})^6 = x^{-4} = \dfrac{1}{x^4}$

33. $(a^{-1/2})^{-2} = a$

35. $(x^{-3/8})^{-4/5} = x^{3/10}$

37. $(a^{1/2} \cdot a)^2 = (a^{3/2})^2 = a^3$

39. $(x^{-1/2} \cdot x^{3/4})^{-2} = (x^{1/4})^{-2} = x^{-1/2} = \dfrac{1}{x^{1/2}}$

41. $(y^{-1/2} \cdot y^{2/3})^{2/3} = (y^{1/6})^{2/3} = y^{1/9}$

43. $(x^{-3}y^6)^{-1/3} = xy^{-2} = \dfrac{x}{y^2}$

45. $(x^{-2}y^{1/3})^{-3/4} = x^{3/2}y^{-1/4} = \dfrac{x^{3/2}}{y^{1/4}}$

47. $\left(\dfrac{x^{1/2}}{y^2}\right)^4 = \dfrac{x^2}{y^8}$

49. $\dfrac{x^{1/4} \cdot x^{-1/2}}{x^{2/3}} = \dfrac{x^{-1/4}}{x^{2/3}} = x^{-11/12} = \dfrac{1}{x^{11/12}}$

51. $\left(\dfrac{y^{2/3} \cdot y^{-5/6}}{y^{1/9}}\right)^9 = \left(\dfrac{y^{-1/6}}{y^{1/9}}\right)^9 = \left(y^{-5/18}\right)^9$

$= y^{-5/2} = \dfrac{1}{y^{5/2}}$

53. $\left(\dfrac{b^2 \cdot b^{-3/4}}{b^{-1/2}}\right)^{-1/2} = \left(\dfrac{b^{5/4}}{b^{-1/2}}\right)^{-1/2} = \left(b^{7/4}\right)^{-1/2}$

$= b^{-7/8} = \dfrac{1}{b^{7/8}}$

55. $(a^{2/3}b^2)^6(a^3b^3)^{1/3} = (a^4b^{12})(ab) = a^5b^{13}$

57. $(16x^{-2}y^4)^{-1/2}(xy^{1/2}) = (16)^{-1/2}(xy^{-2})(xy^{1/2})$

$= (16)^{-1/2}x^2y^{-3/2}$

$= \dfrac{x^2}{16^{1/2}y^{3/2}}$

$= \dfrac{x^2}{4y^{3/2}}$

59. $(x^{-2/3}y^{-3})^3(27x^{-3}y^6)^{-1/3} = (x^{-2}y^{-9})(27)^{-1/3}(xy^{-2})$

$= (27)^{-1/3}x^{-1}y^{-11}$

$= \dfrac{1}{27^{1/3}xy^{11}}$

$= \dfrac{1}{3xy^{11}}$

61. $\dfrac{\left(4a^{4/3}b^{-2}\right)^{-1/2}}{(a^{1/6}b^{-3/2})^2} = \dfrac{\left((4)^{-1/2}a^{-2/3}b\right)}{a^{1/3}b^{-3}} = \dfrac{b^4}{2a}$

63. $\left(\dfrac{x^{1/2}y^{-3/4}}{y^{2/3}}\right)^{-6} = \left(x^{1/2}y^{-17/12}\right)^{-6}$

$= x^{-3}y^{17/2} = \dfrac{y^{17/2}}{x^3}$

65. $\left(\dfrac{b^{-3}}{64a^{-1/2}}\right)^{-2/3} = \dfrac{b^2}{64^{-2/3}a^{1/3}}$

$= \dfrac{64^{2/3}b^2}{a^{1/3}} = \dfrac{16b^2}{a^{1/3}}$

67. $y^{3/2}(y^{1/2} - y^{1/2}) = y^{4/2} - y^{2/2} = y^2 - y$

69. $a^{-1/4}(a^{5/4} - a^{9/4}) = a^{4/4} - a^{8/4} = a - a^2$

71. $x^n \cdot x^{n/2} = x^{3n/2}$

73. $\dfrac{y^{n/2}}{y^{-n}} = y^{3n/2}$

Objective B Exercises

75. False

77. $3^{1/4} = \sqrt[4]{3}$

79. $a^{3/2} = (a^3)^{1/2} = \sqrt{a^3}$

81. $(2t)^{5/2} = \sqrt{(2t)^5} = \sqrt{32t^5}$

83. $-2x^{2/3} = -2(x)^{2/3} = -2\sqrt[3]{x^2}$

85. $(a^2b)^{2/3} = \sqrt[3]{(a^2b)^2} = \sqrt[3]{a^4b^2}$

87. $(a^2b^4)^{3/5} = \sqrt[5]{(a^2b^4)^3} = \sqrt[5]{a^6b^{12}}$

89. $(4x-3)^{3/4} = \sqrt[4]{(4x-3)^3}$

91. $x^{-2/3} = \dfrac{1}{x^{2/3}} = \dfrac{1}{\sqrt[3]{x^2}}$

93. $\sqrt{14} = 14^{1/2}$

95. $\sqrt[3]{x} = x^{1/3}$

97. $\sqrt[3]{x^4} = x^{4/3}$

99. $\sqrt[5]{b^3} = b^{3/5}$

101. $\sqrt[3]{2x^2} = (2x^2)^{1/3}$

103. $-\sqrt{3x^5} = -(3x^5)^{1/2}$

105. $3x\sqrt[3]{y^2} = 3xy^{2/3}$

107. $\sqrt{a^2 - 2} = (a^2 - 2)^{1/2}$

Objective C Exercises

109. Positive

111. Not a real number

113. $\sqrt{x^{16}} = x^8$

115. $-\sqrt{x^8} = -x^4$

117. $\sqrt[3]{x^3y^9} = xy^3$

119. $-\sqrt[3]{x^{15}y^3} = -x^5y$

121. $\sqrt{16a^4b^{12}} = 4a^2b^6$

123. The square root of a negative number is not a real number.

125. $\sqrt[3]{27x^9} = 3x^3$

127. $\sqrt[3]{-64x^9y^{12}} = -4x^3y^4$

129. $-\sqrt[4]{x^8y^{12}} = -x^2y^3$

131. $\sqrt[5]{x^{20}y^{10}} = x^4y^2$

133. $\sqrt[4]{81x^4y^{20}} = 3xy^5$

135. $\sqrt[5]{32a^5b^{10}} = 2ab^2$

Applying the Concepts

137. a) False $\sqrt{(-2)^2} = \sqrt{4} = 2$

b) True

c) True

d) False $\sqrt[n]{a^n + b^n} = (a^n + b^n)^{1/n}$

e) False

$(a^{1/2} + b^{1/2})^2 = a + 2a^{1/2}b^{1/2} + b$

f) False $\sqrt[m]{a^n} = a^{n/m}$

139. No. If $x \geq 0$, the statement is true. However if $x < 0$ the $\sqrt{x^2} = |x|$.

Section 9.2

Objective A Exercises

1. No

3. Yes

5. $\sqrt{x^4y^3z^5} = \sqrt{x^4y^2z^4(yz)}$
$= \sqrt{x^4y^2z^4}\sqrt{yz}$
$= x^2yz^2\sqrt{yz}$

7. $\sqrt{8a^3b^8} = \sqrt{4a^2b^8(2a)}$
$= \sqrt{4a^2b^8}\sqrt{2a}$
$= 2ab^4\sqrt{2a}$

9. $\sqrt{45x^2y^3z^5} = \sqrt{9x^2y^2z^4(5yz)}$
$= \sqrt{9x^2y^2z^4}\sqrt{5yz}$
$= 3xyz^2\sqrt{5yz}$

11. $\sqrt[4]{48x^4y^5z^6} = \sqrt[4]{16x^4y^4z^4(3yz^2)}$
$= \sqrt[4]{16x^4y^4z^4}\sqrt[4]{3yz^2}$
$= 2xyz\sqrt[4]{3yz^2}$

13. $\sqrt[3]{a^{16}b^8} = \sqrt[3]{a^{15}b^6(ab^2)}$
$= \sqrt[3]{a^{15}b^6}\sqrt[3]{ab^2}$
$= a^5b^2\sqrt[3]{ab^2}$

15. $\sqrt[3]{-125x^2y^4} = \sqrt[3]{-125y^3(x^2y)}$
$= \sqrt[3]{-125y^3}\sqrt[3]{x^2y}$
$= -5y\sqrt[3]{x^2y}$

17. $\sqrt[3]{a^4b^5c^6} = \sqrt[3]{a^3b^3c^6(ab^2)}$
$= \sqrt[3]{a^3b^3c^6}\sqrt[3]{ab^2}$
$= abc^2\sqrt[3]{ab^2}$

19. $\sqrt[4]{16x^9y^5} = \sqrt[4]{16x^8y^4(xy)}$
$= \sqrt[4]{16x^8y^4}\sqrt[4]{xy}$
$= 2x^2y\sqrt[4]{xy}$

Objective B Exercises

21. True

23. $2\sqrt{x} - 8\sqrt{x} = -6\sqrt{x}$

25. $\sqrt{8x} - \sqrt{32x} = \sqrt{4 \cdot 2x} - \sqrt{16 \cdot 2x}$
$= \sqrt{4}\sqrt{2x} - \sqrt{16}\sqrt{2x}$
$= 2\sqrt{2x} - 4\sqrt{2x}$
$= -2\sqrt{2x}$

27. $\sqrt{18b} + \sqrt{75b} = \sqrt{9 \cdot 2b} + \sqrt{25 \cdot 3b}$
$= \sqrt{9}\sqrt{2b} + \sqrt{25}\sqrt{3b}$
$= 3\sqrt{2b} + 5\sqrt{3b}$

29. $3\sqrt{8x^2y^3} - 2x\sqrt{32y^3} = 3\sqrt{4 \cdot 2x^2y^3} - 2x\sqrt{16 \cdot 2y^3}$
$= 3\sqrt{4x^2y^2}\sqrt{2y} - 2x\sqrt{16y^2}\sqrt{2y}$
$= 3 \cdot 2xy\sqrt{2y} - 2x \cdot 4y\sqrt{2y}$
$= 6xy\sqrt{2y} - 8xy\sqrt{2y}$
$= -2xy\sqrt{2y}$

31. $2a\sqrt{27ab^5} + 3b\sqrt{3a^3b} = 2a\sqrt{3^3ab^5} + 3b\sqrt{3a^3b}$
$= 2a\sqrt{3^2b^4}\sqrt{3ab} + 3b\sqrt{a^2}\sqrt{3ab}$
$= 2a \cdot 3b^2\sqrt{3ab} + 3ab\sqrt{3ab}$
$= 6ab^2\sqrt{3ab} + 3ab\sqrt{3ab}$

33. $\sqrt[3]{16} - \sqrt[3]{54} = \sqrt[3]{8 \cdot 2} - \sqrt[3]{27 \cdot 2}$
$= \sqrt[3]{8}\sqrt[3]{2} - \sqrt[3]{27}\sqrt[3]{2}$
$= 2\sqrt[3]{2} - 3\sqrt[3]{2}$
$= -\sqrt[3]{2}$

35. $2b\sqrt[3]{16b^2} + \sqrt[3]{128b^5} = 2b\sqrt[3]{8 \cdot 2b^2} + \sqrt[3]{64b^3 \cdot 2b^2}$
$= 2b\sqrt[3]{8}\sqrt[3]{2b^2} + \sqrt[3]{64b^3}\sqrt[3]{2b^2}$
$= 4b\sqrt[3]{2b^2} + 4b\sqrt[3]{2b^2}$
$= 8b\sqrt[3]{2b^2}$

37. $3\sqrt[4]{32a^5} - a\sqrt[4]{162a} = 3\sqrt[4]{16a^4 \cdot 2a} - a\sqrt[4]{81 \cdot 2a}$
$= 3\sqrt[4]{16a^4}\sqrt[4]{2a} - a\sqrt[4]{81}\sqrt[4]{2a}$
$= 3 \cdot 2a\sqrt[4]{2a} - a \cdot 3\sqrt[4]{2a}$
$= 6a\sqrt[4]{2a} - 3a\sqrt[4]{2a}$
$= 3a\sqrt[4]{2a}$

39. $2\sqrt{50} - 3\sqrt{125} + \sqrt{98}$
$= 2\sqrt{25 \cdot 2} - 3\sqrt{25 \cdot 5} + \sqrt{49 \cdot 2}$
$= 2\sqrt{25}\sqrt{2} - 3\sqrt{25}\sqrt{5} + \sqrt{49}\sqrt{2}$
$= 10\sqrt{2} - 15\sqrt{5} + 7\sqrt{2}$
$= 17\sqrt{2} - 15\sqrt{5}$

41. $\sqrt{9b^3} - \sqrt{25b^3} + \sqrt{49b^3}$
$= \sqrt{9b^2 \cdot b} - \sqrt{25b^2 \cdot b} + \sqrt{49b^2 \cdot b}$
$= \sqrt{9b^2}\sqrt{b} - \sqrt{25b^2}\sqrt{b} + \sqrt{49b^2}\sqrt{b}$
$= 3b\sqrt{b} - 5b\sqrt{b} + 7b\sqrt{b}$
$= 5b\sqrt{b}$

43. $2x\sqrt{8xy^2} - 3y\sqrt{32x^3} + \sqrt{4x^3y^3}$
$= 2x\sqrt{4y^2 \cdot 2x} - 3y\sqrt{16x^2 \cdot 2x} + \sqrt{4x^2y^2 \cdot xy}$
$= 2x\sqrt{4y^2}\sqrt{2x} - 3y\sqrt{16x^2}\sqrt{2x} + \sqrt{4x^2y^2}\sqrt{xy}$
$= 4xy\sqrt{2x} - 12xy\sqrt{2x} + 2xy\sqrt{xy}$
$= -8xy\sqrt{2x} + 2xy\sqrt{xy}$

45. $\sqrt[3]{54xy^3} - 5\sqrt[3]{2xy^3} + y\sqrt[3]{128x}$
$= \sqrt[3]{27y^3 \cdot 2x} - 5\sqrt[3]{y^3 \cdot 2x} + y\sqrt[3]{64 \cdot 2x}$
$= \sqrt[3]{27y^3}\sqrt[3]{2x} - 5\sqrt[3]{y^3}\sqrt[3]{2x} + y\sqrt[3]{64}\sqrt[3]{2x}$
$= 3y\sqrt[3]{2x} - 5y\sqrt[3]{2x} + 4y\sqrt[3]{2x}$
$= 2y\sqrt[3]{2x}$

47. $2a\sqrt[4]{32b^5} - 3b\sqrt[4]{162a^4b} + \sqrt[4]{2a^4b^5}$
$= 2a\sqrt[4]{16b^4 \cdot 2b} - 3b\sqrt[4]{81a^4 \cdot 2b} + \sqrt[4]{a^4b^4 \cdot 2b}$
$= 2a\sqrt[4]{16b^4}\sqrt[4]{2b} - 3b\sqrt[4]{81a^4}\sqrt[4]{2b} + \sqrt[4]{a^4b^4}\sqrt[4]{2b}$
$= 4ab\sqrt[4]{2b} - 9ab\sqrt[4]{2b} + ab\sqrt[4]{2b}$
$= -4ab\sqrt[4]{2b}$

Objective C Exercises

49. $\sqrt{8}\sqrt{32} = \sqrt{256} = 16$

51. $\sqrt[3]{4}\sqrt[3]{8} = 2\sqrt[3]{4}$

53. $\sqrt{x^2y^5}\sqrt{xy} = \sqrt{x^3y^6}$
$= \sqrt{x^2y^6 \cdot x} = xy^3\sqrt{x}$

55. $\sqrt{2x^2y}\sqrt{32xy} = \sqrt{64x^3y^2}$
$= \sqrt{64x^2y^2 \cdot x} = 8xy\sqrt{x}$

57. $\sqrt[3]{x^2y}\sqrt[3]{16x^4y^2} = \sqrt[3]{16x^6y^3}$
$= \sqrt[3]{8x^6y^3 \cdot 2} = 2x^2y\sqrt[3]{2}$

59. $\sqrt[4]{12ab^3}\sqrt[4]{4a^5b^2} = \sqrt[4]{48a^6b^5}$
$= \sqrt[4]{16a^4b^4 \cdot 3a^2b} = 2ab\sqrt[4]{3a^2b}$

61. $\sqrt{3}(\sqrt{27} - \sqrt{3}) = \sqrt{81} - \sqrt{9} = 9 - 3 = 6$

63. $\sqrt{x}(\sqrt{x} - \sqrt{2}) = \sqrt{x^2} - \sqrt{2x} = x - \sqrt{2x}$

65. $\sqrt{2x}(\sqrt{8x} - \sqrt{32}) = \sqrt{16x^2} - \sqrt{64x}$
$= \sqrt{16x^2} - \sqrt{64 \cdot x} = 4x - 8\sqrt{x}$

67. $(3-2\sqrt{5})(2+\sqrt{5}) = 6+3\sqrt{5}-4\sqrt{5}-2(\sqrt{5})^2$
$= 6+3\sqrt{5}-4\sqrt{5}-10 = -4-\sqrt{5}$

69. $(-2+\sqrt{7})(3+5\sqrt{7}) = -6-10\sqrt{7}+3\sqrt{7}+5(\sqrt{7})^2$
$= -6-10\sqrt{7}+3\sqrt{7}+35 = 29-7\sqrt{7}$

71. $(6+3\sqrt{2})(4-2\sqrt{2}) = 24-12\sqrt{2}+12\sqrt{2}-6(\sqrt{2})^2$
$= 24-12\sqrt{2}+12\sqrt{2}-12 = 12$

73. $(5-2\sqrt{7})(5+2\sqrt{7}) = 25+10\sqrt{7}-10\sqrt{7}-4(\sqrt{7})^2$
$= 25+10\sqrt{7}-10\sqrt{7}-28 = -3$

75. $(3-\sqrt{2x})(1+5\sqrt{2x})$
$= 3+15\sqrt{2x}-\sqrt{2x}-5(\sqrt{2x})^2$
$= 3+15\sqrt{2x}-\sqrt{2x}-10x$
$= -10x+14\sqrt{2x}+3$

77. $(2+2\sqrt{x})(1+5\sqrt{x})$
$= 2+10\sqrt{x}+2\sqrt{x}+10(\sqrt{x})^2$
$= 2+10\sqrt{x}+2\sqrt{x}+10x$
$= 10x+12\sqrt{2x}+2$

79. $(2+\sqrt{x})^2 = (2+\sqrt{x})(2+\sqrt{x})$
$= 4+2\sqrt{x}+2\sqrt{x}+(\sqrt{x})^2$
$= x+4\sqrt{x}+4$

81. $(4-\sqrt{2x+1})^2 = (4-\sqrt{2x+1})(4-\sqrt{2x+1})$
$= 16-4\sqrt{2x+1}-4\sqrt{2x+1}+(\sqrt{2x+1})^2$
$= 2x+1-8\sqrt{2x+1}+16$
$= 2x-8\sqrt{2x+1}+17$

83. $(\sqrt{6}-5\sqrt{3})(3\sqrt{6}+4\sqrt{3})$
$= 3(\sqrt{6})^2+4\sqrt{18}-15\sqrt{18}-20(\sqrt{3})^2$
$= 18+4\sqrt{9 \cdot 2}-15\sqrt{9 \cdot 2}-60$
$= 18+12\sqrt{2}-45\sqrt{2}-60$
$= -42-33\sqrt{2}$

85. True

Objective D Exercises

87. To rationalize the denominator of a radical expression means to rewrite the expression with no radicals in the denominator. To do this, multiply both the numerator and the denominator by the same expression, one that removes the radicals(s) from the denominator of the original expression.

89. $\sqrt[3]{4x}$

91. $\sqrt{3}+x$

93. $\dfrac{\sqrt{60y^4}}{\sqrt{12y}} = \sqrt{\dfrac{60y^4}{12y}}$
$= \sqrt{5y^3} = \sqrt{y^2 \cdot 5y}$
$= y\sqrt{5y}$

95. $\dfrac{\sqrt{65ab^4}}{\sqrt{5ab}} = \sqrt{\dfrac{65ab^4}{5ab}}$
$= \sqrt{13b^3} = \sqrt{b^2 \cdot 13b}$
$= b\sqrt{13b}$

97. $\dfrac{1}{\sqrt{2}} = \dfrac{1}{\sqrt{2}} \cdot \dfrac{\sqrt{2}}{\sqrt{2}} = \dfrac{\sqrt{2}}{\sqrt{2^2}} = \dfrac{\sqrt{2}}{2}$

99. $\dfrac{2}{\sqrt{3y}} = \dfrac{2}{\sqrt{3y}} \cdot \dfrac{\sqrt{3y}}{\sqrt{3y}} = \dfrac{2\sqrt{3y}}{\sqrt{9y^2}} = \dfrac{2\sqrt{3y}}{3y}$

101. $\dfrac{9}{\sqrt{3a}} = \dfrac{9}{\sqrt{3a}} \cdot \dfrac{\sqrt{3a}}{\sqrt{3a}} = \dfrac{9\sqrt{3a}}{\sqrt{9a^2}}$
$= \dfrac{9\sqrt{3a}}{3a} = \dfrac{3\sqrt{3a}}{a}$

103. $\sqrt{\dfrac{y}{2}} = \dfrac{\sqrt{y}}{\sqrt{2}} = \dfrac{\sqrt{y}}{\sqrt{2}} \cdot \dfrac{\sqrt{2}}{\sqrt{2}} = \dfrac{\sqrt{2y}}{\sqrt{4}} = \dfrac{\sqrt{2y}}{2}$

105. $\dfrac{5}{\sqrt[3]{9}} = \dfrac{5}{\sqrt[3]{9}} \cdot \dfrac{\sqrt[3]{3}}{\sqrt[3]{3}} = \dfrac{5\sqrt[3]{3}}{\sqrt[3]{27}} = \dfrac{5\sqrt[3]{3}}{3}$

107. $\dfrac{5}{\sqrt[3]{3y}} = \dfrac{5}{\sqrt[3]{3y}} \cdot \dfrac{\sqrt[3]{9y^2}}{\sqrt[3]{9y^2}} = \dfrac{5\sqrt[3]{9y^2}}{\sqrt[3]{27y^3}} = \dfrac{5\sqrt[3]{9y^2}}{3y}$

109. $\dfrac{\sqrt{15a^2b^5}}{\sqrt{30a^5b^3}} = \sqrt{\dfrac{15a^2b^5}{30a^5b^3}} = \sqrt{\dfrac{b^2}{2a^3}} = \dfrac{\sqrt{b^2}}{\sqrt{a^2 \cdot 2a}}$
$= \dfrac{b}{a\sqrt{2a}} = \dfrac{b}{a\sqrt{2a}} \cdot \dfrac{\sqrt{2a}}{\sqrt{2a}} = \dfrac{b\sqrt{2a}}{a\sqrt{4a^2}}$
$= \dfrac{b\sqrt{2a}}{2a^2}$

111. $\dfrac{\sqrt{12x^3y}}{\sqrt{20x^4y}} = \sqrt{\dfrac{12x^3y}{20x^4y}} = \sqrt{\dfrac{3}{5x}}$
$= \dfrac{\sqrt{3}}{\sqrt{5x}} = \dfrac{\sqrt{3}}{\sqrt{5x}} \cdot \dfrac{\sqrt{5x}}{\sqrt{5x}} = \dfrac{\sqrt{15x}}{\sqrt{25x^2}}$
$= \dfrac{\sqrt{15x}}{5x}$

113. $\dfrac{-2}{1-\sqrt{2}} = \dfrac{-2}{1-\sqrt{2}} \cdot \dfrac{1+\sqrt{2}}{1+\sqrt{2}}$
$= \dfrac{-2(1+\sqrt{2})}{1^2-(\sqrt{2})^2} = \dfrac{-2-2\sqrt{2}}{1-2} = \dfrac{-2-2\sqrt{2}}{-1}$
$= 2+2\sqrt{2}$

115. $\dfrac{-4}{3-\sqrt{2}} = \dfrac{-4}{3-\sqrt{2}} \cdot \dfrac{3+\sqrt{2}}{3+\sqrt{2}}$
$= \dfrac{-4(3+\sqrt{2})}{3^2-(\sqrt{2})^2} = \dfrac{-12-4\sqrt{2}}{9-2} = \dfrac{-12-4\sqrt{2}}{7}$

117. $\dfrac{5}{2-\sqrt{7}} = \dfrac{5}{2-\sqrt{7}} \cdot \dfrac{2+\sqrt{7}}{2+\sqrt{7}}$
$= \dfrac{5(2+\sqrt{7})}{2^2-(\sqrt{7})^2} = \dfrac{10+5\sqrt{7}}{4-7} = \dfrac{10+5\sqrt{7}}{-3}$
$= -\dfrac{10+5\sqrt{7}}{3}$

119. $\dfrac{-7}{\sqrt{x}-3} = \dfrac{-7}{\sqrt{x}-3} \cdot \dfrac{\sqrt{x}+3}{\sqrt{x}+3}$
$= \dfrac{-7(\sqrt{x}+3)}{(\sqrt{x})^2-3^2} = \dfrac{-(7\sqrt{x}+21)}{x-9}$
$= -\dfrac{7\sqrt{x}+21}{x-9}$

121. $\dfrac{\sqrt{3}+\sqrt{4}}{\sqrt{2}+\sqrt{3}}=\dfrac{\sqrt{3}+\sqrt{2^2}}{\sqrt{2}+\sqrt{3}}=\dfrac{\sqrt{3}+2}{\sqrt{2}+\sqrt{3}}\cdot\dfrac{\sqrt{2}-\sqrt{3}}{\sqrt{2}-\sqrt{3}}$

$=\dfrac{\sqrt{6}-(\sqrt{3})^2+2\sqrt{2}-2\sqrt{3}}{(\sqrt{2})^2-(\sqrt{3})^2}$

$=\dfrac{\sqrt{6}-3+2\sqrt{2}-2\sqrt{3}}{2-3}$

$=\dfrac{\sqrt{6}-3+2\sqrt{2}-2\sqrt{3}}{-1}$

$=-\sqrt{6}+3-2\sqrt{2}+2\sqrt{3}$

123. $\dfrac{2+3\sqrt{5}}{1-\sqrt{5}}=\dfrac{2+3\sqrt{5}}{1-\sqrt{5}}\cdot\dfrac{1+\sqrt{5}}{1+\sqrt{5}}$

$=\dfrac{2+2\sqrt{5}+3\sqrt{5}+3(\sqrt{5})^2}{1^2-(\sqrt{5})^2}$

$=\dfrac{2+5\sqrt{5}+15}{1-5}$

$=\dfrac{17+5\sqrt{4}}{-4}$

$=-\dfrac{17+5\sqrt{4}}{4}$

125. $\dfrac{2\sqrt{a}-\sqrt{b}}{4\sqrt{a}+3\sqrt{b}}=\dfrac{2\sqrt{a}-\sqrt{b}}{4\sqrt{a}+3\sqrt{b}}\cdot\dfrac{4\sqrt{a}-3\sqrt{b}}{4\sqrt{a}-3\sqrt{b}}$

$=\dfrac{8\sqrt{a^2}-6\sqrt{ab}-4\sqrt{ab}+3\sqrt{b^2}}{(4\sqrt{a})^2-(3\sqrt{b})^2}$

$=\dfrac{8a-10\sqrt{ab}+3b}{16a-9b}$

127. $\dfrac{3\sqrt{y}-y}{\sqrt{y}+2y}=\dfrac{3\sqrt{y}-y}{\sqrt{y}+2y}\cdot\dfrac{\sqrt{y}-2y}{\sqrt{y}-2y}$

$=\dfrac{3(\sqrt{y})^2-6y\sqrt{y}-y\sqrt{y}+2y^2}{(\sqrt{y})^2-(2y)^2}$

$=\dfrac{3y-7y\sqrt{y}+2y^2}{y-4y^2}$

$=\dfrac{3-7\sqrt{y}+2y}{1-4y}$

Applying the Concepts

129. a) False $\sqrt[2]{3}\cdot\sqrt[3]{4}=\sqrt[6]{432}$
b) True
c) False $\sqrt[3]{x}\cdot\sqrt[3]{x}=x^{2/3}$
d) False $\sqrt{x}+\sqrt{y}$
e) False $\sqrt[2]{2}+\sqrt[3]{3}$
f) True

131. $\dfrac{\sqrt[4]{(a+b)^3}}{\sqrt{a+b}}=\dfrac{(a+b)^{3/4}}{(a+b)^{1/2}}=(a+b)^{1/4}$

$=\sqrt[4]{a+b}$

Section 9.3

Objective A Exercises

1. $\sqrt[3]{4x}=-2$
$(\sqrt[3]{4x})^3=(-2)^3$
$4x=-8$
$x=-2$
The solution is −2.

3. $\sqrt{3x-2}=5$
$(\sqrt{3x-2})^2=(5)^2$
$3x-2=25$
$3x=27$
$x=9$
The solution is 9.

5. No solution

7. $\sqrt[3]{2x-6}=4$
$(\sqrt[3]{2x-6})^3=(4)^3$
$2x-6=64$
$2x=70$
$x=35$
The solution is 35.

9. $\sqrt[4]{3x}+2=5$

$\sqrt[4]{3x}=3$

$(\sqrt[4]{3x})^4=(3)^4$

$3x=81$

$x=27$

The solution is 27.

11. $\sqrt[3]{2x-3}+5=2$

$\sqrt[3]{2x-3}=-3$

$(\sqrt[3]{2x-3})^3=(-3)^3$

$2x-3=-27$

$2x=-24$

$x=-12$

The solution is −12.

13. $\sqrt{x}+2=x-4$

$\sqrt{x}=x-6$

$(\sqrt{x})^2=(x-6)^2$

$x=x^2-12x+36$

$x^2-13x+36=0$

$(x-9)(x-4)=0$

$x-9=0 \quad x-4=0$

$x=9 \qquad x=4$

Check both solutions in the original equation

$\sqrt{x}+2=x-4$

$\sqrt{9}+2=9-4$

$5=5$

$\sqrt{4}+2=4-4$

$4=0$

The solution is 9.

15. $\sqrt{x}+\sqrt{x-5}=5$

$\sqrt{x}=5-\sqrt{x-5}$

$(\sqrt{x})^2=(5-\sqrt{x-5})^2$

$x=25-10\sqrt{x-5}+x-5$

$0=20-10\sqrt{x-5}$

$-20=-10\sqrt{x-5}$

$2=\sqrt{x-5}$

$(2)^2=(\sqrt{x-5})^2$

$4=x-5$

$x=9$

The solution is 9.

17. $\sqrt{2x+5}-\sqrt{2x}=1$

$\sqrt{2x+5}=1+\sqrt{2x}$

$(\sqrt{2x+5})^2=(1+\sqrt{2x})^2$

$2x+5=1+2\sqrt{2x}+2x$

$5=1+2\sqrt{2x}$

$4=2\sqrt{2x}$

$2=\sqrt{2x}$

$(2)^2=(\sqrt{2x})^2$

$4=2x$

$x=2$

The solution is 2.

19. $\sqrt{2x} - \sqrt{x-1} = 1$

$\sqrt{2x} = 1 + \sqrt{x-1}$

$(\sqrt{2x})^2 = (1 + \sqrt{x-1})^2$

$2x = 1 + 2\sqrt{x-1} + x - 1$

$x = 2\sqrt{x-1}$

$(x)^2 = (2\sqrt{x-1})^2$

$x^2 = 4(x-1)$

$x^2 = 4x - 4$

$x^2 - 4x + 4 = 0$

$(x-2)(x-2) = 0$

$x - 2 = 0 \quad x - 2 = 0$

$x = 2 \quad\quad x = 2$

The solution is 2.

21. $\sqrt{2x+2} + \sqrt{x} = 3$

$\sqrt{2x+2} = 3 - \sqrt{x}$

$(\sqrt{2x+2})^2 = (3 - \sqrt{x})^2$

$2x + 2 = 9 - 6\sqrt{x} + x$

$x + 2 = 9 - 6\sqrt{x}$

$x - 7 = -6\sqrt{x}$

$(x-7)^2 = (-6\sqrt{x})^2$

$x^2 - 14x + 49 = 36x$

$x^2 - 50x + 49 = 0$

$(x-49)(x-1) = 0$

$x - 49 = 0 \quad x - 1 = 0$

$x = 49 \quad\quad x = 1$

Check both solutions in the original equation

$\sqrt{2x+2} + \sqrt{x} = 3$

$\sqrt{2(49)+2} + \sqrt{49} = 3$

$\sqrt{100} + \sqrt{49} = 3$

$10 + 7 = 3$

$17 \neq 3$

$\sqrt{2(1)+2} + \sqrt{1} = 3$

$\sqrt{4} + \sqrt{1} = 3$

$2 + 1 = 3$

$3 = 3$

The solution is 1.

23. $\sqrt{x} < \sqrt{x+5}$.

Therefore $\sqrt{x} - \sqrt{x+5} < 0$ and cannot equal a positive number.

Objective B Exercises

25. Strategy: To find the distance the object will fall substitute the given values for t and g in the equation and solve for d.

Solution: $t = \sqrt{\dfrac{2d}{g}}$

$3 = \sqrt{\dfrac{2d}{5.5}}$

$(3)^2 = \left(\sqrt{\dfrac{2d}{5.5}}\right)^2$

$9 = \dfrac{2d}{5.5}$

$49.5 = 2d$

$24.75 = d$

On the moon, the object will fall 24.75 ft in 3 s.

27. a) **Strategy:** To find the height of the water evaluate the function for $t = 10$.

Solution: $h(t) = (88.18 - 3.18t)^{2/5}$

$h(10) = (88.18 - 3.18(10))^{2/5}$

$h(10) = (88.18 - 31.8)^{2/5}$

$h(10) = (56.38)^{2/5}$

$h(10) = 5.0$

The height of the water is 5.0 ft.

b) **Strategy:** To find how long it will take to empty the take substitute the given value for h in the equation and solve for t.

Solution: $h(t) = (88.18 - 3.18t)^{2/5}$

$0 = (88.18 - 3.18t)^{2/5}$

$0 = ((88.18 - 3.18t)^{2/5})^{5/2}$

$0 = 88.18 - 3.18t$

$3.18t = 88.16$

$t = 27.7$

The tank will empty in 27.7 s.

29. Strategy: To find the length of the pendulum substitute the given value for T in the equation and solve for L.

Solution: $T = 2\pi\sqrt{\frac{L}{32}}$

$3 = 2\pi\sqrt{\frac{L}{32}}$

$\left(\frac{3}{2\pi}\right)^2 = \left(\sqrt{\frac{L}{32}}\right)^2$

$\left(\frac{3}{2\pi}\right)^2 = \frac{L}{32}$

$32\left(\frac{3}{2\pi}\right)^2 = L$

$7.3 = L$

The length of the pendulum is 7.30 ft.

31. Strategy: Find the difference in the widths. Use the Pythagorean Theorem to find the width of the screen of a regular TV and then repeat the process to find the width of the screen for HDTV. Subtract the width of the regular TV from the width of the HDTV.

Solution: $c^2 = a^2 + b^2$

for the regular TV:

$27^2 = 16.2^2 + b^2$

$729 = 262.44 + b^2$

$466.56 = b^2$

$(466.56)^{1/2} = (b^2)^{1/2}$

$21.6 = b$

for the HDTV

$33^2 = 16.2^2 + b^2$

$1089 = 262.44 + b^2$

$826.56 = b^2$

$(826.56)^{1/2} = (b^2)^{1/2}$

$28.75 = b$

$28.75 - 21.6 = 7.15$

The HDTV is approximately 7.15 in. wider.

33. Strategy: Use the Pythagorean Theorem to find out the longest pole that can be placed in the box.

Solution:

Find the length diagonal of the bottom of the box.

$c^2 = a^2 + b^2$

$c^2 = 2^2 + 3^2$

$c^2 = 4 + 9$

$c^2 = 13$

This represents the value one of the legs of the right triangle needed to find the length of the pole.

$c^2 = a^2 + b^2$

$c^2 = 13 + 4^2$

$c^2 = 13 + 16$

$c^2 = 29$

$c = 5.4$

The longest pole can be 5.4 ft.

Applying the Concepts

35. $\sqrt{3x-2} = \sqrt{2x-3} + \sqrt{x-1}$

$(\sqrt{3x-2})^2 = (\sqrt{2x-3} + \sqrt{x-1})^2$

$3x-2 = 2x-3+2\sqrt{2x-3}\cdot\sqrt{x-1}+x-1$

$3x-2 = 3x-4+2\sqrt{2x-3}\cdot\sqrt{x-1}$

$2 = 2\sqrt{2x-3}\cdot\sqrt{x-1}$

$1 = \sqrt{2x-3}\cdot\sqrt{x-1}$

$(1)^2 = (\sqrt{2x-3}\cdot\sqrt{x-1})^2$

$1 = (2x-3)(x-1)$

$1 = 2x^2-5x+3$

$0 = 2x^2-5x+2$

$(2x-1)(x-2) = 0$

$x = \frac{1}{2} \qquad x = 2$

Check both solutions in the original equation.

$\sqrt{3x-2} = \sqrt{2x-3} + \sqrt{x-1}$

$\sqrt{3\left(\frac{1}{2}\right)-2} = \sqrt{2\left(\frac{1}{2}\right)-3} + \sqrt{\left(\frac{1}{2}\right)-1}$

$\sqrt{-\frac{1}{2}} = \sqrt{-1} + \sqrt{-\frac{1}{2}}$

Not real numbers.

$\sqrt{3(2)-2} = \sqrt{2(2)-3} + \sqrt{(2)-1}$

$\sqrt{4} = \sqrt{1} + \sqrt{1}$

$2 = 1+1$

$2 = 2$

The solution is 2.

Section 9.4

Objective A Exercises

1. An imaginary number is a number whose square is a negative number.
A complex number is a number of the form $a + bi$ where a and b are real numbers and $i = \sqrt{-1}$.

3. $i\sqrt{a}$

5. $\sqrt{-25} = i\sqrt{25} = 5i$

7. $\sqrt{-98} = i\sqrt{98} = i\sqrt{49\cdot 2} = 7i\sqrt{2}$

9. $3+\sqrt{-45} = 3+i\sqrt{45} = 3+i\sqrt{9\cdot 5} = 3+3i\sqrt{5}$

11. $-6-\sqrt{-100} = -6-i\sqrt{100} = -6-10i$

13. $-b+\sqrt{b^2-4ac}$

$-4+\sqrt{(4)^2-4(1)(5)} = -4+\sqrt{16-20}$

$= -4+\sqrt{-4} = -4+i\sqrt{4}$

$= -4+2i$

15. $-b+\sqrt{b^2-4ac}$

$-(-4)+\sqrt{(-4)^2-4(2)(10)} = 4+\sqrt{16-80}$

$= 4+\sqrt{-64} = 4+i\sqrt{64}$

$= 4+8i$

17. $-b+\sqrt{b^2-4ac}$

$-(-8)+\sqrt{(-8)^2-4(3)(6)} = 8+\sqrt{64-72}$

$= 8+\sqrt{-8} = 8+i\sqrt{4\cdot 2}$

$= 8+2i\sqrt{2}$

19. $-b+\sqrt{b^2-4ac}$

$-2+\sqrt{(2)^2-4(4)(7)} = -2+\sqrt{4-112}$

$= -2+\sqrt{-108} = -2+i\sqrt{36\cdot 3}$

$= -2+6i\sqrt{3}$

21. $-b+\sqrt{b^2-4ac}$

$-5+\sqrt{(5)^2-4(-2)(-6)} = -5+\sqrt{25-48}$

$= -5+\sqrt{-23}$

$= -5+i\sqrt{23}$

23. $-b+\sqrt{b^2-4ac}$
$-4+\sqrt{(4)^2-4(-3)(-6)}=-4+\sqrt{16-72}$
$=-4+\sqrt{-56}=-4+i\sqrt{4\cdot 14}$
$=-4+2i\sqrt{14}$

Objective B Exercises

25. $(2+4i)+(6-5i)=8-i$

27. $(-2-4i)-(6-8i)=-8+4i$

29. $(8-2i)-(2+4i)=6-6i$

31. $5+(6-4i)=11-4i$

33. $3i-(6+5i)=-6-2i$

35. The real parts of the complex numbers are additive inverses.

Objective C Exercises

37. $(7i)(-9i)=-63i^2=-63(-1)=63$

39. $\sqrt{-2}\sqrt{-8}=i\sqrt{2}\cdot i\sqrt{8}=i^2\sqrt{16}=-\sqrt{16}=-4$

41. $(5+2i)(5-2i)=25-10i+10i-4i^2$
$=25-4i^2=25-4(-1)=29$

43. $2i(6+2i)=12i+4i^2=12i+4(-1)$
$=-4+12i$

45. $-i(4-3i)=-4i+3i^2=-4i+3(-1)$
$=-3-4i$

47. $(5-2i)(3+i)=15+5i-6i-2i^2$
$=15-i-2i^2$
$=15-i-2(-1)$
$=17-i$

49. $(6+5i)(3+2i)=18+12i+15i+10i^2$
$=18+27i+10i^2$
$=18+27i+10(-1)$
$=8+27i$

51. $(2+5i)^2=4+20i+25i^2$
$=4+20i+25(-1)$
$=-21+20i$

53. $\left(\frac{6}{5}+\frac{3}{5}i\right)\left(\frac{2}{3}-\frac{1}{3}i\right)=\frac{4}{5}-\frac{2}{5}i+\frac{2}{5}i-\frac{1}{5}i^2$
$=\frac{4}{5}-\frac{1}{5}i^2$
$=\frac{4}{5}-\frac{1}{5}(-1)$
$=\frac{4}{5}+\frac{1}{5}=1$

55. True

Objective D Exercises

57. $\frac{3}{i}=\frac{3}{i}\cdot\frac{i}{i}=\frac{3i}{i^2}=\frac{3i}{-1}=-3i$

59. $\frac{2-3i}{-4i}=\frac{2-3i}{-4i}\cdot\frac{i}{i}=\frac{2i-3i^2}{-4i^2}$
$=\frac{2i-3(-1)}{-4(-1)}=\frac{2i+3}{4}$
$=\frac{3}{4}+\frac{1}{2}i$

61. $\frac{4}{5+i}=\frac{4}{5+i}\cdot\frac{5-i}{5-i}=\frac{20-4i}{25-i^2}$
$=\frac{20-4i}{25-(-1)}=\frac{20-4i}{26}$
$=\frac{10}{13}-\frac{2}{13}i$

63. $\frac{2}{2-i}=\frac{2}{2-i}\cdot\frac{2+i}{2+i}=\frac{4+2i}{4-i^2}$
$=\frac{4+2i}{4-(-1)}=\frac{4+2i}{5}$
$=\frac{4}{5}+\frac{2}{5}i$

65. $\frac{1-3i}{3+i}=\frac{1-3i}{3+i}\cdot\frac{3-i}{3-i}=\frac{3-i-9i+3i^2}{9-i^2}$

$=\frac{3-10i+3(-1)}{9-(-1)}=\frac{-10i}{10}$

$=-i$

67. $\frac{3i}{1+4i}=\frac{3i}{1+4i}\cdot\frac{1-4i}{1-4i}=\frac{3i-12i^2}{1-16i^2}$

$=\frac{3i-12(-1)}{1-16(-1)}=\frac{3i+12}{17}$

$=\frac{12}{17}+\frac{3}{17}i$

69. $\frac{2-3i}{3+i}=\frac{2-3i}{3+i}\cdot\frac{3-i}{3-i}=\frac{6-2i-9i+3i^2}{9-i^2}$

$=\frac{6-11i+3(-1)}{9-(-1)}=\frac{3-11i}{10}$

$=\frac{3}{10}-\frac{11}{10}i$

71. $\frac{5+3i}{3-i}=\frac{5+3i}{3-i}\cdot\frac{3+i}{3+i}=\frac{15+5i+9i+3i^2}{9-i^2}$

$=\frac{15+14i+3(-1)}{9-(-1)}=\frac{12+14i}{10}$

$=\frac{6+7i}{5}$

$=\frac{6}{5}+\frac{7}{5}i$

Applying the Concepts

73. True

75. a) $2x^2+18=0$

$2(3i)^2+18=0$

$2(9i^2)+18=0$

$2(-9)+18=0$

$0=0$

Yes, $3i$ is a solution.

b) $x^2-6x+10=0$

$(3+i)^2-6(3+i)+10=0$

$9+6i+i^2-18-6i+10=0$

$9+(-1)-8=0$

$0=0$

Yes $3+i$ is a solution.

Concept Review

1. To write a rational exponent as a radical expression, the denominator of the rational exponent is the index of the radical and the numerator of the rational exponent becomes the power of the radicand.

2. If the base of the exponent was a negative number the expression that contained the rational exponent would not always represent a real number.

3. A radical expression is in simplest form when the radicand contains no factor that is a perfect power.

4. To rationalize the denominator of a radical expression that has two terms in the denominator multiply the numerator and the denominator by the conjugate of the denominator.

5. You can add two radical expressions that have the same radicand and the same index.

6. We isolate the radical so that when we raise each side of the equation to the appropriate power, the resulting equation does not contain a radical expression.

7. A solution is an extraneous solution if it is a solution of a rewritten form of the original equation but, when that value is substituted into the original equation the result is an equation that is not true.

8. To add two complex numbers, add the real parts and add the imaginary parts.

9. Rewrite $\sqrt{36}+\sqrt{-36}$ in the form $a + bi$.

$\sqrt{36}+\sqrt{-36}=\sqrt{36}+i\sqrt{36}=6+6i$

10. To find the product of two variables with the same base and fractional exponents, add the exponents. The bases remain unchanged.

Chapter 9 Review Exercises

1. $(16x^{-4}y^{12})^{1/4}(100x^{6}y^{-2})^{1/2}$
$= (16)^{1/4}x^{-1}y^{3}\cdot 100^{1/2}x^{3}y^{-1}$
$= 20x^{2}y^{2}$

2. $\sqrt[4]{3x-5}=2$

$(\sqrt[4]{3x-5})^{4}=2^{4}$

$3x-5=16$

$3x=21$

$x=7$

3. $(6-5i)(4+3i)=24+18i-20i-15i^{2}$
$=24-2i-15(-1)$
$=39-2i$

4. $7y\sqrt[3]{x^{2}}=7x^{2/3}y$

5. $(\sqrt{3}+8)(\sqrt{3}-2)=\sqrt{3^{2}}-2\sqrt{3}+8\sqrt{3}-16$

$=3+6\sqrt{3}-16$

$=6\sqrt{3}-13$

6. $\sqrt{4x+9}+10=11$

$\sqrt{4x+9}=1$

$(\sqrt{4x+9})^{2}=1^{2}$

$4x+9=1$

$4x=-8$

$x=-2$

7. $\dfrac{x^{-3/2}}{x^{7/2}}=x^{-10/2}=x^{-5}=\dfrac{1}{x^{5}}$

8. $\dfrac{8}{\sqrt{3y}}=\dfrac{8}{\sqrt{3y}}\cdot\dfrac{\sqrt{3y}}{\sqrt{3y}}=\dfrac{8\sqrt{3y}}{\sqrt{3^{2}y^{2}}}=\dfrac{8\sqrt{3y}}{3y}$

9. $\sqrt[3]{-8a^{6}b^{12}}=-2a^{2}b^{4}$

10. $\sqrt{50a^{4}b^{3}}-ab\sqrt{18a^{2}b}$

$=\sqrt{25a^{4}b^{2}\cdot 2b}-ab\sqrt{9a^{2}\cdot 2b}$

$=5a^{2}b\sqrt{2b}-3a^{2}b\sqrt{2b}$

$=2a^{2}b\sqrt{2b}$

11. $\dfrac{14}{4-\sqrt{2}}=\dfrac{14}{4-\sqrt{2}}\cdot\dfrac{4+\sqrt{2}}{4+\sqrt{2}}=\dfrac{56+14\sqrt{2}}{16-\sqrt{2^{2}}}$

$=\dfrac{56+14\sqrt{2}}{16-2}=\dfrac{56+14\sqrt{2}}{14}$

$=4+\sqrt{2}$

12. $\dfrac{5+2i}{3i}=\dfrac{5+2i}{3i}\cdot\dfrac{-3i}{-3i}=\dfrac{-15i-6i^{2}}{-9i^{2}}$

$=\dfrac{-15i-6(-1)}{-9(-1)}=\dfrac{-15i+6}{9}$

$=\dfrac{2}{3}-\dfrac{5}{3}i$

13. $\sqrt{18a^{3}b^{6}}=\sqrt{9a^{2}b^{6}\cdot 2a}=3ab^{3}\sqrt{2a}$

14. $(17+8i)-(15-4i)=2+12i$

15. $3x\sqrt[3]{54x^{8}y^{10}}-2x^{2}y\sqrt[3]{16x^{5}y^{7}}$

$=3x\sqrt[3]{27x^{6}y^{9}\cdot 2x^{2}y}-2x^{2}y\sqrt[3]{8x^{3}y^{6}\cdot 2x^{2}y}$

$=9x^{3}y^{3}\sqrt[3]{2x^{2}y}-4x^{3}y^{3}\sqrt[3]{2x^{2}y}$

$=5x^{3}y^{3}\sqrt[3]{2x^{2}y}$

16. $\sqrt[3]{16x^{4}y}\sqrt[3]{4xy^{5}}=\sqrt[3]{64x^{5}y^{6}}$

$=\sqrt[3]{64x^{3}y^{6}\cdot x^{2}}=4xy^{2}\sqrt[3]{x^{2}}$

17. $i(3-7i)=3i-7i^2=3i-7(-1)=7+3i$

18. $\dfrac{(4a^{-2/3}b^4)^{-1/2}}{(a^{-1/6}b^{3/2})^2}=\dfrac{(4)^{-1/2}a^{1/3}b^{-2}}{a^{-1/3}b^3}$

$=\dfrac{a^{2/3}}{2b^5}$

19. $\sqrt[5]{-64a^8b^{12}}=\sqrt[5]{-32a^5b^{10}\cdot 2a^3b^2}$

$=-2ab^2\sqrt[5]{2a^3b^2}$

20. $\dfrac{5+9i}{1-i}=\dfrac{5+9i}{1-i}\cdot\dfrac{1+i}{1+i}=\dfrac{5+5i+9i+9i^2}{1-i^2}$

$=\dfrac{5+14i+9(-1)}{1-(-1)}=\dfrac{-4+14i}{2}$

$=-2+7i$

21. $\sqrt{-12}\sqrt{-6}=i\sqrt{12}\cdot i\sqrt{6}=i^2\sqrt{72}$

$=-1\sqrt{36\cdot 2}=-6\sqrt{2}$

22. $\sqrt{x-5}+\sqrt{x+6}=11$

$\sqrt{x-5}=11-\sqrt{x+6}$

$(\sqrt{x-5})^2=(11-\sqrt{x+6})^2$

$x-5=121-22\sqrt{x+6}+x+6$

$-5=127-22\sqrt{x+6}$

$-132=-22\sqrt{x+6}$

$6=\sqrt{x+6}$

$(6)^2=(\sqrt{x+6})^2$

$36=x+6$

$30=x$

The solution is 30.

23. $\sqrt[4]{81a^8b^{12}}=3a^2b^3$

24. $\dfrac{9}{\sqrt[3]{3x}}=\dfrac{9}{\sqrt[3]{3x}}\cdot\dfrac{\sqrt[3]{9x^2}}{\sqrt[3]{9x^2}}=\dfrac{9\sqrt[3]{9x^2}}{\sqrt[3]{27x^3}}$

$=\dfrac{9\sqrt[3]{9x^2}}{3x}=\dfrac{3\sqrt[3]{9x^2}}{x}$

25. $(-8+3i)-(4-7i)=-12+10i$

26. $(2+\sqrt{2x-1})^2=4+4\sqrt{2x-1}+2x-1$

$=2x+4\sqrt{2x-1}+3$

27. $4x\sqrt{12x^2y}+\sqrt{3x^4y}-x^2\sqrt{27y}$

$=4x\sqrt{4x^2\cdot 3y}+\sqrt{x^4\cdot 3y}-x^2\sqrt{9\cdot 3y}$

$=8x^2\sqrt{3y}+x^2\sqrt{3y}-3x^2\sqrt{3y}$

$=6x^2\sqrt{3y}$

28. $81^{-1/4}=(3^4)^{-1/4}=3^{-1}=\dfrac{1}{3}$

29. $(a^{16})^{-5/8}=a^{-10}=\dfrac{1}{a^{10}}$

30. $-\sqrt{49x^6y^{16}}=-7x^3y^8$

31. $4a^{2/3}=4\sqrt[3]{a^2}$

32. $(9x^2y^4)^{-1/2}(x^6y^6)^{1/3}=9^{-1/2}x^{-1}y^{-2}x^2y^2$

$=9^{-1/2}x^1y^0=9^{-1/2}x$

$=\dfrac{x}{9^{1/2}}=\dfrac{x}{3}$

33. $\sqrt[4]{x^6y^8z^{10}}=\sqrt[4]{x^4y^8z^8\cdot x^2z^2}$

$=xy^2z^2\sqrt[4]{x^2z^2}$

34. $\sqrt{54}+\sqrt{24}=\sqrt{9\cdot 6}+\sqrt{4\cdot 6}$

$=3\sqrt{6}+2\sqrt{6}=5\sqrt{6}$

35. $\sqrt{48x^5y}-x\sqrt{80x^2y}=\sqrt{16x^4\cdot 3xy}-x\sqrt{16x^2\cdot y}$

$=4x^2\sqrt{3xy}-4x^2\sqrt{y}$

36. $\sqrt{32}\sqrt{50}=\sqrt{1600}=40$

37. $\sqrt{3x}(3+\sqrt{3x}) = 3\sqrt{3x} + \sqrt{(3x)^2}$
$= 3\sqrt{3x} + 3x$

38. $\dfrac{\sqrt{125x^6}}{\sqrt{5x^3}} = \sqrt{\dfrac{125x^6}{5x^3}} = \sqrt{25x^3} = 5x\sqrt{x}$

39. $\dfrac{2-3\sqrt{7}}{6-\sqrt{7}} = \dfrac{2-3\sqrt{7}}{6-\sqrt{7}} \cdot \dfrac{6+\sqrt{7}}{6+\sqrt{7}}$
$= \dfrac{12 + 2\sqrt{7} - 18\sqrt{7} - 3(\sqrt{7})^2}{6^2 - (\sqrt{7})^2}$
$= \dfrac{12 - 16\sqrt{7} - 21}{36-7}$
$= \dfrac{-9-16\sqrt{7}}{29}$

40. $\sqrt{-36} = i\sqrt{36} = 6i$

41. $-b + \sqrt{b^2 - 4ac}$
$-(-8) + \sqrt{(-8)^2 - 4(1)(25)} = 8 + \sqrt{64 - 100}$
$= 8 + \sqrt{-36} = 8 + i\sqrt{36}$
$= 8 + 6i$

42. $-b + \sqrt{b^2 - 4ac}$
$-2 + \sqrt{2^2 - 4(1)(9)} = -2 + \sqrt{4 - 36}$
$= -2 + \sqrt{-32} = -2 + i\sqrt{16 \cdot 2}$
$= -2 + 4i\sqrt{2}$

43. $(5 + 2i) + (4 - 3i) = 9 - i$

44. $(3 + 2\sqrt{5})(3 - 2\sqrt{5})$
$= 9 - 6\sqrt{5} + 6\sqrt{5} - 4(\sqrt{5})^2$
$= 9 - 20 = -11$

45. $(3 - 9i) - 7 = -4 - 9i$

46. $(4 - i)^2 = (4 - i)(4 - i)$
$= 16 - 8i + i^2 = 16 - 8i + (-1)$
$= 15 - 8i$

47. $\dfrac{-6}{i} = \dfrac{-6}{i} \cdot \dfrac{i}{i} = \dfrac{-6i}{i^2} = \dfrac{-6i}{-1} = 6i$

48. $\dfrac{7}{2-i} = \dfrac{7}{2-i} \cdot \dfrac{2+i}{2+i} = \dfrac{14+7i}{4-i^2}$
$= \dfrac{14+7i}{4-(-1)} = \dfrac{14+7i}{5}$
$= \dfrac{14}{5} + \dfrac{7}{5}i$

49. $\sqrt{2x-7} + 2 = 5$
$\sqrt{2x-7} = 3$
$\left(\sqrt{2x-7}\right)^2 = 3^2$
$2x - 7 = 9$
$2x = 16$
$x = 8$
The solution is 8.

50. $\sqrt[3]{9x} = -6$
$\left(\sqrt[3]{9x}\right)^3 = (-6)^3$
$9x = -216$
$x = -24$
The solution is −24.

51. Strategy: Use the Pythagorean Theorem to find the width of the rectangle.

Solution: $c^2 = a^2 + b^2$
$13^2 = 12^2 + b^2$
$169 = 144 + b^2$
$25 = b^2$
$(25)^{1/2} = (b^2)^{1/2}$
$5 = b$
The width of the rectangle is 5 in.

52. Strategy: To find the amount of power substitute the given value for v in the equation and solve for P.

Solution: $v = 4.05\sqrt[3]{P}$

$20 = 4.05\sqrt[3]{P}$

$4.94 = \sqrt[3]{P}$

$(4.94)^3 = (\sqrt[3]{P})^3$

$120 \approx P$

The amount of power is 120 watts.

53. Strategy: To find the distance required substitute the given values for v and a in the equation and solve for s.

Solution: $v = \sqrt{2as}$

$88 = \sqrt{2(16s)}$

$(88)^2 = (\sqrt{32s})^2$

$7744 = 32s$

$242 = s$

The distance required is 242 ft. .

54. Strategy: To find the distance use the Pythagorean Theorem.

Solution: $c^2 = a^2 + b^2$

$12^2 = 10^2 + b^2$

$144 = 100 + b^2$

$44 = b^2$

$(44)^{1/2} = (b^2)^{1/2}$

$6.63 = b$

The distance is 6.63 ft.

Chapter 9 Test

1. $\frac{1}{2}\sqrt[4]{x^3} = \frac{1}{2}x^{3/4}$

2. $\sqrt[3]{54x^7y^3} - x\sqrt[3]{128x^4y^3} - x^2\sqrt[3]{2xy^3}$

$= \sqrt[3]{27x^6y^3 \cdot 2x} - x\sqrt[3]{64x^3y^3 \cdot 2x} - x^2\sqrt[3]{y^3 \cdot 2x}$

$= 3x^2y\sqrt[3]{2x} - 4x^2y\sqrt[3]{2x} - x^2y\sqrt[3]{2x}$

$= -2x^2y\sqrt[3]{2x}$

3. $3y^{2/5} = 3\sqrt[5]{y^2}$

4. $(2 + 5i)(4 - 2i) = 8 - 4i + 20i - 10i^2$

$= 8 + 16i - 10(-1)$

$= 18 + 16i$

5. $(3 - 2\sqrt{x})^2 = (3 - 2\sqrt{x})(3 - 2\sqrt{x})$

$= 9 - 12\sqrt{x} + 4(\sqrt{x})^2$

$= 4x - 12\sqrt{x} + 9$

6. $\frac{r^{2/3}r^{-1}}{r^{-1/2}} = \frac{r^{-1/3}}{r^{-1/2}} = r^{1/6}$

7. $\sqrt{x+12} - \sqrt{x} = 2$

$\sqrt{x+12} = 2 + \sqrt{x}$

$(\sqrt{x+12})^2 = (2 + \sqrt{x})^2$

$x + 12 = 4 + 4\sqrt{x} + x$

$12 = 4 + 4\sqrt{x}$

$8 = 4\sqrt{x}$

$2 = \sqrt{x}$

$(2)^2 = (\sqrt{x})^2$

$4 = x$

The solution is 4.

8. $\sqrt[4]{4a^5b^3}\sqrt[4]{8a^3b^7} = \sqrt[4]{32a^8b^{10}}$

$= \sqrt[4]{16a^8b^8 \cdot 2b^2} = 2a^2b^2\sqrt[4]{2b^2}$

9. $\sqrt{3x}(\sqrt{x} - \sqrt{25x}) = \sqrt{3x^2} - \sqrt{75x^2}$

$= \sqrt{x^2 \cdot 3} - \sqrt{25x^2 \cdot 3} = x\sqrt{3} - 5x\sqrt{3}$

$= -4x\sqrt{3}$

10. $(5 - 2i) - (8 - 4i) = -3 + 2i$

11. $\sqrt{32x^4y^7} = \sqrt{16x^4y^6 \cdot 2y} = 4x^2y^3\sqrt{2y}$

12. $(2\sqrt{3}+4)(3\sqrt{3}-1)$

$=6\sqrt{3^2}-2\sqrt{3}+12\sqrt{3}-4$

$=18+10\sqrt{3}-4$

$=14+10\sqrt{3}$

13. $\sqrt{-5}\cdot\sqrt{-20}=i\sqrt{5}\cdot i\sqrt{20}=i^2\sqrt{100}$

$=-1\sqrt{100}=-10$

14. $\dfrac{4-2\sqrt{5}}{2-\sqrt{5}}=\dfrac{4-2\sqrt{5}}{2-\sqrt{5}}\cdot\dfrac{2+\sqrt{5}}{2+\sqrt{5}}$

$=\dfrac{8+4\sqrt{5}-4\sqrt{5}-2(\sqrt{5})^2}{2^2-(\sqrt{5})^2}$

$=\dfrac{8-10}{4-5}=\dfrac{-2}{-1}=2$

15. $\sqrt{18a^3}+a\sqrt{50a}=\sqrt{9a^2\cdot 2a}+a\sqrt{25\cdot 2a}$

$=3a\sqrt{2a}+5a\sqrt{2a}$

$=8a\sqrt{2a}$

16. $(\sqrt{a}-3\sqrt{b})(2\sqrt{a}+5\sqrt{b})$

$=2\sqrt{a^2}+5\sqrt{ab}-6\sqrt{ab}-15\sqrt{b^2}$

$=2a-\sqrt{ab}-15b$

17. $\dfrac{(2x^{1/3}y^{-2/3})^6}{(x^{-4}y^8)^{1/4}}=\dfrac{2^6x^2y^{-4}}{x^{-1}y^2}=\dfrac{64x^3}{y^6}$

18. $\dfrac{10x}{\sqrt[3]{5x^2}}=\dfrac{10x}{\sqrt[3]{5x^2}}\cdot\dfrac{\sqrt[3]{25x}}{\sqrt[3]{25x}}=\dfrac{10x\sqrt[3]{25x}}{\sqrt[3]{125x^3}}$

$=\dfrac{10x\sqrt[3]{25x}}{5x}=2\sqrt[3]{25x}$

19. $\dfrac{2+3i}{1-2i}=\dfrac{2+3i}{1-2i}\cdot\dfrac{1+2i}{1+2i}=\dfrac{2+4i+3i+6i^2}{1-4i^2}$

$=\dfrac{2+7i+6(-1)}{1-4(-1)}=\dfrac{-4+7i}{5}$

$=-\dfrac{4}{5}+\dfrac{7}{5}i$

20. $\sqrt[3]{2x-2}+4=2$

$\sqrt[3]{2x-2}=-2$

$(\sqrt[3]{2x-2})^3=(-2)^3$

$2x-2=-8$

$2x=-6$

$x=-3$

The solution is −3.

21. $\left(\dfrac{4a^4}{b^2}\right)^{-3/2}=\dfrac{(4a^4)^{-3/2}}{(b^2)^{-3/2}}=\dfrac{4^{-3/2}a^{-6}}{b^{-3}}=\dfrac{b^3}{8a^6}$

22. $\sqrt[3]{27a^4b^3c^7}=\sqrt[3]{27a^3b^3c^6\cdot ac}$

$=3abc^2\sqrt[3]{ac}$

23. $\dfrac{\sqrt{32x^5y}}{\sqrt{2xy^3}}=\sqrt{\dfrac{32x^5y}{2xy^3}}=\sqrt{\dfrac{16x^4}{y^2}}=\dfrac{4x^2}{y}$

24. $\dfrac{5x}{\sqrt{5x}}=\dfrac{5x}{\sqrt{5x}}\cdot\dfrac{\sqrt{5x}}{\sqrt{5x}}=\dfrac{5x\sqrt{5x}}{(\sqrt{5x})^2}$

$=\dfrac{5x\sqrt{5x}}{5x}=\sqrt{5x}$

25. Strategy: To find the distance the object has fallen substitute the value for v in the equation and solve for d.

Solution: $v=\sqrt{64d}$

$192=\sqrt{64d}$

$(192)^2=(\sqrt{64d})^2$

$36{,}864=64d$

$576=d$

The object has fallen 576 ft.

Cumulative Review Exercises

1. $2^3 \cdot 3 - 4(3 - 4 \cdot 5) = 2^3 \cdot 3 - 4(3 - 20)$

$= 2^3 \cdot 3 - 4(-17)$

$= 8 \cdot 3 - 4(-17)$

$= 24 + 68$
$= 92$

2. $4a^2b - a^3 = 4(-2)^2(3) - (-2)^3$
$= 4(4)(3) - (-8)$
$= 16(3) + 8$
$= 48 + 8$
$= 56$

3. $-3(4x - 1) - 2(1 - x) = -12x + 3 - 2 + 2x$
$= -10x + 1$

4. $5 - \frac{2}{3}x = 4$

$5 - \frac{2}{3}x - 5 = 4 - 5$

$-\frac{2}{3}x = -1$

$\left(-\frac{3}{2}\right)\left(-\frac{2}{3}x\right) = -1\left(-\frac{3}{2}\right)$

$x = \frac{3}{2}$

The solution is $\frac{3}{2}$.

5. $2[4 - 2(3 - 2x)] = 4(1 - x)$
$2[4 - 6 + 4x] = 4 - 4x$
$2[-2 + 4x] = 4 - 4x$
$-4 + 8x = 4 - 4x$
$-4 + 12x = 4$
$12x = 8$

$x = \frac{2}{3}$

The solution is $\frac{2}{3}$.

6. $6x - 3(2x + 2) > 3 - 3(x + 2)$
$6x - 6x - 6 > 3 - 3x - 6$
$-6 > -3x - 3$
$-3 > -3x$
$1 < x$
$\{x \mid x > 1\}$

7. $2 + |4 - 3x| = 5$
$|4 - 3x| = 3$

$4 - 3x = -3$ $\quad$ $4 - 3x = 3$
$-3x = -7$ $\quad$ $-3x = -1$
$x = \frac{7}{3}$ $\quad$ $x = \frac{1}{3}$

The solutions are $\frac{1}{3}$ and $\frac{7}{3}$.

8. $|2x + 3| \le 9$
$-9 \le 2x + 3 \le 9$
$-9 - 3 \le 2x + 3 - 3 \le 9 - 3$
$-12 \le 2x \le 6$
$-6 \le x \le 3$
$\{x \mid -6 \le x \le 3\}$

9. $A = \frac{1}{2}bh = \frac{1}{2}(25)(15) = 12.5(15) = 187.5$
The area is 187.5 cm^2.

10. $V = L \times W \times H = 3.5 \times 2 \times 2 = 14\ ft^3$
The volume is 14 ft^3.

11. Solve $3x - 2y = -6$ for y.
$3x - 2y = -6$
$-2y = -3x - 6$
$y = \frac{3}{2}x + 3$
The y-intercept is (0, 3)
The slope is $\frac{3}{2}$.

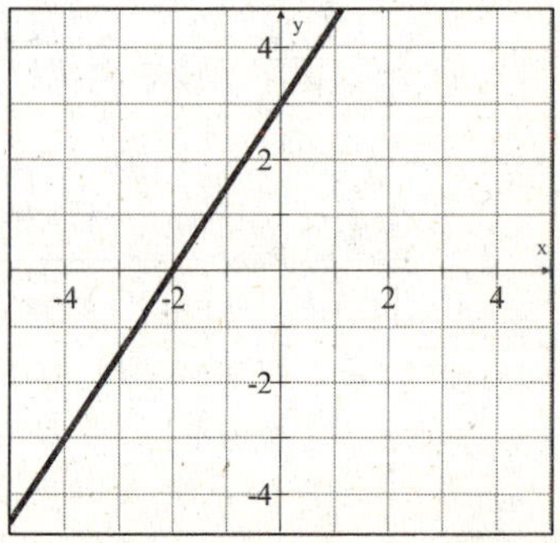

12. $3x + 2y \le 4$
$2y \le -3x + 4$
$y \le -\dfrac{3}{2}x + 2$

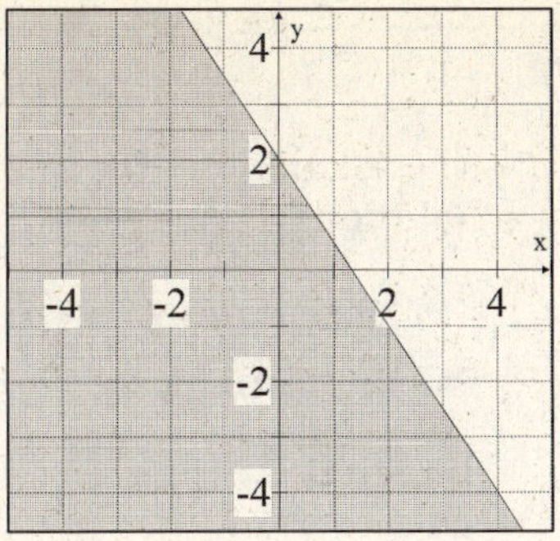

13. Find the slope of the line.
$$m = \frac{y_2 - y_1}{x_2 - x_1} = \frac{2-3}{-1-2} = \frac{-1}{-3} = \frac{1}{3}$$
Use the point-slope formula to find the equation of the line.
$$y - y_1 = m(x - x_1)$$
$$y - 3 = \frac{1}{3}(x-2)$$
$$y - 3 = \frac{1}{3}x - \frac{2}{3}$$
$$y = \frac{1}{3}x + \frac{7}{3}$$
The equation of the line is $y = \frac{1}{3}x + \frac{7}{3}$.

14.
$$\begin{aligned} 2x - y &= 4 \\ \underline{-2x + 3y} &\underline{= 0} \\ 2y &= 4 \\ y &= 2 \end{aligned}$$
$$\begin{aligned} 2x - 2 &= 4 \\ 2x &= 6 \\ x &= 3 \end{aligned}$$
The solution is (3, 2).

15. $(2^{-1}x^2y^{-6})(2^{-1}y^{-4})^{-2}$
$= (2^{-1}x^2y^{-6})(2^2y^8)$
$= 2x^2y^2$

16. $81x^2 - y^2 = (9x + y)(9x - y)$

17. $x^5 + 2x^3 - 3x = x(x^4 + 2x^2 - 3)$
$= x(x^2 + 3)(x^2 - 1)$
$= x(x^2 + 3)(x + 1)(x - 1)$

18. $P = \dfrac{R - C}{n}$
$$P \cdot n = \frac{R-C}{n} \cdot n$$
$$nP = R - C$$
$$nP + C = R$$
$$C = R - nP$$

19. $\left(\dfrac{x^{-2/3}y^{1/2}}{y^{-1/3}}\right)^6 = \dfrac{x^{-4}y^3}{y^{-2}} = \dfrac{y^5}{x^4}$

20. $\sqrt{40x^3} - x\sqrt{90x} = \sqrt{4x^2 \cdot 10x} - x\sqrt{9 \cdot 10x}$
$= 2x\sqrt{10x} - 3x\sqrt{10x}$
$= -x\sqrt{10x}$

21. $(\sqrt{3} - 2)(\sqrt{3} - 5)$
$= \sqrt{3^2} - 5\sqrt{3} - 2\sqrt{3} + 10$
$= 3 - 7\sqrt{3} + 10$
$= 13 - 7\sqrt{3}$

22. $\dfrac{4}{\sqrt{6} - \sqrt{2}} = \dfrac{4}{\sqrt{6} - \sqrt{2}} \cdot \dfrac{\sqrt{6} + \sqrt{2}}{\sqrt{6} + \sqrt{2}}$
$$= \frac{4\sqrt{6} + 4\sqrt{2}}{\sqrt{6^2} - \sqrt{2^2}} = \frac{4\sqrt{6} + 4\sqrt{2}}{6-2} = \frac{4\sqrt{6} + 4\sqrt{2}}{4}$$
$$= \sqrt{6} + \sqrt{2}$$

23. $\dfrac{2i}{3-i} = \dfrac{2i}{3-i} \cdot \dfrac{3+i}{3+i} = \dfrac{6i + 2i^2}{9 - i^2}$
$$= \frac{6i + 2(-1)}{9-(-1)} = \frac{6i - 2}{10}$$
$$= \frac{-1 + 3i}{5}$$
$$= -\frac{1}{5} + \frac{3}{5}i$$

24. $\sqrt[3]{3x-4}+5=1$

$\sqrt[3]{3x-4}=-4$

$(\sqrt[3]{3x-4})^3=(-4)^3$

$3x-4=-64$

$3x=-60$

$x=-20$

25. $\dfrac{BC}{EF}=\dfrac{AC}{DE}$

$\dfrac{8}{12}=\dfrac{18}{DE}$

$8(DE)=12(18)$

$8(DE)=216$

$DE=27$

The length of side DE is 27 m.

26. Strategy: Let x represent amount invested at 3.5%.

	Principal	Rate	Interest
Amount invested at 3.5%	x	0.035	$(0.035)x$
Amount invested at 4.5%	$10{,}000-x$	0.045	$0.045(10{,}000-x)$

The total amount of interest earned is $425.
$0.035x+0.045(10{,}000-x)=425$

Solution: $0.035x+0.045(10{,}000-x)=425$

$0.035x+450-0.045x=425$

$450-0.010x=425$

$-0.010x=-25$

$x=2500$

$2500 must be invested at 3.5%.

27. Strategy: Let x represent the unknown rate of the car.
The unknown rate of the plane is $5x$.

	Distance	Rate	Time
car	25	x	$\frac{25}{x}$
Plane	625	$5x$	$\frac{625}{5x}$

The total time of the trip was 3 h.

$\dfrac{25}{x}+\dfrac{625}{5x}=3$

Solution: $\dfrac{25}{x}+\dfrac{625}{5x}=3$

$5x\left(\dfrac{25}{x}+\dfrac{625}{5x}\right)=5x(3)$

$125+625=15x$

$750=15x$

$50=x$

$250=5x$

The rate of the plane is 250 mph.

28. Strategy: To find the time it takes light to travel from the earth to the moon use the formula $D=RT$. Substitute in the given values for R and D and solve for T.

Solution: $D=RT$

$1.86\times10^5\cdot T=232{,}500$

$T=1.25\times10^0$

$T=1.25$

The time is 1.25 s.

29. $m=\dfrac{y_2-y_1}{x_2-x_1}=\dfrac{400-0}{5000-0}=\dfrac{400}{5000}=0.08$

The annual income is 8% of the investment.

30. Strategy: To find the height if the periscope substitute in the given value for d and solve for h.

Solution: $d=\sqrt{1.5h}$

$7=\sqrt{1.5h}$

$(7)^2=(\sqrt{1.5h})^2$

$49=1.5h$

$32.7\approx h$

The height of the periscope is 32.7 ft.

Chapter 10: Quadratic Equations

Prep Test

1. $\sqrt{18} = \sqrt{9 \cdot 2} = 3\sqrt{2}$

2. $3i$

3. $\dfrac{3x-2}{x-1} - 1 = \dfrac{3x-2}{x-1} - \dfrac{x-1}{x-1}$
$= \dfrac{3x-2-x+1}{x-1}$
$= \dfrac{2x-1}{x-1}$

4. $b^2 - 4ac$
$(-4)^2 - 4(2)(1) = 16 - 8$
$= 8$

5. $4x^2 + 28x + 49 = (2x+7)(2x+7)$
Yes, it is a perfect square trinomial.

6. $4x^2 - 4x + 1 = (2x-1)(2x-1) = (2x-1)^2$

7. $9x^2 - 4 = (3x+2)(3x-2)$

8. $\{x|\ x < -1\} \cap \{x|\ x < 4\}$

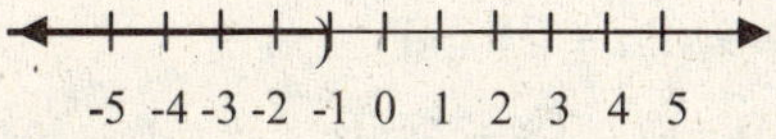

9. $x(x-1) = x + 15$
$x^2 - x = x + 15$
$x^2 - 2x - 15 = 0$
$(x-5)(x+3) = 0$
$x - 5 = 0 \quad x + 3 = 0$
$x = 5 \qquad x = -3$
The solutions are −3 and 5.

10. $\dfrac{4}{x-3} = \dfrac{16}{x}$
$x(x-3)\left(\dfrac{4}{x-3}\right) = x(x-3)\left(\dfrac{16}{x}\right)$
$4x = 16(x-3)$
$4x = 16x - 48$
$-12x = -48$
$x = 4$
The solution is 4.

Section 10.1

Objective A Exercises

1. If $a = 0$ in the quadratic equation there would be no second degree term. The equation would not be a quadratic equation.

3. $2x^2 - 4x - 5 = 0;\ a = 2, b = -4, c = -5$

5. $4x^2 - 5x + 6 = 0;\ a = 4, b = -5, c = 6$

7. $(x-3)(x-5) = 0$
$x - 3 = 0 \quad x - 5 = 0$
$x = 3 \qquad x = 5$
The solutions are 3 and 5.

9. $(x+7)(x+1) = 0$
$x + 7 = 0 \quad x + 1 = 0$
$x = -7 \qquad x = -1$
The solutions are −7 and −1.

11. $x^2 - 4x = 0$
$x(x-4) = 0$
$x = 0 \quad x - 4 = 0$
$x = 4$
The solutions are 0 and 4.

13. $t^2 - 25 = 0$

$(t+5)(t-5) = 0$

$t+5=0 \quad t-5=0$

$t=-5 \qquad t=5$

The solutions are −5 and 5.

15. $s^2 - s - 6 = 0$

$(s-3)(s+2) = 0$

$s-3=0 \quad s+2=0$

$s=3 \qquad s=-2$

The solutions are −2 and 3.

17. $y^2 - 6y + 9 = 0$

$(y-3)(y-3) = 0$

$y-3=0 \quad y-3=0$

$y=3 \qquad y=3$

The solution is 3.

19. $9z^2 - 18z = 0$

$9z(z-2) = 0$

$9z=0 \quad z-2=0$

$z=0 \qquad z=2$

The solutions are 0 and 2.

21. $r^2 - 3r = 10$

$r^2 - 3r - 10 = 0$

$(r-5)(r+2) = 0$

$r-5=0 \quad r+2=0$

$r=5 \qquad r=-2$

The solutions are −2 and 5.

23. $v^2 + 10 = 7v$

$v^2 - 7v + 10 = 0$

$(v-5)(v-2) = 0$

$v-5=0 \quad v-2=0$

$v=5 \qquad v=2$

The solutions are 2 and 5.

25. $2x^2 - 9x - 18 = 0$

$(2x+3)(x-6) = 0$

$2x+3=0 \quad x-6=0$

$2x=-3 \qquad x=6$

$x=-\frac{3}{2}$

The solutions are $-\frac{3}{2}$ and 6.

27. $4z^2 - 9z + 2 = 0$

$(4z-1)(z-2) = 0$

$4z-1=0 \quad z-2=0$

$4z=1 \qquad z=2$

$z=\frac{1}{4}$

The solutions are $\frac{1}{4}$ and 2.

29. $3w^2 + 11w = 4$

$3w^2 + 11w - 4 = 0$

$(3w-1)(w+4) = 0$

$3w-1=0 \quad w+4=0$

$3w=1 \qquad w=-4$

$w=\frac{1}{3}$

The solutions are −4 and $\frac{1}{3}$.

31. $6x^2 = 23x + 18$

$6x^2 - 23x - 18 = 0$

$(2x-9)(3x+2) = 0$

$2x-9=0 \quad 3x+2=0$

$2x=9 \qquad 3x=-2$

$x=\frac{9}{2} \qquad x=-\frac{2}{3}$

The solutions are $-\frac{2}{3}$ and $\frac{9}{2}$.

33. $4-15u-4u^2=0$

$(1-4u)(4+u)=0$

$1-4u=0 \quad 4+u=0$

$-4u=-1 \qquad u=-4$

$u=\frac{1}{4}$

The solutions are $\frac{1}{4}$ and −4.

35. $x+18=x(x-6)$

$x+18=x^2-6x$

$0=x^2-7x-18$

$0=(x-9)(x+2)$

$x-9=0 \quad x+2=0$

$x=9 \qquad x=-2$

The solutions are −2 and 9.

37. $4s(s+3)=s-6$

$4s^2+12s=s-6$

$4s^2+11s+6=0$

$(4s+3)(s+2)=0$

$4s+3=0 \quad s+2=0$

$4s=-3 \qquad s=-2$

$s=-\frac{3}{4}$

The solutions are −2 and $-\frac{3}{4}$.

39. $u^2-2u+4=(2u-3)(u+2)$

$u^2-2u+4=2u^2+u-6$

$0=u^2+3u-10$

$0=(u+5)(u-2)$

$u+5=0 \quad u-2=0$

$u=-5 \qquad u=2$

The solutions are −5 and 2.

41. $(3x-4)(x+4)=x^2-3x-28$

$3x^2+8x-16=x^2-3x-28$

$2x^2+11x+12=0$

$(2x+3)(x+4)=0$

$2x+3=0 \quad x+4=0$

$2x=-3 \qquad x=-4$

$x=-\frac{3}{2}$

The solutions are −4 and $-\frac{3}{2}$.

43. $x^2-9bx+14b^2=0$

$(x-2b)(x-7b)=0$

$x-2b=0 \quad x-7b=0$

$x=2b \qquad x=7b$

The solutions are $2b$ and $7b$.

45. $x^2-6cx-7c^2=0$

$(x+c)(x-7c)=0$

$x+c=0 \quad x-7c=0$

$x=-c \qquad x=7c$

The solutions are $-c$ and $7c$.

47. $2x^2+3bx+b^2=0$

$(2x+b)(x+b)=0$

$2x+b=0 \quad x+b=0$

$2x=-b \qquad x=-b$

$x=-\frac{b}{2}$

The solutions are $-b$ and $-\frac{b}{2}$.

49. $3x^2-14ax+8a^2=0$

$(3x-2a)(x-4a)=0$

$3x-2a=0 \quad x-4a=0$

$3x=2a \qquad x=4a$

$x=\frac{2a}{3}$

The solutions are $4a$ and $\frac{2a}{3}$.

Objective B Exercises

51. $(x-r_1)(x-r_2)=0$
$(x-2)(x-5)=0$
$x^2-7x+10=0$

53. $(x-r_1)(x-r_2)=0$
$[x-(-2)][x-(-4)]=0$
$(x+2)(x+4)=0$
$x^2+6x+8=0$

55. $(x-r_1)(x-r_2)=0$
$(x-6)[x-(-1)]=0$
$(x-6)(x+1)=0$
$x^2-5x-6=0$

57. $(x-r_1)(x-r_2)=0$
$(x-3)[x-(-3)]=0$
$(x-3)(x+3)=0$
$x^2-9=0$

59. $(x-r_1)(x-r_2)=0$
$(x-4)(x-4)=0$
$x^2-8x+16=0$

61. $(x-r_1)(x-r_2)=0$
$(x-0)(x-5)=0$
$x^2-5x=0$

63. $(x-r_1)(x-r_2)=0$
$(x-0)(x-3)=0$
$x^2-3x=0$

65. $(x-r_1)(x-r_2)=0$
$(x-3)\left(x-\frac{1}{2}\right)=0$
$x^2-\frac{7}{2}x+\frac{3}{2}=0$
$2\left(x^2-\frac{7}{2}x+\frac{3}{2}\right)=0\cdot 2$
$2x^2-7x+3=0$

67. $(x-r_1)(x-r_2)=0$
$\left(x-\left(-\frac{3}{4}\right)\right)(x-2)=0$
$\left(x+\frac{3}{4}\right)(x-2)=0$
$x^2-\frac{5}{4}x-\frac{3}{2}=0$
$4\left(x^2-\frac{5}{4}x-\frac{3}{2}\right)=0\cdot 4$
$4x^2-5x-6=0$

69. $(x-r_1)(x-r_2)=0$
$\left(x-\left(-\frac{5}{3}\right)\right)[x-(-2)]=0$
$\left(x+\frac{5}{3}\right)(x+2)=0$
$x^2+\frac{11}{3}x+\frac{10}{3}=0$
$3\left(x^2+\frac{11}{3}x+\frac{10}{3}\right)=0\cdot 3$
$3x^2+11x+10=0$

71. $(x-r_1)(x-r_2)=0$
$\left(x-\left(-\frac{2}{3}\right)\right)\left(x-\frac{2}{3}\right)=0$
$\left(x+\frac{2}{3}\right)\left(x-\frac{2}{3}\right)=0$
$x^2-\frac{4}{9}=0$
$9\left(x^2-\frac{4}{9}\right)=0\cdot 9$
$9x^2-4=0$

73. $(x-r_1)(x-r_2)=0$

$\left(x-\frac{1}{2}\right)\left(x-\frac{1}{3}\right)=0$

$x^2-\frac{5}{6}x+\frac{1}{6}=0$

$6\left(x^2-\frac{5}{6}x+\frac{1}{6}\right)=0\cdot 6$

$6x^2-5x+1=0$

75. $(x-r_1)(x-r_2)=0$

$\left(x-\frac{6}{5}\right)\left(x-\left(-\frac{1}{2}\right)\right)=0$

$\left(x-\frac{6}{5}\right)\left(x+\frac{1}{2}\right)=0$

$x^2-\frac{7}{10}x-\frac{3}{5}=0$

$10\left(x^2-\frac{7}{10}x-\frac{3}{5}\right)=0\cdot 10$

$10x^2-7x-6=0$

77. $(x-r_1)(x-r_2)=0$

$\left(x-\left(-\frac{1}{4}\right)\right)\left(x-\left(-\frac{1}{2}\right)\right)=0$

$\left(x+\frac{1}{4}\right)\left(x+\frac{1}{2}\right)=0$

$x^2+\frac{3}{4}x+\frac{1}{8}=0$

$8\left(x^2+\frac{3}{4}x+\frac{1}{8}\right)=0\cdot 8$

$8x^2+6x+1=0$

79. $(x-r_1)(x-r_2)=0$

$\left(x-\frac{3}{5}\right)\left(x-\left(-\frac{1}{10}\right)\right)=0$

$\left(x-\frac{3}{5}\right)\left(x+\frac{1}{10}\right)=0$

$x^2-\frac{1}{2}x-\frac{3}{50}=0$

$50\left(x^2-\frac{1}{2}x-\frac{3}{50}\right)=0\cdot 50$

$50x^2-25x-3=0$

81. $c=0$

Objective C Exercises

83. $y^2=49$

$\sqrt{y^2}=\sqrt{49}$

$y=\pm\sqrt{49}=\pm 7$

The solutions are −7 and 7.

85. $z^2=-4$

$\sqrt{z^2}=\sqrt{-4}$

$z=\pm\sqrt{-4}=\pm 2i$

The solutions are $-2i$ and $2i$.

87. $s^2-4=0$

$s^2=4$

$\sqrt{s^2}=\sqrt{4}$

$s=\pm\sqrt{4}=\pm 2$

The solutions are −2 and 2.

89. $4x^2 - 81 = 0$

$4x^2 = 81$

$x^2 = \frac{81}{4}$

$\sqrt{x^2} = \sqrt{\frac{81}{4}}$

$x = \pm\sqrt{\frac{81}{4}} = \pm\frac{9}{2}$

The solutions are $-\frac{9}{2}$ and $\frac{9}{2}$.

91. $y^2 + 49 = 0$

$y^2 = -49$

$\sqrt{y^2} = \sqrt{-49}$

$x = \pm\sqrt{-49} = \pm 7i$

The solutions are $-7i$ and $7i$.

93. $v^2 - 48 = 0$

$v^2 = 48$

$\sqrt{v^2} = \sqrt{48}$

$v = \pm\sqrt{48} = \pm 4\sqrt{3}$

The solutions are $-4\sqrt{3}$ and $4\sqrt{3}$.

95. $r^2 - 75 = 0$

$r^2 = 75$

$\sqrt{r^2} = \sqrt{75}$

$r = \pm\sqrt{75} = \pm 5\sqrt{3}$

The solutions are $-5\sqrt{3}$ and $5\sqrt{3}$.

97. $z^2 + 18 = 0$

$z^2 = -18$

$\sqrt{z^2} = \sqrt{-18}$

$z = \pm\sqrt{-18} = \pm 3i\sqrt{2}$

The solutions are $-3i\sqrt{2}$ and $3i\sqrt{2}$.

99. $(x-1)^2 = 36$

$\sqrt{(x-1)^2} = \sqrt{36}$

$x - 1 = \pm\sqrt{36} = \pm 6$

$x - 1 = 6 \quad x - 1 = -6$

$x = 7 \qquad x = -5$

The solutions are −5 and 7.

101. $3(y+3)^2 = 27$

$(y+3)^2 = 9$

$\sqrt{(y+3)^2} = \sqrt{9}$

$y + 3 = \pm\sqrt{9} = \pm 3$

$y + 3 = 3 \quad y + 3 = -3$

$y = 0 \qquad y = -6$

The solutions are −6 and 0.

103. $5(z+2)^2 = 125$

$(z+2)^2 = 25$

$\sqrt{(z+2)^2} = \sqrt{25}$

$z + 2 = \pm\sqrt{25} = \pm 5$

$z + 2 = 5 \quad z + 2 = -5$

$z = 3 \qquad z = -7$

The solutions are −7 and 3.

105. $\left(v - \frac{1}{2}\right)^2 = \frac{1}{4}$

$\sqrt{\left(v - \frac{1}{2}\right)^2} = \sqrt{\frac{1}{4}}$

$v - \frac{1}{2} = \pm\sqrt{\frac{1}{4}} = \pm\frac{1}{2}$

$v - \frac{1}{2} = \frac{1}{2} \quad v - \frac{1}{2} = -\frac{1}{2}$

$v = 1 \qquad v = 0$

The solutions are 1 and 0.

107. $(x+5)^2-6=0$

$(x+5)^2=6$

$\sqrt{(x+5)^2}=\sqrt{6}$

$x+5=\pm\sqrt{6}$

$x+5=\sqrt{6}$ $\quad x+5=-\sqrt{6}$

$x=-5+\sqrt{6}$ $\quad x=-5-\sqrt{6}$

The solutions are $-5-\sqrt{6}$ and $-5+\sqrt{6}$.

109. $(v-3)^2+45=0$

$(v-3)^2=-45$

$\sqrt{(v-3)^2}=\sqrt{-45}$

$v-3=\pm\sqrt{-45}=\pm 3i\sqrt{5}$

$v-3=3i\sqrt{5}$ $\quad v-3=-3i\sqrt{5}$

$v=3+3i\sqrt{5}$ $\quad v=3-3i\sqrt{5}$

The solutions are $3-3i\sqrt{5}$ and $3+3i\sqrt{5}$.

111. $\left(u+\frac{2}{3}\right)^2-18=0$

$\left(u+\frac{2}{3}\right)^2=18$

$\sqrt{\left(u+\frac{2}{3}\right)^2}=\sqrt{18}$

$u+\frac{2}{3}=\pm\sqrt{18}=\pm 3\sqrt{2}$

$u+\frac{2}{3}=3\sqrt{2}$ $\quad u+\frac{2}{3}=-3\sqrt{2}$

$u=-\frac{2}{3}+3\sqrt{2}$ $\quad u=-\frac{2}{3}-3\sqrt{2}$

$u=-\frac{2+9\sqrt{2}}{3}$ $\quad u=-\frac{2-9\sqrt{2}}{3}$

The solutions are $-\frac{2+9\sqrt{2}}{3}$ and $-\frac{2-9\sqrt{2}}{3}$.

113. Two complex solutions

115. Two equal real solutions

Applying the Concepts

117. $(x-r_1)(x-r_2)=0$

$(x-\sqrt{2})[x-(-\sqrt{2})]=0$

$(x-\sqrt{2})(x+\sqrt{2})=0$

$x^2-2=0$

119. $(x-r_1)(x-r_2)=0$

$(x-i)[x-(-i)]=0$

$(x-i)(x+i)=0$

$x^2+1=0$

121. $(x-r_1)(x-r_2)=0$

$(x-2\sqrt{2})[x-(-2\sqrt{2})]=0$

$(x-2\sqrt{2})(x+2\sqrt{2})=0$

$x^2-8=0$

123. $(x-r_1)(x-r_2)=0$

$(x-i\sqrt{2})[x-(-i\sqrt{2})]=0$

$(x-i\sqrt{2})(x+i\sqrt{2})=0$

$x^2+2=0$

125. $(2x-1)^2=(2x+3)^2$

$\sqrt{(2x-1)^2}=\sqrt{(2x+3)^2}$

$2x-1=\pm\sqrt{(2x+3)^2}=\pm(2x+3)$

$2x-1=2x+3$ $\quad 2x-1=-(2x+3)$

$-1=3$ $\quad 4x=-2$

$x=-\frac{1}{2}$

The solution is $-\frac{1}{2}$.

127. $(2x+1)^2=(x-7)^2$

$\sqrt{(2x+1)^2}=\sqrt{(x-7)^2}$

$2x+1=\pm\sqrt{(x-7)^2}=\pm(x-7)$

$2x+1=x-7 \qquad 2x+1=-(x-7)$

$x=-8 \qquad 3x=6$

$x=2$

The solutions are −8 and 2.

129. Because −2, 3 and 5 are solutions, by the Principle of Zero Products we have $(x+2)(x-3)(x-5)=0$. Multiplying the left side gives a cubic polynomial.

Section 10.2

Objective A Exercises

1. $x^2-4x-5=0$

$x^2-4x=5$

$x^2-4x+4=5+4$

$(x-2)^2=9$

$\sqrt{(x-2)^2}=\sqrt{9}$

$x-2=\pm\sqrt{9}=\pm3$

$x-2=3 \qquad x-2=-3$

$x=5 \qquad x=-1$

The solutions are −1 and 5.

3. $v^2+8v-9=0$

$v^2+8v=9$

$v^2+8v+16=9+16$

$(v+4)^2=25$

$\sqrt{(v+4)^2}=\sqrt{25}$

$v+4=\pm\sqrt{25}=\pm5$

$v+4=5 \qquad v+4=-5$

$v=1 \qquad v=-9$

The solutions are −9 and 1.

5. $z^2-6z+9=0$

$z^2-6z=-9$

$z^2-6z+9=-9+9$

$(z-3)^2=0$

$\sqrt{(z-3)^2}=\sqrt{0}$

$z-3=0$

$z=3$

The solution is 3.

7. $r^2+4r-7=0$

$r^2+4r=7$

$r^2+4r+4=7+4$

$(r+2)^2=11$

$\sqrt{(r+2)^2}=\sqrt{11}$

$r+2=\pm\sqrt{11}$

$r+2=\sqrt{11} \qquad r+2=-\sqrt{11}$

$r=-2+\sqrt{11} \qquad r=-2-\sqrt{11}$

The solutions are $-2-\sqrt{11}$ and $-2+\sqrt{11}$

9. $x^2-6x+7=0$

$x^2-6x=-7$

$x^2-6x+9=-7+9$

$(x-3)^2=2$

$\sqrt{(x-3)^2}=\sqrt{2}$

$x-3=\pm\sqrt{2}$

$x-3=\sqrt{2} \qquad x-3=-\sqrt{2}$

$x=3+\sqrt{2} \qquad x=3-\sqrt{2}$

The solutions are $3-\sqrt{2}$ and $3+\sqrt{2}$.

11. $z^2 - 2z + 2 = 0$

$z^2 - 2z = -2$

$z^2 - 2z + 1 = -2 + 1$

$(z-1)^2 = -1$

$\sqrt{(z-1)^2} = \sqrt{-1}$

$z - 1 = \pm i$

$z - 1 = i \quad z - 1 = -i$

$z = 1 + i \quad z = 1 - i$

The solutions are $1+i$ and $1-i$.

13. $s^2 - 5s - 24 = 0$

$s^2 - 5s = 24$

$s^2 - 5s + \frac{25}{4} = 24 + \frac{25}{4}$

$\left(s - \frac{5}{2}\right)^2 = \frac{121}{4}$

$\sqrt{\left(s - \frac{5}{2}\right)^2} = \sqrt{\frac{121}{4}}$

$s - \frac{5}{2} = \pm\frac{11}{2}$

$s - \frac{5}{2} = \frac{11}{2} \quad s - \frac{5}{2} = -\frac{11}{2}$

$s = \frac{16}{2} = 8 \quad s = -\frac{6}{2} = -3$

The solutions are −3 and 8.

15. $x^2 + 5x - 36 = 0$

$x^2 + 5x = 36$

$x^2 + 5x + \frac{25}{4} = 36 + \frac{25}{4}$

$\left(x + \frac{5}{2}\right)^2 = \frac{169}{4}$

$\sqrt{\left(x + \frac{5}{2}\right)^2} = \sqrt{\frac{169}{4}}$

$x + \frac{5}{2} = \pm\frac{13}{2}$

$x + \frac{5}{2} = \frac{13}{2} \quad x + \frac{5}{2} = -\frac{13}{2}$

$x = \frac{8}{2} = 4 \quad x = -\frac{18}{2} = -9$

The solutions are −9 and 4.

17. $p^2 - 3p + 1 = 0$

$p^2 - 3p = -1$

$p^2 - 3p + \frac{9}{4} = -1 + \frac{9}{4}$

$\left(p - \frac{3}{2}\right)^2 = \frac{5}{4}$

$\sqrt{\left(p - \frac{3}{2}\right)^2} = \sqrt{\frac{5}{4}}$

$p - \frac{3}{2} = \pm\frac{\sqrt{5}}{2}$

$p - \frac{3}{2} = \frac{\sqrt{5}}{2} \quad p - \frac{3}{2} = -\frac{\sqrt{5}}{2}$

$p = \frac{3}{2} + \frac{\sqrt{5}}{2} \quad p = \frac{3}{2} - \frac{\sqrt{5}}{2}$

The solutions are $\frac{3+\sqrt{5}}{2}$ and $\frac{3-\sqrt{5}}{2}$.

19. $t^2 - t - 1 = 0$

$t^2 - t = 1$

$t^2 - t + \frac{1}{4} = 1 + \frac{1}{4}$

$\left(t - \frac{1}{2}\right)^2 = \frac{5}{4}$

$\sqrt{\left(t - \frac{1}{2}\right)^2} = \sqrt{\frac{5}{4}}$

$t - \frac{1}{2} = \pm\frac{\sqrt{5}}{2}$

$t - \frac{1}{2} = \frac{\sqrt{5}}{2} \qquad t - \frac{1}{2} = -\frac{\sqrt{5}}{2}$

$t = \frac{1}{2} + \frac{\sqrt{5}}{2} \qquad t = \frac{1}{2} - \frac{\sqrt{5}}{2}$

The solutions are $\frac{1+\sqrt{5}}{2}$ and $\frac{1-\sqrt{5}}{2}$.

21. $y^2 - 6y = 4$

$y^2 - 6y + 9 = 4 + 9$

$(y-3)^2 = 13$

$\sqrt{(y-3)^2} = \sqrt{13}$

$y - 3 = \pm\sqrt{13}$

$y - 3 = \sqrt{13} \qquad y - 3 = -\sqrt{13}$

$y = 3 + \sqrt{13} \qquad y = 3 - \sqrt{13}$

The solutions are $3 - \sqrt{13}$ and $3 + \sqrt{13}$.

23. $x^2 = 8x - 15$

$x^2 - 8x = -15$

$x^2 - 8x + 16 = -15 + 16$

$(x-4)^2 = 1$

$\sqrt{(x-4)^2} = \sqrt{1}$

$x - 4 = \pm 1$

$x - 4 = 1 \qquad x - 4 = -1$

$x = 5 \qquad x = 3$

The solutions are 3 and 5.

25. $v^2 = 4v - 13$

$v^2 - 4v = -13$

$v^2 - 4v + 4 = -13 + 4$

$(v-2)^2 = -9$

$\sqrt{(v-2)^2} = \sqrt{-9}$

$v - 2 = \pm 3i$

$v - 2 = 3i \qquad v - 2 = -3i$

$v = 2 + 3i \qquad v = 2 - 3i$

The solutions are $2 + 3i$ and $2 - 3i$.

27. $p^2 + 6p = -13$

$p^2 + 6p + 9 = -13 + 9$

$(p+3)^2 = -4$

$\sqrt{(p+3)^2} = \sqrt{-4}$

$p + 3 = \pm 2i$

$p + 3 = 2i \qquad p + 3 = -2i$

$p = -3 + 2i \qquad p = -3 - 2i$

The solutions are $-3 - 2i$ and $-3 + 2i$.

29. $y^2 - 2y = 17$

$y^2 - 2y + 1 = 17 + 1$

$(y-1)^2 = 18$

$\sqrt{(y-1)^2} = \sqrt{18}$

$y - 1 = \pm 3\sqrt{2}$

$y - 1 = 3\sqrt{2} \qquad y - 1 = -3\sqrt{2}$

$y = 1 + 3\sqrt{2} \qquad y = 1 - 3\sqrt{2}$

The solutions are $1 - 3\sqrt{2}$ and $1 + 3\sqrt{2}$.

31. $z^2 = z + 4$

$z^2 - z = 4$

$z^2 - z + \frac{1}{4} = 4 + \frac{1}{4}$

$\left(z - \frac{1}{2}\right)^2 = \frac{17}{4}$

$\sqrt{\left(z - \frac{1}{2}\right)^2} = \sqrt{\frac{17}{4}}$

$z - \frac{1}{2} = \pm\frac{\sqrt{17}}{2}$

$z - \frac{1}{2} = \frac{\sqrt{17}}{2} \qquad z - \frac{1}{2} = -\frac{\sqrt{17}}{2}$

$z = \frac{1}{2} + \frac{\sqrt{17}}{2} \qquad z = \frac{1}{2} - \frac{\sqrt{17}}{2}$

The solutions are $\frac{1+\sqrt{17}}{2}$ and $\frac{1-\sqrt{17}}{2}$.

33. $x^2 + 13 = 2x$

$x^2 - 2x = -13$

$x^2 - 2x + 1 = -13 + 1$

$(x-1)^2 = -12$

$\sqrt{(x-1)^2} = \sqrt{-12}$

$x - 1 = \pm 2i\sqrt{3}$

$x - 1 = 2i\sqrt{3} \qquad x - 1 = -2i\sqrt{3}$

$x = 1 + 2i\sqrt{3} \qquad x = 1 - 2i\sqrt{3}$

The solutions are $1 - 2i\sqrt{3}$ and $1 + 2i\sqrt{3}$.

35. $4x^2 - 4x + 5 = 0$

$4x^2 - 4x = -5$

$\frac{1}{4}(4x^2 - 4x) = \frac{1}{4}(-5)$

$x^2 - x = -\frac{5}{4}$

$x^2 - x + \frac{1}{4} = -\frac{5}{4} + \frac{1}{4}$

$\left(x - \frac{1}{2}\right)^2 = -1$

$\sqrt{\left(x - \frac{1}{2}\right)^2} = \sqrt{-1}$

$x - \frac{1}{2} = \pm i$

$x - \frac{1}{2} = i \qquad x - \frac{1}{2} = -i$

$x = \frac{1}{2} + i \qquad x = \frac{1}{2} - i$

The solutions are $\frac{1}{2} - i$ and $\frac{1}{2} + i$.

37. $9x^2 - 6x + 2 = 0$

$9x^2 - 6x = -2$

$\frac{1}{9}(9x^2 - 6x) = \frac{1}{9}(-2)$

$x^2 - \frac{2}{3}x = -\frac{2}{9}$

$x^2 - \frac{2}{3}x + \frac{1}{9} = -\frac{2}{9} + \frac{1}{9}$

$\left(x - \frac{1}{3}\right)^2 = -\frac{1}{9}$

$\sqrt{\left(x - \frac{1}{3}\right)^2} = \sqrt{-\frac{1}{9}}$

$x - \frac{1}{3} = \pm\frac{1}{3}i$

$x - \frac{1}{3} = \frac{1}{3}i \qquad x - \frac{1}{3} = -\frac{1}{3}i$

$x = \frac{1}{3} + \frac{1}{3}i \qquad x = \frac{1}{3} - \frac{1}{3}i$

The solutions are $\frac{1}{3} - \frac{1}{3}i$ and $\frac{1}{3} + \frac{1}{3}i$.

39. $2s^2 = 4s + 5$

$2s^2 - 4s = 5$

$\frac{1}{2}(2s^2 - 4s) = \frac{1}{2}(5)$

$s^2 - 2s = \frac{5}{2}$

$s^2 - 2s + 1 = \frac{5}{2} + 1$

$(s-1)^2 = \frac{7}{2}$

$\sqrt{(s-1)^2} = \sqrt{\frac{7}{2}}$

$s - 1 = \pm\sqrt{\frac{7}{2}} = \pm\frac{\sqrt{14}}{2}$

$s - 1 = \frac{\sqrt{14}}{2} \qquad s - 1 = -\frac{\sqrt{14}}{2}$

$s = 1 + \frac{\sqrt{14}}{2} \qquad s = 1 - \frac{\sqrt{14}}{2}$

The solutions are $\frac{2-\sqrt{14}}{2}$ and $\frac{2+\sqrt{14}}{2}$.

41. $2r^2 = 3 - r$

$2r^2 + r = 3$

$\frac{1}{2}(2r^2 + r) = \frac{1}{2}(3)$

$r^2 + \frac{1}{2}r = \frac{3}{2}$

$r^2 + \frac{1}{2}r + \frac{1}{16} = \frac{3}{2} + \frac{1}{16}$

$\left(r + \frac{1}{4}\right)^2 = \frac{25}{16}$

$\sqrt{\left(r + \frac{1}{4}\right)^2} = \sqrt{\frac{25}{16}}$

$r + \frac{1}{4} = \pm\frac{5}{4}$

$r + \frac{1}{4} = \frac{5}{4} \qquad r + \frac{1}{4} = -\frac{5}{4}$

$r = \frac{4}{4} = 1 \qquad r = -\frac{6}{4} = -\frac{3}{2}$

The solutions are $-\frac{3}{2}$ and 1.

43. $y - 2 = (y-3)(y+2)$

$y - 2 = y^2 - y - 6$

$y^2 - 2y = 4$

$y^2 - 2y + 1 = 4 + 1$

$(y-1)^2 = 5$

$\sqrt{(y-1)^2} = \sqrt{5}$

$y - 1 = \pm\sqrt{5}$

$y - 1 = \sqrt{5} \qquad y - 1 = -\sqrt{5}$

$y = 1 + \sqrt{5} \qquad y = 1 - \sqrt{5}$

The solutions are $1-\sqrt{5}$ and $1+\sqrt{5}$.

45. $6t - 2 = (2t-3)(t-1)$

$6t - 2 = 2t^2 - 5t + 3$

$2t^2 - 11t = -5$

$\frac{1}{2}(2t^2 - 11t) = \frac{1}{2}(-5)$

$t^2 - \frac{11}{2}t = -\frac{5}{2}$

$t^2 - \frac{11}{2}t + \frac{121}{16} = -\frac{5}{2} + \frac{121}{16}$

$\left(t - \frac{11}{4}\right)^2 = \frac{81}{16}$

$\sqrt{\left(t - \frac{11}{4}\right)^2} = \sqrt{\frac{81}{16}}$

$t-\frac{11}{4}=\pm\frac{9}{4}$

$t-\frac{11}{4}=\frac{9}{4}$ $t-\frac{11}{4}=-\frac{9}{4}$

$t=\frac{20}{4}=5$ $t=\frac{2}{4}=\frac{1}{2}$

The solutions are $\frac{1}{2}$ and 5.

47. $(x-4)(x+1)=x-3$

$x^2-3x-4=x-3$

$x^2-4x=1$

$x^2-4x+4=1+4$

$(x-2)^2=5$

$\sqrt{(x-2)^2}=\sqrt{5}$

$x-2=\pm\sqrt{5}$

$x-2=\sqrt{5}$ $x-2=-\sqrt{5}$

$x=2+\sqrt{5}$ $x=2-\sqrt{5}$

The solutions are $2+\sqrt{5}$ and $2-\sqrt{5}$.

49. $z^2+2z=4$

$z^2+2z+1=4+1$

$(z+1)^2=5$

$\sqrt{(z+1)^2}=\sqrt{5}$

$z+1=\pm\sqrt{5}$

$z+1=\sqrt{5}$ $z+1=-\sqrt{5}$

$z=\sqrt{5}-1$ $z=-\sqrt{5}-1$

$z=1.236$ $z=-3.236$

The solutions are -3.326 and 1.236.

51. $2x^2=4x-1$

$2x^2-4x=-1$

$\frac{1}{2}(2x^2-4x)=\frac{1}{2}(-1)$

$x^2-2x=-\frac{1}{2}$

$x^2-2x+1=-\frac{1}{2}+1$

$(x-1)^2=\frac{1}{2}$

$\sqrt{(x-1)^2}=\sqrt{\frac{1}{2}}$

$x-1=\pm\sqrt{\frac{1}{2}}$

$x-1=\sqrt{\frac{1}{2}}$ $x-1=-\sqrt{\frac{1}{2}}$

$x=\sqrt{\frac{1}{2}}+1$ $x=-\sqrt{\frac{1}{2}}+1$

$x=1.707$ $x=0.293$

The solutions are 0.293 and 1.707.

53. $4z^2+2z=1$

$\frac{1}{4}(4z^2+2z)=\frac{1}{4}(1)$

$z^2+\frac{1}{2}z=\frac{1}{4}$

$z^2+\frac{1}{2}z+\frac{1}{16}=\frac{1}{4}+\frac{1}{16}$

$\left(z+\frac{1}{4}\right)^2=\frac{5}{16}$

$\sqrt{\left(z+\frac{1}{4}\right)^2}=\sqrt{\frac{5}{16}}$

$$z+\frac{1}{4}=\pm\frac{\sqrt{5}}{4}$$

$$z+\frac{1}{4}=\frac{\sqrt{5}}{4} \qquad z+\frac{1}{4}=-\frac{\sqrt{5}}{4}$$

$$z=\frac{\sqrt{5}}{4}-\frac{1}{4} \qquad z=-\frac{\sqrt{5}}{4}-\frac{1}{4}$$

$$z=0.309 \qquad z=-0.809$$

The solutions are 0.309 and -0.809.

55. $c \le 4$

Applying the Concepts

57. $x^2-ax-2a^2=0$

$$x^2-ax=2a^2$$

$$x^2-ax+\frac{1}{4}a^2=2a^2+\frac{1}{4}a^2$$

$$\left(x-\frac{1}{2}a\right)^2=\frac{9}{4}a^2$$

$$\sqrt{\left(x-\frac{1}{2}a\right)^2}=\sqrt{\frac{9}{4}a^2}$$

$$x-\frac{1}{2}a=\pm\frac{3}{2}a$$

$$x-\frac{1}{2}a=\frac{3}{2}a \qquad x-\frac{1}{2}a=-\frac{3}{2}a$$

$$x=\frac{4}{2}a=2a \qquad x=-\frac{2}{2}a=-a$$

The solutions are $-a$ and $2a$.

59. $x^2+3ax-10a^2=0$

$$x^2+3ax=10a^2$$

$$x^2+3ax+\frac{9}{4}a^2=10a^2+\frac{9}{4}a^2$$

$$\left(x+\frac{3}{2}a\right)^2=\frac{49}{4}a^2$$

$$\sqrt{\left(x+\frac{3}{2}a\right)^2}=\sqrt{\frac{49}{4}a^2}$$

$$x+\frac{3}{2}a=\pm\frac{7}{2}a$$

$$x+\frac{3}{2}a=\frac{7}{2}a \qquad x+\frac{3}{2}a=-\frac{7}{2}a$$

$$x=\frac{4}{2}a=2a \qquad x=-\frac{10}{2}a=-5a$$

The solutions are $-5a$ and $2a$.

61. Strategy: Using the answer from Exercise 60, we know that it takes the ball 4.43 s to hit the ground. Use this time to find the horizontal distance the ball travels.

Solution: $t=4.43$
$s=44.5t=44.5(4.43)=197.2$ ft

No, the ball will have only gone 197.2 ft when it hits the ground.

Section 10.3

Objective A Exercises

1. The quadratic formula:

$$x=\frac{-b\pm\sqrt{b^2-4ac}}{2a}$$

a is the coefficient of x^2; b is the coefficient of x, and c is the constant term in the quadratic equation $ax^2+bx+c,\ a\neq 0$.

3. $x^2-3x-10=0$
$a=1,\ b=-3,\ c=-10$

$$x=\frac{-b\pm\sqrt{b^2-4ac}}{2a}$$

$$x=\frac{-(-3)\pm\sqrt{(-3)^2-4(1)(-10)}}{2(1)}$$

$$x=\frac{3\pm\sqrt{9+40}}{2}=\frac{3\pm\sqrt{49}}{2}$$

$$x = \frac{3 \pm 7}{2}$$

$$x = \frac{3+7}{2} \qquad x = \frac{3-7}{2}$$

$$x = \frac{10}{2} = 5 \qquad x = -\frac{4}{2} = -2$$

The solutions are −2 and 5.

5. $y^2 + 5y - 36 = 0$

$a = 1, b = 5, c = -36$

$$y = \frac{-b \pm \sqrt{b^2 - 4ac}}{2a}$$

$$y = \frac{-5 \pm \sqrt{5^2 - 4(1)(-36)}}{2(1)}$$

$$y = \frac{-5 \pm \sqrt{25 + 144}}{2} = \frac{-5 \pm \sqrt{169}}{2}$$

$$y = \frac{-5 \pm 13}{2}$$

$$y = \frac{-5+13}{2} \qquad y = \frac{-5-13}{2}$$

$$y = \frac{8}{2} = 4 \qquad y = -\frac{18}{2} = -9$$

The solutions are −9 and 4.

7. $w^2 = 8w + 72$

$w^2 - 8w - 72 = 0$

$a = 1, b = -8, c = -72$

$$w = \frac{-b \pm \sqrt{b^2 - 4ac}}{2a}$$

$$w = \frac{-(-8) \pm \sqrt{(-8)^2 - 4(1)(-72)}}{2(1)}$$

$$w = \frac{8 \pm \sqrt{64 + 288}}{2} = \frac{8 \pm \sqrt{352}}{2}$$

$$w = \frac{8 \pm 4\sqrt{22}}{2}$$

$$w = \frac{8 + 4\sqrt{22}}{2} \qquad w = \frac{8 - 4\sqrt{22}}{2}$$

$$w = 4 + 2\sqrt{22} \qquad w = 4 - 2\sqrt{22}$$

The solutions are $4 - 2\sqrt{22}$ and $4 + 2\sqrt{22}$.

9. $v^2 = 24 - 5v$

$v^2 + 5v - 24 = 0$

$a = 1, b = 5, c = -24$

$$z = \frac{-b \pm \sqrt{b^2 - 4ac}}{2a}$$

$$z = \frac{-5 \pm \sqrt{5^2 - 4(1)(-24)}}{2(1)}$$

$$z = \frac{-5 \pm \sqrt{25 + 96}}{2} = \frac{-5 \pm \sqrt{121}}{2}$$

$$z = \frac{-5 \pm 11}{2}$$

$$z = \frac{-5+11}{2} \qquad z = \frac{-5-11}{2}$$

$$z = \frac{6}{2} = 3 \qquad z = -\frac{16}{2} = -8$$

The solutions are −8 and 3.

11. $2y^2 + 5y - 1 = 0$

$a = 2, b = 5, c = -1$

$$y = \frac{-b \pm \sqrt{b^2 - 4ac}}{2a}$$

$$y = \frac{-5 \pm \sqrt{5^2 - 4(2)(-1)}}{2(2)}$$

$$y = \frac{-5 \pm \sqrt{25 + 8}}{4} = \frac{-5 \pm \sqrt{33}}{4}$$

$$y = \frac{-5 + \sqrt{33}}{4} \qquad y = \frac{-5 - \sqrt{33}}{4}$$

The solutions are $\frac{-5 - \sqrt{33}}{4}$ and $\frac{-5 + \sqrt{33}}{4}$.

13. $8s^2 = 10s + 3$

$8s^2 - 10s - 3 = 0$

$a = 8, b = -10, c = -3$

$$s = \frac{-b \pm \sqrt{b^2 - 4ac}}{2a}$$

$$s = \frac{-(-10) \pm \sqrt{(-10)^2 - 4(8)(-3)}}{2(8)}$$

$$s = \frac{10 \pm \sqrt{100 + 96}}{16} = \frac{10 \pm \sqrt{196}}{16}$$

$$s = \frac{10 \pm 14}{16}$$

$$s = \frac{10 + 14}{16} \qquad s = \frac{10 - 14}{16}$$

$$s = \frac{24}{16} = \frac{3}{2} \qquad s = -\frac{4}{16} = -\frac{1}{4}$$

The solutions are $-\frac{1}{4}$ and $\frac{3}{2}$.

15. $x^2 = 14x - 4$

$x^2 - 14x + 4 = 0$

$a = 1, b = -14, c = 4$

$$x = \frac{-b \pm \sqrt{b^2 - 4ac}}{2a}$$

$$x = \frac{-(-14) \pm \sqrt{(-14)^2 - 4(1)(4)}}{2(1)}$$

$$x = \frac{14 \pm \sqrt{196 - 16}}{2} = \frac{14 \pm \sqrt{180}}{2}$$

$$x = \frac{14 \pm 6\sqrt{5}}{2}$$

$$x = \frac{14 + 6\sqrt{5}}{2} \qquad x = \frac{14 - 6\sqrt{5}}{2}$$

$$x = 7 + 3\sqrt{5} \qquad x = 7 - 3\sqrt{5}$$

The solutions are $7 - 3\sqrt{5}$ and $7 + 3\sqrt{5}$.

17. $2z^2 - 2z - 1 = 0$

$a = 2, b = -2, c = -1$

$$z = \frac{-b \pm \sqrt{b^2 - 4ac}}{2a}$$

$$z = \frac{-(-2) \pm \sqrt{(-2)^2 - 4(2)(-1)}}{2(2)}$$

$$z = \frac{2 \pm \sqrt{4 + 8}}{4} = \frac{2 \pm \sqrt{12}}{4}$$

$$z = \frac{2 \pm 2\sqrt{3}}{4} = \frac{1 \pm \sqrt{3}}{2}$$

$$z = \frac{1 + \sqrt{3}}{2} \qquad z = \frac{1 - \sqrt{3}}{2}$$

The solutions are $\frac{1 - \sqrt{3}}{2}$ and $\frac{1 + \sqrt{3}}{2}$.

19. $z^2 + 2z + 2 = 0$

$a = 1, b = 2, c = 2$

$$z = \frac{-b \pm \sqrt{b^2 - 4ac}}{2a}$$

$$z = \frac{-2 \pm \sqrt{2^2 - 4(1)(2)}}{2(1)}$$

$$z = \frac{-2 \pm \sqrt{4 - 8}}{2} = \frac{-2 \pm \sqrt{-4}}{2}$$

$$z = \frac{-2 \pm 2i}{2} = -1 \pm i$$

$$z = -1 + i \qquad z = -1 - i$$

The solutions are $-1 - i$ and $-1 + i$.

21. $y^2 - 2y + 5 = 0$

$a = 1, b = -2, c = 5$

$$y = \frac{-b \pm \sqrt{b^2 - 4ac}}{2a}$$

$$y = \frac{-(-2) \pm \sqrt{(-2)^2 - 4(1)(5)}}{2(1)}$$

$$y = \frac{2 \pm \sqrt{4 - 20}}{2} = \frac{2 \pm \sqrt{-16}}{2}$$

$$y = \frac{2 \pm 4i}{2} = 1 \pm 2i$$

$y = 1 + 2i \qquad y = 1 - 2i$

The solutions are $1 - 2i$ and $1 + 2i$.

23. $s^2 - 4s + 13 = 0$

$a = 1, b = -4, c = 13$

$$s = \frac{-b \pm \sqrt{b^2 - 4ac}}{2a}$$

$$s = \frac{-(-4) \pm \sqrt{(-4)^2 - 4(1)(13)}}{2(1)}$$

$$s = \frac{4 \pm \sqrt{16 - 52}}{2} = \frac{4 \pm \sqrt{-36}}{2}$$

$$s = \frac{4 \pm 6i}{2} = 2 \pm 3i$$

$s = 2 + 3i \qquad s = 2 - 3i$

The solutions are $2 - 3i$ and $2 + 3i$.

25. $2w^2 - 2w - 5 = 0$

$a = 2, b = -2, c = -5$

$$w = \frac{-b \pm \sqrt{b^2 - 4ac}}{2a}$$

$$w = \frac{-(-2) \pm \sqrt{(-2)^2 - 4(2)(-5)}}{2(2)}$$

$$w = \frac{2 \pm \sqrt{4 + 40}}{4} = \frac{2 \pm \sqrt{44}}{4}$$

$$w = \frac{2 \pm 2\sqrt{11}}{4} = \frac{1 \pm \sqrt{11}}{2}$$

$$w = \frac{1 + \sqrt{11}}{2} \qquad w = \frac{1 - \sqrt{11}}{2}$$

The solutions are $\frac{1 - \sqrt{11}}{2}$ and $\frac{1 + \sqrt{11}}{2}$.

27. $2x^2 + 6x + 5 = 0$

$a = 2, b = 6, c = 5$

$$x = \frac{-b \pm \sqrt{b^2 - 4ac}}{2a}$$

$$x = \frac{-6 \pm \sqrt{6^2 - 4(2)(5)}}{2(2)}$$

$$x = \frac{-6 \pm \sqrt{36 - 40}}{4} = \frac{-6 \pm \sqrt{-4}}{4}$$

$$x = \frac{-6 \pm 2i}{4} = \frac{-3 \pm i}{2}$$

$$x = -\frac{3}{2} + \frac{1}{2}i \qquad x = -\frac{3}{2} - \frac{1}{2}i$$

The solutions are $-\frac{3}{2} - \frac{1}{2}i$ and $-\frac{3}{2} + \frac{1}{2}i$.

29. $4t^2 - 6t + 9 = 0$

$a = 4, b = -6, c = 9$

$$t = \frac{-b \pm \sqrt{b^2 - 4ac}}{2a}$$

$$t = \frac{-(-6) \pm \sqrt{(-6)^2 - 4(4)(9)}}{2(4)}$$

$$t = \frac{6 \pm \sqrt{36 - 144}}{8} = \frac{6 \pm \sqrt{-108}}{8}$$

$$t = \frac{6 \pm 6i\sqrt{3}}{8} = \frac{3 \pm 3i\sqrt{3}}{4}$$

$$t = \frac{3 + 3i\sqrt{3}}{4} \qquad t = \frac{3 - 3i\sqrt{3}}{4}$$

The solutions are $\frac{3}{4} + \frac{3\sqrt{3}}{4}i$ and $\frac{3}{4} - \frac{3\sqrt{3}}{4}i$.

31. $p^2 - 8p + 3 = 0$
$a = 1, b = -8, c = 3$

$$p = \frac{-b \pm \sqrt{b^2 - 4ac}}{2a}$$

$$p = \frac{-(-8) \pm \sqrt{(-8)^2 - 4(1)(3)}}{2(1)}$$

$$p = \frac{8 \pm \sqrt{64 - 12}}{2} = \frac{8 \pm \sqrt{52}}{2}$$

$$p = \frac{8 \pm 2\sqrt{13}}{2} = 4 \pm \sqrt{13}$$

$p = 4 + \sqrt{13}$ $\quad p = 4 - \sqrt{13}$

The solutions are 0.394 and 7.606.

33. $w^2 + 4w = 1$
$w^2 + 4w - 1 = 0$
$a = 1, b = 4, c = -1$

$$w = \frac{-b \pm \sqrt{b^2 - 4ac}}{2a}$$

$$w = \frac{-4 \pm \sqrt{4^2 - 4(1)(-1)}}{2(1)}$$

$$w = \frac{-4 \pm \sqrt{16 + 4}}{2} = \frac{-4 \pm \sqrt{20}}{2}$$

$$r = \frac{-4 \pm 2\sqrt{5}}{2} = -2 \pm \sqrt{5}$$

$r = -2 + \sqrt{5}$ $\quad r = -2 - \sqrt{5}$

The solutions are -4.236 and 0.236.

35. $2y^2 = y + 5$
$2y^2 - y - 5 = 0$
$a = 2, b = -1, c = -5$

$$y = \frac{-b \pm \sqrt{b^2 - 4ac}}{2a}$$

$$y = \frac{-(-1) \pm \sqrt{(-1)^2 - 4(2)(-5)}}{2(2)}$$

$$y = \frac{1 \pm \sqrt{1 + 40}}{4} = \frac{1 \pm \sqrt{41}}{4}$$

$y = \frac{1 + \sqrt{41}}{4}$ $\quad y = \frac{1 - \sqrt{41}}{4}$

The solutions are -1.351 and 1.851.

37. $3y^2 + y + 1 = 0$
$a = 3, b = 1, c = 1$
$b^2 - 4ac = 1^2 - 4(3)(1)$
$= 1 - 12 = -11$
$-11 < 0$
Since the discriminant is less than zero, the equation has two complex number solutions.

39. $4x^2 + 20x + 25 = 0$
$a = 4, b = 20, c = 25$
$b^2 - 4ac = (20)^2 - 4(4)(25)$
$= 400 - 400 = 0$
Since the discriminant is equal to zero, the equation has two equal real number solutions.

41. $3w^2 + 3w - 2 = 0$
$a = 3, b = 3, c = -2$
$b^2 - 4ac = 3^2 - 4(3)(-2)$
$= 9 + 24 = 33$
$33 > 0$
Since the discriminant is greater than zero, the equation has two unequal real number solutions.

43. $\sqrt{4ac}$

Applying the Concepts

45. Strategy: To determine if the arrow reaches a height of 275 ft, use the discriminant to determine if the equation has real number solutions.

Solution: $-16t^2 + 128t = 0$
$a = -16, b = 128, c = 0$

$b^2 - 4ac = (128)^2 - 4(-16)(0)$

$= 16{,}384 + 0 = 16{,}384$

$16{,}384 > 0$

Since the discriminant is greater than zero, the arrow reaches the height of 275 ft.

47. $x^2 - 6x + p = 0$

$x^2 - 6x = -p$

$x^2 - 6x + 9 = -p + 9$

$(x-3)^2 = -p + 9$

$\sqrt{(x-3)^2} = \sqrt{9-p}$

$x - 3 = \pm\sqrt{9-p}$

$x = 3 \pm \sqrt{9-p}$

x will have two real solutions if $9 - p > 0$.
Solving this inequality gives $p < 9$.
The values of p are $\{p | p < 9\}$.

49. $x^2 - 2x + p = 0$
$a = 1, b = -2, c = p$

$$x = \frac{-b \pm \sqrt{b^2 - 4ac}}{2a}$$

$$x = \frac{-(-2) \pm \sqrt{(-2)^2 - 4(1)(p)}}{2(1)}$$

$$x = \frac{2 \pm \sqrt{4 - 4p}}{2}$$

$x = 1 \pm \sqrt{1-p}$

x will have two complex solutions when $1 - p < 0$.
Solving the inequality gives $p > 1$.
The values of p are $(1, \infty)$.

51. $x^2 + ix + 2 = 0$
$a = 1, b = i, c = 2$

$$x = \frac{-b \pm \sqrt{b^2 - 4ac}}{2a}$$

$$x = \frac{-i \pm \sqrt{i^2 - 4(1)(2)}}{2(1)}$$

$$x = \frac{-i \pm \sqrt{-1-8}}{2} = \frac{i \pm \sqrt{-9}}{2}$$

$$x = \frac{-i \pm 3i}{2}$$

$$x = \frac{-i + 3i}{2} \qquad x = \frac{-i - 3i}{2}$$

$x = i \qquad x = -2i$

The solutions are $-2i$ and i.

53. $2x^2 + bx - 2 = 0$
$a = 2, b = b, c = -2$

$$x = \frac{-b \pm \sqrt{b^2 - 4ac}}{2a}$$

$$x = \frac{-b \pm \sqrt{b^2 - 4(2)(-2)}}{2(2)}$$

$$x = \frac{-b \pm \sqrt{b^2 + 16}}{4}$$

If b is real, the quantity $\sqrt{b^2 + 16}$ can never be less than zero, and the quadratic will always have real number solutions.

Section 10.4

Objective A Exercises

1. Yes

3. No

5. $x^4 - 13x^2 + 36 = 0$

$(x^2)^2 - 13(x^2) + 36 = 0$

$u^2 - 13u + 36 = 0$

$(u-4)(u-9) = 0$

$u - 4 = 0 \quad u - 9 = 0$

$u = 4 \quad u = 9$

Replace u with x^2.

$x^2 = 4 \quad x^2 = 9$

$\sqrt{x^2} = \sqrt{4} \quad \sqrt{x^2} = \sqrt{9}$

$x = \pm 2 \quad x = \pm 3$

The solutions are −2, 2, −3 and 3.

7. $z^4 - 6z^2 + 8 = 0$

$(z^2)^2 - 6(z^2) + 8 = 0$

$u^2 - 6u + 8 = 0$

$(u-4)(u-2) = 0$

$u - 4 = 0 \quad u - 2 = 0$

$u = 4 \quad u = 2$

Replace u with z^2.

$z^2 = 4 \quad z^2 = 2$

$\sqrt{z^2} = \sqrt{4} \quad \sqrt{z^2} = \sqrt{2}$

$z = \pm 2 \quad z = \pm\sqrt{2}$

The solutions are $-2, 2, -\sqrt{2}$ and $\sqrt{2}$.

9. $p - 3p^{1/2} + 2 = 0$

$(p^{1/2})^2 - 3(p^{1/2}) + 2 = 0$

$u^2 - 3u + 2 = 0$

$(u-2)(u-1) = 0$

$u - 2 = 0 \quad u - 1 = 0$

$u = 2 \quad u = 1$

Replace u with $p^{1/2}$.

$p^{1/2} = 2 \quad p^{1/2} = 1$

$\left(p^{1/2}\right)^2 = 2^2 \quad \left(p^{1/2}\right)^2 = 1^2$

$p = 4 \quad p = 1$

The solutions are 1 and 4.

11. $x - x^{1/2} - 12 = 0$

$(x^{1/2})^2 - (x^{1/2}) - 12 = 0$

$u^2 - u - 12 = 0$

$(u-4)(u+3) = 0$

$u - 4 = 0 \quad u + 3 = 0$

$u = 4 \quad u = -3$

Replace u with $x^{1/2}$.

$x^{1/2} = 4 \quad x^{1/2} = -3$

$\left(x^{1/2}\right)^2 = 4^2 \quad \left(x^{1/2}\right)^2 = (-3)^2$

$x = 16 \quad x = 9$

9 does not check as a solution. The solution is 16.

13. $z^4 + 3z^2 - 4 = 0$

$(z^2)^2 + 3(z^2) - 4 = 0$

$u^2 + 3u - 4 = 0$

$(u+4)(u-1) = 0$

$u + 4 = 0 \quad u - 1 = 0$

$u = -4 \quad u = 1$

Replace u with z^2.

$z^2 = -4 \quad z^2 = 1$

$\sqrt{z^2} = \sqrt{-4} \quad \sqrt{z^2} = \sqrt{1}$

$z = \pm 2i \quad z = \pm 1$

The solutions are $-1, 1, -2i$ and $2i$.

15. $x^4 + 12x^2 - 64 = 0$

$(x^2)^2 + 12(x^2) - 64 = 0$

$u^2 + 12u - 64 = 0$

$(u+16)(u-4) = 0$

$u + 16 = 0 \quad u - 4 = 0$

$u = -16 \quad u = 4$

Replace u with x^2.

$x^2 = -16 \quad x^2 = 4$

$\sqrt{x^2} = \sqrt{-16} \quad \sqrt{x^2} = \sqrt{4}$

$x = \pm 4i \quad x = \pm 2$

The solutions are $-2, 2, -4i$ and $4i$.

17. $p+2p^{1/2}-24=0$

$(p^{1/2})^2+2(p^{1/2})-24=0$

$u^2+2u-24=0$

$(u+6)(u-4)=0$

$u+6=0 \quad u-4=0$

$u=-6 \qquad u=4$

Replace u with $p^{1/2}$.

$p^{1/2}=-6 \qquad p^{1/2}=4$

$\left(p^{1/2}\right)^2=(-6)^2 \qquad \left(p^{1/2}\right)^2=4^2$

$p=36 \qquad p=16$

36 does not check as a solution. The solution is 16.

19. $y^{2/3}-9y^{1/3}+8=0$

$(y^{1/3})^2-9(y^{1/3})+8=0$

$u^2-9u+8=0$

$(u-8)(u-1)=0$

$u-8=0 \quad u-1=0$

$u=8 \qquad u=1$

Replace u with $y^{1/3}$.

$y^{1/3}=8 \qquad y^{1/3}=1$

$(y^{1/3})^3=8^3 \qquad (y^{1/3})^3=1^3$

$y=512 \qquad y=1$

The solutions are 1 and 512.

21. $9w^4-13w^2+4=0$

$9(w^2)^2-13(w^2)+4=0$

$9u^2-13u+4=0$

$(9u-4)(u-1)=0$

$9u-4=0 \quad u-1=0$

$9u=4 \qquad u=1$

$u=\frac{4}{9}$

Replace u with w^2.

$w^2=\frac{4}{9} \qquad w^2=1$

$\sqrt{w^2}=\sqrt{\frac{4}{9}} \qquad \sqrt{w^2}=\sqrt{1}$

$w=\pm\frac{2}{3} \qquad w=\pm 1$

The solutions are -1, 1, $-\frac{2}{3}$ and $\frac{2}{3}$.

Objective B Exercises

23. Exercises 28, 29, 30, 34, 35, 36, 38, 39, 42

25. $\sqrt{x+1}+x=5$

$\sqrt{x+1}=5-x$

$\left(\sqrt{x+1}\right)^2=(5-x)^2$

$x+1=25-10x+x^2$

$0=x^2-11x+24$

$(x-3)(x-8)=0$

$x-3=0 \quad x-8=0$

$x=3 \qquad x=8$

8 does not check as a solution.
The solution is 3.

27. $x=\sqrt{x}+6$

$x-6=\sqrt{x}$

$(x-6)^2=(\sqrt{x})^2$

$x^2-12x+36=x$

$x^2-13x+36=0$

$(x-9)(x-4)=0$

$x-9=0 \quad x-4=0$

$x=9 \qquad x=4$

4 does not check as a solution.
The solution is 9.

29. $\sqrt{3w+3} = w+1$

$\left(\sqrt{3w+3}\right)^2 = (w+1)^2$

$3w+3 = w^2 + 2w + 1$

$0 = w^2 - w - 2$

$(w-2)(w+1) = 0$

$w-2=0 \quad w+1=0$

$w = 2 \qquad w = -1$

The solutions are −1 and 2.

31. $\sqrt{4y+1} - y = 1$

$\sqrt{4y+1} = 1 + y$

$\left(\sqrt{4y+1}\right)^2 = (1+y)^2$

$4y+1 = 1 + 2y + y^2$

$0 = y^2 - 2y$

$y(y-2) = 0$

$y = 0 \quad y - 2 = 0$

$y = 2$

The solutions are 0 and 2.

33. $\sqrt{10x+5} - 2x = 1$

$\sqrt{10x+5} = 1 + 2x$

$\left(\sqrt{10x+5}\right)^2 = (1+2x)^2$

$10x + 5 = 1 + 4x + 4x^2$

$0 = 4x^2 - 6x - 4$

$2(2x+1)(x-2) = 0$

$2x+1=0 \quad x-2=0$

$2x = -1 \qquad x = 2$

$x = -\frac{1}{2}$

The solutions are $-\frac{1}{2}$ and 2.

35. $\sqrt{p+11} = 1 - p$

$\left(\sqrt{p+11}\right)^2 = (1-p)^2$

$p + 11 = 1 - 2p + p^2$

$0 = p^2 - 3p - 10$

$(p-5)(p+2) = 0$

$p-5=0 \quad p+2=0$

$p = 5 \qquad p = -2$

5 does not check as a solution.
The solution is −2.

37. $\sqrt{x-1} - \sqrt{x} = -1$

$\sqrt{x-1} = \sqrt{x} - 1$

$\left(\sqrt{x-1}\right)^2 = (\sqrt{x} - 1)^2$

$x - 1 = x - 2\sqrt{x} + 1$

$2\sqrt{x} = 2$

$\sqrt{x} = 1$

$\left(\sqrt{x}\right)^2 = 1^2$

$x = 1$

The solution is 1.

39. $\sqrt{2x-1} = 1 - \sqrt{x-1}$

$\left(\sqrt{2x-1}\right)^2 = (1 - \sqrt{x-1})^2$

$2x - 1 = 1 - 2\sqrt{x-1} + x - 1$

$2\sqrt{x-1} = -x + 1$

$\left(2\sqrt{x-1}\right)^2 = (-x+1)^2$

$4(x-1) = x^2 - 2x + 1$

$4x - 4 = x^2 - 2x + 1$

$0 = x^2 - 6x + 5$

$(x-5)(x-1) = 0$

$x-5=0 \quad x-1=0$

$x = 5 \qquad x = 1$

5 does not check as a solution.
The solution is 1.

41. $\sqrt{t+3}+\sqrt{2t+7}=1$

$\sqrt{t+3}=1-\sqrt{2t+7}$

$\left(\sqrt{t+3}\right)^2=(1-\sqrt{2t+7})^2$

$t+3=1-2\sqrt{2t+7}+2t+7$

$2\sqrt{2t+7}=t+5$

$\left(2\sqrt{2t+7}\right)^2=(t+5)^2$

$4(2t+7)=t^2+10t+25$

$8t+28=t^2+10t+25$

$0=t^2+2t-3$

$(t+3)(t-1)=0$

$t+3=0 \quad t-1=0$

$t=-3 \quad t=1$

1 does not check as a solution.
The solution is −3.

Objective C Exercises

43. $y+2$

45. $x=\dfrac{10}{x-9}$

$(x-9)x=\left(\dfrac{10}{x-9}\right)(x-9)$

$x^2-9x=10$

$x^2-9x-10=0$

$(x-10)(x+1)=0$

$x-10=0 \quad x+1=0$

$x=10 \quad x=-1$

The solutions are −1 and 10.

47. $\dfrac{y-1}{y+2}+y=1$

$(y+2)\left(\dfrac{y-1}{y+2}+y\right)=1(y+2)$

$y-1+y(y+2)=y+2$

$y-1+y^2+2y=y+2$

$y^2+2y-3=0$

$(y+3)(y-1)=0$

$y+3=0 \quad y-1=0$

$y=-3 \quad y=1$

The solutions are −3 and 1.

49. $\dfrac{3r+2}{r+2}-2r=1$

$(r+2)\left(\dfrac{3r+2}{r+2}-2r\right)=1(r+2)$

$3r+2-2r(r+2)=r+2$

$3r+2-2r^2-4r=r+2$

$-2r^2-2r=0$

$-2r(r+1)=0$

$-2r=0 \quad r+1=0$

$r=0 \quad r=-1$

The solutions are −1 and 0.

51. $\dfrac{2}{2x+1}+\dfrac{1}{x}=3$

$x(2x+1)\left(\dfrac{2}{2x+1}+\dfrac{1}{x}\right)=3x(2x+1)$

$2x+2x+1=6x^2+3x$

$0=6x^2-x-1$

$(2x-1)(3x+1)=0$

$2x-1=0 \quad 3x+1=0$

$2x=1 \quad 3x=-1$

$x=\dfrac{1}{2} \quad x=-\dfrac{1}{3}$

The solutions are $-\dfrac{1}{3}$ and $\dfrac{1}{2}$.

53. $\dfrac{16}{z-2}+\dfrac{16}{z+2}=6$

$(z-2)(z+2)\left(\dfrac{16}{z-2}+\dfrac{16}{z+2}\right)=6(z-2)(z+2)$

$16(z+2)+16(z-2)=6(z^2-4)$

$16z+32+16z-32=6z^2-24$

$0=6z^2-32z-24$

$2(3z^2-16z-12)=0$

$2(3z+2)(z-6)=0$

$3z+2=0 \quad z-6=0$

$3z=-2 \qquad z=6$

$z=-\dfrac{2}{3}$

The solutions are $-\dfrac{2}{3}$ and 6.

55. $\dfrac{t}{t-2}+\dfrac{2}{t-1}=4$

$(t-2)(t-1)\left(\dfrac{t}{t-2}+\dfrac{2}{t-1}\right)=4(t-2)(t-1)$

$t(t-1)+2(t-2)=4(t^2-3t+2)$

$t^2-t+2t-4=4t^2-12t+8$

$0=3t^2-13t+12$

$(3t-4)(t-3)=0$

$3t-4=0 \quad t-3=0$

$3t=4 \qquad t=3$

$t=\dfrac{4}{3}$

The solutions are $\dfrac{4}{3}$ and 3.

Applying the Concepts

57. $(x^2-7)^{1/2}=(x-1)^{1/2}$

$\left((x^2-7)^{1/2}\right)^2=\left((x-1)^{1/2}\right)^2$

$x^2-7=x-1$

$x^2-x-6=0$

$(x-3)(x+2)=0$

$x-3=0 \quad x+2=0$

$x=3 \qquad x=-2$

The solutions are −2 and 3.

59. $\left(\sqrt{x}+3\right)^2-4\sqrt{x}-17=0$

Let $u=\sqrt{x}+3$.

$u^2-4(u-3)-17=0$

$u^2-4u+12-17=0$

$u^2-4u-5=0$

$(u-5)(u+1)=0$

$u-5=0 \quad u+1=0$

$u=5 \qquad u=-1$

Replace u with $\sqrt{x}+3$.

$\sqrt{x}+3=5 \quad \sqrt{x}+3=-1$

$\sqrt{x}=2 \qquad \sqrt{x}=-4$

$\left(\sqrt{x}\right)^2=2^2 \quad \left(\sqrt{x}\right)^2=(-4)^2$

$x=4 \qquad x=16$

The solution is 4.

Section 10.5

Objective A Exercises

1. It must be true that $x-3>0$ and $x-5>0$ or the $x-3<0$ and $x-5<0$. In other words, either both factors are positive or both factors are negative.

3. a) $x=2$ True
b) $x=-2$ False
c) $x=-3$ False

5. a) $x<6$ False
b) $-5\le x\le 6$ True

7. $(x-4)(x+2)>0$

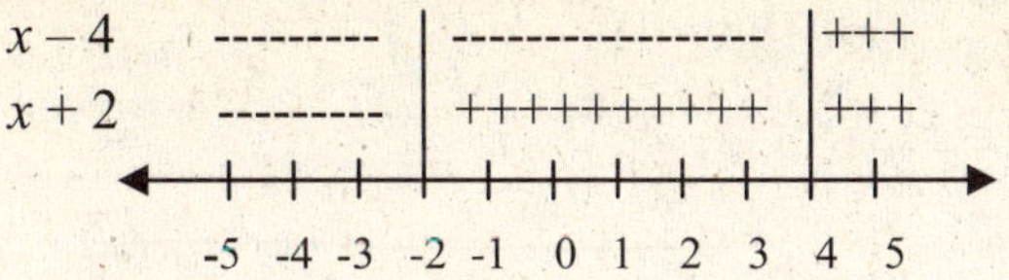

$\{x \mid x<-2 \text{ or } x>4\}$

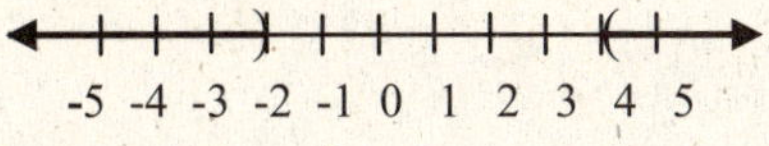

9. $x^2-3x+2\geq 0$

$(x-2)(x-1)\geq 0$

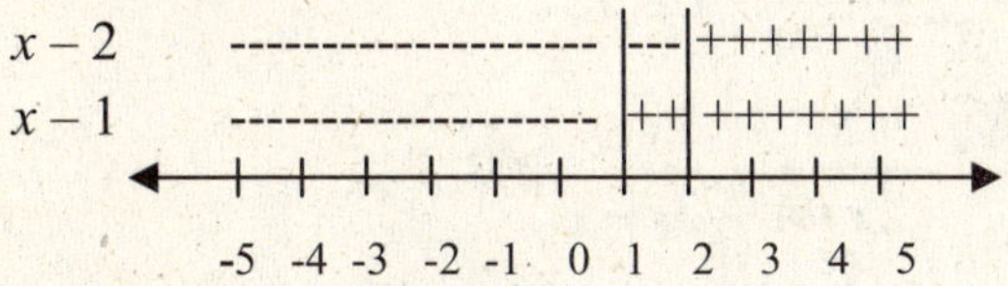

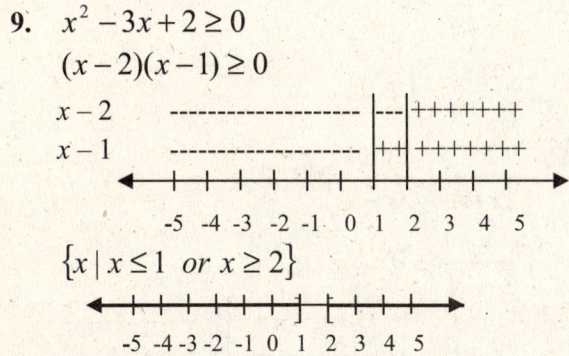

$\{x \mid x\leq 1 \text{ or } x\geq 2\}$

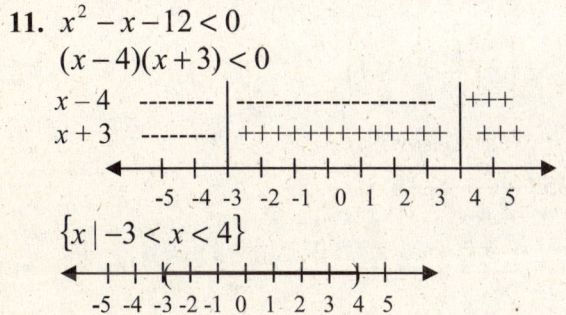

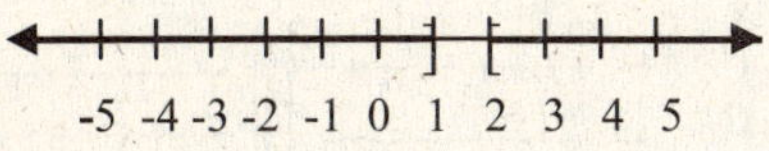

11. $x^2-x-12<0$

$(x-4)(x+3)<0$

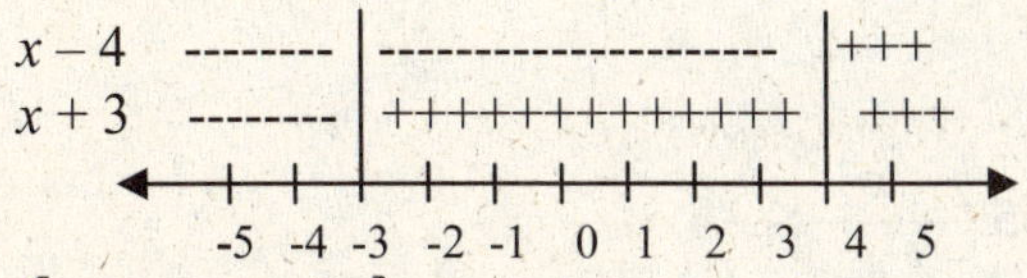

$\{x \mid -3<x<4\}$

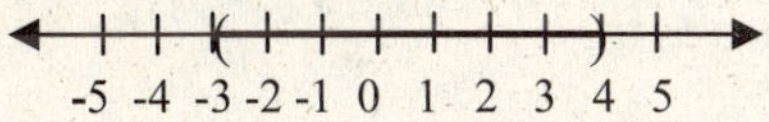

13. $(x-1)(x+2)(x-3)<0$

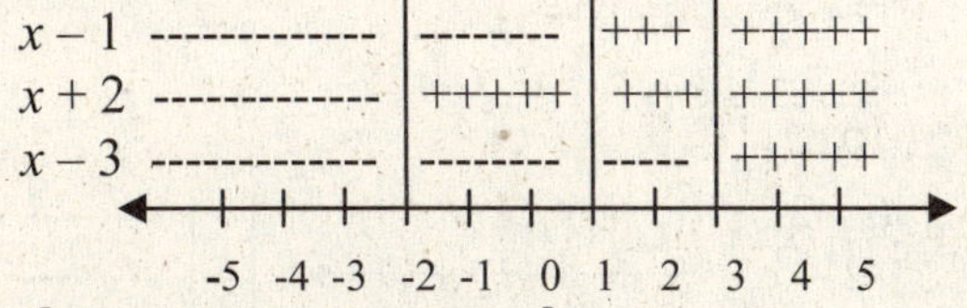

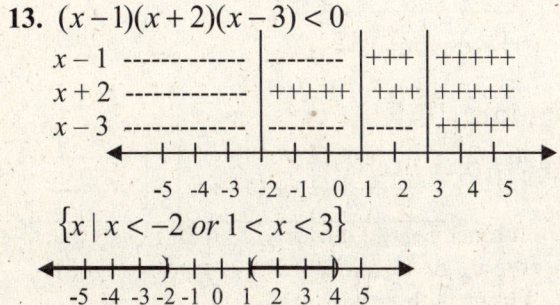

$\{x \mid x<-2 \text{ or } 1<x<3\}$

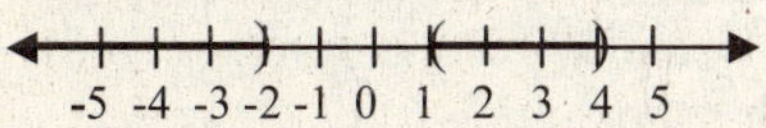

15. $(x+4)(x-2)(x-1)\geq 0$

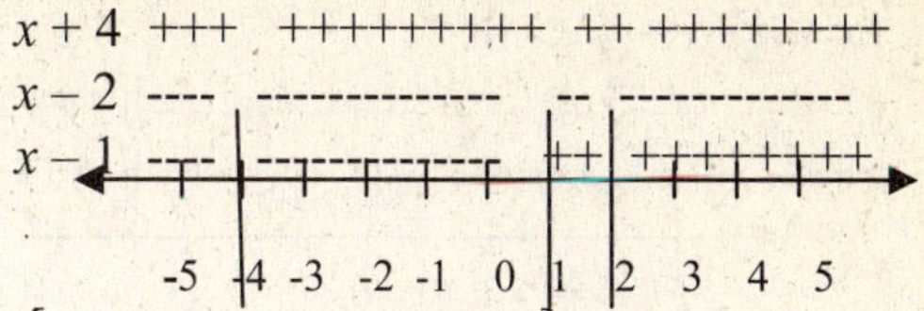

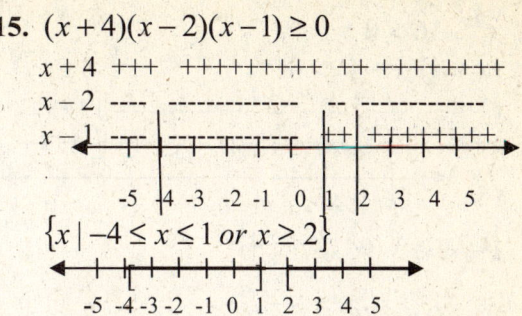

$\{x \mid -4\leq x\leq 1 \text{ or } x\geq 2\}$

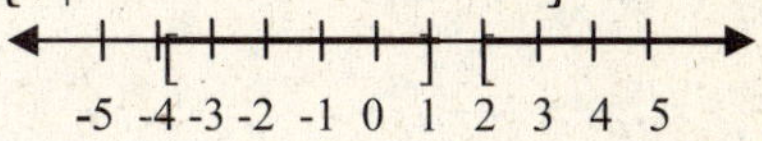

17. $\dfrac{x-4}{x+2}>0$

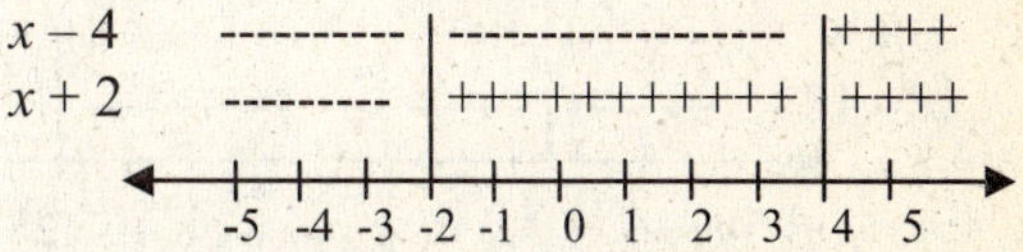

$\{x \mid x<-2 \text{ or } x>4\}$

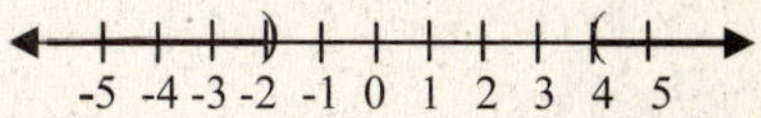

19. $\dfrac{x-3}{x+1}\leq 0$

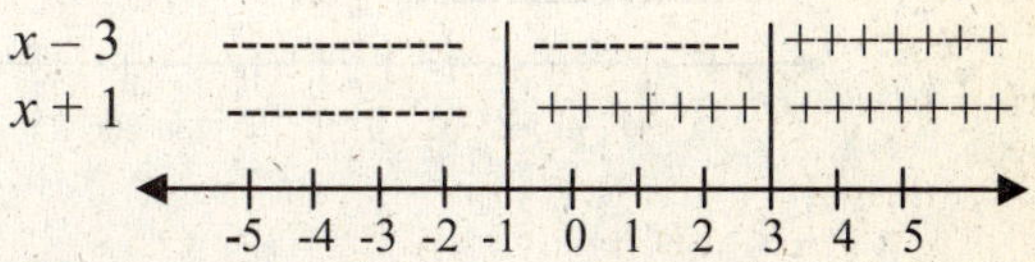

$\{x \mid -1<x\leq 3\}$

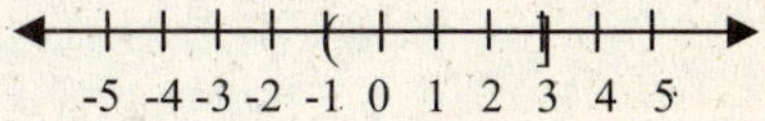

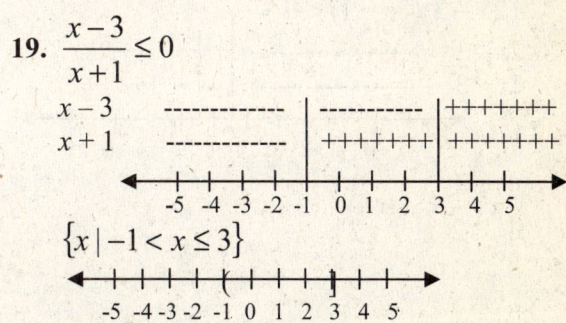

21. $\dfrac{(x-1)(x+2)}{x-3}\leq 0$

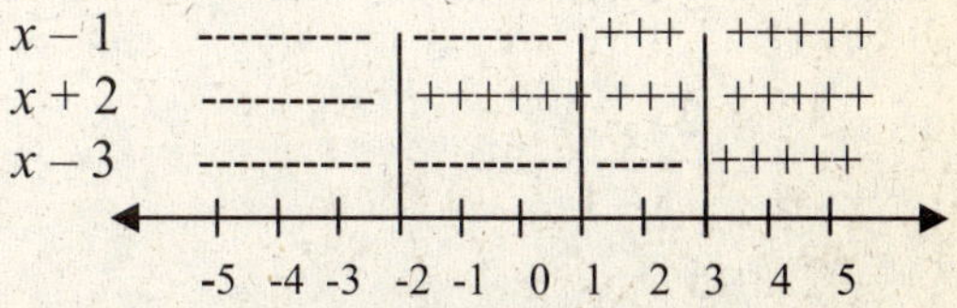

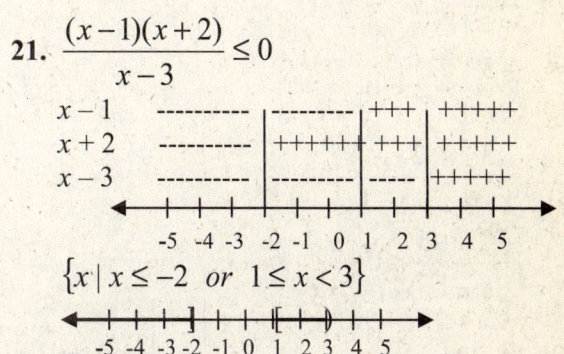

$\{x \mid x\leq -2 \text{ or } 1\leq x<3\}$

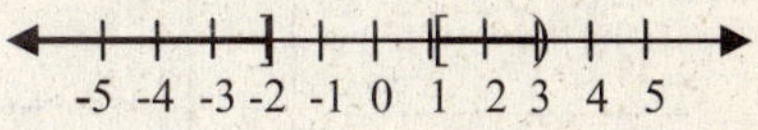

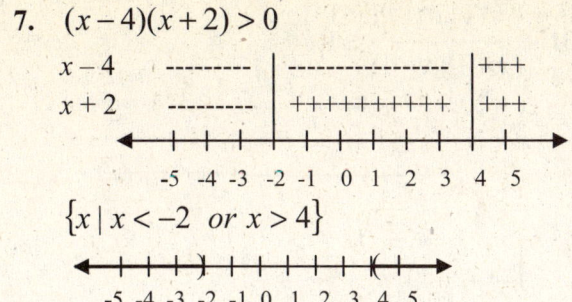

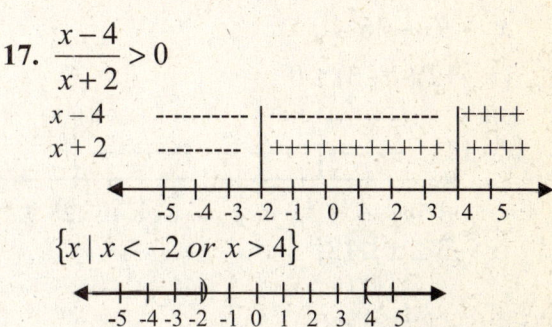

23. $x^2 - 16 > 0$

$(x+4)(x-4) > 0$

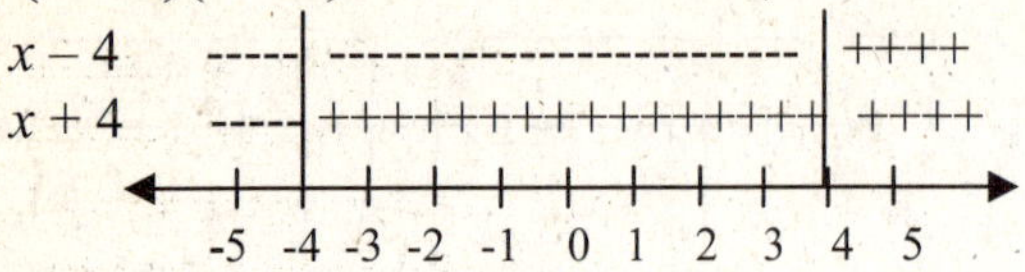

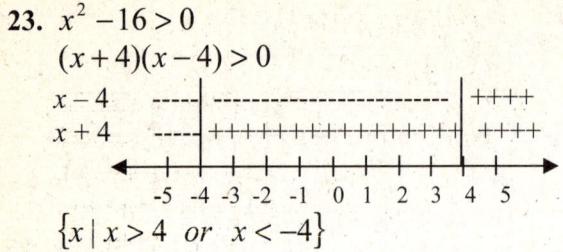

$\{x \mid x > 4 \text{ or } x < -4\}$

25. $x^2 - 9x \le 36$

$x^2 - 9x - 36 \le 0$

$(x-12)(x+3) \le 0$

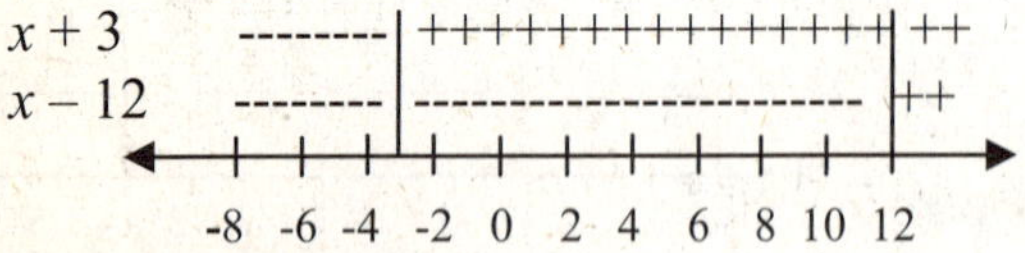

$\{x \mid -3 \le x \le 12\}$

27. $4x^2 - 8x + 3 < 0$

$(2x-3)(2x-1) < 0$

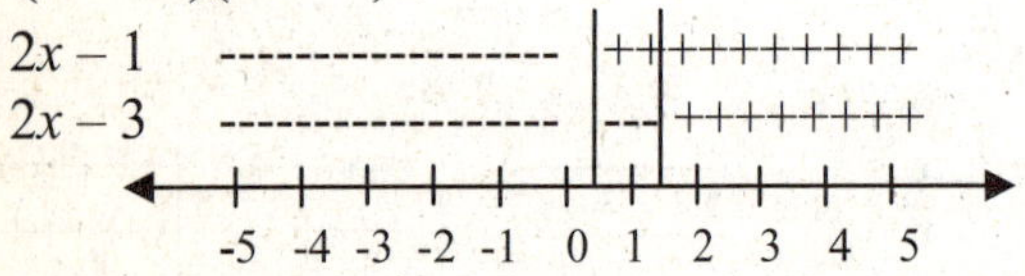

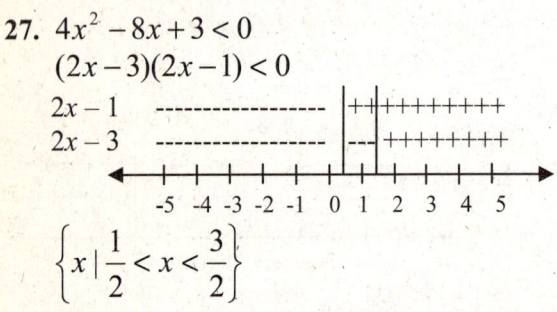

$\left\{x \mid \frac{1}{2} < x < \frac{3}{2}\right\}$

29. $\frac{3}{x-1} < 2$

$\frac{3}{x-1} - 2 < 0$

$\frac{3}{x-1} - \frac{2x-2}{x-1} < 0$

$\frac{-2x+5}{x-1} < 0$

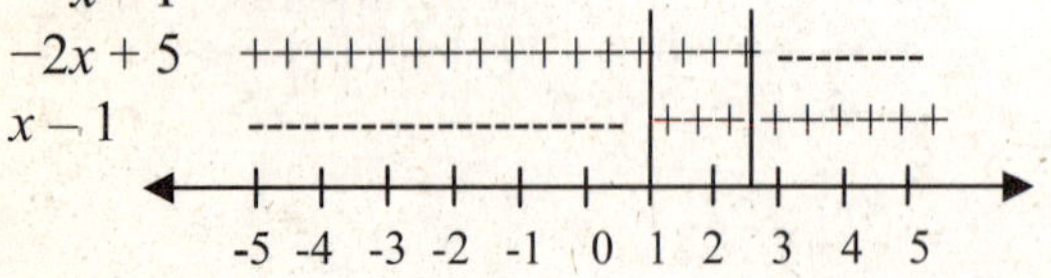

$\left\{x \mid x < 1 \text{ or } x > \frac{5}{2}\right\}$

31. $\frac{x-2}{(x+1)(x-1)} \le 0$

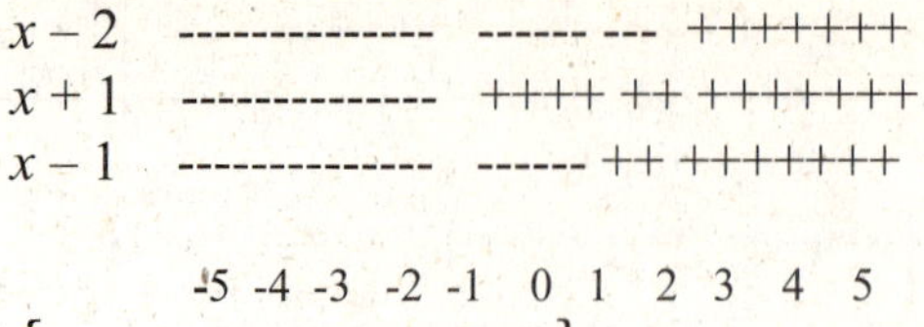

$\{x \mid x < -1 \text{ or } 1 < x \le 2\}$

33. $\frac{x}{2x-1} \ge 1$

$\frac{x}{2x-1} - 1 \ge 0$

$\frac{x}{2x-1} - \frac{2x-1}{2x-1} \ge 0$

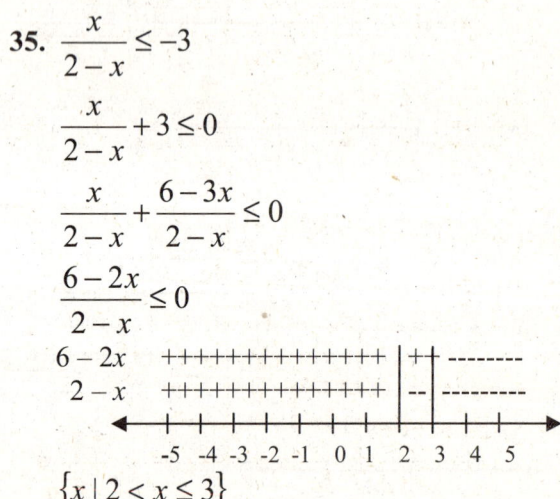

$\frac{-x+1}{2x-1} \ge 0$

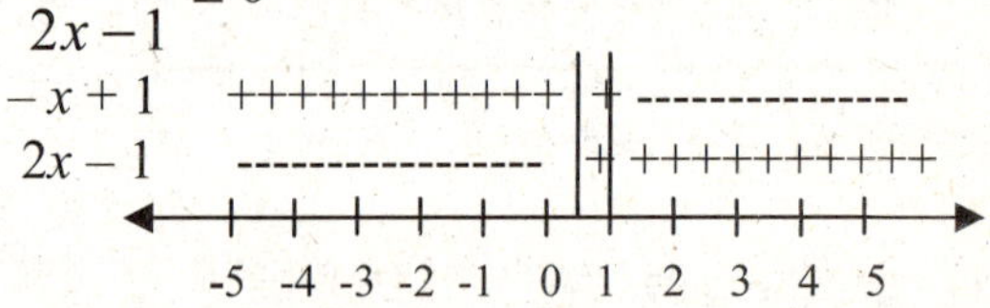

$\left\{x \mid \frac{1}{2} < x \le 1\right\}$

35. $\frac{x}{2-x} \le -3$

$\frac{x}{2-x} + 3 \le 0$

$\frac{x}{2-x} + \frac{6-3x}{2-x} \le 0$

$\frac{6-2x}{2-x} \le 0$

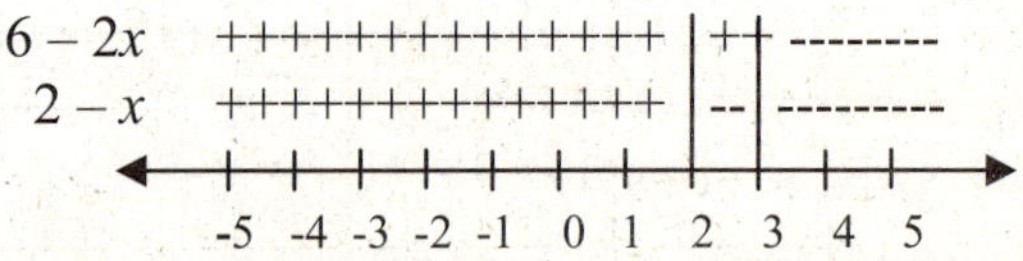

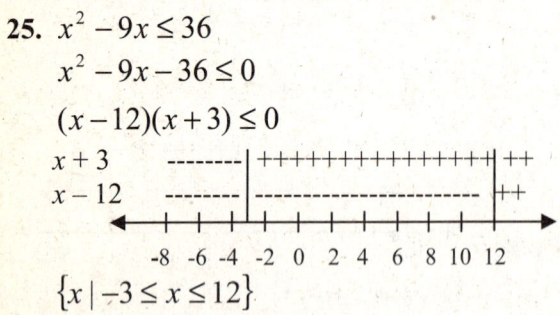

$\{x \mid 2 < x \le 3\}$

37. $\dfrac{3}{x-5} > \dfrac{1}{x+1}$

$\dfrac{3}{x-5} - \dfrac{1}{x+1} > 0$

$\dfrac{3(x+1)}{(x-5)(x+1)} - \dfrac{(x-5)}{(x-5)(x+1)} > 0$

$\dfrac{3x+3-x+5}{(x-5)(x+1)} > 0$

$\dfrac{2x+8}{(x-5)(x+1)} > 0$

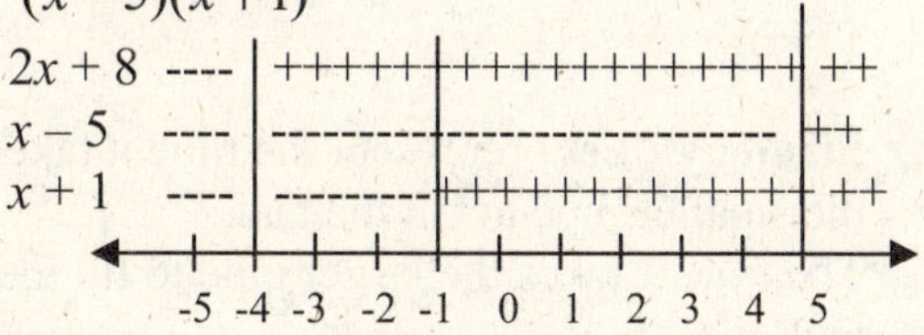

$\{x \mid x > 5 \text{ or } -4 < x < -1\}$

Applying the Concepts

39. $(x-1)(x+3)(x-2)(x-4) \ge 0$

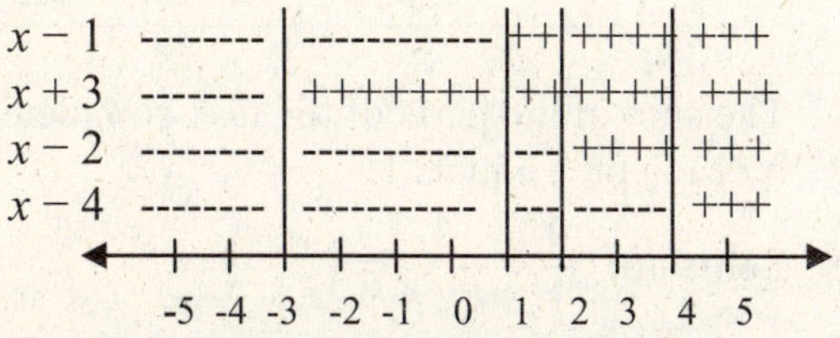

$\{x \mid x \le -3 \text{ or } 1 \le x \le 2 \text{ or } x \ge 4\}$

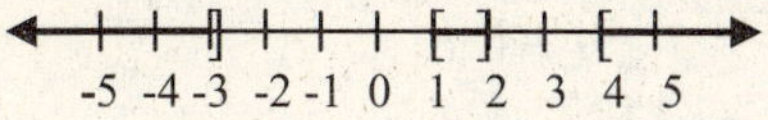

41. $(x^2+2x-3)(x^2+3x+2) \ge 0$

$(x-1)(x+3)(x+1)(x+2) \ge 0$

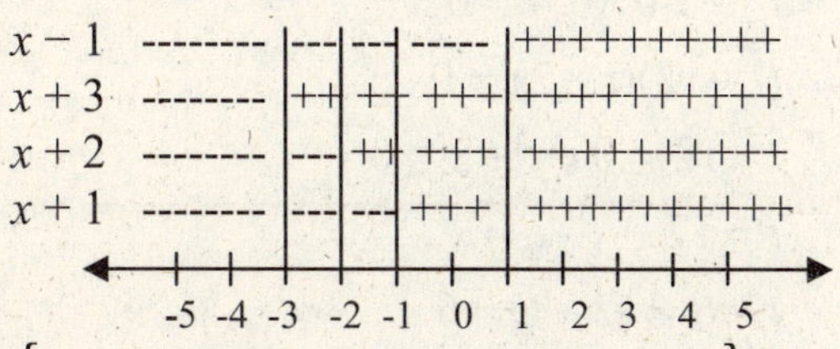

$\{x \mid x \le -3 \text{ or } -2 \le x \le -1 \text{ or } x \ge 1\}$

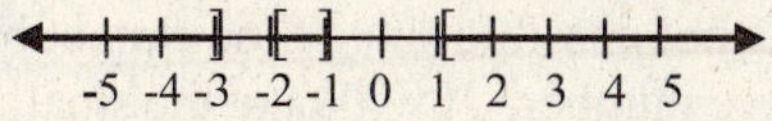

43. $\dfrac{x^2(3-x)(2x+1)}{(x+4)(x+2)} \ge 0$

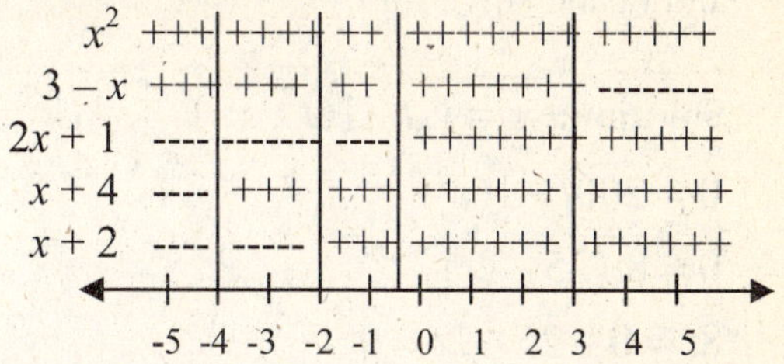

$\left\{x \mid -4 < x < -2 \text{ or } -\dfrac{1}{2} \le x \le 3\right\}$

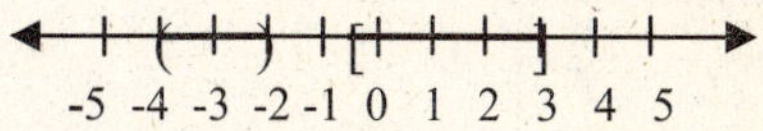

Section 10.6

Objective A Exercises

1. **Strategy:** To find the maximum safe speed, substitute for *d* and solve for *v*.

Solution: $d = 0.04v^2 + 0.5v$

$60 = 0.04v^2 + 0.5v$

$0 = 0.04v^2 + 0.5v - 60$

$v = \dfrac{-b \pm \sqrt{b^2 - 4ac}}{2a}$

$v = \dfrac{-0.5 \pm \sqrt{(0.5)^2 - 4(0.04)(-60)}}{2(0.04)}$

$v = \dfrac{-0.5 \pm \sqrt{9.85}}{0.08}$

$v = 33$

$v = -45$

Since the speed cannot be a negative number the maximum speed is 33 mph.

3. **Strategy:** To find the time it takes for the projectile to return to earth, substitute for s and v_0 and solve for t.

Solution: $s = v_0 t - 16t^2$

$0 = 200t - 16t^2$

$0 = 8t(25 - 2t)$

$8t = 0 \quad 25 - 2t = 0$

$t = 0 \qquad t = 12.5$

The projectile will take 12.5 s to return to earth.

5. a) **Strategy:** The maximum height occurs in the middle of the arch (when $x = 0$).

Solution: $h(x) = -\frac{3}{64}x^2 + 27$

$h(0) = -\frac{3}{64}(0)^2 + 27 = 27$

The maximum height of the arch is 27 ft.

b) **Strategy:** Let $x = 8$ in the function $h(x)$.

Solution: $h(x) = -\frac{3}{64}x^2 + 27$

$h(8) = -\frac{3}{64}(8)^2 + 27 = 24$

The height of the arch 8 ft to the right of the center is 24 ft.

c) **Strategy:** Let $h = 8$ in the height equation.

Solution: $h(x) = -\frac{3}{64}x^2 + 27$

$8 = -\frac{3}{64}x^2 + 27$

$-19 = -\frac{3}{64}x^2$

$x^2 = \frac{1216}{3}$

$x = \pm 20.13$

Since height cannot be a negative number the arch is 8 ft tall approximately 20.13 ft from the center.

7. **Strategy**: Substitute the given value for H and V and solve for the length of the side. In a square base the $L = W$.

Solution: $V = LWH$

Let x represent the length of the side of the square base. $V = x^2 H$

$971{,}199 = x^2(31)$

$x^2 = 31329$

$x = 177$

The length of a side of the square base is 177 m.

9. **Strategy:** Let t represent the time it takes the smaller pipe to fill the tank.
The time it takes the larger pipe to fill the tank is $t - 6$.

	Rate	Time	Part
Smaller pipe	$\frac{1}{t}$	4	$\frac{4}{t}$
Larger pipe	$\frac{1}{t-6}$	4	$\frac{4}{t-6}$

The sum of the parts of the task completed by each pipe equals 1.

Solution: $\frac{4}{t} + \frac{4}{t-6} = 1$

$t(t-6)\left(\frac{4}{t} + \frac{4}{t-6}\right) = 1t(t-6)$

$4(t-6) + 4t = t^2 - 6t$

$4t - 24 + 4t = t^2 - 6t$

$t^2 - 14t + 24 = 0$

$(t-12)(t-2) = 0$

$t - 12 = 0 \quad t - 2 = 0$

$t = 12 \qquad t = 2$

$t - 6 = 12 - 6 = 6$

$t - 6 = 2 - 6 = -4$

$t = 2$ is not possible since time cannot be a negative number. It will take the smaller pipe 12 min and the larger pipe 6 min.

11. Strategy: Let t represent the time it takes the faster computer working alone.

	Rate	Time	Part
Slower computer	$\frac{1}{t+4}$	3	$\frac{3}{t+4}$
Faster computer	$\frac{1}{t}$	1	$\frac{1}{t}$

The sum of the parts of the task completed by each computer equals 1.

Solution: $\frac{1}{t}+\frac{3}{t+4}=1$

$$t(t+4)\left(\frac{1}{t}+\frac{3}{t+4}\right)=1t(t+4)$$

$$t+4+3t=t^2+4t$$

$$4t+4=t^2+4t$$

$$t^2-4=0$$

$$(t+2)(t-2)=0$$

$$t+2=0 \quad t-2=0$$

$$t=-2 \qquad t=2$$

Time cannot be a negative number. It will take the faster computer 2 h working alone.

13. Strategy: Let t represent the time it takes the experienced carpenter.
The time it takes the apprentice is $t+2$.

	Rate	Time	Part
Experienced carpenter	$\frac{1}{t}$	2	$\frac{2}{t}$
Apprentice carpenter	$\frac{1}{t+2}$	4	$\frac{4}{t+2}$

The sum of the parts of the task completed by each computer equals 1.

Solution: $\frac{2}{t}+\frac{4}{t+2}=1$

$$t(t+2)\left(\frac{2}{t}+\frac{4}{t+2}\right)=1t(t+2)$$

$$2(t+2)+4t=t^2+2t$$

$$2t+4+4t=t^2+2t$$

$$t^2-4t-4=0$$

$$t=\frac{-b\pm\sqrt{b^2-4ac}}{2a}$$

$$t=\frac{-(-4)\pm\sqrt{(-4)^2-4(1)(-4)}}{2(1)}$$

$$t=\frac{4\pm\sqrt{32}}{2}=2\pm2\sqrt{2}$$

$$t=4.8$$

$$t=-0.8$$

Time cannot be a negative number. It will take the apprentice carpenter $t+2=6.8$ h working alone.

15. Strategy: Let r represent the rate of the wind.

	Distance	Rate	Time
With wind	4000	$1320+r$	$\frac{4000}{1320+r}$
Against wind	4000	$1320-r$	$\frac{4000}{1320-r}$

It took 0.5 h less time to make the return trip.

Solution: $\frac{4000}{1320-r}-\frac{4000}{1320+r}=0.5$

$$(1320-r)(1320+r)\left(\frac{4000}{1320-r}-\frac{4000}{1320+r}\right)=0.5(1320-r)(1320+r)$$

$$4000(1320+r)-4000(1320-r)=0.5(1{,}742{,}400-r^2)$$

$5{,}280{,}000 + 4000r - 5{,}280{,}000 + 4000r = 871{,}200 - 0.5r^2$

$8000r = 871{,}200 - 0.5r^2$

$0.5r^2 + 8000r - 871{,}200 = 0$

$$r = \frac{-b \pm \sqrt{b^2 - 4ac}}{2a}$$

$$r = \frac{-(8000) \pm \sqrt{(8000)^2 - 4(0.5)(-871{,}200)}}{2(0.5)}$$

$$r = \frac{-8000 \pm \sqrt{65{,}742{,}400}}{1}$$

$r = 108$

$r = -16{,}108$

Since the rate cannot be a negative number. The rate of the wind was approximately 108 mph.

17. Strategy: Let r represent the rate of the jet stream.

	Distance	Rate	Time
With jet stream	3660	$630 + r$	$\frac{3660}{630+r}$
Against jet stream	3660	$630 - r$	$\frac{3660}{630-r}$

It took 1.75 h less time to make the trip flying with the jet stream.

Solution: $\dfrac{3660}{630-r} - \dfrac{3660}{630+r} = 1.75$

$$(630+r)(630-r)\left(\frac{3660}{630-r} - \frac{3660}{630+r}\right) = 1.75(630+r)(630-r)$$

$3660(630+r) - 3660(630-r) = 1.75(396{,}900 - r^2)$

$2{,}305{,}800 + 3660r - 2{,}305{,}800 + 3660r = 694{,}575 - 1.75r^2$

$7320r = 694{,}575 - 1.75r^2$

$1.75r^2 + 7320r - 694{,}575 = 0$

$$r = \frac{-b \pm \sqrt{b^2 - 4ac}}{2a}$$

$$r = \frac{-(7320) \pm \sqrt{(7320)^2 - 4(1.75)(-694{,}575)}}{2(1.75)}$$

$$r = \frac{-7320 \pm \sqrt{58{,}444{,}425}}{3.5}$$

$r = 93$

$r = -4276$

The rate cannot be a negative number. The rate of the jet stream is 93 mph.

19. Strategy: Let x represent the width of the rectangle.
The length of the rectangle is $x + 111$.
The area of the rectangle is 104,000 mi^2.
Solution: $A = LW$
$104{,}000 = x(x+111)$
$104{,}000 = x^2 + 111x$
$0 = x^2 + 111x - 104{,}000$
$$x = \frac{-b \pm \sqrt{b^2 - 4ac}}{2a}$$
$$x = \frac{-111 \pm \sqrt{(111)^2 - 4(1)(-104{,}000)}}{2(1)}$$
$$x = \frac{-111 \pm \sqrt{428{,}321}}{2}$$
$x = 272$
$x = -383$
The width cannot be a negative number.
$x + 111 = 272 + 111 = 383$
The width is 272 mi.
The length is 383 mi.

21. Strategy: Let x represent the height of the triangle.
The base of the triangle is $5x - 1$.
The area of the triangle is 21 cm^2.
Solution: $A = \frac{1}{2}bh$
$21 = \frac{1}{2}(5x-1)(x)$
$42 = 5x^2 - x$
$0 = 5x^2 - x - 42$
$0 = (5x+14)(x-3)$
$5x + 14 = 0 \quad x - 3 = 0$
$5x = -14 \quad x = 3$
$x = -\frac{14}{5}$
Since the height cannot be negative, $-\frac{14}{5}$ is not a solution.
$5x - 1 = 5(3) - 1 = 14$
The height is 3 cm.
The base is 14 cm.

23. Strategy: Let x represent a side of the square base of the box.
The volume of the box is 49,000 cm^3.

Solution: $V = LWH$
$49{,}000 = (x)(x)(10)$
$49{,}000 = 10x^2$
$x^2 = 4900$
$x = 70$
The side of the original square is 20 cm more than side x.
$x + 20 = 70 + 20 = 90$
The dimensions of the original square base is 90 cm by 90 cm.

25. Strategy: Let x represent the width of the rectangle.
The length of the rectangle is $40 - x$.
The area of the rectangle is 300 ft^2.

Solution: $A = LW$
$300 = x(40 - x)$
$300 = 40x - x^2$
$x^2 - 40x + 300 = 0$
$(x-10)(x-30) = 0$
$x - 10 = 0 \quad x - 30 = 0$
$x = 10 \quad x = 30$
$40 - x = 40 - 10 = 30$
$40 - x = 40 - 30 = 10$
The dimensions of the rectangle are 10 ft by 30 ft.

Applying the Concepts

27. Strategy: To find the radius of the cone substitute 11.25π for A and 6 for s in the equation $A = \pi r^2 + \pi rs$ and solve for r.

Solution: $A = \pi r^2 + \pi rs$
$11.25\pi = \pi r^2 + \pi r(6)$
$0 = \pi r^2 + 6\pi r - 11.25\pi$
$a = \pi, b = 6\pi, c = -11.25\pi$

$$r = \frac{-b \pm \sqrt{b^2 - 4ac}}{2a}$$

$$r = \frac{-6\pi \pm \sqrt{(6\pi)^2 - 4(\pi)(-11.25\pi)}}{2(\pi)}$$

$$r = \frac{-6\pi \pm \sqrt{36\pi^2 + 45\pi^2}}{2\pi}$$

$$r = \frac{-6\pi \pm \sqrt{81\pi^2}}{2\pi}$$

$$r = \frac{-6\pi \pm 9\pi}{2\pi}$$

$r = 1.5$

$r = -7.5$

The radius cannot be a negative number.
The radius of the cone is 1.5 in.

Concept Review

1. Write the quadratic equation by using the equation $(x - r_1)(x - r_2) = 0$ where r_1 and r_2 are solutions to the equation.

2. The symbol ± means plus or minus.

3. $x^2 - 18x$

$$x^2 - 18x + \left(\frac{1}{2}(-18)\right)^2$$

$$= x^2 - 18x + 81$$

4. The approximate values of $4 + 3\sqrt{2}$ and $4 - 3\sqrt{2}$ can be found as follows:

$4 + 3\sqrt{2} = 4 + 3(1.4142) = 8.2426$

$4 - 3\sqrt{2} = 4 - 3(1.4142) = -0.2426$

5. To solve a quadratic by completing the square:
1) Write the equation in the form $ax^2 + bx = -c$.

2) If $a \neq 1$, multiply both sides of the equation by $\frac{1}{a}$.

3) Complete the square on $x^2 + \frac{b}{a}x$. Add the number that completes the square to both sides of the equation.
4) Factor the perfect square trinomial.
5) Take the square root of each side of the equation.
6) Solve the resulting equation for x.
7) Check the solutions.

6. The quadratic formula $x = \frac{-b \pm \sqrt{b^2 - 4ac}}{2a}$ gives the solutions for the quadratic equation.

7. The discriminant $b^2 - 4ac$ tells us if the quadratic equation has one real number solution, two unequal real number solutions or two complex number solutions.

8. It is important to check the solution to a radical equation because when both sides of an equation are squared the resulting equation may have a solution that is not a solution of the original equation.

9. A quadratic inequality is one that can be written in the form $a^2 + bx + c > 0$ or $a^2 + bx + c < 0$. The symbols $\geq$ or $\leq$ can also be used.

10. In a quadratic inequality a $>$ or $<$ symbol indicates that endpoints are not included in the solution set. The symbols $\geq$ or $\leq$ indicate that the endpoints are included in the solution set.

11. In a rational inequality, check that no element of the solution would result a denominator of zero when substituted for the variable in the original inequality.

Chapter 10 Review Exercises

1. $2x^2 - 3x = 0$

$x(2x-3) = 0$

$2x - 3 = 0 \quad x = 0$

$2x = 3$

$x = \dfrac{3}{2}$

The solutions are 0 and $\dfrac{3}{2}$.

2. $6x^2 + 9cx = 6c^2$

$6x^2 + 9cx - 6c^2 = 0$

$3(2x^2 + 3cx - 2c^2) = 0$

$3(2x - c)(x + 2c) = 0$

$2x - c = 0 \quad x + 2c = 0$

$2x = c \qquad x = -2c$

$x = \dfrac{c}{2}$

The solutions are $-2c$ and $\dfrac{c}{2}$.

3. $x^2 = 48$

$\sqrt{x^2} = \sqrt{48}$

$x = \pm\sqrt{48} = \pm 4\sqrt{3}$

The solutions are $4\sqrt{3}$ and $-4\sqrt{3}$.

4. $\left(x + \dfrac{1}{2}\right)^2 + 4 = 0$

$\left(x + \dfrac{1}{2}\right)^2 = -4$

$\sqrt{\left(x + \dfrac{1}{2}\right)^2} = \sqrt{-4}$

$x + \dfrac{1}{2} = \pm\sqrt{-4} = \pm 2i$

$x + \dfrac{1}{2} = 2i \quad x + \dfrac{1}{2} = -2i$

$x = 2i - \dfrac{1}{2} \quad x = -2i - \dfrac{1}{2}$

The solutions are $2i - \dfrac{1}{2}$ and $-2i - \dfrac{1}{2}$.

5. $x^2 + 4x + 3 = 0$

$x^2 + 4x = -3$

$x^2 + 4x + 4 = -3 + 4$

$(x+2)^2 = 1$

$\sqrt{(x+2)^2} = \sqrt{1}$

$x + 2 = \pm 1$

$x + 2 = 1 \quad x + 2 = -1$

$x = -1 \qquad x = -3$

The solutions are -1 and -3.

6. $7x^2 - 14x + 3 = 0$

$7x^2 - 14x = -3$

$\dfrac{1}{7}(7x^2 - 14x) = \dfrac{1}{7}(-3)$

$x^2 - 2x = -\dfrac{3}{7}$

$x^2 - 2x + 1 = -\dfrac{3}{7} + 1$

$(x-1)^2 = \dfrac{4}{7}$

$\sqrt{(x-1)^2} = \sqrt{\dfrac{4}{7}}$

$x - 1 = \pm\sqrt{\dfrac{4}{7}} = \pm\dfrac{2\sqrt{7}}{7}$

$x - 1 = \dfrac{2\sqrt{7}}{7} \quad x - 1 = -\dfrac{2\sqrt{7}}{7}$

$x = 1 + \dfrac{2\sqrt{7}}{7} \quad x = 1 - \dfrac{2\sqrt{7}}{7}$

The solutions are $\dfrac{7 + 2\sqrt{7}}{7}$ and $\dfrac{7 - 2\sqrt{7}}{7}$.

7. $12x^2 - 25x + 12 = 0$

$a = 12, b = -25, c = 12$

$$x = \frac{-b \pm \sqrt{b^2 - 4ac}}{2a}$$

$$x = \frac{-(-25) \pm \sqrt{(-25)^2 - 4(12)(12)}}{2(12)}$$

$$x = \frac{25 \pm \sqrt{625 - 576}}{24}$$

$$x = \frac{25 \pm \sqrt{49}}{24}$$

$$x = \frac{25 \pm 7}{24}$$

$$x = \frac{25+7}{24} = \frac{32}{24} = \frac{4}{3}$$

$$x = \frac{25-7}{24} = \frac{18}{24} = \frac{3}{4}$$

The solutions are $\frac{4}{3}$ and $\frac{3}{4}$.

8. $x^2 - x + 8 = 0$

$a = 1, b = -1, c = 8$

$$x = \frac{-b \pm \sqrt{b^2 - 4ac}}{2a}$$

$$x = \frac{-(-1) \pm \sqrt{(-1)^2 - 4(1)(8)}}{2(1)}$$

$$x = \frac{1 \pm \sqrt{1 - 32}}{2}$$

$$x = \frac{1 \pm \sqrt{-31}}{2}$$

$$x = \frac{1 \pm i\sqrt{31}}{2}$$

The solutions are $\frac{1 + i\sqrt{31}}{2}$ and $\frac{1 - i\sqrt{31}}{2}$.

9. $(x - r_1)(x - r_2) = 0$

$(x - 0)(x - (-3)) = 0$

$x(x + 3) = 0$

$x^2 + 3x = 0$

10. $(x - r_1)(x - r_2) = 0$

$$\left(x - \frac{3}{4}\right)\left(x - (-\frac{2}{3})\right) = 0$$

$$\left(x - \frac{3}{4}\right)\left(x + \frac{2}{3}\right) = 0$$

$$x^2 - \frac{1}{12}x - \frac{1}{2} = 0$$

$$12\left(x^2 - \frac{1}{12}x - \frac{1}{2}\right) = 0(12)$$

$$12x^2 - x - 6 = 0$$

11. $x^2 - 2x + 8 = 0$

$x^2 - 2x = -8$

$x^2 - 2x + 1 = -8 + 1$

$(x - 1)^2 = -7$

$\sqrt{(x-1)^2} = \sqrt{-7}$

$x - 1 = \pm\sqrt{-7} = \pm i\sqrt{7}$

$x = 1 \pm i\sqrt{7}$

The solutions are $1 + i\sqrt{7}$ and $1 - i\sqrt{7}$.

12. $(x - 2)(x + 3) = x - 10$

$x^2 + x - 6 = x - 10$

$x^2 = -4$

$\sqrt{x^2} = \sqrt{-4}$

$x = \pm\sqrt{-4}$

$x = \pm 2i$

The solutions are $2i$ and $-2i$.

13. $3x(x-3)=2x-4$

$3x^2-9x=2x-4$

$3x^2-11x+4=0$

$a=3, b=-11, c=4$

$x=\dfrac{-b\pm\sqrt{b^2-4ac}}{2a}$

$x=\dfrac{-(-11)\pm\sqrt{(-11)^2-4(3)(4)}}{2(3)}$

$x=\dfrac{11\pm\sqrt{121-48}}{6}$

$x=\dfrac{11\pm\sqrt{73}}{6}$

The solutions are $\dfrac{11+\sqrt{73}}{6}$ and $\dfrac{11-\sqrt{73}}{6}$.

14. $3x^2-5x+3=0$

$a=3, b=-5, c=3$

b^2-4ac

$(-5)^2-4(3)(3)=25-36=-11$

$-11>0$

Since the discriminant is less than zero the equation has two complex number solutions.

15. $(x+3)(2x-5)<0$

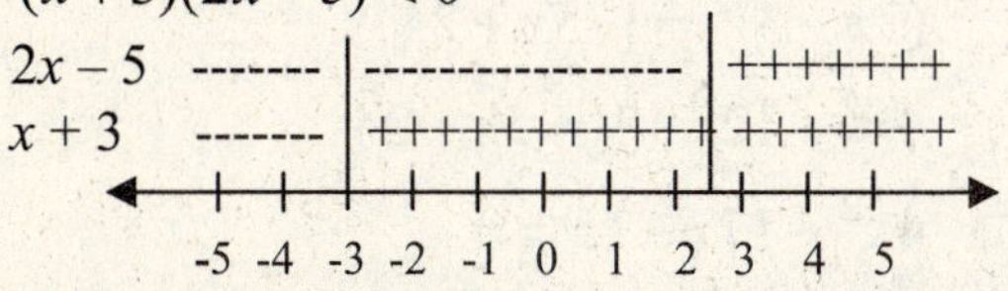

$\left\{x\,|\,-3<x<\dfrac{5}{2}\right\}$

16. $(x-2)(x+4)(2x+3)\le 0$

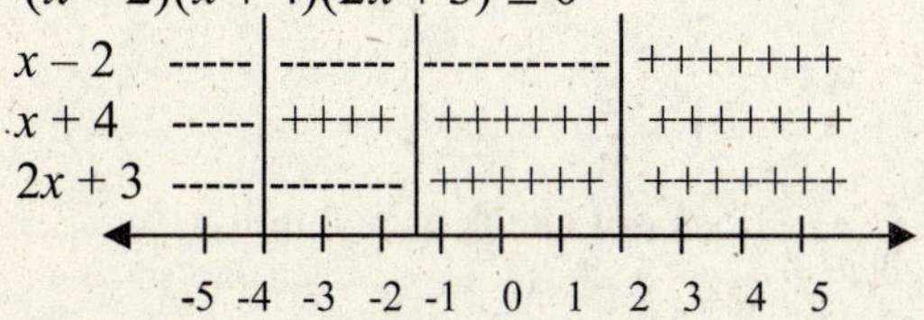

$\left\{x\,|\,x\le -4 \text{ or } -\dfrac{3}{2}\le x\le 2\right\}$

17. $x^{2/3}+x^{1/3}-12=0$

$\left(x^{1/3}\right)^2+x^{1/3}-12=0$

$u^2+u-12=0$

$(u+4)(u-3)=0$

$u+4=0 \quad u-3=0$

$u=-4 \qquad u=3$

Replace u with $x^{1/3}$.

$x^{1/3}=-4 \qquad x^{1/3}=3$

$\left(x^{1/3}\right)^3=(-4)^3 \quad \left(x^{1/3}\right)^3=3^3$

$x=-64 \qquad x=27$

The solutions are −64 and 27.

18. $2(x-1)+3\sqrt{x-1}-2=0$

$2(\sqrt{x-1})^2+3\sqrt{x-1}-2=0$

$2u^2+3u-2=0$

$(2u-1)(u+2)=0$

$2u-1=0 \qquad u+2=0$

$2u=1 \qquad u=-2$

$u=\dfrac{1}{2}$

Replace u with $\sqrt{x-1}$.

$\sqrt{x-1}=\dfrac{1}{2} \qquad \sqrt{x-1}=-2$

$\left(\sqrt{x-1}\right)^2=\left(\dfrac{1}{2}\right)^2 \quad \left(\sqrt{x-1}\right)=(-2)^2$

$x-1=\dfrac{1}{4} \qquad x-1=4$

$x=\dfrac{5}{4} \qquad x=5$

5 does not check as a solution.

The solution is $\dfrac{5}{4}$.

19. $3x = \dfrac{9}{x-2}$

$3x(x-2) = \dfrac{9}{x-2}(x-2)$

$3x^2 - 6x = 9$

$3x^2 - 6x - 9 = 0$

$3(x^2 - 2x - 3) = 0$

$3(x-3)(x+1) = 0$

$x - 3 = 0 \quad x + 1 = 0$

$x = 3 \qquad x = -1$

The solutions are -1 and 3.

20. $\dfrac{3x+7}{x+2} + x = 3$

$\dfrac{3x+7}{x+2} = 3 - x$

$(x+2)\left(\dfrac{3x+7}{x+2}\right) = (3-x)(x+2)$

$3x + 7 = 3x + 6 - x^2 - 2x$

$x^2 + 2x + 1 = 0$

$(x+1)^2 = 0$

$\sqrt{(x+1)^2} = \sqrt{0}$

$x + 1 = 0$

$x = -1$

The solution is -1.

21. $\dfrac{x-2}{2x-3} \geq 0$

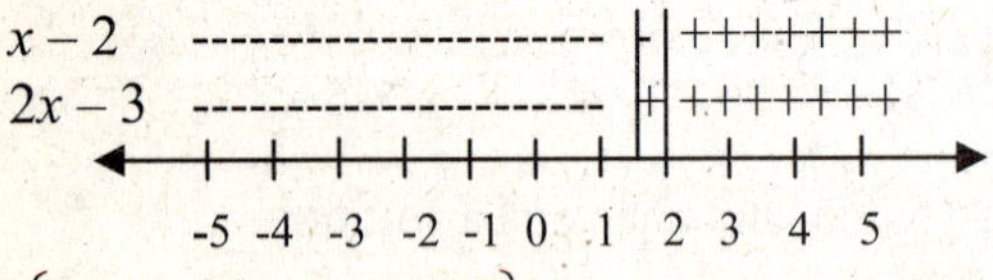

$\left\{x \mid x < \dfrac{3}{2} \text{ or } x \geq 2\right\}$

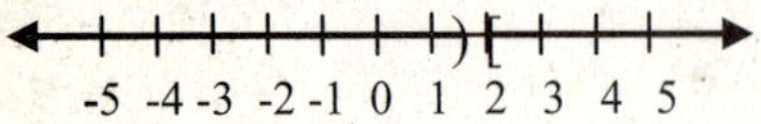

22. $\dfrac{(2x-1)(x+3)}{x-4} \leq 0$

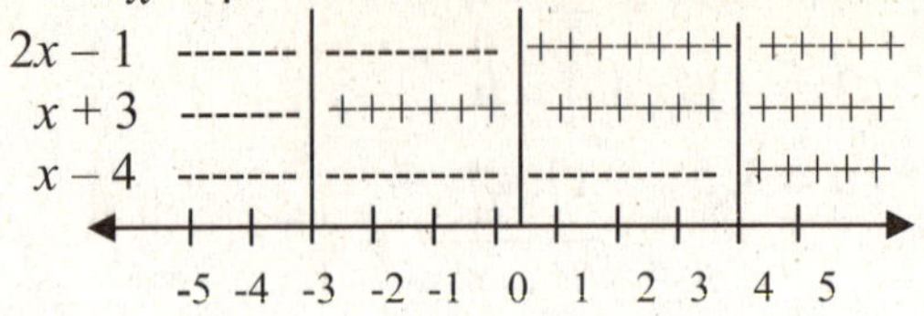

$\left\{x \mid x \leq -3 \text{ or } \dfrac{1}{2} \leq x < 4\right\}$

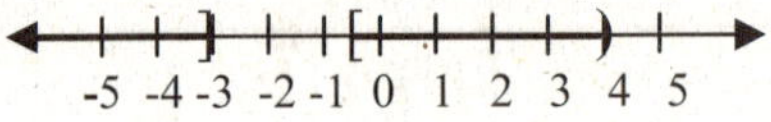

23. $x = \sqrt{x} + 2$

$x - 2 = \sqrt{x}$

$(x-2)^2 = \left(\sqrt{x}\right)^2$

$x^2 - 4x + 4 = x$

$x^2 - 5x + 4 = 0$

$(x-4)(x-1) = 0$

$x - 4 = 0 \quad x - 1 = 0$

$x = 4 \qquad x = 1$

1 does not check as a solution.
The solution is 4.

24. $2x = \sqrt{5x+24} + 3$

$2x - 3 = \sqrt{5x+24}$

$(2x-3)^2 = \left(\sqrt{5x+24}\right)^2$

$4x^2 - 12x + 9 = 5x + 24$

$4x^2 - 17x - 15 = 0$

$(4x+3)(x-5) = 0$

$4x + 3 = 0 \quad x - 5 = 0$

$4x = -3 \qquad x = 5$

$x = -\dfrac{3}{4}$

$-\dfrac{3}{4}$ does not check as a solution.

The solution is 5.

25. $\dfrac{x-2}{2x+3}-\dfrac{x-4}{x}=2$

$(2x+3)(x)\left(\dfrac{x-2}{2x+3}-\dfrac{x-4}{x}\right)=2(2x+3)(x)$

$x(x-2)-(x-4)(2x+3)=2x(2x+3)$

$x^2-2x-2x^2-3x+8x+12=4x^2+6x$

$0=5x^2+3x-12$

$a=5, b=3, c=-12$

$x=\dfrac{-b\pm\sqrt{b^2-4ac}}{2a}$

$x=\dfrac{-3\pm\sqrt{3^2-4(5)(-12)}}{2(5)}$

$x=\dfrac{-3\pm\sqrt{9+240}}{10}$

$x=\dfrac{-3\pm\sqrt{249}}{10}$

The solutions are $\dfrac{-3+\sqrt{249}}{10}$ and $\dfrac{-3-\sqrt{249}}{10}$.

26. $1-\dfrac{x+4}{2-x}=\dfrac{x-3}{x+2}$

$(2-x)(x+2)\left(1-\dfrac{x+4}{2-x}\right)=(2-x)(x+2)\dfrac{x-3}{x+2}$

$(2-x)(x+2)-(x+2)(x+4)=(2-x)(x-3)$

$4-x^2-x^2-6x-8=-x^2+5x-6$

$x^2+11x-2=0$

$a=1, b=11, c=-2$

$x=\dfrac{-b\pm\sqrt{b^2-4ac}}{2a}$

$x=\dfrac{-11\pm\sqrt{11^2-4(1)(-2)}}{2(1)}$

$x=\dfrac{-11\pm\sqrt{121+8}}{2}$

$x=\dfrac{-11\pm\sqrt{129}}{2}$

The solutions are $\dfrac{-11+\sqrt{129}}{2}$ and $\dfrac{-11-\sqrt{129}}{2}$.

27. $(x-r_1)(x-r_2)=0$

$\left(x-\dfrac{1}{3}\right)(x-(-3))=0$

$\left(x-\dfrac{1}{3}\right)(x+3)=0$

$x^2+\dfrac{8}{3}x-1=0$

$3\left(x^2+\dfrac{8}{3}x-1\right)=0(3)$

$3x^2+8x-3=0$

28. $2x^2 + 9x = 5$

$2x^2 + 9x - 5 = 0$

$(2x-1)(x+5) = 0$

$2x - 1 = 0 \quad x + 5 = 0$

$2x = 1 \qquad x = -5$

$x = \frac{1}{2}$

The solutions are -5 and $\frac{1}{2}$.

29. $2(x+1)^2 - 36 = 0$

$2(x+1)^2 = 36$

$(x+1)^2 = 18$

$(x+1)^2 = \left(\sqrt{18}\right)^2$

$x + 1 = \pm\sqrt{18} = \pm 3\sqrt{2}$

$x = -1 \pm 3\sqrt{2}$

The solutions are $-1 + 3\sqrt{2}$ and $-1 - 3\sqrt{2}$.

30. $x^2 + 6x + 10 = 0$

$a = 1, b = 6, c = 10$

$x = \frac{-b \pm \sqrt{b^2 - 4ac}}{2a}$

$x = \frac{-6 \pm \sqrt{6^2 - 4(1)(10)}}{2(1)}$

$x = \frac{-6 \pm \sqrt{36 - 40}}{2} = \frac{-6 \pm \sqrt{-4}}{2}$

$x = \frac{-6 \pm 2i}{2}$

$x = -3 \pm i$

The solutions are $-3 + i$ and $-3 - i$.

31. $\frac{2}{x-4} + 3 = \frac{x}{2x-3}$

$(2x-3)(x-4)\left(\frac{2}{x-4} + 3\right) = (2x-3)(x-4)\frac{x}{2x-3}$

$2(2x-3) + 3(x-4)(2x-3) = x(x-4)$

$4x - 6 + 6x^2 - 33x + 36 = x^2 - 4x$

$5x^2 - 25x + 30 = 0$

$5(x^2 - 5x + 6) = 0$

$5(x-3)(x-2) = 0$

$x - 3 = 0 \quad x - 2 = 0$

$x = 3 \qquad x = 2$

The solutions are 2 and 3.

32. $x^4 - 28x^2 + 75 = 0$

$\left(x^2\right)^2 - 28x^2 + 75 = 0$

$u^2 - 28u + 75 = 0$

$(u-25)(u-3) = 0$

$u - 25 = 0 \quad u - 3 = 0$

$u = 25 \qquad u = 3$

Replace u with x^2.

$x^2 = 25 \qquad x^2 = 3$

$\sqrt{x^2} = \sqrt{25} \quad \sqrt{x^2} = \sqrt{3}$

$x = \pm 5 \qquad x = \pm\sqrt{3}$

The solutions are -5, 5, $\sqrt{3}$ and $-\sqrt{3}$.

33. $\sqrt{2x-1}+\sqrt{2x}=3$

$\sqrt{2x-1}=3-\sqrt{2x}$

$(\sqrt{2x-1})^2=\left(3-\sqrt{2x}\right)^2$

$2x-1=9-6\sqrt{2x}-2x$

$-10=-6\sqrt{2x}$

$5=3\sqrt{2x}$

$5^2=\left(3\sqrt{2x}\right)^2$

$25=18x$

$x=\frac{25}{18}$

The solution is $\frac{25}{18}$.

34. $2x^{2/3}+3x^{1/3}-2=0$

$2\left(x^{1/3}\right)^2+3x^{1/3}-2=0$

$2u^2+3u-2=0$

$(2u-1)(u+2)=0$

$2u-1=0 \quad u+2=0$

$2u=1 \qquad u=-2$

$u=\frac{1}{2}$

Replace u with $x^{1/3}$.

$x^{1/3}=\frac{1}{2} \qquad x^{1/3}=-2$

$\left(x^{1/3}\right)^3=\left(\frac{1}{2}\right)^3 \quad \left(x^{1/3}\right)^3=(-2)^3$

$x=\frac{1}{8} \qquad x=-8$

The solutions are -8 and $\frac{1}{8}$.

35. $\sqrt{3x-2}+4=3x$

$\sqrt{3x-2}=3x-4$

$\left(\sqrt{3x-2}\right)^2=(3x-4)^2$

$3x-2=9x^2-24x+16$

$9x^2-27x+18=0$

$9(x^2-3x+2)=0$

$(x-2)(x-1)=0$

$x-2=0 \quad x-1=0$

$x=2 \qquad x=1$

1 does not check as a solution.
The solution is 2.

36. $x^2-10x+7=0$

$x^2-10x=-7$

$x^2-10x+25=-7+25$

$(x-5)^2=18$

$\sqrt{(x-5)^2}=\sqrt{18}$

$x-5=\pm 3\sqrt{2}$

$x=5\pm 3\sqrt{2}$

The solutions are $5+3\sqrt{2}$ and $5-3\sqrt{2}$.

37. $\frac{2x}{x-4}+\frac{6}{x+1}=11$

$(x-4)(x+1)\left(\frac{2x}{x-4}+\frac{6}{x+1}\right)=11(x-4)(x+1)$

$2x(x+1)+6(x-4)=11(x-4)(x+1)$

$2x^2+2x+6x-24=11x^2-33x-44$

$0=9x^2-41x-20$

$0=(9x+4)(x-5)$

$9x+4=0 \quad x-5=0$

$9x=-4 \qquad x=5$

$x=-\frac{4}{9}$

The solutions are $-\frac{4}{9}$ and 5.

38. $9x^2 - 3x = 1$

$9x^2 - 3x - 1 = 0$

$a = 9, b = -3, c = -1$

$$x = \frac{-b \pm \sqrt{b^2 - 4ac}}{2a}$$

$$x = \frac{-(-3) \pm \sqrt{(-3)^2 - 4(9)(-1)}}{2(9)}$$

$$x = \frac{3 \pm \sqrt{9+36}}{18} = \frac{3 \pm \sqrt{45}}{18}$$

$$x = \frac{3 \pm 3\sqrt{5}}{18}$$

$$x = \frac{1 \pm \sqrt{5}}{6}$$

The solutions are $\frac{1+\sqrt{5}}{6}$ and $\frac{1-\sqrt{5}}{6}$.

39. $2x = 4 - 3\sqrt{x-1}$

$2x - 4 = 3\sqrt{x-1}$

$(2x-4)^2 = \left(3\sqrt{x-1}\right)^2$

$4x^2 - 16x + 16 = 9x - 9$

$4x^2 - 25x + 25 = 0$

$(4x-5)(x-5) = 0$

$4x - 5 = 0 \quad x - 5 = 0$

$4x = 5 \qquad x = 5$

$x = \frac{5}{4}$

$\frac{5}{4}$ does not check as a solution.

The solution is 5.

40. $1 - \frac{x+3}{3-x} = \frac{x-4}{x+3}$

$$(3-x)(x+3)\left(1 - \frac{x+3}{3-x}\right) = (3-x)(x+3)\frac{x-4}{x+3}$$

$(3-x)(x+3) - (x+3)(x+3) = (3-x)(x-4)$

$3x + 9 - x^2 - 3x - x^2 - 6x - 9 = -x^2 + 7x - 12$

$x^2 + 13x - 12 = 0$

$a = 1, b = 13, c = -12$

$$x = \frac{-b \pm \sqrt{b^2 - 4ac}}{2a}$$

$$x = \frac{-13 \pm \sqrt{13^2 - 4(1)(-12)}}{2(1)}$$

$$x = \frac{-13 \pm \sqrt{169+48}}{2}$$

$$x = \frac{-13 \pm \sqrt{217}}{2}$$

The solutions are $\frac{-13+\sqrt{217}}{2}$ and $-\frac{13-\sqrt{217}}{2}$.

41. $2x^2 - 5x = 6$

$2x^2 - 5x - 6 = 0$

$a = 2, b = -5, c = -6$

$b^2 - 4ac = (-5)^2 - 4(2)(-6) = 73$

$73 > 0$

Since the discriminant is greater than zero the equation has two unequal real number solutions.

42. $x^2 - 3x \le 10$

$x^2 - 3x - 10 \le 0$

$(x-5)(x+2) \le 0$

The zeros are −2 and 5. The factors have opposite signs between the zeros. The solution set is $\{x \mid -2 \le x \le 5\}$.

43. Strategy: Let r represent the rate of the rowing in calm water.

	Distance	Rate	Time
With current	16	$r+2$	$\frac{16}{r+2}$
Against current	16	$r-2$	$\frac{16}{r-2}$

The total time traveled was 6 h.

Solution: $\frac{16}{r+2}+\frac{16}{r-2}=6$

$(r-2)(r+2)\left(\frac{16}{r+2}+\frac{16}{r-2}\right)=6(r-2)(r+2)$

$16(r-2)+16(r+2)=6r^2-24$

$16r-32+16r+32=6r^2-24$

$0=6r^2-32r-24$

$0=2(3r^2-16r-12)$

$0=(3r+2)(r-6)$

$3r+2=0 \quad r-6=0$

$3r=-2 \quad r=6$

$r=-\frac{2}{3}$

The rate cannot be a negative number. The rowing rate in calm water is 6 mph.

44. Strategy: Let x represent the width of the rectangle.
The length of the rectangle is $2x+2$.
The area of the rectangle is 60 cm^2.

Solution: $A=LW$

$60=x(2x+2)$

$60=2x^2+2x$

$0=2x^2+2x-60$

$0=2(x^2+x-30)$

$0=2(x+6)(x-5)$

$x+6=0 \quad x-5=0$

$x=-6 \quad x=5$

The width cannot be a negative number.
$2x+2=2(5)+2=12$
The width is 5 cm.
The length is 12 cm.

45. Strategy: Let x represent the first integer.
The second consecutive even integer: $x+2$.
The third consecutive even integer: $x+4$.
The sum of the squares of the three consecutive even integers is 56.

Solution: $x^2+(x+2)^2+(x+4)^2=56$

$x^2+x^2+4x+4+x^2+8x+16=56$

$3x^2+12x+20=56$

$3x^2+12x-36=0$

$3(x^2+4x-12)=0$

$3(x+6)(x-2)=0$

$x+6=0 \quad x-2=0$

$x=-6 \quad x=2$

$x=-6, x+2=-4, x+4=-2$

$x=2, x+2=4, x+4=6$

The integers are −6, −4 and −2 or 2, 4 and 6.

46. Strategy: Let t represent the time it takes the new computer to print the payroll.
The time it takes the older computer to print the payroll is $t+12$.

	Rate	Time	Part
New computer	$\frac{1}{t}$	8	$\frac{8}{t}$
Older computer	$\frac{1}{t+12}$	8	$\frac{8}{t+12}$

The sum of the parts of the task completed equals 1.

Solution: $\frac{8}{t}+\frac{8}{t+12}=1$

$t(t+12)\left(\frac{8}{t}+\frac{8}{t+12}\right)=1t(t+12)$

$8(t+12)+8t=t^2+12t$

$8t+96+8t=t^2+12t$

$t^2 - 4t - 96 = 0$

$(t-12)(t+8) = 0$

$t-12=0 \quad t+8=0$

$t=12 \quad t=-8$

$t=-8$ is not possible since time cannot be a negative number. Working alone the new computer can print the payroll in 12 min.

47. Strategy: Let r represent the rate of the first car.
The rate of the second car is $r+10$.

	Distance	Rate	Time
1st car	200	r	$\frac{200}{r}$
2nd car	200	$r+10$	$\frac{200}{r+10}$

The second car's time is one hour less than the first car's time.

Solution: $\frac{200}{r+10} = \frac{200}{r} - 1$

$(r)(r+10)\left(\frac{200}{r+10}\right) = (r)(r+10)\left(\frac{200}{r} - 1\right)$

$200r = 200(r+10) - r(r+10)$

$200r = 200r + 2000 - r^2 - 10r$

$0 = r^2 - 10r - 2000$

$0 = (r-40)(r+50)$

$r-40=0 \quad r+50=0$

$r=40 \quad r=-50$

The rate cannot be a negative number.
The rate of the first car is 40 mph.
The rate of the second car is 50 mph.

Chapter 10 Test

1. $3x^2 + 10x = 8$

$3x^2 + 10x - 8 = 0$

$(3x-2)(x+4) = 0$

$(3x-2)(x+4) = 0$

$3x-2=0 \quad x+4=0$

$3x=2 \quad x=-4$

$x = \frac{2}{3}$

The solutions are -4 and $\frac{2}{3}$.

2. $6x^2 - 5x - 6 = 0$

$(2x-3)(3x+2) = 0$

$2x-3=0 \quad 3x+2=0$

$2x=3 \quad 3x=-2$

$x=\frac{3}{2} \quad x=-\frac{2}{3}$

The solutions are $\frac{3}{2}$ and $-\frac{2}{3}$.

3. $(x-r_1)(x-r_2)=0$

$(x-3)(x-(-3))=0$

$(x-3)(x+3)=0$

$x^2-9=0$

4. $(x-r_1)(x-r_2)=0$

$\left(x-\frac{1}{2}\right)(x-(-4))=0$

$\left(x-\frac{1}{2}\right)(x+4)=0$

$x^2+\frac{7}{2}x-2=0$

$2\left(x^2+\frac{7}{2}x-2\right)=0(2)$

$2x^2+7x-4=0$

5. $3(x-2)^2-24=0$

$3(x-2)^2=24$

$(x-2)^2=8$

$\sqrt{(x-2)^2}=\sqrt{8}$

$x-2=\pm 2\sqrt{2}$

$x=2\pm 2\sqrt{2}$

The solutions are $2+2\sqrt{2}$ and $2-2\sqrt{2}$.

6. $x^2-6x-2=0$

$x^2-6x=2$

$x^2-6x+9=2+9$

$(x-3)^2=11$

$\sqrt{(x-3)^2}=\sqrt{11}$

$x-3=\pm\sqrt{11}$

$x=3\pm\sqrt{11}$

The solutions are $3+\sqrt{11}$ and $3-\sqrt{11}$.

7. $3x^2-6x=2$

$\frac{1}{3}(3x^2-6x)=2\left(\frac{1}{3}\right)$

$x^2-2x=\frac{2}{3}$

$x^2-2x+1=\frac{2}{3}+1$

$(x-1)^2=\frac{5}{3}$

$\sqrt{(x-1)^2}=\sqrt{\frac{5}{3}}$

$x-1=\pm\frac{\sqrt{15}}{3}$

$x=1\pm\frac{\sqrt{15}}{3}$

$x=\frac{3\pm\sqrt{15}}{3}$

The solutions are $\frac{3+\sqrt{15}}{3}$ and $\frac{3-\sqrt{15}}{3}$.

8. $2x^2-2x=1$

$2x^2-2x-1=0$

$a=2, b=-2, c=-1$

$x=\frac{-b\pm\sqrt{b^2-4ac}}{2a}$

$x=\frac{-(-2)\pm\sqrt{(-2)^2-4(2)(-1)}}{2(2)}$

$x=\frac{2\pm\sqrt{4+8}}{4}=\frac{2\pm\sqrt{12}}{4}$

$x=\frac{2\pm 2\sqrt{3}}{4}$

$x=\frac{1\pm\sqrt{3}}{2}$

The solutions are $\frac{1+\sqrt{3}}{2}$ and $\frac{1-\sqrt{3}}{2}$.

9. $x^2+4x+12=0$

$a=1, b=4, c=12$

$x=\frac{-b\pm\sqrt{b^2-4ac}}{2a}$

$x=\frac{-4\pm\sqrt{4^2-4(1)(12)}}{2(1)}$

$x=\frac{-4\pm\sqrt{16-48}}{2}=\frac{-4\pm\sqrt{-32}}{2}$

$x=\frac{-4\pm 4i\sqrt{2}}{2}$

$x=-2\pm 2i\sqrt{2}$

The solutions are $-2+2i\sqrt{2}$ and $-2-2i\sqrt{2}$.

10. $2x + 7x^{1/2} - 4 = 0$

$2(x^{1/2})^2 + 7x^{1/2} - 4 = 0$

$2u^2 + 7u - 4 = 0$

$(2u - 1)(u + 4) = 0$

$2u - 1 = 0 \quad u + 4 = 0$

$2u = 1 \qquad u = -4$

$u = \frac{1}{2}$

Replace u with $x^{1/2}$.

$x^{1/2} = \frac{1}{2} \qquad x^{1/2} = -4$

$(x^{1/2})^2 = \left(\frac{1}{2}\right)^2 \quad (x^{1/2})^2 = (-4)^2$

$x = \frac{1}{4} \qquad x = 16$

16 does not check as a solution.

The solution is $\frac{1}{4}$.

11. $x^4 - 4x^2 + 3 = 0$

$(x^2)^2 - 4x^2 + 3 = 0$

$u^2 - 4u + 3 = 0$

$(u - 1)(u - 3) = 0$

$u - 1 = 0 \quad u - 3 = 0$

$u = 1 \qquad u = 3$

Replace u with x^2.

$x^2 = 1 \qquad x^2 = 3$

$\sqrt{x^2} = \sqrt{1} \quad \sqrt{x^2} = \sqrt{3}$

$x = \pm 1 \qquad x = \pm\sqrt{3}$

The solutions are -1, 1, $\sqrt{3}$ and $-\sqrt{3}$.

12. $\sqrt{2x+1} + 5 = 2x$

$\sqrt{2x+1} = 2x - 5$

$\left(\sqrt{2x+1}\right)^2 = (2x - 5)^2$

$2x + 1 = 4x^2 - 20x + 25$

$4x^2 - 22x + 24 = 0$

$2(2x^2 - 11x + 12) = 0$

$2(2x - 3)(x - 4) = 0$

$2x - 3 = 0 \quad x - 4 = 0$

$2x = 3 \qquad x = 4$

$x = \frac{3}{2}$

$\frac{3}{2}$ does not check as a solution.

The solution is 4.

13. $\sqrt{x-2} = \sqrt{x} - 2$

$(\sqrt{x-2})^2 = \left(\sqrt{x} - 2\right)^2$

$x - 2 = x - 4\sqrt{x} + 4$

$-6 = -4\sqrt{x}$

$3 = 2\sqrt{x}$

$3^2 = \left(2\sqrt{x}\right)^2$

$9 = 4x$

$x = \frac{9}{4}$

$\frac{9}{4}$ does not check as a solution.

There is no solution.

14. $\dfrac{2x}{x-3}+\dfrac{5}{x-1}=1$

$(x-3)(x-1)\left(\dfrac{2x}{x-3}+\dfrac{5}{x-1}\right)=1(x-3)(x-1)$

$2x(x-1)+5(x-3)=1(x-3)(x-1)$

$2x^2-2x+5x-15=x^2-4x+3$

$x^2+7x-18=0$

$(x+9)(x-2)=0$

$x+9=0 \quad x-2=0$

$x=-9 \qquad x=2$

The solutions are −9 and 2.

15. $(x-2)(x+4)(x-4)<0$

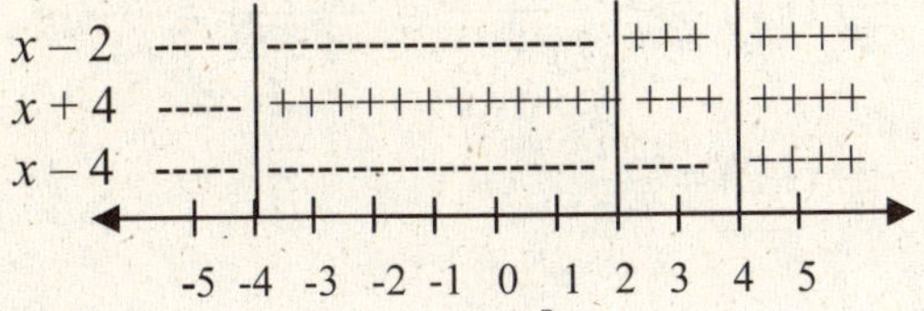

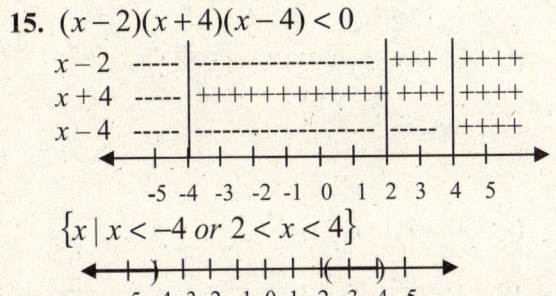

$\{x \mid x<-4 \text{ or } 2<x<4\}$

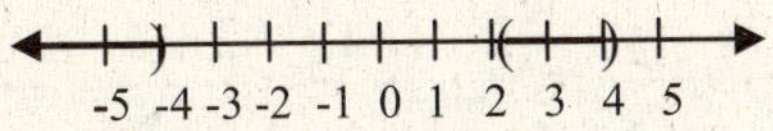

16. $\dfrac{2x-3}{x+4}\le 0$

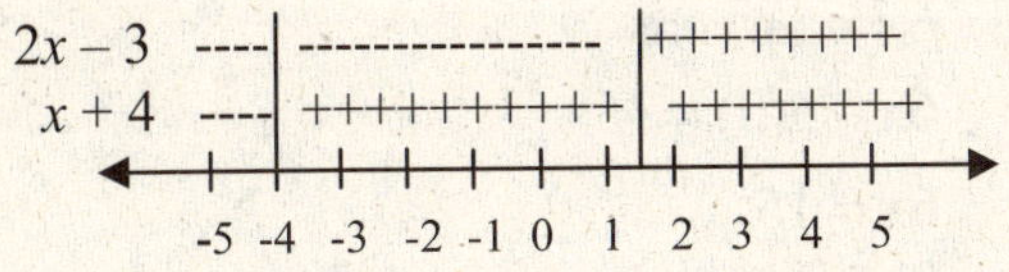

$\left\{x \mid -4<x\le \dfrac{3}{2}\right\}$

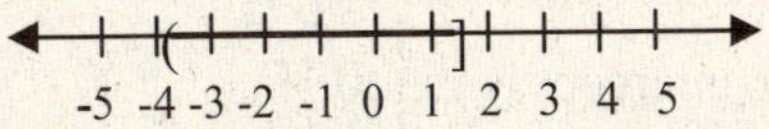

17. $9x^2+24x=-16$

$9x^2+24x+16=0$

$a=9, b=24, c=16$

$b^2-4ac=(24)^2-4(9)(16)=0$

Since the discriminant is equal to zero the equation has two equal real number solutions.

18. **Strategy:** To find the time when the ball hits the basket substitute 10 ft for h in the equation and solve for t.

Solution: $h=-16t^2+32t+6.5$

$10=-16t^2+32t+6.5$

$0=-16t^2+32t-3.5$

$0=16t^2-32t+3.5$

$a=16, b=-32, c=3.5$

$t=\dfrac{-b\pm\sqrt{b^2-4ac}}{2a}$

$t=\dfrac{-(-32)\pm\sqrt{(-32)^2-4(16)(3.5)}}{2(16)}$

$t=\dfrac{32\pm\sqrt{800}}{32}$

$t=1.88$

$t=0.12$

We need to find the time it takes to reach the basket after the ball has reached its peak. This occurs 1.88 s after the ball has been released.

19. **Strategy:** Let t represent the time it takes Cora to stain a bookcase.
The time it takes Clive to stain a bookcase is $t+6$.

	Rate	Time	Part
Cora	$\frac{1}{t}$	4	$\frac{4}{t}$
Clive	$\frac{1}{t+6}$	4	$\frac{4}{t+6}$

The sum of the parts of the task completed equals 1.

Solution: $\dfrac{4}{t}+\dfrac{4}{t+6}=1$

$t(t+6)\left(\dfrac{4}{t}+\dfrac{4}{t+6}\right)=1t(t+6)$

$4(t+6)+4t=t^2+6t$

$4t+24+4t=t^2+6t$

$t^2 - 2t - 24 = 0$

$(t-6)(t+4) = 0$

$t-6=0 \quad t+4=0$

$t=6 \quad\quad t=-4$

$t=-4$ is not possible since time cannot be a negative number. Working alone it will take Cora 6 h to stain the bookcase.

20. Strategy: Let r represent the rate of the canoe in calm water.

	Distance	Rate	Time
With current	6	$r+2$	$\frac{6}{r+2}$
Against current	6	$r-2$	$\frac{6}{r-2}$

The total time traveled was 4 h.

Solution: $\frac{6}{r+2} + \frac{6}{r-2} = 4$

$(r-2)(r+2)\left(\frac{6}{r+2} + \frac{6}{r-2}\right) = 4(r-2)(r+2)$

$6(r-2) + 6(r+2) = 4r^2 - 16$

$6r - 12 + 6r + 12 = 4r^2 - 16$

$0 = 4r^2 - 12r - 16$

$0 = 4(r^2 - 3r - 4)$

$0 = 4(r-4)(r+1)$

$r-4=0 \quad r+1=0$

$r=4 \quad\quad r=-1$

The rate cannot be a negative number. The rate of the canoe in calm water is 4 mph.

Cumulative Review Exercises

1. $2a^2 - b^2 \div c^2$

$2(3)^2 - (-4)^2 \div (-2)^2 = 2(9) - 16 \div 4$

$= 18 - 16 \div 4 = 18 - 4$

$= 14$

2. $|3x-2| < 8$

$-8 < 3x - 2 < 8$

$-8 + 2 < 3x - 2 + 2 < 8 + 2$

$-6 < 3x < 10$

$\frac{1}{3}\cdot(-6) < \frac{1}{3}\cdot(3x) < \frac{1}{3}\cdot 10$

$-2 < x < \frac{10}{3}$

$\left\{x \mid -2 < x < \frac{10}{3}\right\}$

3. $V = \pi r^2 h = \pi(3)^2(6) = \pi(9)(6) = 54\pi$

The volume is 54π m^3.

4. $f(x) = \frac{2x-3}{x^2-1}$

$f(-2) = \frac{2(-2)-3}{(-2)^2-1} = \frac{-4-3}{4-1} = \frac{-7}{3}$

$f(-2) = -\frac{7}{3}$

5. $(3, -4)$ and $(-1, 2)$

$m = \frac{y_2 - y_1}{x_2 - x_1} = \frac{2-(-4)}{-1-3} = \frac{2+4}{-4} = \frac{6}{-4}$

$m = -\frac{3}{2}$

6. $6x - 5y = 15$

$6x - 5(0) = 15$

$6x = 15$

$x = \frac{15}{6} = \frac{5}{2}$

The x-intercept is $\left(\frac{5}{2}, 0\right)$.

$6(0) - 5y = 15$

$-5y = 15$

$y = -3$

The y-intercept is $(0, -3)$.

7. $x - y = 1$

$y = x - 1$

$m = 1$ and $(1, 2)$

$y - y_1 = m(x - x_1)$

$y - 2 = 1(x - 1)$

$y - 2 = x - 1$

$y = x + 1$

8. (1) $x + y + z = 2$

(2) $-x + 2y - 3z = -9$

(3) $x - 2y - 2z = -1$

Eliminate x and y. Add equations (2) and (3).

$$\begin{array}{r} -x + 2y - 3z = -9 \\ \underline{x - 2y - 2z = -1} \\ -5z = -10 \\ z = 2 \end{array}$$

Eliminate y and z. Multiply equation (1) by 2 and add to equation (3).

$2(x + y + z = 2)$

$2x + 2y + 2z = 4$

$$\begin{array}{r} 2x + 2y + 2z = 4 \\ \underline{x - 2y - 2z = -1} \\ 3x \qquad\quad = 3 \\ x = 1 \end{array}$$

Replace x with 1 and z with 2 in equation (1)

$x + y + z = 2$

$1 + y + 2 = 2$

$y + 3 = 2$

$y = -1$

The solution is $(1, -1, 2)$.

9. Solve each inequality for y.

$x + y \le 3 \quad 2x - y < 4$

$y \le 3 - x \quad -y < 4 - 2x$

$y > 2x - 4$

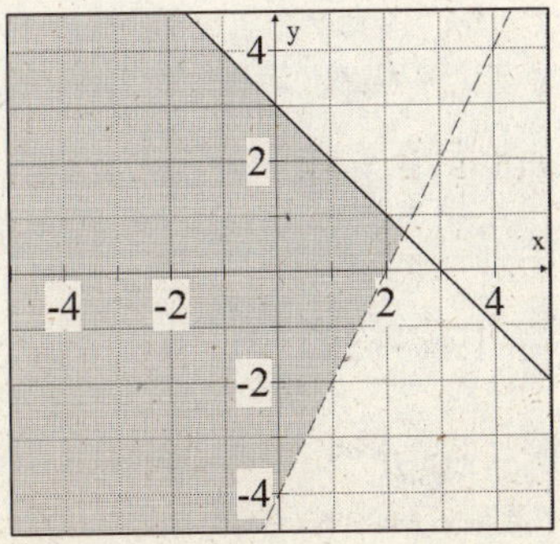

10. $\dfrac{BC}{EF} = \dfrac{\textit{height of triangle ABC}}{\textit{height of triangle DEF}}$

$\dfrac{12}{24} = \dfrac{8}{h}$

$12h = 24(8)$

$12h = 192$

$h = 16$

The height of triangle DEF is 16 cm.

11.

$$\begin{array}{r} x^2 - 3x - 4 \\ 3x - 4\overline{)\,3x^3 - 13x^2 + 0x + 10} \\ \underline{3x^3 - 4x^2} \\ -9x^2 + 0x \\ \underline{-9x^2 + 12x} \\ -12x + 10 \\ \underline{-12x + 16} \\ -6 \end{array}$$

$(3x^3 - 13x^2 + 10) \div (3x - 4) = x^2 - 3x - 4 + \dfrac{-6}{3x - 4}$

12. $-3x^3y + 6x^2y^2 - 9xy^3 = -3xy(x^2 - 2xy + 3y^2)$

13. $6x^2 - 7x - 20 = (2x - 5)(3x + 4)$

14. $\dfrac{x^2 + 2x + 1}{8x^2 + 8x} \cdot \dfrac{4x^3 - 4x^2}{x^2 - 1}$

$= \dfrac{(x+1)(x+1)}{8x(x+1)} \cdot \dfrac{4x^2(x-1)}{(x+1)(x-1)}$

$= \dfrac{(x+1)(x+1) \cdot 4x^2(x-1)}{8x(x+1) \cdot (x+1)(x-1)}$

$= \dfrac{x}{2}$

15. $\dfrac{x}{x+2}-\dfrac{4x}{x+3}=1$

$(x+2)(x+3)\left(\dfrac{x}{x+2}-\dfrac{4x}{x+3}\right)=1(x+2)(x+3)$

$x(x+3)-4x(x+2)=1(x+2)(x+3)$

$x^2+3x-4x^2-8x=x^2+5x+6$

$0=4x^2+10x+6$

$0=2(2x^2+5x+3)$

$(2x+3)(x+1)=0$

$2x+3=0 \quad x+1=0$

$2x=-3 \qquad x=-1$

$x=-\dfrac{3}{2}$

The solutions are -1 and $-\dfrac{3}{2}$.

16. $S=\dfrac{n}{2}(a+b)$

$2S=n(a+b)$

$2S=na+nb$

$2S-na=nb$

$\dfrac{2S-na}{n}=b$

17. $a^{-1/2}(a^{1/2}-a^{3/2})=a^0-a^1$
$=1-a$

18. $-2i(7-4i)=-14i+8i^2=-8-14i$

19. $\sqrt{3x+1}-1=x$

$\sqrt{3x+1}=x+1$

$\left(\sqrt{3x+1}\right)^2=(x+1)^2$

$3x+1=x^2+2x+1$

$0=x^2-x$

$0=x(x-1)$

$x=0 \qquad x-1=0$

$x=1$

Both 0 and 1 check as solutions. The solutions are 0 and 1.

20. $x^4-6x^2+8=0$

$\left(x^2\right)^2-6x^2+8=0$

$u^2-6u+8=0$

$(u-4)(u-2)=0$

$u-4=0 \quad u-2=0$

$u=4 \qquad u=2$

Replace u with x^2.

$x^2=4 \qquad x^2=2$

$\sqrt{x^2}=\sqrt{4} \quad \sqrt{x^2}=\sqrt{2}$

$x=\pm 2 \qquad x=\pm\sqrt{2}$

The solutions are -2, 2, $\sqrt{2}$ and $-\sqrt{2}$.

21. Strategy: Let P represent the length of the piston rod, T the tolerance and m the given length. Solve the absolute value inequality $|m-p|\le T$ for m.

Solution: $|m-p|\le T$

$\left|m-9\dfrac{3}{8}\right|\le\dfrac{1}{64}$

$-\dfrac{1}{64}\le m-9\dfrac{3}{8}\le\dfrac{1}{64}$

$-\dfrac{1}{64}+9\dfrac{3}{8}\le m-9\dfrac{3}{8}+9\dfrac{3}{8}\le\dfrac{1}{64}+9\dfrac{3}{8}$

$9\dfrac{23}{64}\le m\le 9\dfrac{25}{64}$

The lower limit is $9\dfrac{23}{64}$ in.

The upper limit is $9\dfrac{25}{64}$ in.

22. Strategy: The base of the triangle is $x+8$. The height of the triangle is $2x-4$.

Solution: $A=\dfrac{1}{2}bh$

$A=\dfrac{1}{2}(x+8)(2x-4)$

$A=\dfrac{1}{2}(2x^2+12x-32)$

$A=x^2+6x-16\ ft^2$

The area is $(x^2+6x-16)$ ft^2.

23. (0, 250) and (30, 0)

$$m = \frac{y_2 - y_1}{x_2 - x_1} = \frac{250 - 0}{0 - 30} = \frac{250}{-30} = -\frac{25}{3}$$

$$m = -\frac{25000}{3}$$

The building depreciates $\frac{\$25000}{3}$ or about $8333 each year.

24. Strategy: To find the distance use the Pythagorean Theorem. The hypotenuse is the length of the ladder. The distance from the bottom of the ladder to the base of the house is one leg and the distance along the house from the ground to the top of the ladder is the unknown leg.

Solution: $a^2 + b^2 = c^2$

$$8^2 + b^2 = 17^2$$
$$64 + b^2 = 289$$
$$b^2 = 225$$

$$b = \pm 15$$

The solution $b = -15$ is not possible since the distance cannot be a negative number. The distance is 15 ft.

25. $2x^2 + 4x + 3 = 0$

$a = 2, b = 4, c = 3$

$b^2 - 4ac = 4^2 - 4(2)(3) = 16 - 24 = -8$

$-8 < 0$

Since the discriminant is less than zero the equation has two complex number solutions.

Chapter 11: Functions and Relations

Prep Test

1. $-\dfrac{b}{2a}$

$-\dfrac{(-4)}{2(2)}=-\dfrac{-4}{4}=-(-1)=1$

2. $y=-x^2+2x+1$
$y=-(-2)^2+2(-2)+1=-4-4+1=-7$

3. $f(x)=x^2-3x+2$
$f(-4)=(-4)^2-3(-4)+2$
$f(-4)=16+12+2$
$f(-4)=30$

4. $p(r)=r^2-5$
$p(2+h)=(2+h)^2-5$
$=4+4h+h^2-5$
$=h^2+4h-1$

5. $0=3x^2-7x-6$
$0=(3x+2)(x-3)$
$0=3x+2 \quad 0=x-3$
$-2=3x \qquad 3=x$
$-\dfrac{2}{3}=x$

The solutions are $-\dfrac{2}{3}$ and 3.

6. $0=x^2-4x+1$
$a=1 \quad b=-4 \quad c=1$

$x=\dfrac{-b\pm\sqrt{b^2-4ac}}{2a}$

$x=\dfrac{-(-4)\pm\sqrt{(-4)^2-4(1)(1)}}{2(1)}$

$x=\dfrac{4\pm\sqrt{16-4}}{2}=\dfrac{4\pm\sqrt{12}}{2}=\dfrac{4\pm2\sqrt{3}}{2}$

$x=\dfrac{4}{2}\pm\dfrac{2\sqrt{3}}{2}=2\pm\sqrt{3}$

The solutions are $2+\sqrt{3}$ and $2-\sqrt{3}$.

7. $x=2y+4$
$2y+4=x$
$2y=x-4$
$\left(\dfrac{1}{2}\right)2y=\left(\dfrac{1}{2}\right)(x-4)$
$y=\dfrac{1}{2}x-2$

8. Domain: $\{-2, 3, 4, 6\}$
Range: $\{4, 5, 6\}$
Yes the relation is a function.

9.

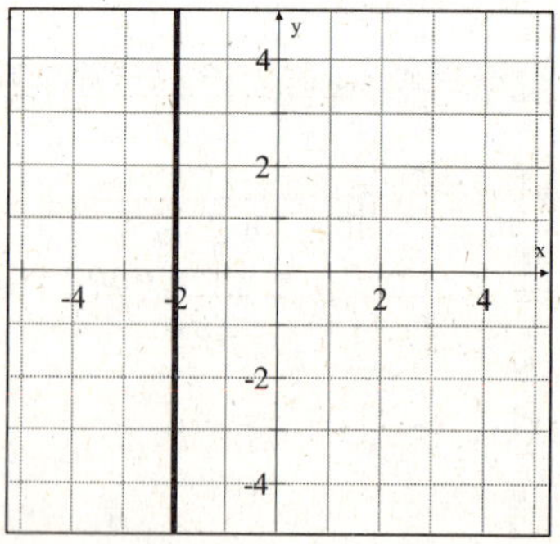

Section 11.1

Objective A Exercises

1. A quadratic function is a function of the form $f(x) = ax^2 + bx + c$, $a \neq 0$.

3. The vertex of a parabola is the point with the smallest y-coordinate or the largest y-coordinate. When $a > 0$, the parabola opens up and the vertex of the parabola is the point with the smallest y-coordinate. When $a < 0$, the parabola opens down and the vertex of the parabola is the point with the largest y-coordinate.

5. -5

7. $x = 7$

9. $y = x^2 - 2x - 4$

$$-\frac{b}{2a} = -\frac{-2}{2(1)} = -\frac{-2}{2} = -(-1) = 1$$

$$y = (1)^2 - 2(1) - 4 = -5$$

Vertex: $(1, -5)$
Axis of symmetry: $x = 1$

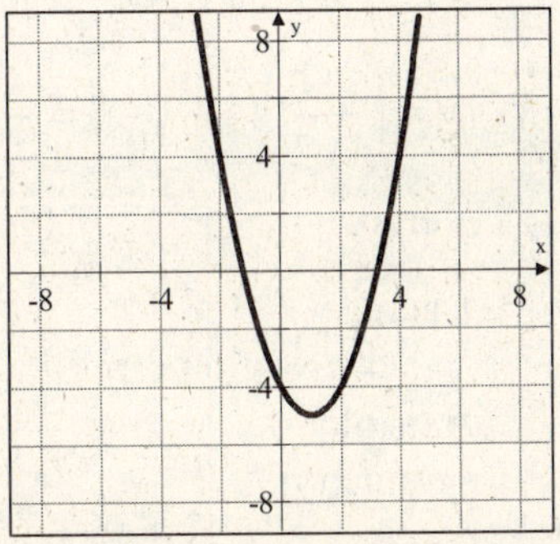

11. $y = -x^2 + 2x - 3$

$$-\frac{b}{2a} = -\frac{2}{2(-1)} = -\frac{2}{-2} = -(-1) = 1$$

$$y = -(1)^2 + 2(1) - 3 = -2$$

Vertex: $(1, -2)$
Axis of symmetry: $x = 1$

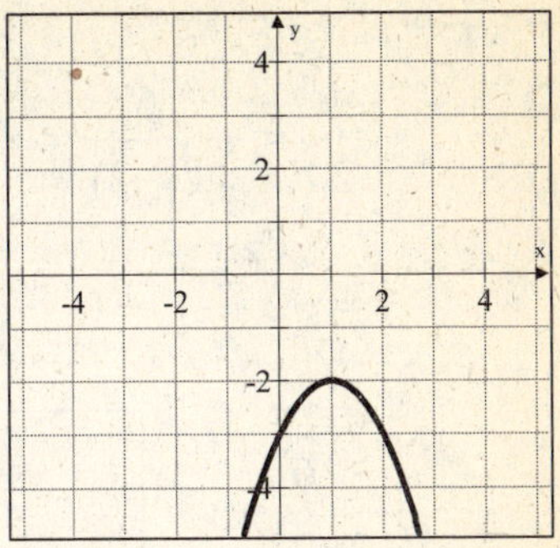

13. $f(x) = x^2 - x - 6$

$$-\frac{b}{2a} = -\frac{(-1)}{2(1)} = -\frac{-1}{2} = \frac{1}{2}$$

$$y = \left(\frac{1}{2}\right)^2 - \frac{1}{2} - 6 = -\frac{25}{4}$$

Vertex: $\left(\frac{1}{2}, -\frac{25}{4}\right)$

Axis of symmetry: $x = \frac{1}{2}$

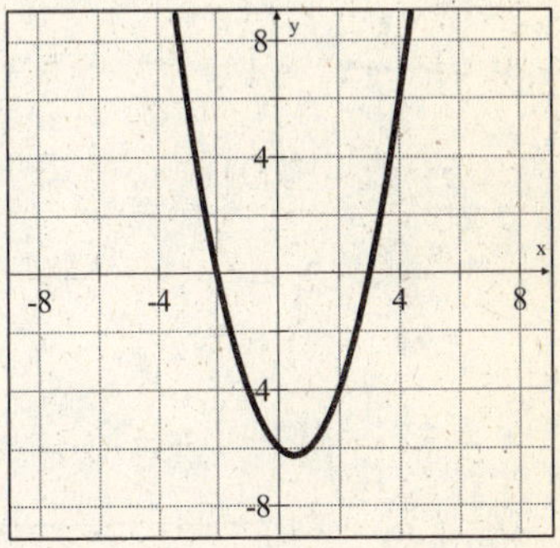

15. $F(x) = x^2 - 3x + 2$

$$-\frac{b}{2a} = -\frac{(-3)}{2(1)} = -\frac{-3}{2} = \frac{3}{2}$$

$$y = \left(\frac{3}{2}\right)^2 - 3\left(\frac{3}{2}\right) + 2 = -\frac{1}{4}$$

Vertex: $\left(\frac{3}{2}, -\frac{1}{4}\right)$

Axis of symmetry: $x = \frac{3}{2}$

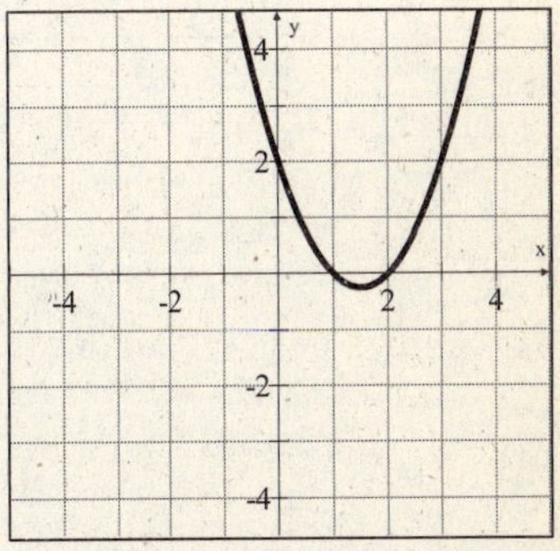

17. $y = -2x^2 + 6x$

$$-\frac{b}{2a} = -\frac{6}{2(-2)} = -\frac{6}{-4} = \frac{3}{2}$$

$$y = -2\left(\frac{3}{2}\right)^2 + 6\left(\frac{3}{2}\right) = \frac{9}{2}$$

Vertex: $\left(\frac{3}{2}, \frac{9}{2}\right)$

Axis of symmetry: $x = \frac{3}{2}$

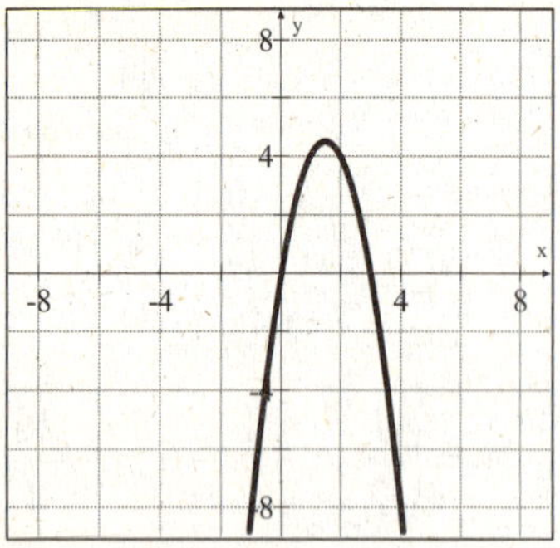

19. $y = -\frac{1}{4}x^2 - 1$

$$-\frac{b}{2a} = -\frac{(0)}{2\left(-\frac{1}{4}\right)} = -\frac{0}{-\frac{1}{2}} = 0$$

$$y = -\left(\frac{1}{4}\right)0^2 - 1 = -1$$

Vertex: $(0, -1)$

Axis of symmetry: $x = 0$

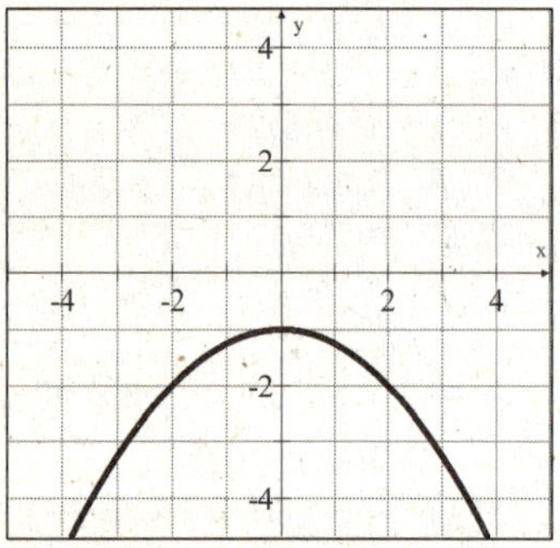

21. $P(x) = -\frac{1}{2}x^2 + 2x - 3$

$$-\frac{b}{2a} = -\frac{2}{2\left(-\frac{1}{2}\right)} = -\frac{2}{-1} = -(-2) = 2$$

$$y = \left(-\frac{1}{2}\right)2^2 + 2(2) - 3 = -1$$

Vertex: $(2, -1)$

Axis of symmetry: $x = 2$

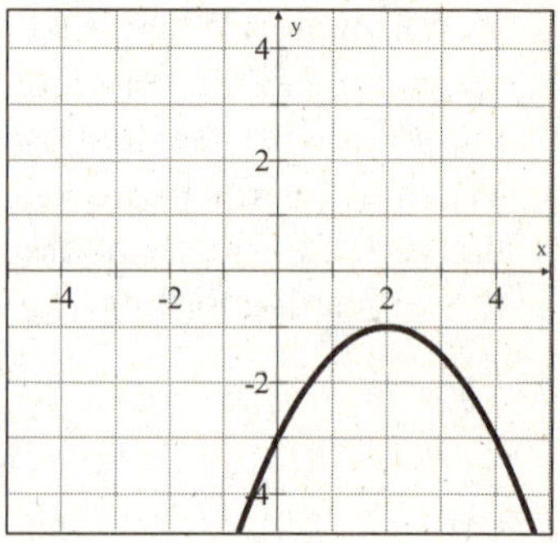

23. $y = -\frac{1}{2}x^2 + x - 3$

$$-\frac{b}{2a} = -\frac{1}{2\left(-\frac{1}{2}\right)} = -\frac{1}{-1} = -(-1) = 1$$

$$y = \left(-\frac{1}{2}\right)1^2 + 1 - 3 = -\frac{5}{2}$$

Vertex: $\left(1, -\frac{5}{2}\right)$

Axis of symmetry: $x = 1$

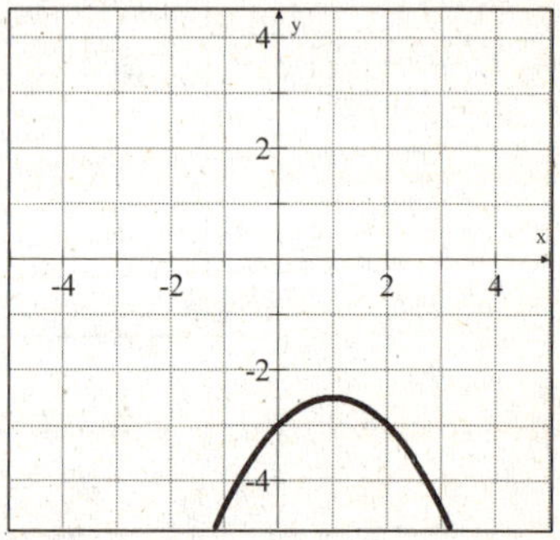

25. Domain: $\{x \mid x \text{ is all real numbers}\}$
Range: $\{y \mid y \geq 2\}$

27. Domain: $\{x \mid x \text{ is all real numbers}\}$
Range: $\{y \mid y \geq -5\}$

29. Domain: $\{x \mid x \text{ is all real numbers}\}$
Range: $\{y \mid y \leq 0\}$

31. Domain: $\{x|\ x \text{ is all real numbers}\}$
Range: $\{y \mid y \geq -7\}$

Objective B Exercises

33. a) A y-intercept of the graph of a parabola is a point at which the graph crosses the y-axis. It is a point at which $x = 0$.

b) The graph of a parabola has one y-intercept.

35. a) The zeros are −4 and 1.
b) The solutions are −4 and 1.

37. $y = x^2 - 9$
$0 = x^2 - 9$
$0 = (x+3)(x-3)$
$x+3=0 \qquad x-3=0$
$x=-3 \qquad x=3$
The x-intercepts are (−3, 0) and (3, 0).

39. $y = 3x^2 + 6x$
$0 = 3x^2 + 6x$
$0 = 3x(x+2)$
$3x = 0 \qquad x+2=0$
$x = 0 \qquad x=-2$
The x-intercepts are (0, 0) and (−2, 0).

41. $y = x^2 - 2x - 8$
$0 = x^2 - 2x - 8$
$0 = (x-4)(x+2)$
$x-4=0 \qquad x+2=0$
$x=4 \qquad x=-2$
The x-intercepts are (4, 0) and (−2, 0).

43. $y = 2x^2 - 5x - 3$
$0 = 2x^2 - 5x - 3$
$0 = (x-3)(2x+1)$
$x-3=0 \qquad 2x+1=0$
$x=3 \qquad 2x=-1$
$x = -\frac{1}{2}$
The x-intercepts are (3, 0) and $\left(-\frac{1}{2}, 0\right)$.

45. $y = x^2 + 4x - 3$
$0 = x^2 + 4x - 3$
$a=1 \quad b=4 \quad c=-3$
$$x = \frac{-b \pm \sqrt{b^2 - 4ac}}{2a}$$
$$x = \frac{-4 \pm \sqrt{(4)^2 - 4(1)(-3)}}{2(1)}$$
$$x = \frac{-4 \pm \sqrt{16+12}}{2} = \frac{-4 \pm \sqrt{28}}{2} = \frac{-4 \pm 2\sqrt{7}}{2}$$
$x = -2 \pm \sqrt{7}$
The x-intercepts are $(-2+\sqrt{7}, 0)$ and $(-2-\sqrt{7}, 0)$.

47. $y = -x^2 - 4x - 5$
$0 = -x^2 - 4x - 5$
$a=-1 \quad b=-4 \quad c=-5$
$$x = \frac{-b \pm \sqrt{b^2 - 4ac}}{2a}$$
$$x = \frac{-(-4) \pm \sqrt{(-4)^2 - 4(-1)(-5)}}{2(-1)}$$
$$x = \frac{4 \pm \sqrt{16-20}}{-2} = \frac{4 \pm \sqrt{-4}}{-2} = \frac{4 \pm 2i}{-2}$$
$x = -2 \pm i$
There are no real solutions. The parabola has no x-intercepts.

49. $f(x) = x^2 - 4x - 5$

$x^2 - 4x - 5 = 0$

$(x-5)(x+1) = 0$

$x - 5 = 0 \qquad x + 1 = 0$

$x = 5 \qquad x = -1$

The zeros are 5 and −1.

51. $f(x) = 3x^2 - 2x - 8$

$3x^2 - 2x - 8 = 0$

$(3x+4)(x-2) = 0$

$3x + 4 = 0 \qquad x - 2 = 0$

$3x = -4 \qquad x = 2$

$x = -\frac{4}{3}$

The zeros are $-\frac{4}{3}$ and 2.

53. $h(x) = 4x^2 - 4x + 1$

$4x^2 - 4x + 1 = 0$

$(2x-1)(2x-1) = 0$

$2x - 1 = 0 \qquad 2x - 1 = 0$

$2x = 1 \qquad 2x = 1$

$x = \frac{1}{2} \qquad x = \frac{1}{2}$

The zero is $\frac{1}{2}$.

55. $f(x) = -3x^2 + 4x$

$-3x^2 + 4x = 0$

$x(-3x+4) = 0$

$-3x + 4 = 0 \qquad x = 0$

$-3x = -4$

$x = \frac{4}{3}$

The zeros are $\frac{4}{3}$ and 0.

57. $f(x) = -3x^2 + 12$

$-3x^2 + 12 = 0$

$-3(x^2 - 4) = 0$

$-3(x+2)(x-2) = 0$

$x + 2 = 0 \qquad x - 2 = 0$

$x = -2 \qquad x = 2$

The zeros are −2 and 2.

59. $f(x) = 2x^2 - 54$

$2x^2 - 54 = 0$

$2(x^2 - 27) = 0$

$x^2 - 27 = 0$

$x^2 = 27$

$x = \pm\sqrt{27} = \pm 3\sqrt{3}$

The zeros are $3\sqrt{3}$ and $-3\sqrt{3}$.

61. $f(x) = x^2 - 2x - 17$

$x^2 - 2x - 17 = 0$

$a = 1 \quad b = -2 \quad c = -17$

$$x = \frac{-b \pm \sqrt{b^2 - 4ac}}{2a}$$

$$x = \frac{-(-2) \pm \sqrt{(-2)^2 - 4(1)(-17)}}{2(1)}$$

$$x = \frac{2 \pm \sqrt{4+68}}{2} = \frac{2 \pm \sqrt{72}}{2} = \frac{2 \pm 6\sqrt{2}}{2}$$

$x = 1 \pm 3\sqrt{2}$

The zeros are $1 + 3\sqrt{2}$ and $1 - 3\sqrt{2}$.

63. $f(x) = x^2 + 4x + 5$

$x^2 + 4x + 5 = 0$

$a = 1 \quad b = 4 \quad c = 5$

$$x = \frac{-b \pm \sqrt{b^2 - 4ac}}{2a}$$

$$x = \frac{-4 \pm \sqrt{(4)^2 - 4(1)(5)}}{2(1)}$$

$$x = \frac{-4 \pm \sqrt{16 - 20}}{2} = \frac{-4 \pm \sqrt{-4}}{2} = \frac{-4 \pm 2i}{2}$$

$x = -2 \pm i$

The zeros are $-2 + i$ and $-2 - i$.

65. $f(x) = x^2 + 4x + 13$

$x^2 + 4x + 13 = 0$

$a = 1 \quad b = 4 \quad c = 13$

$$x = \frac{-b \pm \sqrt{b^2 - 4ac}}{2a}$$

$$x = \frac{-4 \pm \sqrt{(4)^2 - 4(1)(13)}}{2(1)}$$

$$x = \frac{-4 \pm \sqrt{16 - 52}}{2} = \frac{-4 \pm \sqrt{-36}}{2} = \frac{-4 \pm 6i}{2}$$

$x = -2 \pm 3i$

The zeros are $-2 + 3i$ and $-2 - 3i$.

67. $y = -x^2 - x + 3$

$a = -1 \quad b = -1 \quad c = 3$

$b^2 - 4ac$

$(-1)^2 - 4(-1)(3) = 1 + 12 = 13$

$13 > 0$

Since the discriminant is greater than zero the parabola has two x-intercepts.

69. $y = x^2 - 10x + 25$

$a = 1 \quad b = -10 \quad c = 25$

$b^2 - 4ac$

$(-10)^2 - 4(1)(25) = 100 - 100 = 0$

Since the discriminant is equal to zero the parabola has one x-intercept.

71. $y = -2x^2 + x - 1$

$a = -2 \quad b = 1 \quad c = -1$

$b^2 - 4ac$

$(1)^2 - 4(-2)(-1) = 1 - 8 = -7$

$-7 < 0$

Since the discriminant is less than zero the parabola has no x-intercepts.

73. $y = 4x^2 - x - 2$

$a = 4 \quad b = -1 \quad c = -2$

$b^2 - 4ac$

$(-1)^2 - 4(4)(-2) = 1 + 32 = 33$

$33 > 0$

Since the discriminant is greater than zero the parabola has two x-intercepts.

75. $y = 2x^2 + x + 4$

$a = 2 \quad b = 1 \quad c = 4$

$b^2 - 4ac$

$(1)^2 - 4(2)(4) = 1 - 32 = -31$

$-31 < 0$

Since the discriminant is less than zero the parabola has no x-intercepts.

77. $y = 4x^2 + 2x - 5$

$a = 4 \quad b = 2 \quad c = -5$

$b^2 - 4ac$

$(2)^2 - 4(4)(-5) = 4 + 80 = 84$

$84 > 0$

Since the discriminant is greater than zero the parabola has two x-intercepts.

79. a) $a > 0$
b) $a = 0$
c) $a < 0$

Objective C Exercises

81. To find the minimum or maximum value of a quadratic function find the x-coordinate of the vertex. Then evaluate the function at the value of this x-coordinate.

83. a) Since $a > 0$, the parabola opens up. The function has a minimum value.
b) Since $a < 0$, the parabola opens down. The function has a maximum value.
c) Since $a > 0$, the parabola opens up. The function has a minimum value.

85. $f(x) = 2x^2 + 4x$

$$x = -\frac{b}{2a} = -\frac{4}{2(2)} = -1$$

$f(x) = 2x^2 + 4x$

$f(-1) = 2(-1)^2 + 4(-1) = 2 - 4 = -2$

Since $a > 0$, the function has a minimum value. The minimum value of the function is -2.

87. $f(x) = -2x^2 + 4x - 5$

$$x = -\frac{b}{2a} = -\frac{4}{2(-2)} = 1$$

$f(x) = -2x^2 + 4x - 5$

$f(1) = -2(1)^2 + 4(1) - 5 = -2 + 4 - 5 = -3$

Since $a < 0$, the function has a maximum value. The maximum value of the function is -3.

89. $f(x) = -2x^2 - 3x$

$$x = -\frac{b}{2a} = -\frac{-3}{2(-2)} = -\frac{3}{4}$$

$f(x) = -2x^2 - 3x$

$$f\left(-\frac{3}{4}\right) = -2\left(-\frac{3}{4}\right)^2 - 3\left(-\frac{3}{4}\right) = -\frac{9}{8} + \frac{9}{4}$$

$$= \frac{9}{8}$$

Since $a < 0$, the function has a maximum value. The maximum value of the function is $\frac{9}{8}$.

91. $f(x) = 3x^2 + 3x - 2$

$$x = -\frac{b}{2a} = -\frac{3}{2(3)} = -\frac{1}{2}$$

$f(x) = 3x^2 + 3x - 2$

$$f\left(-\frac{1}{2}\right) = 3\left(-\frac{1}{2}\right)^2 + 3\left(-\frac{1}{2}\right) - 2 = \frac{3}{4} - \frac{3}{2} - 2$$

$$= -\frac{11}{4}$$

Since $a > 0$, the function has a minimum value. The minimum value of the function is $-\frac{11}{4}$.

93. $f(x) = -x^2 - x + 2$

$$x = -\frac{b}{2a} = -\frac{-1}{2(-1)} = -\frac{1}{2}$$

$f(x) = -x^2 - x + 2$

$$f\left(-\frac{1}{2}\right) = -\left(-\frac{1}{2}\right)^2 - \left(-\frac{1}{2}\right) + 2 = -\frac{1}{4} + \frac{1}{2} + 2$$

$$= \frac{9}{4}$$

Since $a < 0$, the function has a maximum value. The maximum value of the function is $\frac{9}{4}$.

95. $f(x) = 3x^2 + 5x + 2$

$$x = -\frac{b}{2a} = -\frac{5}{2(3)} = -\frac{5}{6}$$

$f(x) = 3x^2 + 5x + 2$

$$f\left(-\frac{5}{6}\right) = 3\left(-\frac{5}{6}\right)^2 + 5\left(-\frac{5}{6}\right) + 2 = \frac{25}{12} - \frac{25}{6} + 2$$

$$= -\frac{1}{12}$$

Since $a > 0$, the function has a minimum

value. The minimum value of the function is $-\frac{1}{12}$.

Objective D Exercises

97. Strategy: To find the price that will give the maximum revenue find the *P*-coordinate of the vertex.

Solution:

$$P=-\frac{b}{2a}=-\frac{125}{2\left(-\frac{1}{4}\right)}=250$$

A price of $250 will give the maximum revenue.

99. Strategy: To find the time it takes the plane to reach its maximum height find the *t*-coordinate of the vertex. To find the maximum height evaluate the function at the *t*-coordinate of the vertex.

Solution:

$$t=-\frac{b}{2a}=-\frac{119}{2(-1.42)}\approx 42$$

$$h(t)=-1.42t^2+119t+6000$$

$$h(42)=-1.42(42)^2+119(42)+6000$$

$$=-2504.88+4998+6000=8493.12$$

The maximum height of the plane is about 8500 m.

101. Strategy: To find the distance from one end of the bridge where the cable is at its minimum height find the *x*-coordinate of the vertex. To find the minimum height evaluate the function at the *x*-coordinate of the vertex.

Solution:

$$x=-\frac{b}{2a}=-\frac{-0.8}{2(0.25)}=1.6$$

$$h(x)=0.25x^2-0.8x+25$$

$$h(1.6)=0.25(1.6)^2-0.8(1.6)+25$$

$$=0.64-1.28+25=24.36$$

The cable is at its minimum height 1.6 ft from the end of the bridge.
The minimum height is 24.36 ft.

103. Strategy: Let x represent one number.
The other number is $20-x$.
Their product is $x(20-x)$.
To find one number find the *x*-coordinate of the vertex. To find the second number evaluate $20-x$ at the *x*-coordinate of the vertex.

Solution:

$$x(20-x)=20x-x^2$$

$$x=-\frac{b}{2a}=-\frac{20}{2(-1)}=10$$

$$20-x=20-10=10$$

The two numbers are 10 and 10.

105. Strategy: Let x represent one number.
The other number is $x-14$.
Their product is $x(x-14)$.
To find one number find the *x*-coordinate of the vertex. To find the second number evaluate $x-14$ at the *x*-coordinate of the vertex.

Solution:

$$x(x-14)=x^2-14x$$

$$x=-\frac{b}{2a}=-\frac{-14}{2(1)}=7$$

$$x-14=7-14=-7$$

The two numbers are 7and −7.

107. Strategy: Let x represent width of the rectangular corral. The length is $200-2x$.
The area is $x(200-2x)$.
To find the width find the *x*-coordinate of the vertex. To find the length evaluate $200-2x$ at the *x*-coordinate of the vertex.

Solution:

$$x(200-2x)=200x-2x^2$$

$$x=-\frac{b}{2a}=-\frac{200}{2(-2)}=50$$

$$200-2x=200-2(50)=100$$

The width is 50 ft and the length is 100 ft.

109. **Strategy**: Let x represent width of the ball fields. The length is $\frac{2100-3x}{2}$.

The area is $x\left(\frac{2100-3x}{2}\right)$.

To find the width find the x-coordinate of the vertex. To find the length evaluate $\frac{2100-3x}{2}$ at the x-coordinate of the vertex.

Solution:

$$x\left(\frac{2100-3x}{2}\right)=x(1050-1.5x)=1050x-1.5x^2$$

$$x=-\frac{b}{2a}=-\frac{1050}{2(-1.5)}=350$$

$$\frac{2100-3x}{2}=\frac{2100-1050}{2}=\frac{1050}{2}=525$$

The dimensions are 350 ft by 525 ft.

Applying the Concepts

111. **Strategy**: To find the root:
Let $x=4$, $a=2$, $b=-5$ and $c=k$ in the quadratic formula.
Substitute the value for k in the original equation and factor to solve.

Solution:

$$2x^2-5x+k=0$$

$$a=2 \quad b=-5 \quad c=k \quad x=4$$

$$x=\frac{-b\pm\sqrt{b^2-4ac}}{2a}$$

$$4=\frac{-(-5)\pm\sqrt{(-5)^2-4(2)(k)}}{2(2)}$$

$$4=\frac{5\pm\sqrt{25-8k}}{4}$$

$$16=5\pm\sqrt{25-8k}$$

$$11=\pm\sqrt{25-8k}$$

$$(11)^2=\left(\pm\sqrt{25-8k}\right)^2$$

$$121=25-8k$$

$$96=-8k$$

$$k=-12$$

$$2x^2-5x-12=0$$

$$(2x+3)(x-4)=0$$

$$2x+3=0 \quad x-4=0$$

$$2x=-3 \qquad x=4$$

$$x=-\frac{3}{2}$$

The other root is $-\frac{3}{2}$.

113. To find the root substitute $x=-2$, $y=0$ and $x=3$, $y=0$ into $f(x)=mx^2+nx+1$. Use these two equations to find the relationship between m and n.

Solution:

$$0=m(-2)^2+n(-2)+1 \qquad 0=m(3)^2+n(3)+1$$

$$0=4m-2n+1 \qquad 0=9m+3n+1$$

$$4m-2n+1=9m+3n+1$$

$$4m-2n=9m+3n$$

$$-5n=5m$$

$$n=-m$$

n and m are opposites.
Substituting $-n$ for m and $-m$ for n we get

$$f(x)=mx^2+nx+1=(-n)x^2+(-m)x+1$$

$$=-nx^2-mx+1$$

$$=nx^2+mx-1=g(x)$$

Therefore since $f(x)=g(x)$ their roots are the same. $g(x)$ has roots -2 and 3.

Section 11.2

Objective A Exercises

1. A vertical line intersects the function no more than once. Yes, the graph is a function.

3. A vertical line intersects the relation more than once. No, the graph is not a function.

5. A vertical line intersects the function no more than once. Yes, the graph is a function.

7. No

9. $f(x) = 1 - x^3$

Domain: $\{x \mid x \in \text{real numbers}\}$

Range: $\{y \mid y \in \text{real numbers}\}$

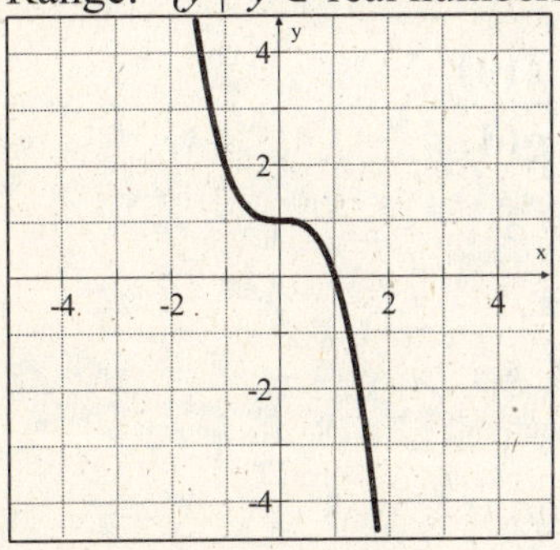

11. $f(x) = 3|2 - x|$

Domain: $\{x \mid x \in \text{real numbers}\}$

Range: $\{y \mid y \geq 0\}$

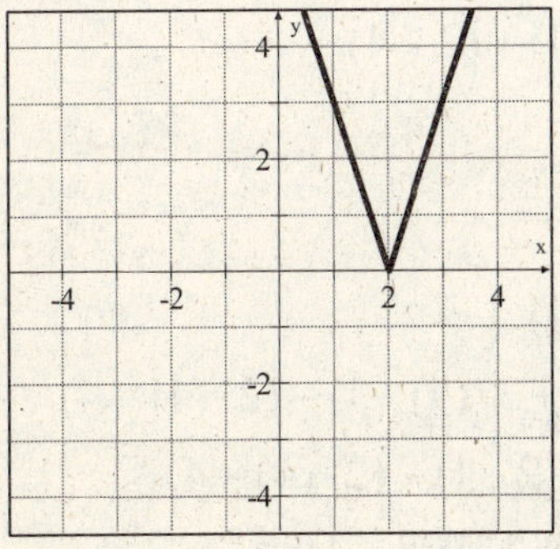

13. $f(x) = \sqrt{4 - x}$

Domain: $\{x \mid x \leq 4\}$

Range: $\{y \mid y \geq 0\}$

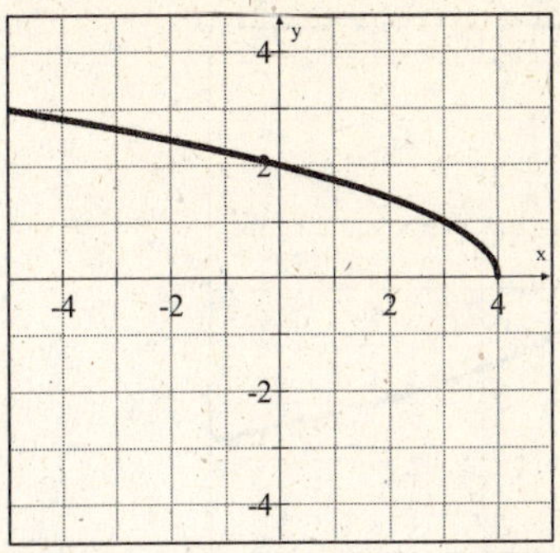

15. $f(x) = x^3 + 4x^2 + 4x$

Domain: $\{x \mid x \in \text{real numbers}\}$

Range: $\{y \mid y \in \text{real numbers}\}$

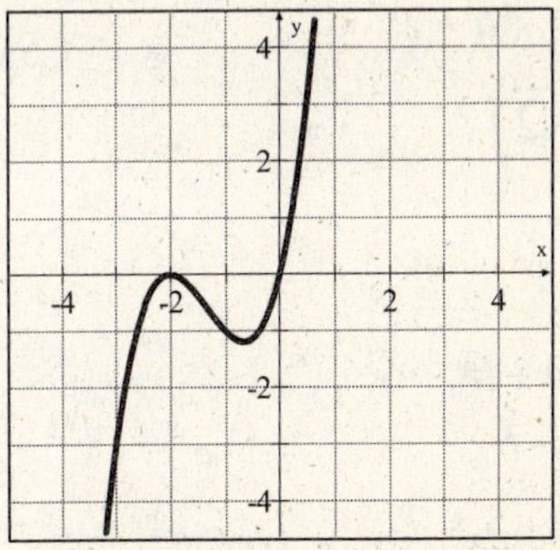

17. $f(x) = |2x + 2|$

Domain: $\{x \mid x \in \text{real numbers}\}$

Range: $\{y \mid y \geq 0\}$

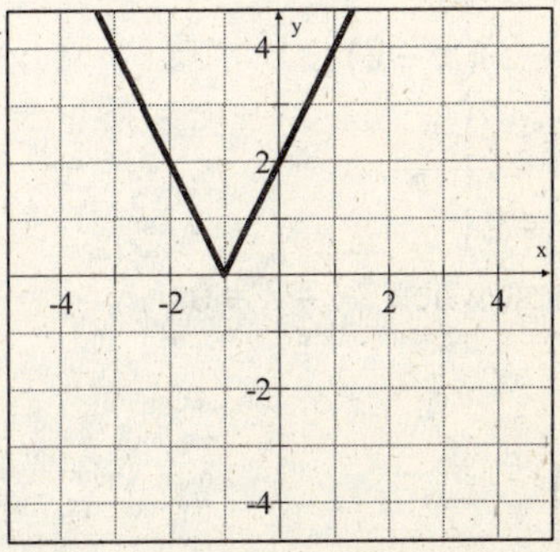

19. $f(x) = -\sqrt{x+2}$

Domain: $\{x \mid x \geq -2\}$
Range: $\{y \mid y \leq 0\}$

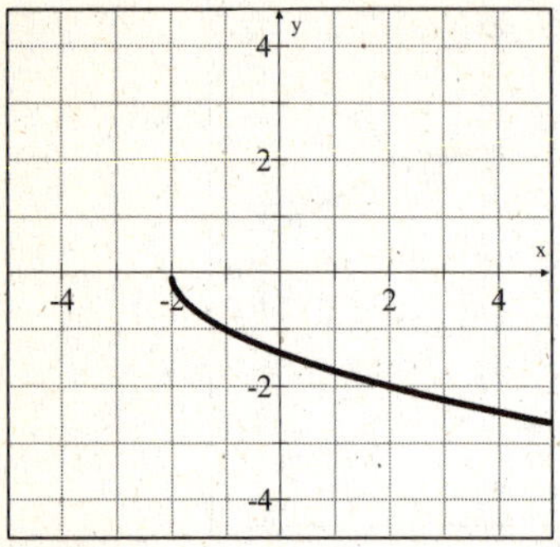

Applying the Concepts

21. $f(x) = \sqrt{x-2}$

$f(a) = 4 = \sqrt{a-2}$

$4^2 = \left(\sqrt{a-2}\right)^2$

$16 = a - 2$

$18 = a$

23. $f(a,b) = a + b$

$g(a,b) = a \cdot b$

$f(2,5) = 2 + 5 = 7$

$g(2,5) = 2 \cdot 5 = 10$

$f(2,5) + g(2,5) = 7 + 10 = 17$

25. $f(14) = 8$

27. $f(x) = (x+2)(x-2)$

$\{x \mid -2 < x < 2\}$

29. $f(x) = |2x - 2|$

$f(x)$ is smallest when $2x - 2 = 0$.
$x = 1$

Section 11.3

Objective A Exercises

1. $f(x) = x^2 + 4 \qquad g(x) = \sqrt{x+4}$

a) Yes
b) Yes
c) Yes
d) No

3. $f(2) - g(2) = (2 \cdot 2^2 - 3) - (-2 \cdot 2 + 4)$

$= (2 \cdot 4 - 3) - (-4 + 4)$

$= (8 - 3) - (0)$

$= 5$

5. $f(0) + g(0) = (2 \cdot 0^2 - 3) + (-2 \cdot 0 + 4)$

$= (0 - 3) + (0 + 4)$

$= -3 + 4$

$= 1$

7. $(f \cdot g)(2) = f(2) \cdot g(2)$

$= (2 \cdot 2^2 - 3) \cdot (-2 \cdot 2 + 4)$

$= (2 \cdot 4 - 3) \cdot (-4 + 4)$

$= (8 - 3) \cdot (0)$

$= 0$

9. $\left(\dfrac{f}{g}\right)(4) = \dfrac{f(4)}{g(4)}$

$= \dfrac{2 \cdot (4)^2 - 3}{-2 \cdot (4) + 4}$

$= \dfrac{2 \cdot 16 - 3}{-8 + 4}$

$= \dfrac{29}{-4} = -\dfrac{29}{4}$

11. $f(1) + g(1) = (2 \cdot 1^2 + 3 \cdot 1 - 1) + (2 \cdot 1 - 4)$

$= (2 \cdot 1 + 3 - 1) + (2 \cdot 1 - 4)$

$= (2 + 3 - 1) + (2 - 4)$

$= 4 - 2$

$= 2$

13. $f(4) - g(4)$

$= (2 \cdot (4)^2 + 3 \cdot (4) - 1) - (2 \cdot (4) - 4)$

$= (2 \cdot 16 + 12 - 1) - (2 \cdot (4) - 4)$

$= (32 + 12 - 1) - (8 - 4)$

$= 43 - 4$

$= 39$

15. $(f \cdot g)(1) = f(1) \cdot g(1)$
$= (2 \cdot (1)^2 + 3 \cdot (1) - 1) \cdot (2 \cdot (1) - 4)$
$= (2 \cdot 1 + 3 - 1) \cdot (2 \cdot (1) - 4)$
$= (2 + 3 - 1) \cdot (2 - 4)$
$= 4 \cdot (-2)$
$= -8$

17. $\left(\frac{f}{g}\right)(2) = \frac{f(2)}{g(2)}$
$= \frac{2 \cdot (2)^2 + 3 \cdot (2) - 1}{2 \cdot (2) - 4}$
$= \frac{2 \cdot 4 + 6 - 1}{4 - 4}$
$= \frac{13}{0}$
Undefined

19. $f(2) - g(2)$
$= (2^2 + 3 \cdot (2) - 5) - (2^3 - 2 \cdot (2) + 3)$
$= (4 + 6 - 5) - (8 - 4 + 3)$
$= 5 - 7$
$= -2$

21. $\left(\frac{f}{g}\right)(-2) = \frac{f(-2)}{g(-2)}$
$= \frac{(-2)^2 + 3 \cdot (-2) - 5}{(-2)^3 - 2 \cdot (-2) + 3}$
$= \frac{4 - 6 - 5}{-8 + 4 + 3}$
$= \frac{-7}{-1}$
$= 7$

Objective B Exercises

23. The expression $(f \circ g)(-2)$ means to evaluate the function f at $g(-2)$.

25. $f(-3) = 5$

27. $f(x) = 2x - 3 \qquad g(x) = 4x - 1$
$f(0) = 2(0) - 3$
$= 0 - 3 = -3$
$g(-3) = 4(-3) - 1$
$= -12 - 1 = -13$
$g[f(0)] = -13$

29. $f(x) = 2x - 3 \qquad g(x) = 4x - 1$
$f(-2) = 2(-2) - 3$
$= -4 - 3 = -7$
$g(-7) = 4(-7) - 1$
$= -28 - 1 = -29$
$g[f(-2)] = -29$

31. $f(x) = 2x - 3 \qquad g(x) = 4x - 1$
$g(2x - 3) = 4(2x - 3) - 1$
$= 8x - 12 - 1$
$= 8x - 13$
$g[f(x)] = 8x - 13$

33. $h(x) = 2x + 4 \qquad f(x) = \frac{1}{2}x + 2$
$h(0) = 2(0) + 4$
$= 0 + 4 = 4$
$f(4) = \frac{1}{2}(4) + 2$
$= 2 + 2 = 4$
$f[h(0)] = 4$

35. $h(x) = 2x + 4 \qquad f(x) = \frac{1}{2}x + 2$
$h(-1) = 2(-1) + 4$
$= -2 + 4 = 2$
$f(2) = \frac{1}{2}(2) + 2$
$= 1 + 2 = 3$
$f[h(-1)] = 3$

37. $h(x) = 2x + 4 \qquad f(x) = \frac{1}{2}x + 2$

$f(2x+4) = \frac{1}{2}(2x+4) + 2$

$= x + 2 + 2$

$= x + 4$

$f[h(x)] = x + 4$

39. $g(x) = x^2 + 3 \qquad h(x) = x - 2$

$g(0) = 0^2 + 3 = 3$

$h(3) = 3 - 2 = 1$

$h[g(0)] = 1$

41. $g(x) = x^2 + 3 \qquad h(x) = x - 2$

$g(-2) = (-2)^2 + 3 = 4 + 3 = 7$

$h(7) = 7 - 2 = 5$

$h[g(-2)] = 5$

43. $g(x) = x^2 + 3 \qquad h(x) = x - 2$

$h(x^2 + 3) = x^2 + 3 - 2$

$= x^2 + 1$

$h[g(x)] = x^2 + 1$

45. $f(x) = x^2 + x + 1 \qquad h(x) = 3x + 2$

$f(0) = (0)^2 + 0 + 1 = 1$

$h(1) = 3(1) + 2 = 3 + 2 = 5$

$h[f(0)] = 5$

47. $f(x) = x^2 + x + 1 \qquad h(x) = 3x + 2$

$f(-2) = (-2)^2 + (-2) + 1 = 4 - 2 + 1 = 3$

$h(3) = 3(3) + 2 = 9 + 2 = 11$

$h[f(-2)] = 11$

49. $f(x) = x^2 + x + 1 \qquad h(x) = 3x + 2$

$h(x^2 + x + 1) = 3(x^2 + x + 1) + 2$

$= 3x^2 + 3x + 3 + 2$

$= 3x^2 + 3x + 5$

$h[f(x)] = 3x^2 + 3x + 5$

51. $f(x) = x - 2 \qquad g(x) = x^3$

$g(-1) = (-1)^3 = -1$

$f(-1) = -1 - 2 = -3$

$f[g(-1)] = -3$

53. $f(x) = x - 2 \qquad g(x) = x^3$

$f(-1) = -1 - 2 = -3$

$g(-3) = (-3)^3 = -27$

$g[f(-1)] = -27$

55. $f(x) = x - 2 \qquad g(x) = x^3$

$g(x-2) = (x-2)^3$

$= (x-2)(x-2)(x-2)$

$= x^3 - 6x^2 + 12x - 8$

$g[f(x)] = x^3 - 6x^2 + 12x - 8$

57. a) **Strategy:** Selling price equals the cost plus the mark-up. If the cost is x and the mark-up is 60% then $S = x + 0.60x$.

If $M(x) = \dfrac{50x + 10{,}000}{x}$ is the cost per camera, then $(S \circ M)(x)$ is the selling price per camera. Find $(S \circ M)(x)$.

Solution:

$(S \circ M)(x) = S(M(x))$

$= M(x) + 0.60(M(x)) = 1.60(M(x))$

$= 1.60\left(\dfrac{50x + 10{,}000}{x}\right)$

$= \dfrac{80x + 16{,}000}{x}$

$S(M(x)) = 80 + \dfrac{16{,}000}{x}$

b) $(S \circ M)(5000) = 80 + \dfrac{16000}{5000}$

$= 80 + 3.2$

$= \$83.20$

c) When 5000 digital cameras are manufactured, the camera store sells each camera for $83.20.

59. a) $I(n) = 12{,}500n \quad n(m) = 4m$

$(I \circ n)(m) = I(n(m))$

$= 12{,}500(4m)$

$= 50{,}000m$

b) $(I \circ n)(3) = 50{,}000(3) = \$150{,}000$

c) The garage's income from conversions done during a 3 month period is $150,000.

61. rebate: $r(p) = p - 1500$
discounted price: $d(p) = 0.90p$

a) If the dealer takes the rebate first and then the discount we are finding $d(r(p)) = 0.90(p - 1500) = 0.90p - 1350$.

b) If the dealer takes the discount first and then the rebate we are finding $r(d(p)) = 0.90p - 1500$.

c) As a buyer you would prefer the dealer to use $r(d(p))$ since the cost would be less.

Applying the Concepts

63. $f(1) = 2$ and $g(2) = 0$
$g[f(1)] = g(2) = 0$

65. $(f \circ g)(3) = f(g(3))$
$g(3) = 5$ and $f(5) = -2$
$(f \circ g)(3) = f(g(3)) = -2$

67. $g(0) = -4$ and $f(-4) = 7$
$f[g(0)] = f(-4) = 7$

69. $g(x) = x^2 - 1$

$g(3+h) - g(3) = (3+h)^2 - 1 - (3^2 - 1)$

$= 9 + 6h + h^2 - 1 - 8$

$g(3+h) - g(3) = h^2 + 6h$

71. $g(x) = x^2 - 1$

$\dfrac{g(1+h) - g(1)}{h} = \dfrac{(1+h)^2 - 1 - (1^2 - 1)}{h}$

$= \dfrac{1 + 2h + h^2 - 1 - 0}{h}$

$= \dfrac{2h + h^2}{h} = 2 + h$

$\dfrac{g(1+h) - g(1)}{h} = 2 + h$

73. $g(x) = x^2 - 1$

$\dfrac{g(a+h) - g(a)}{h} = \dfrac{(a+h)^2 - 1 - (a^2 - 1)}{h}$

$= \dfrac{a^2 + 2ah + h^2 - 1 - a^2 + 1}{h}$

$= \dfrac{2ah + h^2}{h} = 2a + h$

$\dfrac{g(a+h) - g(a)}{h} = 2a + h$

75. $f(x) = 2x \quad g(x) = 3x - 1 \quad h(x) = x - 2$
$f(1) = 2(1) = 2$
$h(2) = 2 - 2 = 0$
$g(0) = 3(0) - 1 = 0 - 1 = -1$
$g(h[f(1)]) = -1$

77. $f(x) = 2x \quad g(x) = 3x - 1 \quad h(x) = x - 2$
$g(0) = 3(0) - 1 = 0 - 1 = -1$
$h(-1) = -1 - 2 = -3$
$f(-3) = 2(-3) = -6$
$f(h[g(0)]) = -6$

79. $f(x) = 2x \quad g(x) = 3x - 1 \quad h(x) = x - 2$
$h(x) = x - 2$
$f(x-2) = 2(x-2) = 2x - 4$
$g(2x-4) = 3(2x-4) - 1 = 6x - 13$
$g(f[h(x)]) = 6x - 13$

Section 11.4

Objective A Exercises

1. A function is a 1-1 function if, for any *a* and *b* in the domain of *f*, $f(a) = f(b)$ implies that $a = b$.

3. a) Yes
b) No

5. Yes, the graph represents a 1-1 function.

7. No, the graph is not a 1-1 function.

9. Yes, the graph represents a 1-1 function.

11. No, the graph is not a 1-1 function.

13. No, the graph is not a 1-1 function.

15. No, the graph is not a 1-1 function.

Objective B Exercises

17. The coordinates of each ordered pair of the inverse of a function are in the reverse order of the coordinates of the ordered pairs of the original function.

19. No

21. a) $f^{-1}(5) = 4$
b) $f^{-1}(4) = 3$
c) $f^{-1}(7) = 6$

23. Inverse function: $\{(0,1),(3,2),(8,3),(15,4)\}$

25. No inverse because the numbers 5 and -5 would each be paired with two different values in the range.

27. Inverse function:
$\{(-2,0),(5,-1),(3,3),(6,-4)\}$

29. No inverse because the number 3 would be paired with three different values of the range.

31. $f(x) = 4x - 8$
$y = 4x - 8$
$x = 4y - 8$
$x + 8 = 4y$
$\frac{1}{4}x + 2 = y$
$f^{-1}(x) = \frac{1}{4}x + 2$

33. $f(x) = 2x + 4$
$y = 2x + 4$
$x = 2y + 4$
$x - 4 = 2y$
$\frac{1}{2}x - 2 = y$
$f^{-1}(x) = \frac{1}{2}x - 2$

35. $f(x) = \frac{1}{2}x - 1$

$y = \frac{1}{2}x - 1$

$x = \frac{1}{2}y - 1$

$x + 1 = \frac{1}{2}y$

$2x + 2 = y$

$f^{-1}(x) = 2x + 2$

37. $f(x) = -2x + 2$

$y = -2x + 2$

$x = -2y + 2$

$x - 2 = -2y$

$-\frac{1}{2}x + 1 = y$

$f^{-1}(x) = -\frac{1}{2}x + 1$

39. $f(x) = \frac{2}{3}x + 4$

$y = \frac{2}{3}x + 4$

$x = \frac{2}{3}y + 4$

$x - 4 = \frac{2}{3}y$

$\frac{3}{2}x - 6 = y$

$f^{-1}(x) = \frac{3}{2}x - 6$

41. $f(x) = -\frac{1}{3}x + 1$

$y = -\frac{1}{3}x + 1$

$x = -\frac{1}{3}y + 1$

$x - 1 = -\frac{1}{3}y$

$-3x + 3 = y$

$f^{-1}(x) = -3x + 3$

43. $f(x) = 2x - 5$

$y = 2x - 5$

$x = 2y - 5$

$x + 5 = 2y$

$\frac{1}{2}x + \frac{5}{2} = y$

$f^{-1}(x) = \frac{1}{2}x + \frac{5}{2}$

45. $f(x) = 5x - 2$

$y = 5x - 2$

$x = 5y - 2$

$x + 2 = 5y$

$\frac{1}{5}x + \frac{2}{5} = y$

$f^{-1}(x) = \frac{1}{5}x + \frac{2}{5}$

47. $f(x) = 6x - 3$

$y = 6x - 3$

$x = 6y - 3$

$x + 3 = 6y$

$\frac{1}{6}x + \frac{1}{2} = y$

$f^{-1}(x) = \frac{1}{6}x + \frac{1}{2}$

49. $f(x) = 3x - 5$

$y = 3x - 5$

$x = 3y - 5$

$x + 5 = 3y$

$\frac{1}{3}x + \frac{5}{3} = y$

$f^{-1}(x) = \frac{1}{3}x + \frac{5}{3}$

$f^{-1}(0) = \frac{1}{3}(0) + \frac{5}{3}$

$f^{-1}(0) = \frac{5}{3}$

51. $f(x) = 3x - 5$

$y = 3x - 5$

$x = 3y - 5$

$x + 5 = 3y$

$\frac{1}{3}x + \frac{5}{3} = y$

$f^{-1}(x) = \frac{1}{3}x + \frac{5}{3}$

$f^{-1}(4) = \frac{1}{3}(4) + \frac{5}{3}$

$f^{-1}(4) = \frac{9}{3} = 3$

53. Using the vertical line test the graph is a function. Using the horizontal line test the graph is 1-1 and therefore does have an inverse.

55. $f(g(x)) = f\left(\frac{x}{4}\right) = 4\left(\frac{x}{4}\right) = x$

$g(f(x)) = g(4x) = \frac{4x}{4} = x$

Yes, the functions are inverses of each other.

57. $f(h(x)) = f\left(\frac{1}{3x}\right) = 3\left(\frac{1}{3x}\right) = \frac{1}{x}$

$h(f(x)) = h(3x) = \frac{1}{3(3x)} = \frac{1}{9x}$

No, the functions are not inverses of each other.

59. $f(g(x)) = f(3x + 2)$

$= \frac{1}{3}(3x + 2) - \frac{2}{3} = x + \frac{2}{3} - \frac{2}{3} = x$

$g(f(x)) = g\left(\frac{1}{3}x - \frac{2}{3}\right)$

$= 3\left(\frac{1}{3}x - \frac{2}{3}\right) + 2 = x - 2 + 2 = x$

Yes, the functions are inverses of each other.

61. $f(g(x)) = f(2x + 3)$

$= \frac{1}{2}(2x + 3) - \frac{3}{2} = x + \frac{3}{2} - \frac{3}{2} = x$

$g(f(x)) = g\left(\frac{1}{2}x - \frac{3}{2}\right)$

$= 2\left(\frac{1}{2}x - \frac{3}{2}\right) + 3 = x - 3 + 3 = x$

Yes, the functions are inverses of each other.

63. $f(x) = \frac{x}{16}$

$y = \frac{x}{16}$

$x = \frac{y}{16}$

$16x = y$

$f^{-1}(x) = 16x$

The inverse function converts pounds to ounces.

65. $f(x) = x + 30$

$y = x + 30$

$x = y + 30$

$x - 30 = y$

$f^{-1}(x) = x - 30$

The inverse function converts a dress size in France to a dress size in the United States.

67. $f(x) = 90x + 65$

$y = 90x + 65$

$x = 90y + 65$

$x - 65 = 90y$

$\frac{1}{90}x - \frac{13}{18} = y$

$f^{-1}(x) = \frac{1}{90}x - \frac{13}{18}$

The inverse function gives the training intensity percent for a given target heart rate.

Applying the Concepts

69.

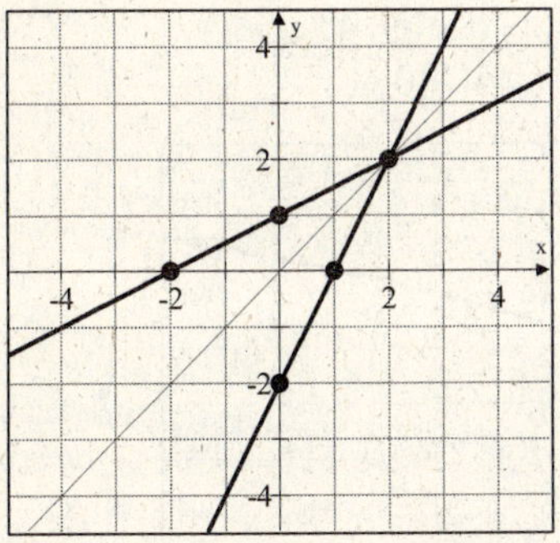

71.

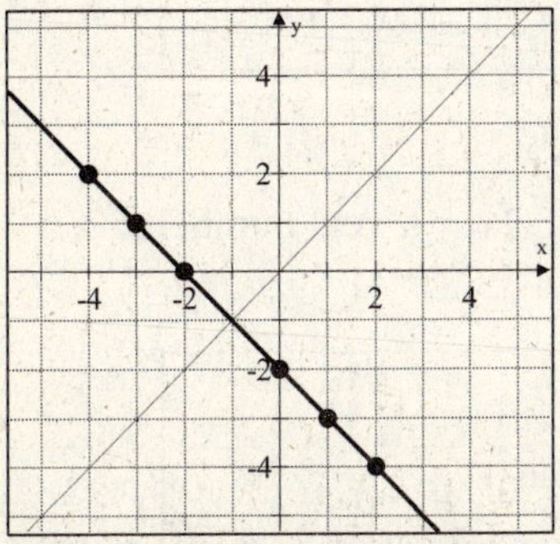

The inverse is the same graph.

73.

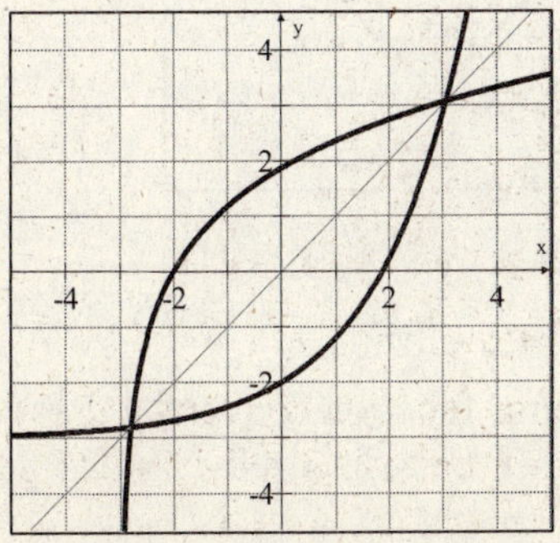

75. Inverse of the function:

Grade	Score
A	90-100
B	80-89
C	70-79
D	60-69
F	0-59

No, the inverse of the grading scale is not a function because each grade is paired with more than one score.

77. A constant function is defined as $y = b$ where b is a constant value. The inverse of this function would be $x = a$ where a is a constant. This is not a function.

Concept Review

1. Given a quadratic equation $y = ax^2 + bx + c$ the axis of symmetry is $x = -\frac{b}{2a}$ and the coordinates of the vertex are $\left(-\frac{b}{2a}, f\left(-\frac{b}{2a}\right)\right)$.

2. You cannot accurately graph a parabola given a vertex and axis of symmetry because this does not give enough information about how wide or narrow the parabola is.

3. If you are given the vertex of a parabola you can determine the range of the function. If the parabola opens up, the range is all y values greater than or equal to the y-coordinate of the vertex. If the parabola opens down, the range is all y values less than or equal to the y-coordinate of the vertex.

4. The discriminant tells us the number of x-intercepts of a parabola.
$b^2 - 4ac = 0$ indicates the parabola has one x-intercept.
$b^2 - 4ac > 0$ indicates the parabola has two x-intercepts.
$b^2 - 4ac < 0$ indicates the parabola has no x-intercept.

5. The coefficient of the x^2 term determines how wide or narrow the parabola is. The larger the absolute value of the coefficient of the x^2 term the narrower the parabola; the smaller the absolute value of the coefficient of the x^2 term the wider the parabola is.

6. The vertical line test states that a graph defines a function if any vertical line intersects the graph at no more than one point.

7. The graph of the absolute value of a linear polynomial is V-shaped.

8. The four basic operations on functions are:

$$(f+g)(x) = f(x) + g(x)$$
$$(f-g)(x) = f(x) - g(x)$$
$$(f \cdot g)(x) = f(x) \cdot g(x)$$
$$\left(\frac{f}{g}\right)(x) = \frac{f(x)}{g(x)},\ g(x) \neq 0$$

9. The notation $(f \circ g)(x) = f[g(x)]$.

10. The horizontal line test states that a graph represents a 1-1 function if any horizontal line intersects the graph at no more than one point.

11. For a function to have an inverse, the function must be 1-1.

Chapter 11 Review Exercises

1. Yes, the graph is a function. It passes the vertical line test.

2. Yes, the graph is a 1-1 function. It passes the horizontal line test.

3. $f(x) = 3x^3 - 2$
Domain: $\{x \mid x \in \text{real numbers}\}$
Range: $\{y \mid y \in \text{real numbers}\}$

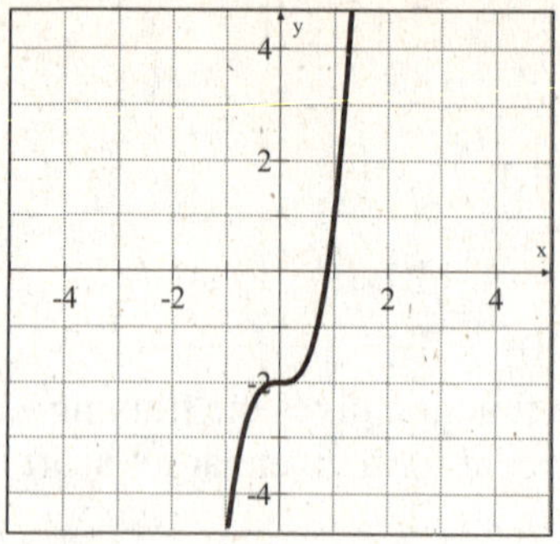

4. $f(x) = \sqrt{x+4}$
Domain: $\{x \mid x \geq -4\}$
Range: $\{y \mid y \geq 0\}$

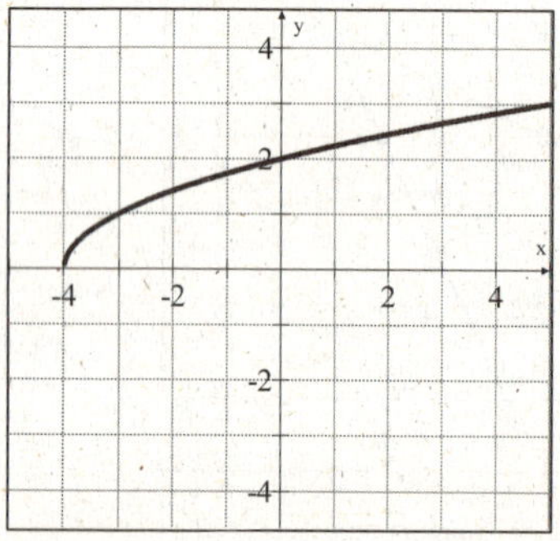

5. $f(x) = |x| - 3$
Domain: $\{x \mid x \in \text{real numbers}\}$
Range: $\{y \mid y \geq -3\}$

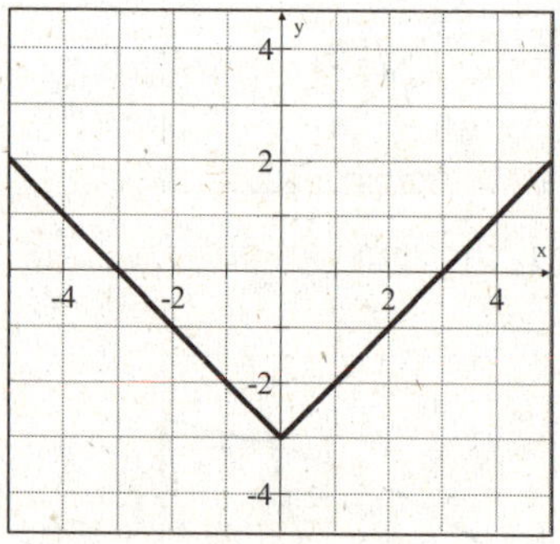

6. $y = x^2 - 2x + 3$

$$-\frac{b}{2a} = -\frac{-2}{2(1)} = -\frac{-2}{2} = -(-1) = 1$$

$y = 1^2 - 2(1) + 3 = 2$

Vertex: $(1,2)$
Axis of symmetry: $x = 1$

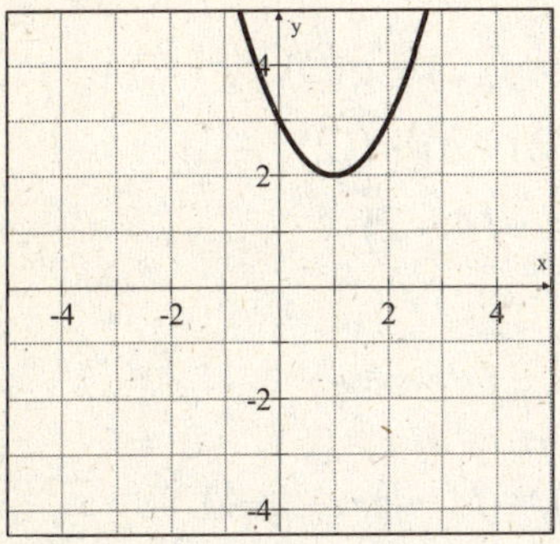

7. $y = -3x^2 + 4x + 6$
$a = -3 \quad b = 4 \quad c = 6$

$b^2 - 4ac$

$(4)^2 - 4(-3)(6) = 16 + 72 = 88$

$88 > 0$
Since the discriminant is greater than zero the parabola has two x-intercepts.

8. $f(x) = 2x^2 + x + 5$
$a = 2 \quad b = 1 \quad c = 5$

$b^2 - 4ac$

$(1)^2 - 4(2)(5) = 1 - 40 = -39$

$-39 < 0$
Since the discriminant is less than zero the parabola has no x-intercepts.

9. $y = 3x^2 + 9x$

$0 = 3x^2 + 9x$

$0 = 3x(x + 3)$

$3x = 0 \quad x + 3 = 0$

$x = 0 \quad x = -3$
The x-intercepts are (0,0) and (−3, 0).

10. $f(x) = x^2 - 6x + 7$

$0 = x^2 - 6x + 7$

$a = 1 \quad b = -6 \quad c = 7$

$$x = \frac{-b \pm \sqrt{b^2 - 4ac}}{2a}$$

$$x = \frac{-(-6) \pm \sqrt{(-6)^2 - 4(1)(7)}}{2(1)}$$

$$x = \frac{6 \pm \sqrt{36 - 28}}{2} = \frac{6 \pm \sqrt{8}}{2} = \frac{6 \pm 2\sqrt{2}}{2}$$

$x = 3 \pm \sqrt{2}$

The x-intercepts are $(3 + \sqrt{2}, 0)$ and $(3 - \sqrt{2}, 0)$.

11. $f(x) = 2x^2 - 7x - 15$

$2x^2 - 7x - 15 = 0$

$(2x + 3)(x - 5) = 0$

$2x + 3 = 0 \quad x - 5 = 0$

$2x = -3 \quad x = 5$

$$x = -\frac{3}{2}$$

The zeros are $-\frac{3}{2}$ and 5.

12. $f(x) = x^2 - 2x + 10$

$x^2 - 2x + 10 = 0$

$a = 1 \quad b = -2 \quad c = 10$

$$x = \frac{-b \pm \sqrt{b^2 - 4ac}}{2a}$$

$$x = \frac{-(-2) \pm \sqrt{(-2)^2 - 4(1)(10)}}{2(1)}$$

$$x = \frac{2 \pm \sqrt{4 - 40}}{2} = \frac{2 \pm \sqrt{-36}}{2} = \frac{2 \pm 6i}{2}$$

$x = 1 \pm 3i$
The zeros are $1 + 3i$ and $1 - 3i$.

13. $f(x) = -2x^2 + 4x + 1$

$x = -\frac{b}{2a} = -\frac{4}{2(-2)} = 1$

$f(x) = -2x^2 + 4x + 1$

$f(1) = -2(1)^2 + 4(1) + 1 = -2 + 4 + 1 = 3$

The maximum value of the function is 3.

14. $f(x) = x^2 - 7x + 8$

$x = -\frac{b}{2a} = -\frac{-7}{2(1)} = -\frac{7}{2}$

$f(x) = x^2 - 7x + 8$

$f\left(\frac{7}{2}\right) = \left(\frac{7}{2}\right)^2 - 7\left(\frac{7}{2}\right) + 8 = \frac{49}{4} - \frac{49}{2} + 8 = -\frac{17}{4}$

The minimum value of the function is $-\frac{17}{4}$.

15. $f(x) = x^2 + 4 \quad g(x) = 4x - 1$

$g(0) = 4(0) - 1 = -1$

$f(-1) = (-1)^2 + 4 = 1 + 4 = 5$

$f[g(0)] = 5$

16. $f(x) = 6x + 8 \quad g(x) = 4x + 2$

$f(-1) = 6(-1) + 8 = -6 + 8 = 2$

$g(2) = 4(2) + 2 = 8 + 2 = 10$

$g[f(-1)] = 10$

17. $f(x) = 3x^2 - 4 \quad g(x) = 2x + 1$

$f(g(x)) = f(2x + 1)$

$= 3(2x+1)^2 - 4 = 3(2x+1)(2x+1) - 4$

$= 3(4x^2 + 4x + 1) - 4$

$= 12x^2 + 12x + 3 - 4 = 12x^2 + 12x - 1$

$f[g(x)] = 12x^2 + 12x - 1$

18. $f(x) = 2x^2 + x - 5 \quad g(x) = 3x - 1$

$g(f(x)) = g(2x^2 + x - 5)$

$= 3(2x^2 + x - 5) - 1$

$= 6x^2 + 3x - 15 - 1$

$= 6x^2 + 3x - 16$

$g[f(x)] = 6x^2 + 3x - 16$

19. $(f + g)(2) = f(2) + g(2)$

$= ((2)^2 + 2(2) - 3) + ((2)^2 - 2)$

$= (4 + 4 - 3) + (4 - 2)$

$= 5 + 2$

$= 7$

20. $(f - g)(-4) = f(-4) - g(-4)$

$= ((-4)^2 + 2(-4) - 3) - ((-4)^2 - 2)$

$= (16 - 8 - 3) - (16 - 2)$

$= 5 - 14$

$= -9$

21. $(f \cdot g)(-4) = f(-4) \cdot g(-4)$

$= ((-4)^2 + 2(-4) - 3) \cdot ((-4)^2 - 2)$

$= (16 - 8 - 3) \cdot (16 - 2)$

$= 5 \cdot 14$

$= 70$

22. $\left(\frac{f}{g}\right)(3) = \frac{f(3)}{g(3)}$

$= \frac{(3)^2 + 2(3) - 3}{(3)^2 - 2}$

$= \frac{9 + 6 - 3}{9 - 2}$

$= \frac{12}{7}$

23. $f(x) = -6x + 4$

$y = -6x + 4$

$x = -6y + 4$

$x - 4 = -6y$

$-\frac{1}{6}x + \frac{2}{3} = y$

$f^{-1}(x) = -\frac{1}{6}x + \frac{2}{3}$

24. $f(x) = \frac{2}{3}x - 12$

$y = \frac{2}{3}x - 12$

$x = \frac{2}{3}y - 12$

$x + 12 = \frac{2}{3}y$

$\frac{3}{2}x + 18 = y$

$f^{-1}(x) = \frac{3}{2}x + 18$

25. $f(g(x)) = f(-4x + 5)$

$= -\frac{1}{4}(-4x + 5) + \frac{5}{4} = x - \frac{5}{4} + \frac{5}{4} = x$

$g(f(x)) = g\left(-\frac{1}{4}x + \frac{5}{4}\right)$

$= -4\left(-\frac{1}{4}x + \frac{5}{4}\right) + 5 = x - 5 + 5 = x$

Yes, the functions are inverses of each other.

26. $f(g(x)) = f(2x + 1)$

$= \frac{1}{2}(2x + 1) = x + \frac{1}{2}$

$g(f(x)) = g\left(\frac{1}{2}x\right)$

$= 2\left(\frac{1}{2}x\right) + 1 = x + 1$

No, the functions are not inverses of each other.

27. $p(x) = 0.4x + 15$

$p = 0.4x + 15$

$x = 0.4p + 15$

$x - 15 = 0.4p$

$2.5x - 37.5 = p$

$p^{-1}(x) = 2.5x - 37.5$

The inverse function gives the diver's depth below the surface of the water for a given pressure on the diver.

28. **Strategy**: To find the number of gloves for a maximum profit find the x-coordinate of the vertex. To find the maximum profit evaluate the function at the x-coordinate of the vertex.

Solution:

$x = -\frac{b}{2a} = -\frac{100}{2(-1)} = -\frac{100}{-2} = -(-50) = 50$

$P(x) = -x^2 + 100x + 2500$

$P(50) = -(50)^2 + 100(50) + 2500$

$= -2500 + 5000 + 2500 = 5000$

The company should make 50 baseball gloves each month to maximize profit. The maximum profit is \$5000.

29. Strategy: Let x represent width of the rectangle.
The length is $14 - x$.
The area is $x(14 - x)$.
To find the width find the x-coordinate of the vertex. To find the length evaluate $14 - x$ at the x-coordinate of the vertex.

Solution:

$x(14 - x) = 14x - x^2$

$x = -\frac{b}{2a} = -\frac{14}{2(-1)} = -\frac{14}{-2} = -(-7) = 7$

$14 - x = 14 - 7 = 7$

The dimensions are 7 ft by 7 ft.

Chapter 11 Test

1. $f(x) = x^2 - 6x + 4$

$$-\frac{b}{2a} = -\frac{-6}{2(1)} = -\frac{-6}{2} = -(-3) = 3$$

$$y = 3^2 - 6(3) + 4 = -5$$

Vertex: $(3,-5)$
Axis of symmetry: $x = 3$

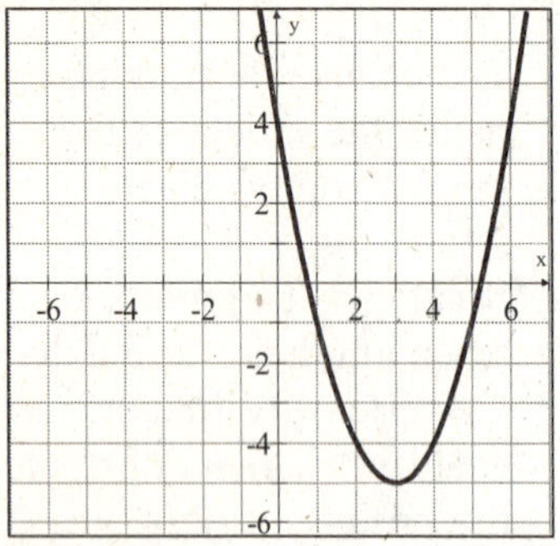

2. $f(x) = \left|\frac{1}{2}x\right| - 2$

Domain: $\{x | x \in \text{real numbers}\}$
Range: $\{y | y \geq -2\}$

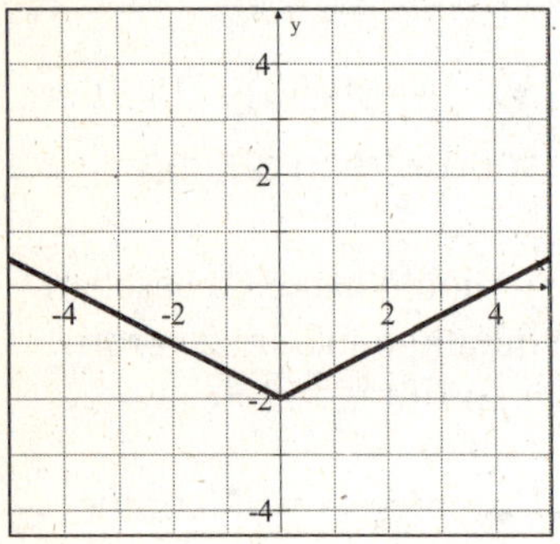

3. $f(x) = -\sqrt{3-x}$

Domain: $\{x | x \leq 3\}$
Range: $\{y | y \leq 0\}$

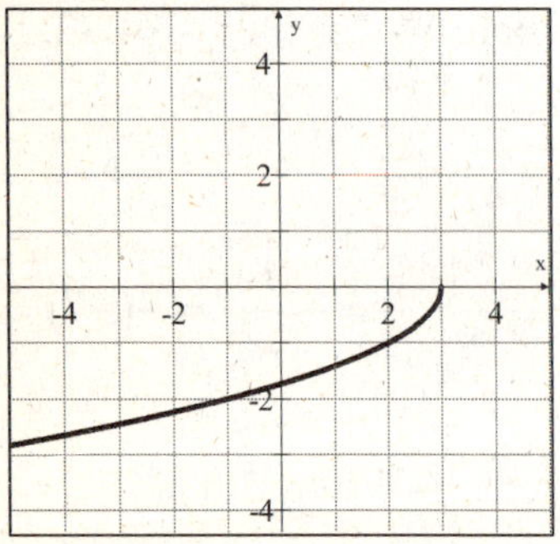

4. $f(x) = x^3 - 3x + 2$

Domain: $\{x | x \in \text{real numbers}\}$
Range: $\{y | y \in \text{real numbers}\}$

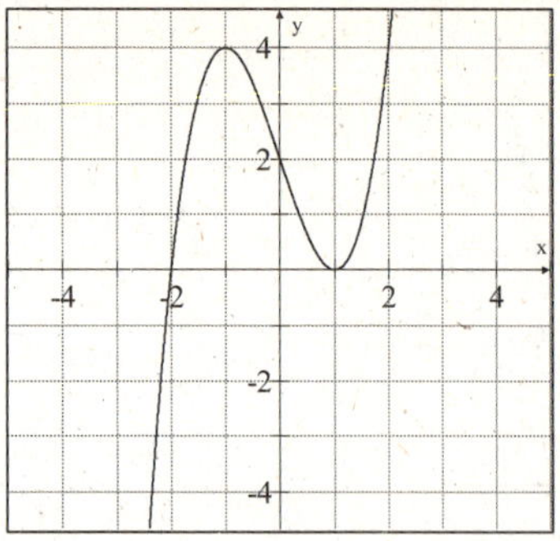

5. $y = 3x^2 - 4x + 6$

$a = 3 \quad b = -4 \quad c = 6$

$b^2 - 4ac$

$(-4)^2 - 4(3)(6) = 16 - 72 = -56$

$-56 < 0$

Since the discriminant is less than zero the parabola has no x-intercepts.

6. $y = 3x^2 - 7x - 6$

$0 = 3x^2 - 7x - 6$

$0 = (3x + 2)(x - 3)$

$3x + 2 = 0 \quad x - 3 = 0$

$3x = -2 \qquad x = 3$

$x = -\frac{2}{3}$

The x-intercepts are $(-\frac{2}{3},0)$ and $(3, 0)$.

7. $f(x) = -x^2 + 8x - 7$

$$x = -\frac{b}{2a} = -\frac{8}{2(-1)} = -(-4) = 4$$

$f(x) = -x^2 + 8x - 7$

$f(4) = -(4)^2 + 8(4) - 7 = -16 + 32 - 7 = 9$

The maximum value of the function is 9.

8. $f(x) = 2x^2 + 4x - 5$
Domain: $\{x| x \in \text{real numbers}\}$
Range: $\{y| y \geq -7\}$

9. $f(x) = x^2 + 2x - 3 \quad g(x) = x^3 - 1$
$$(f - g)(2) = f(2) - g(2)$$
$$= (2^2 + 2(2) - 3) - (2^3 - 1)$$
$$= (4 + 4 - 3) - (8 - 1)$$
$$= 5 - 7$$
$$= -2$$

10. $f(x) = x^3 + 1 \quad g(x) = 2x - 3$
$$(f \cdot g)(-3) = f(-3) \cdot g(-3)$$
$$= ((-3)^3 + 1) \cdot (2(-3) - 3)$$
$$= (-27 + 1) \cdot (-6 - 3)$$
$$= (-26) \cdot (-9)$$
$$= 234$$

11. $f(x) = 4x - 5 \quad g(x) = x^2 + 3x + 4$
$$\left(\frac{f}{g}\right)(-2) = \frac{f(-2)}{g(-2)}$$
$$= \frac{4(-2) - 5}{(-2)^2 + 3(-2) + 4}$$
$$= \frac{-8 - 5}{4 - 6 + 4} = \frac{-13}{2}$$
$$= -\frac{13}{2}$$

12. $f(x) = x^2 + 4 \quad g(x) = 2x^2 + 2x + 1$
$$(f - g)(-4) = f(-4) - g(-4)$$
$$= ((-4)^2 + 4) - (2(-4)^2 + 2(-4) + 1)$$
$$= (16 + 4) - (32 - 8 + 1)$$
$$= 20 - 25$$
$$= -5$$

13. $f(x) = 2x - 7 \quad g(x) = x^2 - 2x - 5$
$$g(2) = 2^2 - 2(2) - 5 = 4 - 4 - 5 = -5$$
$$f(-5) = 2(-5) - 7 = -10 - 7 = -17$$
$$f[g(2)] = -17$$

14. $f(x) = x^2 + 1 \quad g(x) = x^2 + x + 1$
$$f(-2) = (-2)^2 + 1 = 4 + 1 = 5$$
$$g(5) = (5)^2 + 5 + 1 = 25 + 5 + 1 = 31$$
$$g[f(-2)] = 31$$

15. $f(x) = x^2 - 1 \quad g(x) = 3x + 2$
$$g(f(x)) = g(x^2 - 1)$$
$$= 3(x^2 - 1) + 2 = 3x^2 - 3 + 2$$
$$= 3x^2 - 1$$
$$g[f(x)] = 3x^2 - 1$$

16. $f(x) = 2x^2 - 7 \quad g(x) = x - 1$
$$f(g(x)) = f(x - 1)$$
$$= 2(x - 1)^2 - 7 = 2(x - 1)(x - 1) - 7$$
$$= 2(x^2 - 2x + 1) - 7$$
$$= 2x^2 - 4x + 2 - 7$$
$$= 2x^2 - 4x - 5$$
$$f[g(x)] = 2x^2 - 4x - 5$$

17. No, inverse because the numbers 4 and 5 would be paired with two different values of the range.

18. Inverse function: $\{(6, 2), (5, 3), (4, 4), (3, 5)\}$

19. $f(x) = 4x - 2$
$$y = 4x - 2$$
$$x = 4y - 2$$
$$x + 2 = 4y$$
$$\frac{1}{4}x + \frac{1}{2} = y$$
$$f^{-1}(x) = \frac{1}{4}x + \frac{1}{2}$$

20. $f(x) = \frac{1}{4}x - 4$

$y = \frac{1}{4}x - 4$

$x = \frac{1}{4}y - 4$

$x + 4 = \frac{1}{4}y$

$4x + 16 = y$

$f^{-1}(x) = 4x + 16$

21. $f(g(x)) = f(2x - 4)$

$= \frac{1}{2}(2x - 4) + 2 = x - 2 + 2 = x$

$g(f(x)) = g\left(\frac{1}{2}x + 2\right)$

$= 2\left(\frac{1}{2}x + 2\right) - 4 = x + 4 - 4 = x$

Yes, the functions are inverses of each other.

22. $f(g(x)) = f\left(\frac{3}{2}x + 3\right)$

$= \frac{2}{3}\left(\frac{3}{2}x + 3\right) - 2 = x + 2 - 2 = x$

$g(f(x)) = g\left(\frac{2}{3}x - 2\right)$

$= \frac{3}{2}\left(\frac{2}{3}x - 2\right) + 3 = x - 3 + 3 = x$

Yes, the functions are inverses of each other.

23. No, the graph is not a 1-1 function. It does not pass the horizontal line test.

24. $C(x) = 1.25x + 5$

$C = 1.25x + 5$

$x = 1.25C + 5$

$x - 5 = 1.25C$

$0.8x - 4 = C$

$C^{-1}(x) = 0.8x - 4$

The inverse function gives the number of miles to a certain location for a given cost.

25. Strategy: To find the number of speakers for a minimum production cost find the *x*-coordinate of the vertex. To find the minimum cost evaluate the function at the *x*-coordinate of the vertex.

Solution:

$x = -\frac{b}{2a} = -\frac{-50}{2} = -\frac{-50}{2} = -(-25) = 25$

$C(x) = x^2 - 50x + 675$

$C(25) = (25)^2 - 50(25) + 675$

$= 625 - 1250 + 675 = 50$

The company should make 25 speakers each day to minimize production costs.
The minimum daily production cost is \$50.

26. Strategy: Let *x* represent one number. The other number is $28 - x$.
Their product is $x(28 - x)$.
To find one number find the *x*-coordinate of the vertex. To find the second number evaluate $28 - x$ at the *x*-coordinate of the vertex.

Solution:

$x(28 - x) = 28x - x^2$

$x = -\frac{b}{2a} = -\frac{28}{2(-1)} = -\frac{28}{-2} = -(-14) = 14$

$28 - x = 28 - 14 = 14$

The two numbers are 14 and 14.

27. Strategy: Let *x* represent width of the rectangle. The length is 100 - *x*.
The area is $x(100 - x)$.
To find the width find the *x*-coordinate of the vertex. To find the length evaluate $100 - x$ at the *x*-coordinate of the vertex.

Solution:

$x(100 - x) = 100x - x^2$

$x = -\frac{b}{2a} = -\frac{100}{2(-1)} = -\frac{100}{-2} = -(-50) = 50$

$100 - x = 100 - 50 = 50$

$A = l \cdot w = 50 \cdot 50 = 2500$

The dimensions of the rectangle are 50 cm by 50 cm. The area is 2500 cm ^{2}.

Cumulative Review Exercises

1. $-3a + \left|\dfrac{3b - ab}{3b - c}\right|$

$-3(2) + \left|\dfrac{3(2) - (2)(2)}{3(2) - (-2)}\right|$

$= -6 + \left|\dfrac{6-4}{6+2}\right| = -6 + \left|\dfrac{2}{8}\right| = -6 + \left|\dfrac{1}{4}\right|$

$= -6 + \dfrac{1}{4} = -\dfrac{23}{4}$

2. 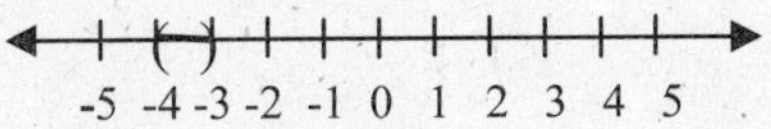

3. $\dfrac{3x-1}{6} - \dfrac{5-x}{4} = \dfrac{5}{6}$

$12\left(\dfrac{3x-1}{6} - \dfrac{5-x}{4}\right) = 12\left(\dfrac{5}{6}\right)$

$2(3x-1) - 3(5-x) = 10$

$6x - 2 - 15 + 3x = 10$

$9x - 17 = 10$

$9x = 27$

$x = 3$

The solution is 3.

4. $4x - 2 < -10$ or $3x - 1 > 8$

$4x - 2 < -10 \qquad 3x - 1 > 8$

$4x < -8 \qquad 3x > 9$

$x < -2 \qquad x > 3$

$\{x \mid x < -2\}$ or $\{x \mid x > 3\}$

$\{x \mid x < -2\} \cup \{x \mid x > 3\}$

$= \{x \mid x < -2 \text{ or } x > 3\}$

5. $|8 - 2x| \geq 0$

$8 - 2x \leq 0 \qquad 8 - 2x \geq 0$

$-2x \leq -8 \qquad -2x \geq -8$

$x \geq 4 \qquad x \leq 4$

$\{x \mid x \geq 4\}$ or $\{x \mid x \leq 4\}$

$\{x \mid x \geq 4\} \cup \{x \mid x \leq 4\}$

$= \{x \mid x \in \text{real numbers}\}$

6. $\left(\dfrac{3a^3b}{2a}\right)^2 \left(\dfrac{a^2}{-3b^2}\right)^3 = \left(\dfrac{3a^2b}{2}\right)^2 \left(\dfrac{a^2}{-3b^2}\right)^3$

$= \left(\dfrac{3^2a^4b^2}{2^2}\right)\left(\dfrac{a^6}{(-3)^3b^6}\right) = \dfrac{9a^{10}b^2}{4(-27)b^6}$

$= \dfrac{9a^{10}}{-108b^4} = -\dfrac{a^{10}}{12b^4}$

7. $(x-4)(2x^2 + 4x - 1)$

$= x(2x^2 + 4x - 1) - 4(2x^2 + 4x - 1)$

$= 2x^3 + 4x^2 - x - 8x^2 - 16x + 4$

$= 2x^3 - 4x^2 - 17x + 4$

8. $6x - 2y = -3$

$4x + y = 5$

$6x - 2y = -3$

$\underline{8x + 2y = 10}$

$14x = 7$

$x = \dfrac{1}{2}$

$6\left(\dfrac{1}{2}\right) - 2y = -3$

$3 - 2y = -3$

$-2y = -6$

$y = 3$

The solution is $\left(\dfrac{1}{2}, 3\right)$.

9. $x^3y + x^2y^2 - 6xy^3 = xy(x^2 + xy - 6y^2)$

$= xy(x + 3y)(x - 2y)$

10. $(b+2)(b-5) = 2b + 14$

$b^2 - 3b - 10 = 2b + 14$

$b^2 - 5b - 24 = 0$

$(b-8)(b+3) = 0$

$b - 8 = 0 \qquad b + 3 = 0$

$b = 8 \qquad b = -3$

The solutions are −3 and 8.

11. $x^2 - 2x > 15$

$x^2 - 2x - 15 > 0$

$(x-5)(x+3) > 0$

$\{x \mid x < -3 \text{ or } x > 5\}$

12. $\dfrac{x^2+4x-5}{2x^2-3x+1}-\dfrac{x}{2x-1}$

$=\dfrac{(x+5)(x-1)}{(2x-1)(x-1)}-\dfrac{x}{2x-1}$

$=\dfrac{x+5}{2x-1}-\dfrac{x}{2x-1}=\dfrac{x+5-x}{2x-1}$

$=\dfrac{5}{2x-1}$

13. $\dfrac{5}{x^2+7x+12}=\dfrac{9}{x+4}-\dfrac{2}{x+3}$

$\dfrac{5}{(x+4)(x+3)}=\dfrac{9}{x+4}\cdot\dfrac{x+3}{x+3}-\dfrac{2}{x+3}\cdot\dfrac{x+4}{x+4}$

$\dfrac{5}{(x+4)(x+3)}=\dfrac{9x+27}{(x+4)(x+3)}-\dfrac{2x+8}{(x+4)(x+3)}$

$5=(9x+27)-(2x+8)$

$5=7x+19$

$-14=7x$

$-2=x$

The solution is −2.

14. $\dfrac{4-6i}{2i}=\dfrac{4-6i}{2i}\cdot\dfrac{i}{i}=\dfrac{4i-6i^2}{2i^2}$

$=\dfrac{4i+6}{-2}=-3-2i$

15. $f(x)=\dfrac{1}{4}x^2$

$-\dfrac{b}{2a}=-\dfrac{0}{2\left(\dfrac{1}{4}\right)}=-\dfrac{0}{\dfrac{1}{4}}=0$

$y=\left(\dfrac{1}{4}\right)0^2=0$

Vertex: (0,0)

Axis of symmetry: $x=0$

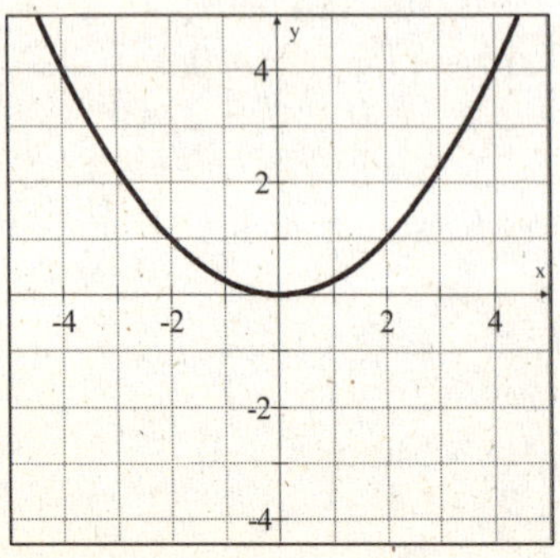

16. $3x-4y\geq 8$

$-4y\geq -3x+8$

$y\leq \dfrac{3}{4}x-2$

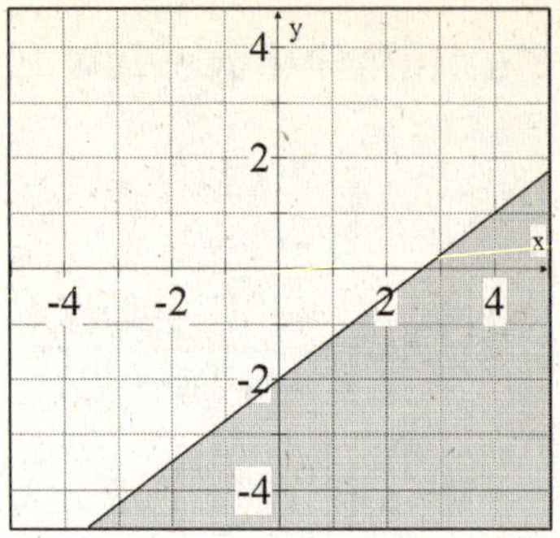

17. $m=\dfrac{y_2-y_1}{x_2-x_1}=\dfrac{-6-4}{2-(-3)}=\dfrac{-10}{5}=-2$

$y-y_1=m(x-x_1)$

$y-4=-2(x-(-3))$

$y-4=-2(x+3)$

$y-4=-2x-6$

$y=-2x-2$

18. The product of the slopes of perpendicular lines is −1.

$2x-3y=6$

$-3y=-2x+6$

$y=\dfrac{2}{3}x-2$

$m_1\cdot m_2=-1$

$\dfrac{2}{3}\cdot m_2=-1$

$m_2=-\dfrac{3}{2}$

$y-y_1=m(x-x_1)$

$y-1=-\dfrac{3}{2}(x-(-3))$

$y-1=-\dfrac{3}{2}(x+3)$

$y-1=-\dfrac{3}{2}x-\dfrac{9}{2}$

$y=-\dfrac{3}{2}x-\dfrac{7}{2}$

19. $3x^2 = 3x - 1$

$3x^2 - 3x + 1 = 0$

$a = 3 \quad b = -3 \quad c = 1$

$$x = \frac{-b \pm \sqrt{b^2 - 4ac}}{2a}$$

$$x = \frac{-(-3) \pm \sqrt{(-3)^2 - 4(3)(1)}}{2(3)}$$

$$x = \frac{3 \pm \sqrt{9-12}}{6} = \frac{3 \pm \sqrt{-3}}{6} = \frac{3 \pm i\sqrt{3}}{6}$$

$$x = \frac{1}{2} \pm \frac{i\sqrt{3}}{6}$$

The zeros are $\frac{1}{2} + \frac{i\sqrt{3}}{6}$ and $\frac{1}{2} - \frac{i\sqrt{3}}{6}$.

20. $\sqrt{8x+1} = 2x - 1$

$\left(\sqrt{8x+1}\right)^2 = (2x-1)^2$

$8x + 1 = 4x^2 - 4x + 1$

$0 = 4x^2 - 12x$

$0 = 4x(x-3)$

$4x = 0 \qquad x - 3 = 0$

$x = 0 \qquad x = 3$

Check both solutions in the original equation:

$\sqrt{8(0)+1} = 2(0) - 1$

$\sqrt{1} = -1$

$1 \neq -1$

$\sqrt{8(3)+1} = 2(3) - 1$

$\sqrt{25} = 5$

$5 = 5$

The solution is 3.

21. $f(x) = 2x^2 - 3$

$$x = -\frac{b}{2a} = -\frac{0}{2(2)} = -\frac{0}{4} = 0$$

$f(x) = 2x^2 - 3$

$f(0) = 2(0)^2 - 3 = 0 - 3$

$= -3$

Since $a > 0$, the function has a minimum value. The minimum value of the function is -3.

22. $f(x) = |3x - 4|$

$f(0) = |3(0) - 4| = |-4| = 4$

$f(1) = |3(1) - 4| = |-1| = 1$

$f(2) = |3(2) - 4| = |2| = 2$

$f(3) = |3(3) - 4| = |5| = 5$

The range is $\{1, 2, 4, 5\}$.

23. Yes

24. $\sqrt[3]{5x-2} = 2$

$\left(\sqrt[3]{5x-2}\right)^3 = 2^3$

$5x - 2 = 8$

$5x = 10$

$x = 2$

The solution is 2.

25. $g(x) = 3x - 5 \quad h(x) = \frac{1}{2}x + 4$

$h(2) = \frac{1}{2}(2) + 4 = 1 + 4 = 5$

$g(5) = 3(5) - 5 = 15 - 5 = 10$

$g[h(2)] = 10$

26. $f(x) = -3x + 9$

$y = -3x + 9$

$x = -3y + 9$

$x - 9 = -3y$

$-\frac{1}{3}x + 3 = y$

$f^{-1}(x) = -\frac{1}{3}x + 3$

27. Strategy: Let x represent the cost per pound of the mixture.

	Amount	Cost	Value
\$4.50 tea	30	4.50	30(4.50)
\$3.60 tea	45	3.60	45(3.60)
Mixture	75	x	$75x$

The sum of the values before mixing is equal to the value after mixing.

Solution:
$30(4.50) + 45(3.60) = 75x$
$135 + 162 = 75x$
$297 = 75x$
$x = 3.96$
The cost per pound of the mixture is \$3.96.

28. Strategy: Let x represent the number of pounds of 80% copper alloy.

	Amount	Percent	Quantity
80%	x	0.80	$0.80x$
20%	50	0.20	0.20(50)
40%	$50 + x$	0.40	$0.40(50 + x)$

The sum of the quantities before mixing is equal to the quantity after mixing.

Solution:
$0.80x + 0.20(50) = 0.40(50 + x)$
$0.80x + 10 = 20 + 0.40x$
$0.40x + 10 = 20$
$0.40x = 10$
$x = 25$
25 lb of the 80% copper alloy must be used.

29. Strategy: Let x represent the additional amount of insecticide.
The total amount of insecticide is $x + 6$.
To find the additional amount of insecticide write and solve a proportion.

Solution: $\frac{6}{16} = \frac{x+6}{28}$.

$\frac{3}{8} = \frac{x+6}{28}$

$\frac{3}{8} \cdot 56 = \frac{x+6}{28} \cdot 56$

$21 = 2x + 12$

$9 = 2x$

$4.5 = x$

An additional 4.5 oz of insecticide are required.

30. Strategy: Let x represent the time it takes for the smaller pipe to fill the tank.
The time it takes the larger pipe to fill the tank is $x - 8$.

	Rate	Time	Part
Smaller pipe	$\frac{1}{t}$	3	$\frac{3}{t}$
Larger pipe	$\frac{1}{t-8}$	3	$\frac{3}{t-8}$

The sum of the parts of the task completed must equal 1.

Solution:

$\frac{3}{t} + \frac{3}{t-8} = 1$

$t(t-8)\left(\frac{3}{t} + \frac{3}{t-8}\right) = 1(t(t-8))$

$3(t-8) + 3t = t^2 - 8t$

$3t - 24 + 3t = t^2 - 8t$

$6t - 24 = t^2 - 8t$

$0 = t^2 - 14t + 24$

$0 = (t-2)(t-12)$

$t - 2 = 0 \quad t - 12 = 0$

$t = 2 \quad t = 12$

The solution $t = 2$ is not possible since the time for the larger pipe would then be a negative number.
$t - 8 = 2 - 8 = -6$.
It takes the larger pipe $t - 8 = 12 - 8 = 4$ min to fill the tank.

31. Strategy: Write the basic direct variation equation replacing the variable with the given values. Solve for *k*.
Write the direct variation equation replacing *k* with its value. Substitute 40 for *f* and solve for *d*.

Solution:

$d = kf$ $\qquad d = \frac{3}{5}f$

$30 = k(50)$ $\qquad = \frac{3}{5}(40)$

$\frac{3}{5} = k$ $\qquad = 24$

A force of 40 lb will stretch the spring 24 in.

32. Strategy: Write the basic inverse variation equation replacing the variable with the given values. Solve for *k*.
Write the inverse variation equation replacing *k* with its value. Substitute 1.5 for *L* and solve for *f*.

Solution:

$f = \frac{k}{L}$ $\qquad f = \frac{120}{L}$

$60 = \frac{k}{2}$ $\qquad = \frac{120}{1.5}$

$120 = k$ $\qquad = 80$

The frequency is 80 vibrations/min.

Chapter 12: Exponential and Logarithmic Functions

Prep Test

1. $3^{-2} = \frac{1}{3^2} = \frac{1}{9}$

2. $\left(\frac{1}{2}\right)^{-4} = \left(\frac{2}{1}\right)^{4} = 2^4 = 16$

3. $\frac{1}{8} = \frac{1}{2^3} = 2^{-3}$

4. $f(x) = x^4 + x^3$
$f(-1) = (-1)^4 + (-1)^3 = 1 + (-1) = 0$
$f(3) = (3)^4 + (3)^3 = 81 + 27 = 108$

5. $3x + 7 = x - 5$
$2x + 7 = -5$
$2x = -12$
$x = -6$
The solution is -6.

6. $16 = x^2 - 6x$
$0 = x^2 - 6x - 16$
$0 = (x - 8)(x + 2)$
$x - 8 = 0 \quad x + 2 = 0$
$x = 8 \quad x = -2$
The solutions are -2 and 8.

7. $A(1 + r)^n$
$5000(1 + 0.04)^6 = 5000(1.04)^6$
$= 6326.60$

8. $f(x) = x^2 - 1$
$x = -\frac{b}{2a} = \frac{0}{2(1)} = \frac{0}{2} = 0$
$f(0) = (0)^2 - 1 = -1$

Vertex:
$(0, -1)$.
Axis of symmetry:
$x = 0$.

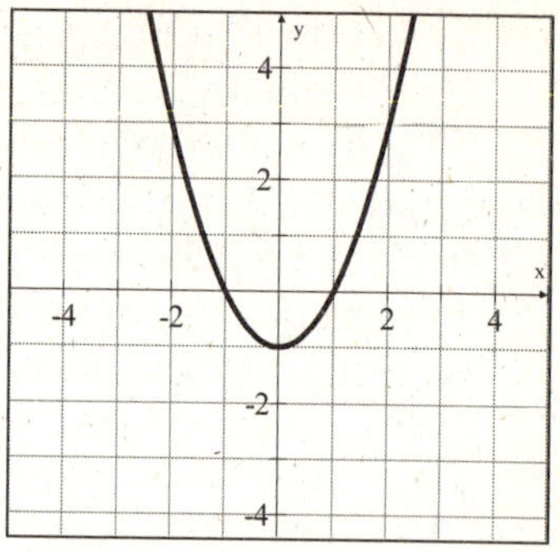

Section 12.1

Objective A Exercises

1. An exponential function with base b is defined by $f(x) = b^x$, $b > 0$, $b \neq 1$, and x is any real number.

3. $f(x) = b^x$, $b > 0$, $b \neq 1$
(iii) cannot be the base.

5. $f(x) = 3^x$
a) $f(2) = 3^2 = 9$
b) $f(0) = 3^0 = 1$
c) $f(-2) = 3^{-2} = \frac{1}{3^2} = \frac{1}{9}$

7. $g(x) = 2^{x+1}$
a) $g(3) = 2^{3+1} = 2^4 = 16$
b) $g(1) = 2^{1+1} = 2^2 = 4$
c) $g(-3) = 2^{-3+1} = 2^{-2} = \frac{1}{2^2} = \frac{1}{4}$

9. $P(x)=\left(\frac{1}{2}\right)^{2x}$

a) $P(0)=\left(\frac{1}{2}\right)^{2(0)}=\left(\frac{1}{2}\right)^{0}=1$

b) $P\left(\frac{3}{2}\right)=\left(\frac{1}{2}\right)^{2(3/2)}=\left(\frac{1}{2}\right)^{3}=\frac{1}{8}$

c) $P(-2)=\left(\frac{1}{2}\right)^{2(-2)}=\left(\frac{1}{2}\right)^{-4}=2^4=16$

11. $G(x)=e^{x/2}$

a) $G(4)=e^{4/2}=e^2=7.3891$

b) $G(-2)=e^{-2/2}=e^{-1}=\frac{1}{e^1}=0.3679$

c) $G\left(\frac{1}{2}\right)=e^{(1/2)/2}=e^{1/4}=1.2840$

13. $H(r)=e^{-r+3}$

a) $H(-1)=e^{-(-1)+3}=e^4=54.5982$

b) $H(3)=e^{-3+3}=e^0=1$

c) $H(5)=e^{-5+3}=e^{-2}=\frac{1}{e^2}=0.1353$

15. $F(x)=2^{x^2}$

a) $F(2)=2^{2^2}=2^4=16$

b) $F(-2)=2^{(-2)^2}=2^4=16$

c) $F\left(\frac{3}{4}\right)=2^{(3/4)^2}=2^{9/16}=1.4768$

17. $f(x)=e^{-x^2/2}$

a) $f(-2)=e^{-(-2)^2/2}=e^{-2}=\frac{1}{e^2}=0.1353$

b) $f(2)=e^{-2^2/2}=e^{-2}=\frac{1}{e^2}=0.1353$

c)

$f(-3)=e^{-(-3)^2/2}=e^{-9/2}=\frac{1}{e^{9/2}}=0.0111$

19. $f(a)>f(b)$

Objective B Exercises

21. $f(x)=3^x$

x	y
0	1
-1	⅓
1	3

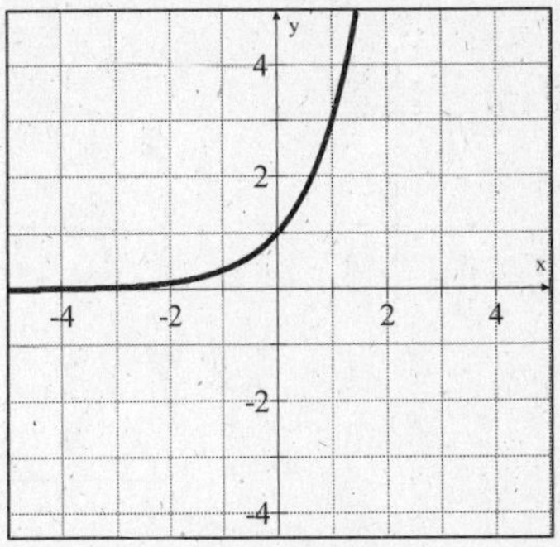

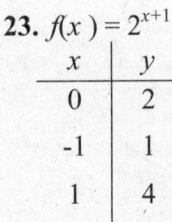

23. $f(x)=2^{x+1}$

x	y
0	2
-1	1
1	4

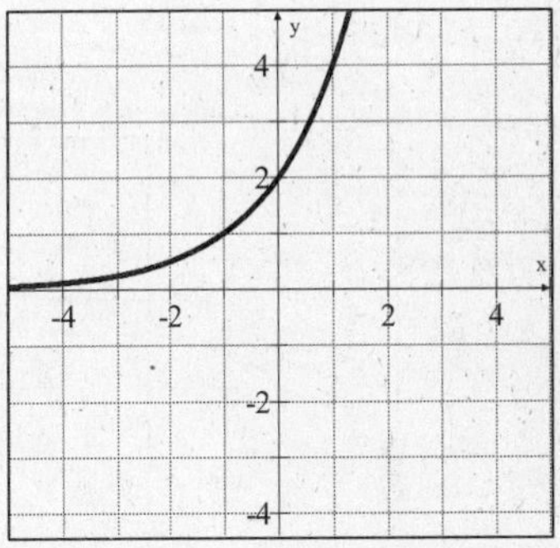

25. $f(x)=\left(\frac{1}{3}\right)^x$

x	y
0	1
-1	3
1	⅓

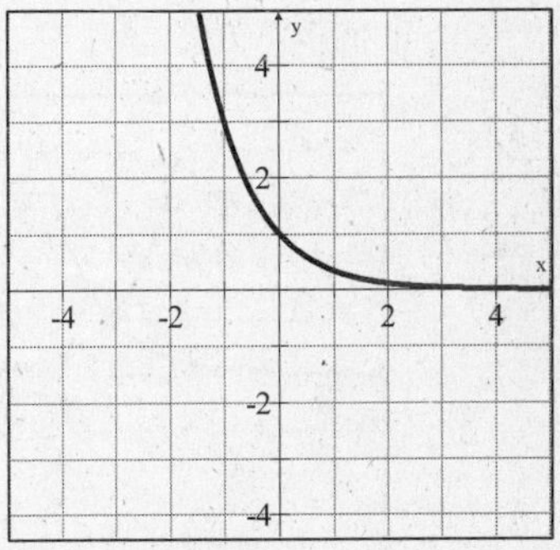

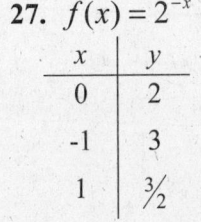

27. $f(x)=2^{-x}+1$

x	y
0	2
-1	3
1	$\frac{3}{2}$

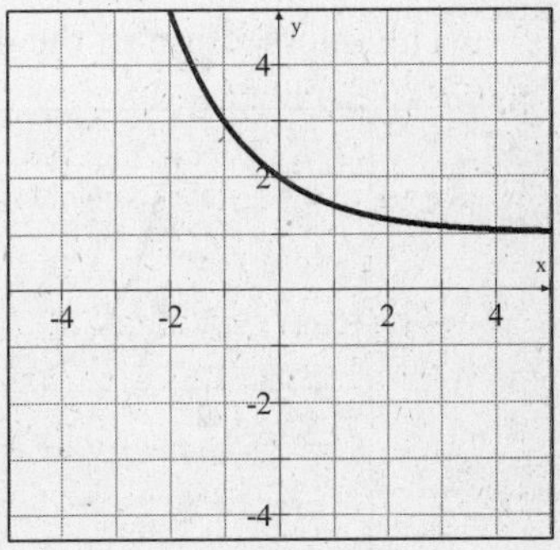

29. $f(x)=\left(\dfrac{1}{3}\right)^{-x}$

x	y
0	1
-1	$\frac{1}{3}$
1	3

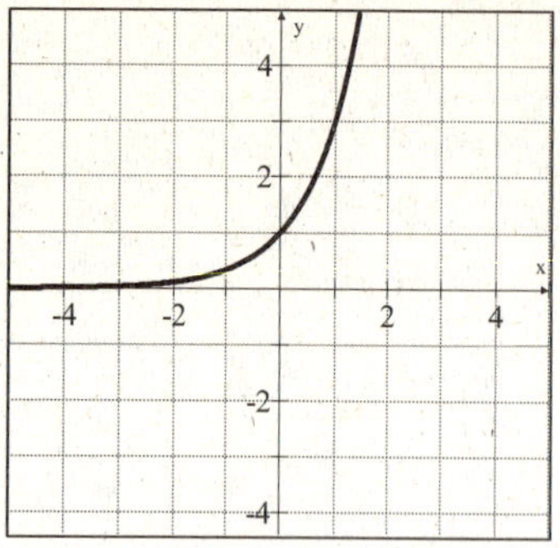

31. $f(x)=\left(\dfrac{1}{2}\right)^{-x}+2$

x	y
0	3
-1	$\frac{5}{2}$
1	4

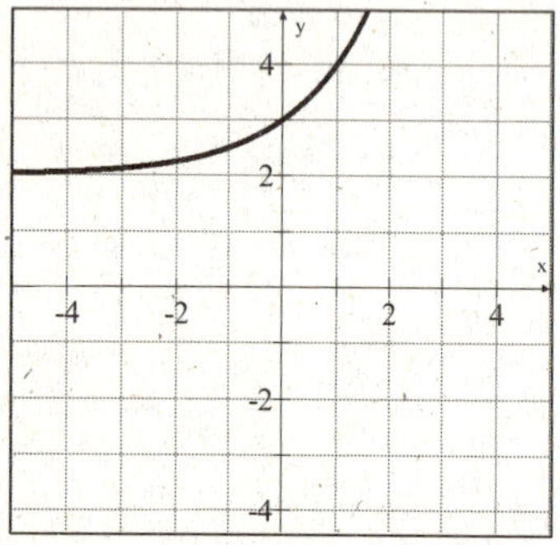

33.

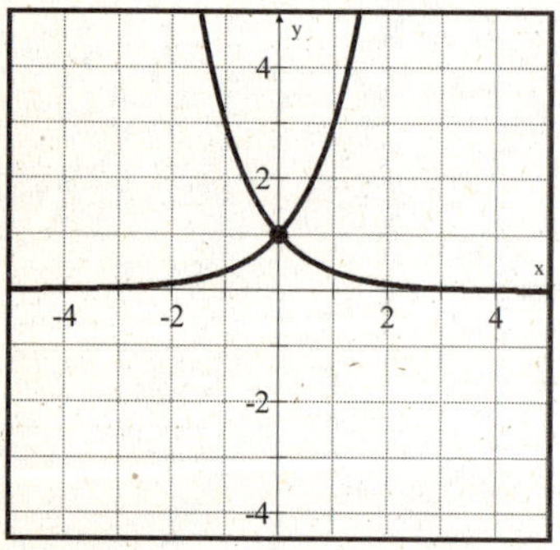

The intersection of the graphs is (0, 1).

35.

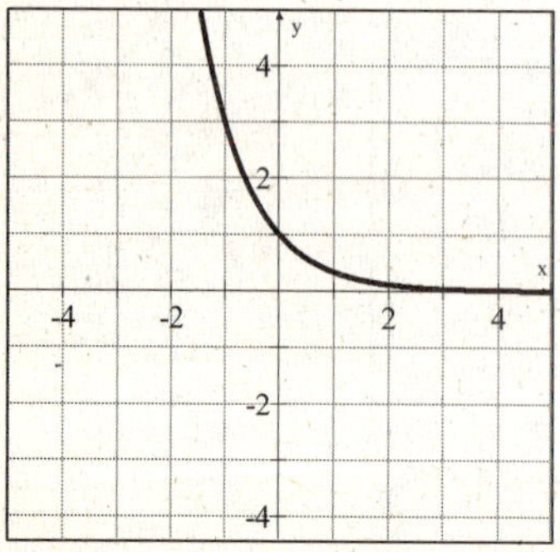

There is no x-intercept. The y-intercept is (0,1).

37. (i) and (iii) have the same graphs.
(ii) and (iv) have the same graphs.

Applying the Concepts

39. $P(x)=\left(\sqrt{3}\right)^{x}$

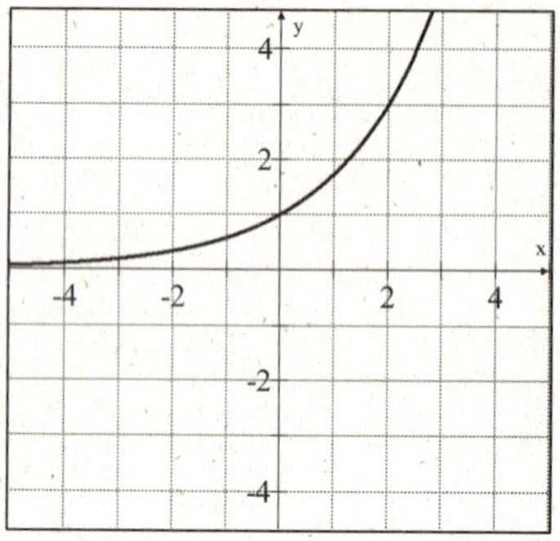

41. $f(x)=\pi^{x}$

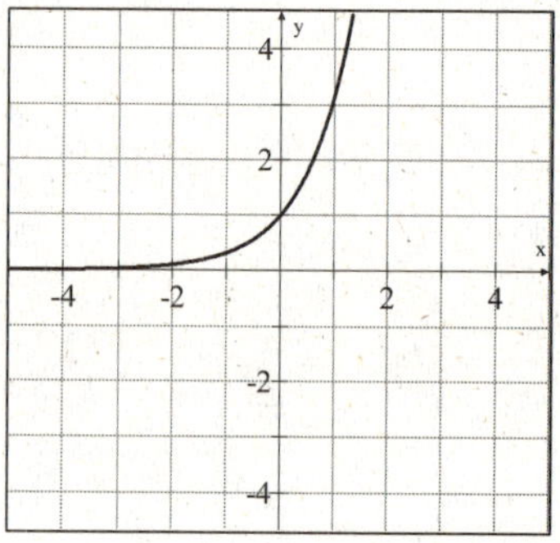

43. a)

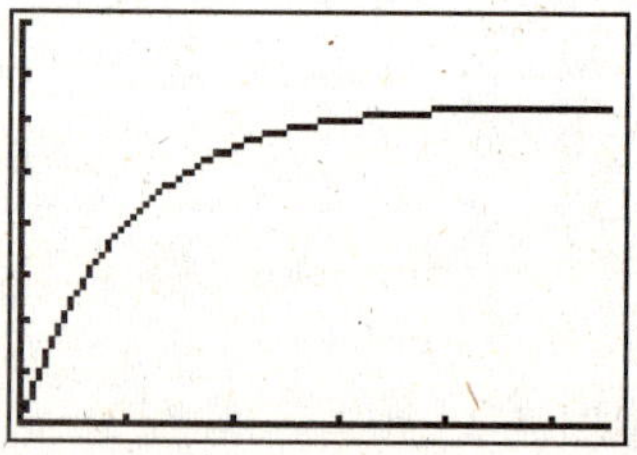

Xmin = 0, Xmax = 5.5,Xscl = 1
Ymin = 0, Ymax = 40, Yscl = 5

b) The point (2, 27.7) means that after 2 s the object will be falling at a speed of 27.7 ft/s.

Section 12.2

Objective A Exercises

1. a) A common logarithm is a logarithm with a base of 10.

b) $\log 4z$

3. $\log_5 25 = 2$

5. $\log_4 \frac{1}{16} = -2$

7. $\log_{10} x = y$

9. $\log_a w = x$

11. $3^2 = 9$

13. $10^{-2} = 0.01$

15. $e^y = x$

17. $b^v = u$

19. $\log_3 81 = x$
$3^x = 81$
$x = 4$
$\log_3 81 = 4$

21. $\log_2 128 = x$
$2^x = 128$
$x = 7$
$\log_2 128 = 7$

23. $\log 100 = x$
$10^x = 100$
$x = 2$
$\log 100 = 2$

25. $\ln e^3 = x$
$3 \ln e = x$
$3(1) = x$
$x = 3$
$\ln e^3 = 3$

27. $\log_8 1 = x$
$8^x = 1$
$x = 0$
$\log_8 1 = 0$

29. $\log_5 625 = x$
$5^x = 625$
$x = 4$
$\log_5 625 = 4$

31. $\log_3 x = 2$
$3^2 = x$
$x = 9$

33. $\log_4 x = 3$
$4^3 = x$
$x = 64$

35. $\log_7 x = -1$
$7^{-1} = x$
$x = \frac{1}{7}$

37. $\log_6 x = 0$
$6^0 = x$
$x = 1$

39. $\log x = 2.5$
$10^{2.5} = x$
$x = 316.23$

41. $\log x = -1.75$
$10^{-1.75} = x$
$x = 0.02$

43. $\ln x = 2$

$e^2 = x$

$x = 7.39$

45. $\ln x = -\frac{1}{2}$

$e^{-1/2} = x$

$x = 0.61$

47. $x > 1$

Objective B Exercises

49. $\log_b(xy) = \log_b(x) + \log_b(y)$

51. False

53. Logarithm of One Property

$\log_{12} 1 = 0$

55. Logarithm of the Base Property

$\ln e = 1$

57. Inverse Property of Logarithms

$\log_3 3^x = x$

59. Inverse Property of Logarithms

$e^{\ln v} = v$

61. Inverse Property of Logarithms

$2^{\log_2(x^2+1)} = x^2 + 1$

63. Inverse Property of Logarithms

$\log_5 5^{x^2-x-1} = x^2 - x - 1$

65. $\log_8(xz) = \log_8 x + \log_8 z$

67. $\log_3 x^5 = 5\log_3 x$

69. $\log_b\left(\frac{r}{s}\right) = \log_b r - \log_b s$

71. $\log_3(x^2y^6) = \log_3 x^2 + \log_3 y^6$

$= 2\log_3 x + 6\log_3 y$

73. $\log_7\left(\frac{u^3}{v^4}\right) = \log_7 u^3 - \log_7 v^4$

$= 3\log_7 u - 4\log_7 v$

75. $\log_2(rs)^2 = 2\log_2(rs) = 2[\log_2 r + \log_2 s]$

77. $\ln(x^2yz) = \ln x^2 + \ln y + \ln z$

$= 2\ln x + \ln y + \ln z$

79. $\log_5\left(\frac{xy^2}{z^4}\right) = \log_5 xy^2 - \log_5 z^4$

$= \log_5 x + \log_5 y^2 - \log_5 z^4$

$= \log_5 x + 2\log_5 y - 4\log_5 z$

81. $\log_8\left(\frac{x^2}{yz^2}\right) = \log_8 x^2 - \log_8 yz^2$

$= \log_8 x^2 - (\log_8 y + \log_8 z^2)$

$= \log_8 x^2 - \log_8 y - \log_8 z^2$

$= 2\log_8 x - \log_8 y - 2\log_8 z$

83. $\log_4 \sqrt{x^3y} = \log_4(x^3y)^{1/2} = \frac{1}{2}\log_4(x^3y)$

$= \frac{1}{2}[\log_4 x^3 + \log_4 y]$

$= \frac{1}{2}[3\log_4 x + \log_4 y]$

$= \frac{3}{2}\log_4 x + \frac{1}{2}\log_4 y$

85. $\log_7 \sqrt{\frac{x^3}{y}} = \log_7\left(\frac{x^3}{y}\right)^{1/2} = \frac{1}{2}\log_7 \frac{x^3}{y}$

$= \frac{1}{2}[\log_7 x^3 - \log_7 y]$

$= \frac{1}{2}[3\log_7 x - \log_7 y]$

$= \frac{3}{2}\log_7 x - \frac{1}{2}\log_7 y$

87. $\log_3\left(\frac{t}{\sqrt{x}}\right) = \log_3\left(\frac{t}{x^{1/2}}\right)$
$= \log_3 t - \log_3 x^{1/2}$
$= \log_3 t - \frac{1}{2}\log_3 x$

89. $\log_3 x^3 + \log_3 y^2 = \log_3(x^3y^2)$

91. $\ln x^4 - \ln y^2 = \ln\left(\frac{x^4}{y^2}\right)$

93. $3\log_7 x = \log_7 x^3$

95. $3\ln x + 4\ln y = \ln x^3 + \ln y^4 = \ln(x^3y^4)$

97. $2(\log_4 x + \log_4 y) = 2\log_4(xy)$
$= \log_4(xy)^2$
$= \log_4(x^2y^2)$

99. $2\log_3 x - \log_3 y + 2\log_3 z$
$= \log_3 x^2 - \log_3 y + \log_3 z^2$
$= \log_3\left(\frac{x^2}{y}\right) + \log_3 z^2$
$= \log_3\left(\frac{x^2z^2}{y}\right)$

101. $\ln x - (2\ln y + \ln z) = \ln x - (\ln y^2 + \ln z)$
$= \ln x - \ln(y^2z)$
$= \ln\left(\frac{x}{y^2z}\right)$

103. $\frac{1}{2}(\log_6 x - \log_6 y) = \frac{1}{2}\log_6\left(\frac{x}{y}\right)$
$= \log_6\left(\frac{x}{y}\right)^{1/2}$
$= \log_6\sqrt{\frac{x}{y}}$

105. $2(\log_4 s - 2\log_4 t + \log_4 r)$
$= 2(\log_4 s - \log_4 t^2 + \log_4 r)$
$= 2(\log_4 \frac{s}{t^2} + \log_4 r)$
$= 2\log_4\left(\frac{sr}{t^2}\right)$
$= \log_4\left(\frac{sr}{t^2}\right)^2$
$= \log_4 \frac{s^2r^2}{t^4}$

107. $\ln x - 2(\ln y + \ln z)$
$= \ln x - 2\ln(yz)$
$= \ln x - \ln(yz)^2$
$= \ln\left(\frac{x}{(yz)^2}\right)$
$= \ln\frac{x}{y^2z^2}$

109. $\frac{1}{2}(3\log_4 x - 2\log_4 y + \log_4 z)$
$= \frac{1}{2}(\log_4 x^3 - \log_4 y^2 + \log_4 z)$
$= \frac{1}{2}(\log_4\left(\frac{x^3}{y^2}\right) + \log_4 z)$
$= \log_4\left(\frac{x^3z}{y^2}\right)^{1/2}$
$= \log_4\sqrt{\frac{x^3z}{y^2}}$

111. $\frac{1}{2}\log_2 x - \frac{2}{3}\log_2 y + \frac{1}{2}\log_2 z$

$= \log_2 x^{1/2} - \log_2 y^{2/3} + \log_2 z^{1/2}$

$= \log_2\left(\frac{x^{1/2}}{y^{2/3}}\right) + \log_5 z^{1/2}$

$= \log_2\left(\frac{x^{1/2}z^{1/2}}{y^{2/3}}\right)$

$= \log_2 \frac{\sqrt{xy}}{\sqrt[3]{y^2}}$

Objective C Exercises

113. $\log_8 6 = \frac{\log_{10} 6}{\log_{10} 8} = 0.8617$

115. $\log_5 30 = \frac{\log_{10} 30}{\log_{10} 5} = 2.1133$

117. $\log_3 0.5 = \frac{\log_{10} 0.5}{\log_{10} 3} = -0.6309$

119. $\log_7 1.7 = \frac{\log_{10} 1.7}{\log_{10} 7} = 0.2727$

121. $\log_5 15 = \frac{\log_{10} 15}{\log_{10} 5} = 1.6826$

123. $\log_{12} 120 = \frac{\log_{10} 120}{\log_{10} 12} = 1.9266$

125. $\log_4 2.55 = \frac{\log_{10} 2.55}{\log_{10} 4} = 0.6752$

127. $\log_5 67 = \frac{\log_{10} 67}{\log_{10} 5} = 2.6125$

129. $\log_5 x = \frac{\log_{10} x}{\log_{10} 5}$

Applying the Concepts

131. a) False

$3^{-2} = \frac{1}{3^2} = \frac{1}{9} \neq -9$

b) True

c) False

$\log x^{-1} = -\log x \neq \frac{1}{\log x}$

d) True

e) False

$\log(x \cdot y) = \log x + \log y \neq \log x \cdot \log y$

f) True

Section 12.3

Objective A Exercises

1. They have the same graph.

3. $f(x) = \log_4 x$

$y = \log_4 x$ is equivalent to $x = 4^y$.

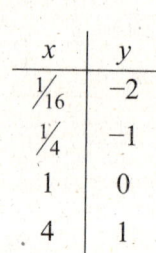

x	y
$\frac{1}{16}$	-2
$\frac{1}{4}$	-1
1	0
4	1

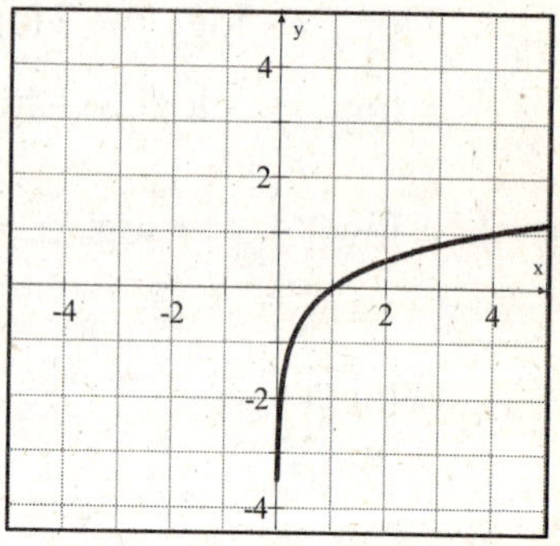

5. $f(x) = \log_3(2x-1)$

$y = \log_3(2x-1)$ is equivalent to

$2x - 1 = 3^y$ or $x = \frac{1}{2}(3^y + 1)$

x	y
$\frac{2}{3}$	-1
1	0
2	1
5	2

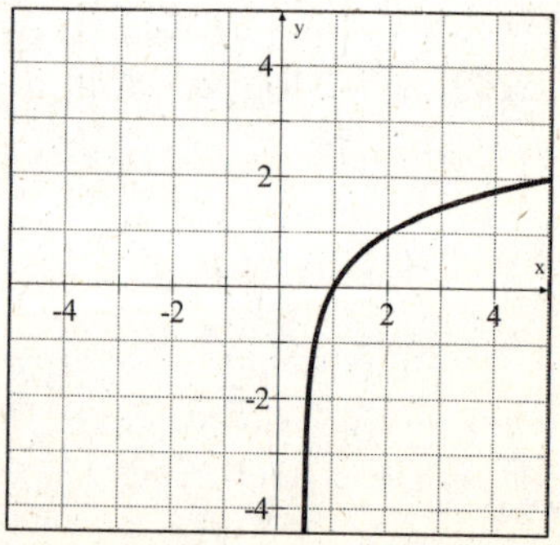

7. $f(x) = 3\log_2 x$

$y = 3\log_2 x$ is equivalent to $x = 2^{y/3}$

x	y
½	−3
1	0
2	3
4	6

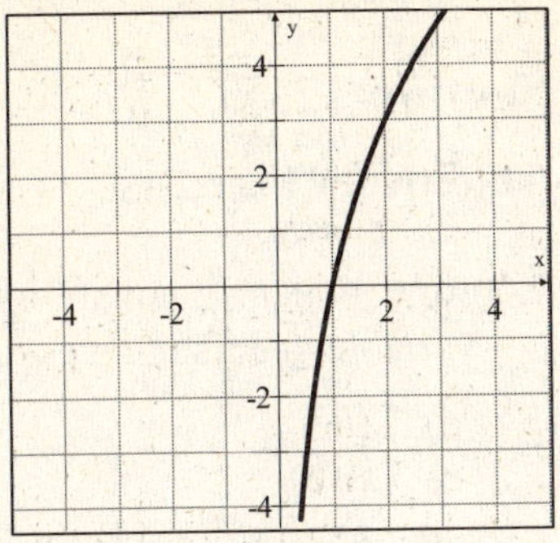

9. $f(x) = -\log_2 x$

$y = -\log_2 x$ is equivalent to $x = 2^{-y}$

x	y
2	−1
1	0
½	1
¼	2

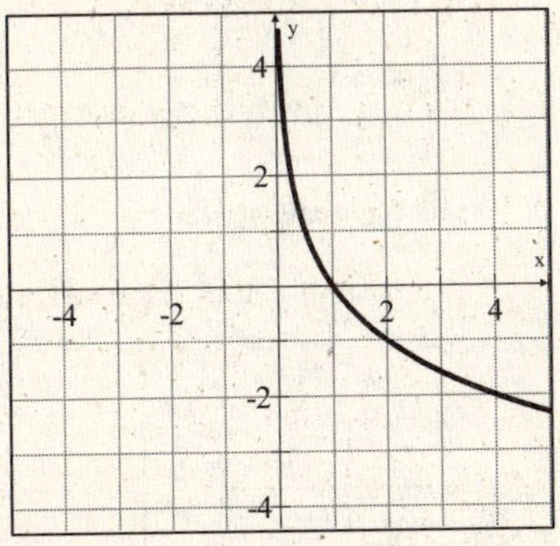

11. $f(x) = \log_2(x-1)$

$y = \log_2(x-1)$ is equivalent to $x - 1 = 2^y$ or $x = 2^y + 1$

x	y
3⁄2	−1
2	0
3	1
5	2

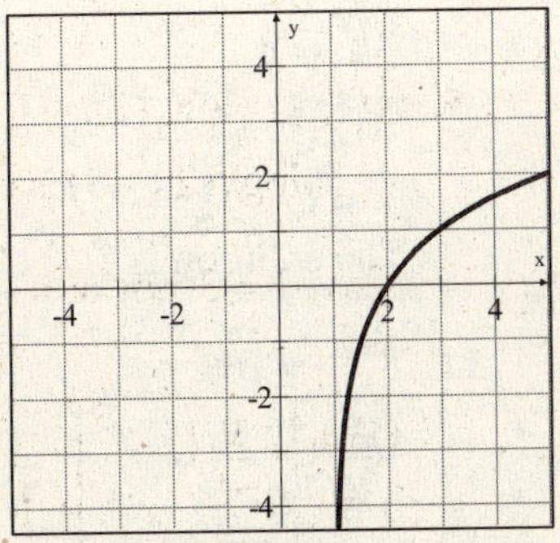

13. $f(x) = -\log_2(x-1)$

$y = -\log_2(x-1)$ is equivalent to $x - 1 = 2^{-y}$ or $x = 2^{-y} + 1$

x	y
3	−1
2	0
3⁄2	1
5⁄4	2

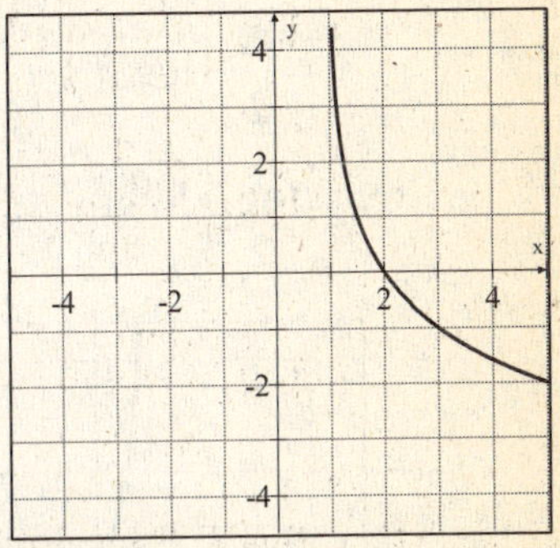

15. They intersect at the point (1,0).

Applying the Concepts

17. $f(x) = x - \log_2(1-x)$

$y = x - \log_2(1-x)$

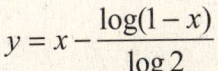

$$y = x - \frac{\log(1-x)}{\log 2}$$

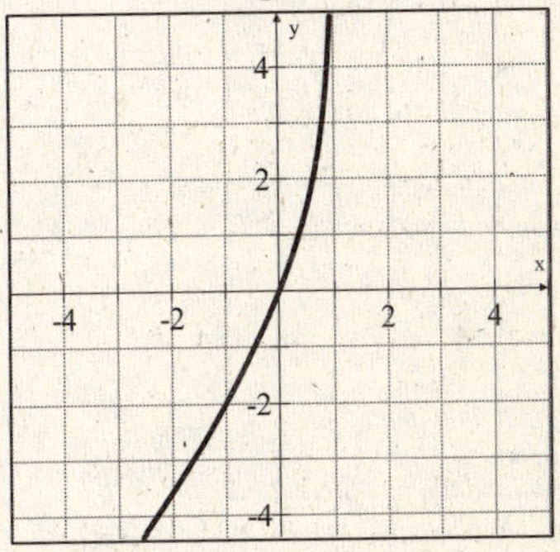

19. $f(x) = \frac{x}{2} - 2\log_2(x+1)$

$y = \frac{x}{2} - 2\log_2(x+1)$

$y = \frac{x}{2} - \log_2(x+1)^2$

$y = \frac{x}{2} - \frac{\log(x+1)^2}{\log 2}$

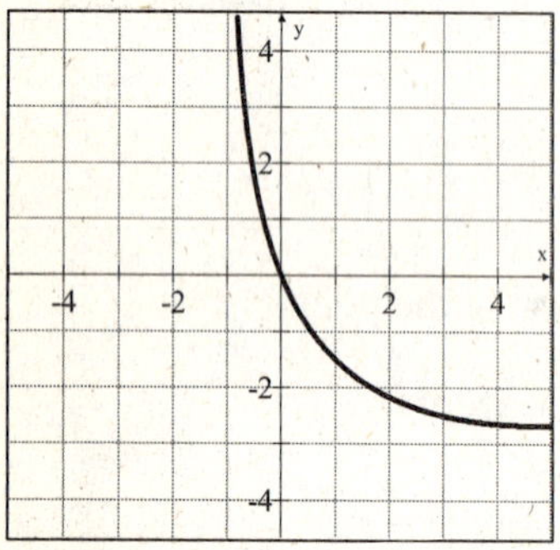

21. $f(x) = x^2 - 10\ln(x-1)$

$y = x^2 - 10\ln(x-1)$

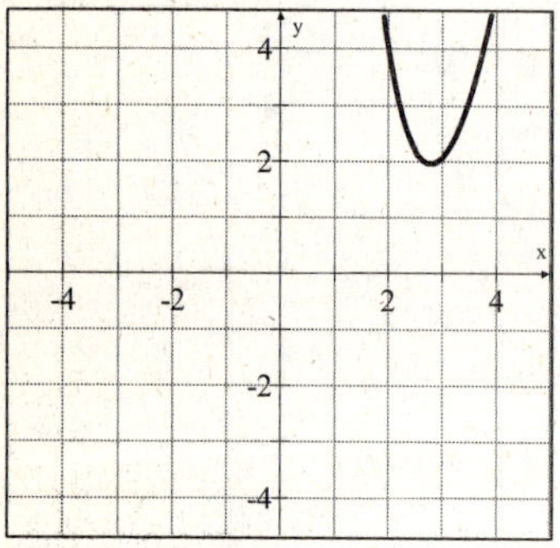

23. a) $M = 5\log s - 5$

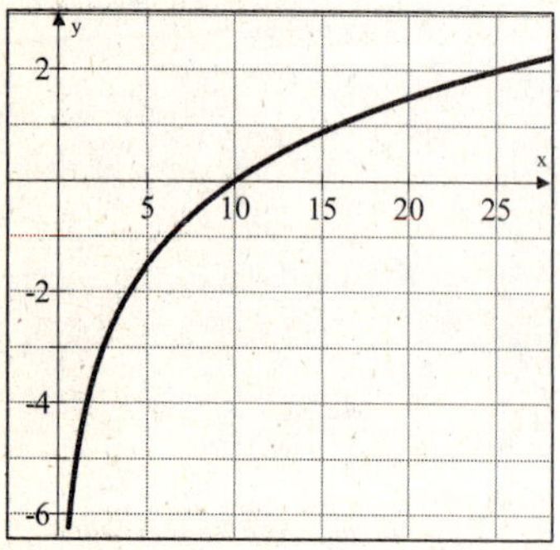

b) The point (25.1, 2) means that a star that is 25.1 parsecs from Earth has a distance modulus of 2.

Section 12.4

Objective A Exercises

1. An exponential equation is one in which a variable occurs in an exponent.

3. $x < 0$

5. $5^{4x-1} = 5^{x-2}$

$4x - 1 = x - 2$

$3x - 1 = -2$

$3x = -1$

$x = -\frac{1}{3}$

The solution is $-\frac{1}{3}$.

7. $8^{x-4} = 8^{5x+8}$

$x - 4 = 5x + 8$

$-4 = 4x + 8$

$4x = -12$

$x = -3$

The solution is −3.

9. $9^x = 3^{x+1}$

$3^{2x} = 3^{x+1}$

$2x = x + 1$

$x = 1$

The solution is 1.

11. $8^{x+2} = 16^x$

$(2^3)^{x+2} = 2^{4x}$

$2^{3x+6} = 2^{4x}$

$3x + 6 = 4x$

$x = 6$

The solution is 6.

13. $16^{2-x} = 32^{2x}$

$(2^4)^{2-x} = (2^5)^{2x}$

$2^{8-4x} = 2^{10x}$

$8 - 4x = 10x$

$8 = 14x$

$x = \frac{8}{14} = \frac{4}{7}$

The solution is $\frac{4}{7}$.

15. $25^{3-x} = 125^{2x-1}$

$(5^2)^{3-x} = (5^3)^{2x-1}$

$5^{6-2x} = 5^{6x-3}$

$6 - 2x = 6x - 3$

$6 = 8x - 3$

$9 = 8x$

$x = \frac{9}{8}$

The solution is $\frac{9}{8}$.

17. $5^x = 6$

$\log 5^x = \log 6$

$x \log 5 = \log 6$

$x = \frac{\log 6}{\log 5}$

$x = 1.1133$

The solution is 1.1133.

19. $8^{x/4} = 0.4$

$\log 8^{x/4} = \log 0.4$

$\frac{x}{4} \log 8 = \log 0.4$

$\frac{x}{4} = \frac{\log 0.4}{\log 8}$

$x = 4 \cdot \frac{\log 0.4}{\log 8}$

$x = -1.7626$

The solution is -1.7626.

21. $2^{3x} = 5$

$\log 2^{3x} = \log 5$

$3x \log 2 = \log 5$

$3x = \frac{\log 5}{\log 2}$

$3x = 2.3219$

$x = 0.7740$

The solution is 0.7740.

23. $2^{-x} = 7$

$\log 2^{-x} = \log 7$

$-x \log 2 = \log 7$

$-x = \frac{\log 7}{\log 2}$

$-x = 2.8074$

$x = -2.8074$

The solution is -2.8074.

25. $2^{x-1} = 6$

$\log 2^{x-1} = \log 6$

$(x-1) \log 2 = \log 6$

$x - 1 = \frac{\log 6}{\log 2}$

$x = \frac{\log 6}{\log 2} + 1$

$x = 3.5850$

The solution is 3.5850.

27. $3^{2x-1} = 4$

$\log 3^{2x-1} = \log 4$

$(2x-1) \log 3 = \log 4$

$2x - 1 = \frac{\log 4}{\log 3}$

$2x - 1 = 1.2619$

$2x = 2.2619$

$x = 1.1309$

The solution is 1.1309.

29. $\left(\frac{1}{2}\right)^{x+1} = 3$

$\log\left(\frac{1}{2}\right)^{x+1} = \log 3$

$(x+1)\log\left(\frac{1}{2}\right) = \log 3$

$x+1 = \frac{\log 3}{\log \frac{1}{2}}$

$x = \frac{\log 3}{\log \frac{1}{2}} - 1$

$x = -2.5850$

The solution is −2.5850.

31. $3 \cdot 2^x = 7$

$\log(3 \cdot 2^x) = \log 7$

$\log 3 + \log 2^x = \log 7$

$\log 3 + x\log 2 = \log 7$

$x\log 2 = \log 7 - \log 3$

$x = \frac{\log 7 - \log 3}{\log 2}$

$x = 1.2224$

The solution is 1.2224.

33. $7 = 10\left(\frac{1}{2}\right)^{x/8}$

$\log 7 = \log 10\left(\frac{1}{2}\right)^{x/8}$

$\log 7 = \log 10 + \log\left(\frac{1}{2}\right)^{x/8}$

$\log 7 = \log 10 + \frac{x}{8}\log\frac{1}{2}$

$\log 7 - \log 10 = \frac{x}{8}\log\frac{1}{2}$

$\frac{\log 7 - \log 10}{\log \frac{1}{2}} = \frac{x}{8}$

$0.5146 = \frac{x}{8}$

$4.1166 = x$

The solution is 4.1166.

35. $15 = 12(e)^{0.05x}$

$\ln 15 = \ln 12(e)^{0.05x}$

$\ln 15 = \ln 12 + \ln(e)^{0.05x}$

$\ln 15 = \ln 12 + 0.05x$

$\ln 15 - \ln 12 = 0.05x$

$0.2231 = 0.05x$

$4.4629 = x$

The solution is 4.4629.

Objective B Exercises

37. $\log x = \log(1-x)$

$x = 1 - x$

$2x = 1$

$x = \frac{1}{2}$

The solution is $\frac{1}{2}$.

39. $\ln(3x+2) = \ln(5x+4)$

$3x + 2 = 5x + 4$

$-2x = 2$

$x = -1$

When we substitute $x = -1$ in either side of the equation we get a logarithm of a negative number.

Because the logarithm of a negative number is not a real number there is no solution.

41. $\log_2(8x) - \log_2(x^2 - 1) = \log_2 3$

$\log_2 \dfrac{8x}{x^x - 1} = \log_2 3$

$\dfrac{8x}{x^2 - 1} = 3$

$(x^2 - 1)\dfrac{8x}{x^2 - 1} = 3(x^2 - 1)$

$8x = 3x^2 - 3$

$0 = 3x^2 - 8x - 3$

$0 = (3x + 1)(x - 3)$

$3x + 1 = 0 \qquad x - 3 = 0$

$3x = -1 \qquad x = 3$

$x = -\dfrac{1}{3}$

$-\dfrac{1}{3}$ does not check as a solution.

The solution is 3.

43. $\log_9 x + \log_9(2x - 3) = \log_9 2$

$\log_9(x(2x - 3)) = \log_9 2$

$x(2x - 3) = 2$

$2x^2 - 3x = 2$

$2x^2 - 3x - 2 = 0$

$(2x + 1)(x - 2) = 0$

$2x + 1 = 0 \qquad x - 2 = 0$

$2x = -1 \qquad x = 2$

$x = -\dfrac{1}{2}$

$-\dfrac{1}{2}$ does not check as a solution.

The solution is 2.

45. $\log_2(2x - 3) = 3$

$2x - 3 = 2^3$

$2x - 3 = 8$

$2x = 11$

$x = \dfrac{11}{2}$

The solution is $\dfrac{11}{2}$.

47. $\ln(3x + 2) = 4$

$3x + 2 = e^4$

$3x + 2 = 54.5982$

$3x = 52.5982$

$x = 17.5327$

The solution is 17.5327.

49. $\log_2(x + 1) + \log_2(x + 3) = 3$

$\log_2((x + 1)(x + 3)) = 3$

$\log_2(x^2 + 4x + 3) = 3$

$x^2 + 4x + 3 = 2^3$

$x^2 + 4x + 3 = 8$

$x^2 + 4x - 5 = 0$

$(x + 5)(x - 1) = 0$

$x + 5 = 0 \qquad x - 1 = 0$

$x = -5 \qquad x = 1$

-5 does not check as a solution.

The solution is 1.

51. $\log_5(2x) - \log_5(x-1) = 1$

$\log_5 \dfrac{2x}{x-1} = 1$

$\dfrac{2x}{x-1} = 5^1$

$(x-1)\dfrac{2x}{x-1} = 5(x-1)$

$2x = 5x - 5$

$-3x = -5$

$x = \dfrac{-5}{-3} = \dfrac{5}{3}$

The solution is $\dfrac{5}{3}$.

53. $\log_8(6x) = \log_8 2 + \log_8(x-4)$

$\log_8(6x) = \log_8(2(x-4))$

$6x = 2x - 8$

$4x = -8$

$x = -2$

-2 does not check as a solution. The equation has no solution.

55. $x - 2 < x$ and therefore $\log(x-2) < \log x$. This means that $\log(x-2) - \log x < 0$ and could not equal the positive number 3.

Applying the Concepts

57. $3^{x+1} = 2^{x-2}$

$\log 3^{x+1} = \log 2^{x-2}$

$(x+1)\log 3 = (x-2)\log 2$

$x + 1 = \dfrac{\log 2}{\log 3}(x-2)$

$x + 1 = 0.6309297536(x-2)$

$x + 1 = 0.6309297536x - 1.261859507$

$0.3690702464x = -2.261859507$

$x = -6.1285$

The solution is −6.1285.

59. $7^{2x-1} = 3^{2x+3}$

$\log 7^{2x-1} = \log 3^{2x+3}$

$(2x-1)\log 7 = (2x+3)\log 3$

$2x - 1 = \dfrac{\log 3}{\log 7}(2x+3)$

$2x - 1 = 0.5645750341(2x+3)$

$2x - 1 = 1.129150068x + 1.693725102$

$0.870849932x = 2.693725102$

$x = 3.0932$

The solution is 3.09352.

61. a) $s = 312.5\ln\dfrac{e^{0.32t} + e^{-0.32t}}{2}$

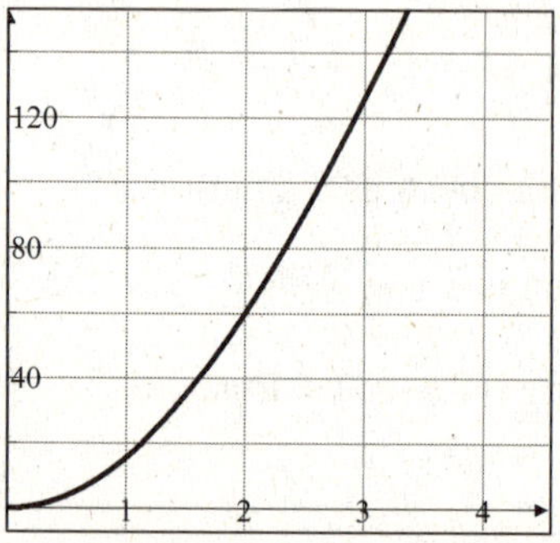

b) Use the graphing calculator to find t when $s = 100$.

$t = 2.64$

It will take 2.64s for the object to fall 100 ft.

Section 12.5

Objective A Exercises

1. Strategy: To find the value of the investment use the compound interest formula.

$P = 1000,\ n = 8,\ i = \dfrac{8\%}{4} = \dfrac{0.08}{4} = 0.02$

Solution: $A = P(1+i)^n$

$A = 1000(1+0.02)^8$

$A = 1000(1.02)^8$

$A = 1171.66$

The value of the investment after 2 years is \$1172.

3. **Strategy**: To find how many years it will take for the investment to be worth $15,000 solve the compound interest formula for n.

$A = 15{,}000,\ P = 5000,\ i = \dfrac{6\%}{12} = \dfrac{0.06}{12} = 0.005$

Solution: $A = P(1+i)^n$

$15000 = 5000(1+0.005)^n$

$3 = (1.005)^n$

$\log 3 = \log(1.005)^n$

$\log 3 = n \log 1.005$

$\dfrac{\log 3}{\log 1.005} = n$

$n \approx 220$

$\dfrac{n}{12} = \dfrac{220}{12} \approx 18$

The investment will be worth $15,000 in approximately 18 years.

5. a) **Strategy**: To find the technetium level use the exponential decay formula.

$A_0 = 30,\ k = 6,\ t = 3$

Solution: $A = A_0(0.5)^{t/k}$

$A = 30(0.5)^{3/6}$

$A = 21.2$

The technetium level is 21.2 mg after 3 h.

b) **Strategy**: To find out how long it will take the technetium level to reach 20 mg use the exponential decay formula.

$A_0 = 30,\ A = 20,\ k = 6$

Solution: $A = A_0(0.5)^{t/k}$

$20 = 30(0.5)^{t/6}$

$\dfrac{2}{3} = 0.5^{t/6}$

$\log\dfrac{2}{3} = \log 0.5^{t/6}$

$\log\dfrac{2}{3} = \dfrac{t}{6}\log 0.5$

$\dfrac{6\log\frac{2}{3}}{\log 0.5} = t$

$t = 3.5$

The technetium level is 20 mg after 3.5 h.

7. **Strategy**: To find the half life use the exponential decay formula.

$A_0 = 25,\ A = 18.95,\ t = 1$

Solution: $A = A_0(0.5)^{t/k}$

$18.95 = 25(0.5)^{1/k}$

$0.758 = 0.5^{1/k}$

$\log 0.758 = \log 0.5^{1/k}$

$\log 0.758 = \dfrac{1}{k}\log 0.5$

$k \log 0.758 = \log 0.5$

$k = \dfrac{\log 0.5}{\log 0.758}$

$k = 2.5$

The half life is 2.5 years.

9. **Strategy**: To determine the intensity of the earthquake use the Richter scale equation.

$M = 8.9$

Solution: $M = \log\dfrac{I}{I_0}$

$8.9 = \log\dfrac{I}{I_0}$

$10^{8.9} = \dfrac{I}{I_0}$

$I = 10^{8.9} I_0$

$I = 794{,}328{,}235 I_0$

The intensity of the earthquake was $794{,}328{,}235 I_0$.

11. Strategy: To determine the how many times stronger the Honshu earthquake was use the Richter scale equation.

$M_1 = 6.9 \quad M_2 = 6.4$

Solution: $M = \log\frac{I}{I_0} = \log I - \log I_0$

$6.9 = \log I_1 - \log I_0$

$6.4 = \log I_2 - \log I_0$

Subtract the equations.

$0.5 = \log I_1 - \log I_2$

$0.5 = \log\frac{I_1}{I_2}$

$10^{0.5} = \frac{I_1}{I_2}$

$I_1 = 10^{0.5} I_2$

$I_1 = 3.16 I_2$

The Honshu earthquake was 3.2 times stronger than the Quetta earthquake.

13. Strategy: To determine the magnitude of the earthquake for the seismogram given use the given equation.

$A = 23 \quad t = 24$

Solution: $M = \log A + 3\log 8t - 2.92$

$M = \log 23 + 3\log 8(24) - 2.92$

$M = \log 23 + 3\log 192 - 2.92$

$M = 5.29$

The magnitude of the earthquake was 5.3.

15. Strategy: To determine the magnitude of the earthquake for the seismogram given use the given equation.

$A = 28 \quad t = 28$

Solution: $M = \log A + 3\log 8t - 2.92$

$M = \log 28 + 3\log 8(28) - 2.92$

$M = \log 28 + 3\log 224 - 2.92$

$M = 5.58$

The magnitude of the earthquake was 5.6.

17. Strategy: To find the pH replace H^+ with its given value and solve for pH.

Solution: $pH = -\log(H^+)$

$pH = -\log(3.98\times10^{-9})$

$pH = 8.4$

The pH of baking soda is 8.4.

19. Strategy: To find the hydrogen ion concentration replace pH with its given value and solve for H^+.

Solution: $pH = -\log(H^+)$

$5.3 < -\log(H^+) < 6.6$

$-5.3 > \log(H^+) > -6.6$

$10^{-5.3} > H^+ > 10^{-6.6}$

$5\times10^{-6} > H^+ > 2.5\times10^{-7}$

The range of hydrogen ion concentration for peanuts is 2.5×10^{-7} to 5.0×10^{-6}.

21. Strategy: To find the number of decibels replace I with its given value in the equation and solve for D.

Solution: $D = 10(\log I + 16)$

$D = 10(\log(630) + 16)$

$D = 10(18.7993)$

$D = 187.993$

The blue whale sounds emit 188 decibels.

23. Strategy: To find the intensity replace D with its given value in the equation and solve for I.

Solution: $D = 10(\log I + 16)$

$25 = 10(\log(I) + 16)$

$25 = 10\log I + 160$

$-135 = 10\log I$

$-13.5 = \log I$

$10^{-13.5} = I$

$I = 3.16\times10^{-14}$

The intensity is 3.16×10^{-14} watts/cm^2.

25. Strategy: To find the percent solve the equation for P.
$d = 0.005 \quad k = 20$

Solution: $\log P = -kd$

$\log P = -20(0.005)$

$\log P = -0.1$

$P = 10^{-.1}$

$P = 0.7943$

79.4% of the light will pass through the glass.

27. Strategy: To find the thickness of copper needed replace I and I_o with the given values then solve for x. $I = 0.25 \quad I_0 = 1$

Solution: $I = I_0 e^{-3.2x}$

$0.25 = e^{-3.2x}$

$\ln 0.25 = \ln e^{-3.2x}$

$-1.39 = -3.2x$

$x = 0.43$

The thickness of the copper is 0.4 cm.

29. a) **Strategy**: Evaluate the given function at $x = 375$ ft.

$$f(x) = \left(\frac{0.5774v + 155.3}{v}\right)x + 565.3\ln\left(\frac{v - 0.2747x}{v}\right) + 3.5$$

Solution:

$$f(375) = \left(\frac{0.5774(160) + 155.3}{160}\right)375 + 565.3\ln\left(\frac{160 - 0.2747(375)}{160}\right) + 3.5$$

$$f(375) = \left(\frac{247.684}{160}\right)375 + 565.3\ln\left(\frac{56.9875}{160}\right) + 3.5$$

$$f(375) = 580.5094 - 583.5829 + 3.5$$

$$f(375) = 0.4265$$

The ball will hit 0.43ft from the bottom of the fence.

b) **Strategy**: Increase the speed by 4% so that $v = 166.4$ ft/s.

$$f(375) = \left(\frac{0.5774(166.4) + 155.3}{166.4}\right)375 + 565.3\ln\left(\frac{166.4 - 0.2747(375)}{166.4}\right) + 3.5$$

$$f(375) = 566.5100 - 545.5868 + 3.5$$

$$f(375) = 24.4$$

The height of the ball is 24.4 feet so it will clear the 15 ft fence by approximately 9 ft.

31. Strategy: Determine the value for x for which $h(x) = 0$ (or when it hits the ground).

$$h(x) = \left(\frac{-21.33}{v^2}\right)x^2 + 0.5774x + 3.5$$

Solution:

$$0 = \left(\frac{-21.33}{v^2}\right)x^2 + 0.5774x + 3.5$$

$x = 699$

Use the graphing calculator to determine the ball hits the ground at $x = 699$ ft. if air resistance is ignored.

This is 324 ft. further than the ball would travel if we did not ignore air resistance.

Applying the Concepts

33. $A = A_0(0.5)^{t/713,000,000}$

$\frac{A}{A_0} = (0.5)^{4,280,000,000/713,000,000}$

$\frac{A}{A_0} = (0.5)^{4280/713}$

$\frac{A}{A_0} = 0.016$

35. a) **Strategy:** To find the value of the investments after 3 years use the given equation.

Solution: $A = A_0e^{rt}$

$A = 5000e^{(0.06)(3)}$

$A = 5000e^{0.18}$

$A = 5986.09$

The value of the investment will be worth \$5986.09.

b) **Strategy:** To find the interest rate needed to grow an investment from \$1000 to \$1250 in 2 years use the given equation.

Solution: $A = A_0e^{rt}$

$1250 = 1000e^{2r}$

$1.25 = e^{2r}$

$\ln 1.25 = \ln e^{2r}$

$0.2231 = 2r$

$r = 0.112$

The rate must be 11.2%.

Concept Review

1. You know that a function is an exponential function if the base is a constant and the exponent contains a variable.

2. If the base were negative the value of the function would be a complex number for some values of the variable. To avoid this the base is always a positive number. $b \neq 1$ because if $b = 1$ the value of the exponential function b^x would always be 1.

3. The number e is an irrational number approximately equal to 2.71828183.

4. Because a logarithm is the inverse function of the exponential function it has the same restriction as the base.

5. The relationship between logarithmic functions and exponential functions is that $y = \log_b x$ is equivalent to $x = b^y$.

6. $\log_b(xy) = \log_b x + \log_b y$

7. $\log_b\left(\frac{x}{y}\right) = \log_b x - \log_b y$

8. We can rewrite a logarithm with any base into a common logarithm or natural logarithm.

9. The domain of the logarithmic function is $\{x \mid x > 0\}$.

10. The compound interest formula is $A = P(1+i)^n$ where P is the original amount invested, i is the interest rate per compounding period, n is the total number of compounding periods and A is the value of the investment after n periods.

11. The Richter scale magnitude of an earthquake uses logarithms to convert the intensity of the shock waves I into a number $M = \log\left(\frac{I}{I_0}\right)$.

Chapter 12 Review Exercises

1. $f(2) = e^{2-2} = e^0 = 1$

2. $5^2 = 25$

3. $f(x) = 3^{-x} + 2$

x	y
0	3
-1	5
1	$7/3$
2	$19/9$

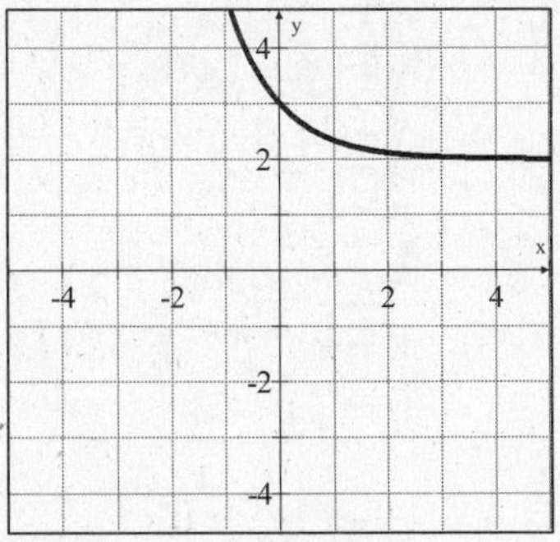

4. $f(x) = \log_3(x-1)$

$y = \log_3(x-1)$ is equivalent to

$x - 1 = 3^y$ or $x = 3^y + 1$

x	y
$4/3$	-1
2	0
4	1

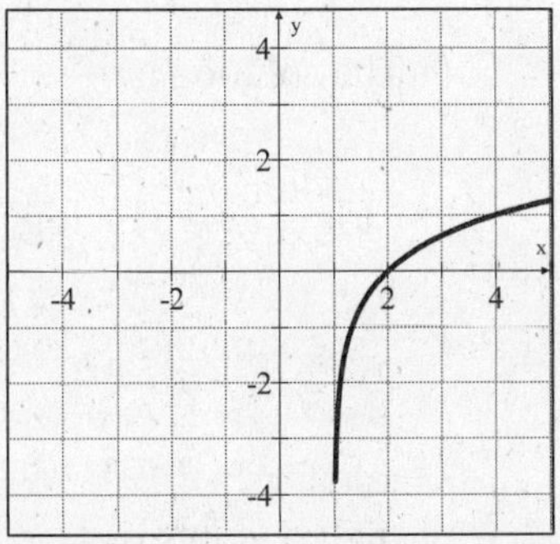

5. $\log_3 \sqrt[5]{x^2y^4} = \log_3(x^2y^4)^{1/5} = \frac{1}{5}\log_3(x^2y^4)$

$= \frac{1}{5}[\log_3 x^2 + \log_3 y^4]$

$= \frac{1}{5}[2\log_3 x + 4\log_3 y]$

$= \frac{2}{5}\log_3 x + \frac{4}{5}\log_3 y$

6. $2\log_3 x - 5\log_3 y = \log_3 x^2 - \log_3 y^5$

$= \log_3\left(\frac{x^2}{y^5}\right)$

7. $27^{2x+4} = 81^{x-3}$

$(3^3)^{2x+4} = (3^4)^{x-3}$

$6x + 12 = 4x - 12$

$2x = -24$

$x = -12$

The solution is -12.

8. $\log_5 \frac{7x+2}{3x} = 1$

Rewrite in exponential form.

$5^1 = \frac{7x+2}{3x}$

$15x = 7x + 2$

$8x = 2$

$x = \frac{1}{4}$

The solution is $\frac{1}{4}$.

9. $\log_6 22 = \frac{\log_{10} 22}{\log_{10} 6} = 1.7251$

10. $\log_2 x = 5$

Rewrite in exponential form.

$2^5 = x$

$x = 32$

The solution is 32.

11. $\log_3(x+2) = 4$

Rewrite in exponential form.

$3^4 = x + 2$

$81 = x + 2$

$x = 79$

The solution is 79.

12. $\log_{10} x = 3$

Rewrite in exponential form.

$10^3 = x$

$1x = 1000$

The solution is 1000.

13. $\frac{1}{3}(\log_7 x + 4\log_7 y) = \frac{1}{3}(\log_7 x + \log_7 y^4)$

$\frac{1}{3}(\log_7 (xy^4)) = \log_7 (xy^4)^{1/3}$

$= \log_7 \sqrt[3]{xy^4}$

14. $\log_8 \sqrt{\frac{x^5}{y^3}} = \log_8 \left(\frac{x^5}{y^3}\right)^{1/2}$

$= \frac{1}{2}(\log_8 x^5 - \log_8 y^3)$

$= \frac{5}{2}\log_8 x - \frac{3}{2}\log_8 y$

15. $\log_2 32 = 5$

16. $\log_3 1.6 = \frac{\log_{10} 1.6}{\log_{10} 3} = 0.4278$

17. $3^{x+2} = 5$

$\log 3^{x+2} = \log 5$

$(x+2)\log 3 = \log 5$

$x + 2 = \frac{\log 5}{\log 3}$

$x = \frac{\log 5}{\log 3} - 2$

$x = -0.535$

The solution is -0.535.

18. $f(-3) = \left(\frac{2}{3}\right)^{-3+2} = \left(\frac{2}{3}\right)^{-1} = \frac{3}{2}$

19. $\log_2 (x+3) - \log_2 (x-1) = 3$

$\log_2 \frac{x+3}{x-1} = 3$

$\frac{x+3}{x-1} = 2^3$

$\frac{x+3}{x-1} = 8$

$(x-1)\frac{x+3}{x-1} = 8(x-1)$

$x + 3 = 8x - 8$

$3 = 7x - 8$

$11 = 7x$

$x = \frac{11}{7}$

The solution is $\frac{11}{7}$.

20. $\log_3 (2x+3) + \log_3 (x-2) = 2$

$\log_3 ((2x+3)(x-2)) = 2$

$2x^2 - x - 6 = 3^2$

$2x^2 - x - 6 = 9$

$2x^2 - x - 15 = 0$

$(2x+5)(x-3) = 0$

$2x + 5 = 0 \quad x - 3 = 0$

$2x = -5 \qquad x = 3$

$x = -\frac{5}{2}$

$-\frac{5}{2}$ does not check as a solution.

The solution is 3.

21. $f(x) = \left(\frac{2}{3}\right)^{x+1}$

x	y
-1	1
-2	$\frac{3}{2}$
0	$\frac{2}{3}$
1	$\frac{4}{9}$

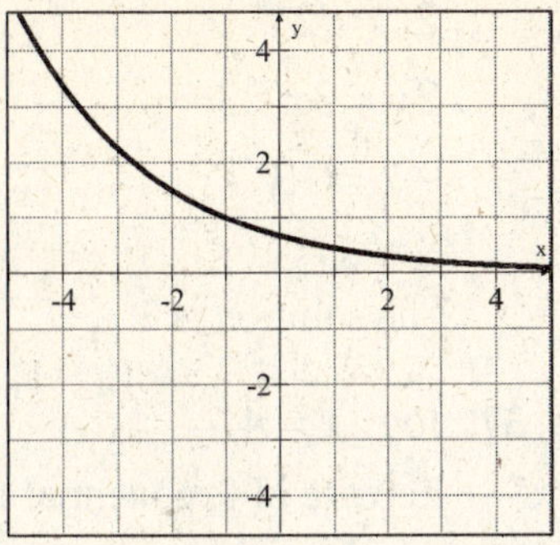

22. $f(x) = \log_2(2x-1)$

$y = \log_2(2x-1)$ is equivalent to

$2x - 1 = 2^y$ or $x = \dfrac{2^y + 1}{2}$

x	y
$\frac{3}{4}$	-1
1	0
$\frac{3}{2}$	1

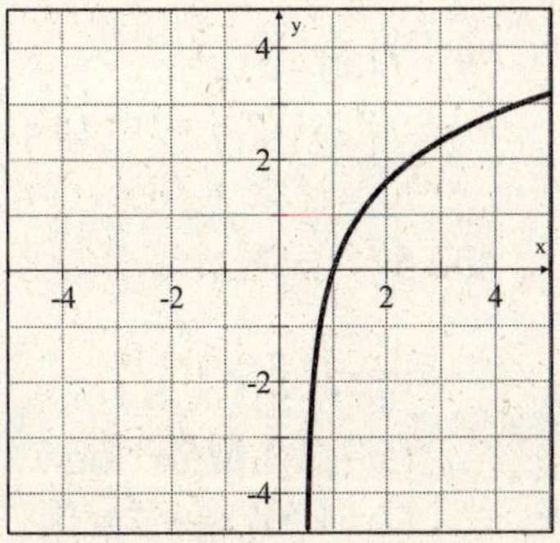

23. $\log_6 36 = x$

$6^x = 36$

$x = 2$

24. $\frac{1}{3}(\log_2 x - \log_2 y) = \frac{1}{3}\log_2\left(\frac{x}{y}\right)$

$= \log_2\left(\frac{x}{y}\right)^{1/3} = \log_2 \sqrt[3]{\frac{x}{y}}$

25. $9^{2x} = 3^{x+3}$

$(3^2)^{2x} = 3^{x+3}$

$3^{4x} = 3^{x+3}$

$4x = x + 3$

$3x = 3$

$x = 1$

The solution is 1.

26. $5 \cdot 3^{x/2} = 12$

$3^{x/2} = 2.4$

$\log 3^{x/2} = \log 2.4$

$\frac{x}{2}\log 3 = \log 2.4$

$\frac{x}{2} = \frac{\log 2.4}{\log 3}$

$\frac{x}{2} = 0.7969$

$x = 1.5938$

The solution is 1.5938.

27. $\log_5 x = -1$

Rewrite in exponential form.

$5^{-1} = x$

$\frac{1}{5} = x$

The solution is $\frac{1}{5}$.

28. $\log_3 81 = 4$

29. $\log x + \log(x-2) = \log 15$

$\log(x(x-2)) = \log 15$

$x(x-2) = 15$

$x^2 - 2x - 15 = 0$

$(x-5)(x+3) = 0$

$x - 5 = 0 \quad x + 3 = 0$

$x = 5 \qquad x = -3$

-3 does not check as a solution.

The solution is 5.

30. $\log_5 \sqrt[3]{x^2 y} = \log_3 (x^2 y)^{1/3} = \frac{1}{3}\log_3 (x^2 y)$

$= \frac{1}{3}[\log_3 x^2 + \log_3 y]$

$= \frac{1}{3}[2\log_3 x + \log_3 y]$

$= \frac{2}{3}\log_3 x + \frac{1}{3}\log_3 y$

31. $6e^{-2x} = 17$

$\ln 6e^{-2x} = \ln 17$

$\ln 6 + \ln e^{-2x} = \ln 17$

$-2x = \ln 17 - \ln 6$

$-2x = 1.0415$

$x = -0.5207$

32. $f(-3) = 7^{-3+2} = 7^{-1} = \frac{1}{7}$

33. $\log_2 16 = x$

$2^x = 16$

$x = 4$

34. $\log_6 x = \log_6 2 + \log_6 (2x - 3)$

$\log_6 x = \log_6 (2(2x - 3))$

$x = 2(2x - 3)$

$x = 4x - 6$

$-3x = -6$

$x = 2$

The solution is 2.

35. $\log_2 5 = x$

$x = \frac{\log 5}{\log 2} = 2.3219$

36. $4^x = 8^{x-1}$

$(2^2)^x = (2^3)^{x-1}$

$2^{2x} = 2^{3x-3}$

$2x = 3x - 3$

$-x = -3$

$x = 3$

The solution is 3.

37. $\log_5 x = 4$

Rewrite in exponential form

$x = 5^4 = 625$

38. $3\log_b x - 7\log_b y = \log_b x^3 - \log_b y^7$

$= \log_b \left(\frac{x^3}{y^7}\right)$

39. $f(x) = 5^{-x-1}$

$f(-2) = 5^{-(-2)-1} = 5^{2-1} = 5^1 = 5$

40. $5^{x-2} = 7$

$\log 5^{x-2} = \log 7$

$(x-2)\log 5 = \log 7$

$x - 2 = \frac{\log 7}{\log 5}$

$x = \frac{\log 7}{\log 5} + 2$

$x = 3.2091$

41. **Strategy**: To find the value of the investment use the compound interest formula.

$P = 4000, n = 24, i = \frac{8\%}{12} = \frac{0.08}{12} = 0.00\overline{6}$

Solution: $A = P(1+i)^n$

$A = 4000(1 + 0.00\overline{6})^{24}$

$A = 4000(1.00\overline{6})^{24}$

$A \approx 4691.55$

The value of the investment after 2 years is \$4692.

42. Strategy: To determine the magnitude of the earthquake using the Richter scale equation.

Solution: $M = \log\dfrac{I}{I_0}$

$M = \log\dfrac{199{,}526{,}232 I_0}{I_0}$

$M = \log 199{,}526{,}232$

$M = 8.3$

The magnitude of the earthquake was 8.3.

43. Strategy: To find the half life use the exponential decay formula.

$A_0 = 25,\ A = 15,\ t = 20$

Solution: $A = A_0(0.5)^{t/k}$

$15 = 25(0.5)^{20/k}$

$0.6 = 0.5^{20/k}$

$\log 0.6 = \log 0.5^{20/k}$

$\log 0.6 = \dfrac{20}{k}\log 0.5$

$k\log 0.6 = 20\log 0.5$

$k = \dfrac{20\log 0.5}{\log 0.6}$

$k = 27.14$

The half life is 27 days.

44. Strategy: To find the number of decibels replace I with its given value in the equation and solve for D.

Solution: $D = 10(\log I + 16)$

$D = 10(\log(5\times 10^{-6}) + 16)$

$D = 10(10.6990)$

$D = 106.99$

The sound emitted from a busy street corner is 107 decibels.

Chapter 12 Test

1. $f(0) = \left(\dfrac{2}{3}\right)^0 = 1$

2. $f(-2) = 3^{-2+1} = 3^{-1} = \dfrac{1}{3}$

3. $f(x) = 2^x - 3$

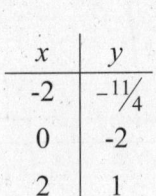

x	y
-2	$-11/4$
0	-2
2	1

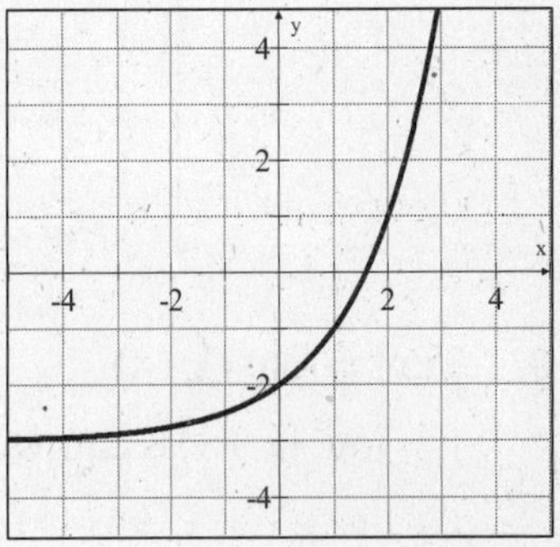

4. $f(x) = 2^x + 2$

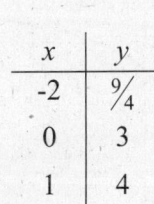

x	y
-2	$9/4$
0	3
1	4

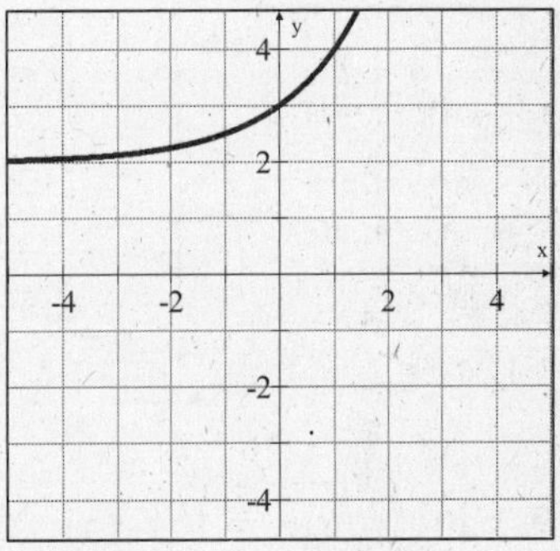

5. $\log_4 16 = x$

$4^x = 16$

$x = 2$

6. $\log_3 x = -2$

$x = 3^{-2} = \dfrac{1}{3^2} = \dfrac{1}{9}$

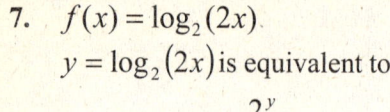

7. $f(x) = \log_2(2x)$

$y = \log_2(2x)$ is equivalent to

$2x = 2^y$ or $x = \dfrac{2^y}{2}$

x	y
$\frac{1}{4}$	-1
$\frac{1}{2}$	0
1	1
2	2

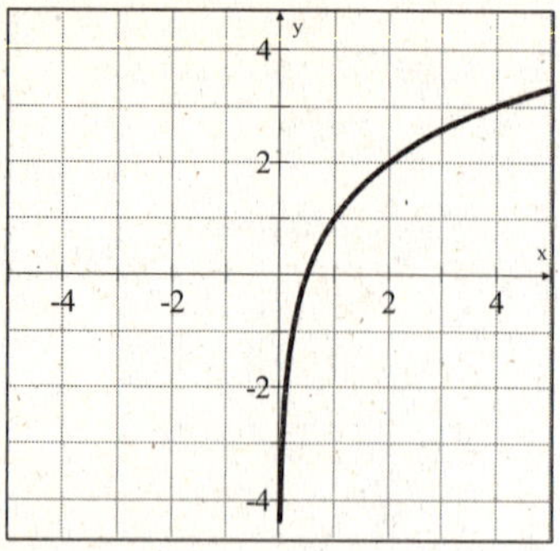

8. $f(x) = \log_3(x+1)$

$y = \log_3(x+1)$ is equivalent to

$x+1 = 3^y$ or $x = 3^y - 1$

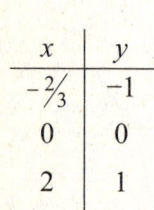

x	y
$-\frac{2}{3}$	-1
0	0
2	1

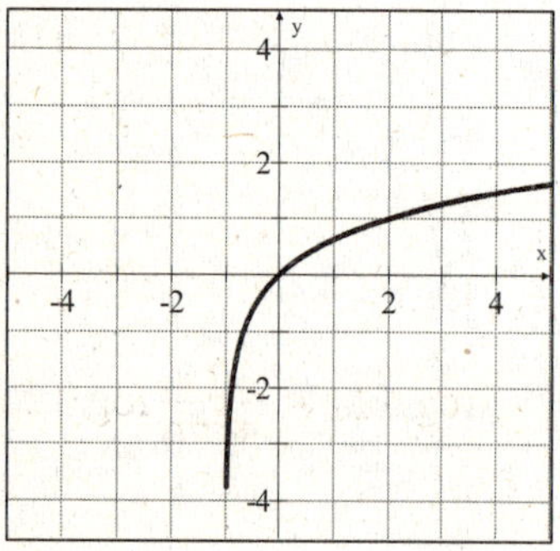

9. $\log_6 \sqrt{xy^3} = \log_6 (xy^3)^{1/2} = \frac{1}{2}\log_3 (xy^3)$

$= \frac{1}{2}[\log_6 x + \log_6 y^3]$

$= \frac{1}{2}[\log_6 x + 3\log_6 y]$

$= \frac{1}{2}\log_6 x + \frac{3}{2}\log_6 y$

10. $\frac{1}{2}(\log_3 x - \log_3 y) = \frac{1}{2}\log_3\left(\frac{x}{y}\right)$

$= \log_3 \sqrt{\frac{x}{y}}$

11. $\ln\frac{x}{\sqrt{z}} = \ln x - \ln\sqrt{z} = \ln x - \ln z^{1/2}$

$= \ln x - \frac{1}{2}\ln z$

12. $3\ln x - \ln y - \frac{1}{2}\ln z = \ln x^3 - \ln y - \ln z^{1/2}$

$= \ln\frac{x^3}{y} - \ln z^{1/2} = \ln\frac{x^3}{yz^{1/2}}$

$= \ln\frac{x^3}{y\sqrt{z}}$

13. $3^{7x+1} = 3^{4x-5}$

$7x + 1 = 4x - 5$

$3x = -6$

$x = -2$

The solution is -2.

14. $8^x = 2^{x-6}$

$(2^3)^x = 2^{x-6}$

$3x = x - 6$

$2x = -6$

$x = -3$

The solution is -3.

15. $3^x = 17$

$\log 3^x = \log 17$

$x\log 3 = \log 17$

$x = \dfrac{\log 17}{\log 3}$

$x = 2.5789$

The solution is 2.5789.

16. $\log x + \log(x-4) = \log 12$

$\log(x(x-4)) = \log 12$

$x(x-4) = 12$

$x^2 - 4x - 12 = 0$

$(x-6)(x+2) = 0$

$x - 6 = 0 \quad x + 2 = 0$

$x = 6 \qquad x = -2$

-2 does not check as a solution.
The solution is 6.

17. $\log_6 x + \log_6 (x-1) = 1$

$\log_6 (x(x-1)) = 1$

$x(x-1) = 6^1$

$x^2 - x - 6 = 0$

$(x-3)(x+2) = 0$

$x-3=0 \quad x+2=0$

$x=3 \qquad x=-2$

-2 does not check as a solution.
The solution is 3.

18. $\log_5 9 = x$

$x = \dfrac{\log 9}{\log 5} = 1.3652$

19. $\log_3 19 = x$

$x = \dfrac{\log 19}{\log 3} = 2.6801$

20. $5^{2x-5} = 9$

$\log 5^{2x-5} = \log 9$

$(2x-5)\log 5 = \log 9$

$2x-5 = \dfrac{\log 9}{\log 5}$

$2x-5 = 1.3652$

$2x = 6.3652$

$x = 3.1826$

The solution is 3.1826.

21. $2e^{x/4} = 9$

$\ln 2e^{x/4} = \ln 9$

$\ln 2 + \ln e^{x/4} = \ln 9$

$\dfrac{x}{4} = \ln 9 - \ln 2$

$\dfrac{x}{4} = 1.5041$

$x = 6.0163$

22. $\log_5 (30x) - \log_5 (x+1) = 2$

$\log_5 \dfrac{30x}{x+1} = 2$

$\dfrac{30x}{x+1} = 5^2$

$\dfrac{30x}{x+1} = 25$

$(x+1)\dfrac{30x}{x+1} = 25(x+1)$

$30x = 25x + 25$

$5x = 25$

$x = 5$

The solution is 5.

23. Strategy: To find the approximate age of the shard use the given equation.
$A_0 = 250,\ A = 170$

Solution: $A = A_0 (0.5)^{t/5570}$

$170 = 250(0.5)^{t/5570}$

$0.68 = 0.5^{t/5570}$

$\log 0.68 = \log 0.5^{t/5570}$

$\log 0.68 = \dfrac{t}{5570} \log 0.5$

$\dfrac{5570 \log 0.68}{\log 0.5} = t$

$t = 3099$

The shard is approximately 3099 years old.

24. Strategy: To find the intensity replace D with its given value in the equation and solve for I.

Solution: $D = 10(\log I + 16)$

$75 = 10(\log(I) + 16)$

$75 = 10\log I + 160$

$-85 = 10\log I$

$-8.5 = \log I$

$10^{-8.5} = I$

$I = 3.16\times 10^{-9}$

The intensity is 3.16×10^{-9} watts/cm^2.

25. Strategy: To find out the half life of a radioactive material use the exponential decay formula.

$A_0 = 10, A = 9, t = 5$

Solution: $A = A_0(0.5)^{t/k}$

$9 = 10(0.5)^{5/k}$

$0.9 = 0.5^{5/k}$

$\log 0.9 = \log 0.5^{5/k}$

$\log 0.9 = \frac{5}{k}\log 0.5$

$k = \frac{5\log 0.5}{\log 0.9}$

$k = 32.9$

The half life is 33 h.

Cumulative Review Exercises

1. $4 - 2[x - 3(2 - 3x) - 4x] = 2x$

$4 - 2[x - 6 + 9x - 4x] = 2x$

$4 - 2[6x - 6] = 2x$

$4 - 12x + 12 = 2x$

$16 = 14x$

$x = \frac{16}{14} = \frac{8}{7}$

The solution is $\frac{8}{7}$.

2. $2x - y = 5$

$-y = -2x + 5$

$y = 2x - 5$

$m = 2$ and $(2, -2)$

$y - y_1 = m(x - x_1)$

$y - (-2) = 2(x - 2)$

$y + 2 = 2x - 4$

$y = 2x - 6$

3. $4x^{2n} + 7x^n + 3 = (4x^n + 3)(x^n + 1)$

4. $\dfrac{1 - \frac{5}{x} + \frac{6}{x^2}}{1 + \frac{1}{x} - \frac{6}{x^2}} = \dfrac{1 - \frac{5}{x} + \frac{6}{x^2}}{1 + \frac{1}{x} - \frac{6}{x^2}} \cdot \dfrac{x^2}{x^2}$

$= \dfrac{x^2 - 5x + 6}{x^2 + x - 6} = \dfrac{(x-2)(x-3)}{(x-2)(x+3)}$

$= \dfrac{x-3}{x+3}$

5. $\dfrac{\sqrt{xy}}{\sqrt{x} - \sqrt{y}} = \dfrac{\sqrt{xy}}{\sqrt{x} - \sqrt{y}} \cdot \dfrac{\sqrt{x} + \sqrt{y}}{\sqrt{x} + \sqrt{y}}$

$= \dfrac{\sqrt{x^2y} + \sqrt{xy^2}}{\sqrt{x^2} - \sqrt{y^2}}$

$= \dfrac{x\sqrt{y} - y\sqrt{x}}{x - y}$

6. $x^2 - 4x - 6 = 0$

$x^2 - 4x + 4 = 6 + 4$

$(x-2)^2 = 10$

$\sqrt{(x-2)^2} = \sqrt{10}$

$x - 2 = \pm\sqrt{10}$

$x = 2 \pm \sqrt{10}$

The solutions are $2 + \sqrt{10}$ and $2 - \sqrt{10}$.

7. $(x - r_1)(x - r_2) = 0$

$\left(x - \frac{1}{3}\right)(x - (-3)) = 0$

$\left(x - \frac{1}{3}\right)(x + 3) = 0$

$x^2 + \frac{8}{3}x - 1 = 0$

$3x^2 + 8x - 3 = 0$

8. $2x - y < 3$ $\quad x + y < 1$
$-y < -2x + 3$ $\quad y < -x + 1$
$y > 2x - 3$

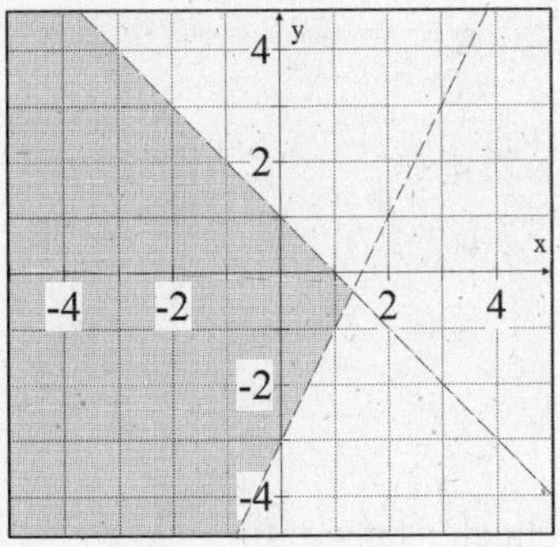

9. (1) $3x - y + z = 3$
(2) $x + y + 4z = 7$
(3 $3x - 2y + 3z = 8$
Eliminate y. Add equations (1) and (2).
$3x - y + z = 3$
$x + y + 4z = 7$
(4) $4x + 5z = 10$
Multiply equation (2) by 2 and add to equation (3).
$2(x + y + 4z) = 2(7)$
$3x - 2y + 3z = 8$

$2x + 2y + 8z = 14$
$3x - 2y + 3z = 8$
(5) $5x + 11z = 22$
Multiply equation (4) by 5 and equation (5) by −4, and add.
$5(4x + 5z) = 5(10)$
$-4(5x + 11z) = -4(22)$

$20x + 25z = 50$
$-20x - 44z = -88$
$-19z = -38$
$z = 2$
Replace z with 2 in equation (4).
$4x + 5(2) = 10$
$4x = 0$
$x = 0$
Replace x with 0 and z with 2 in equation (1).
$3(0) - y + 2 = 3$
$-y + 2 = 3$
$-y = 1$
$y = -1$
The solution is (0, −1, 2).

10. $\dfrac{x-4}{2-x} - \dfrac{1-6x}{2x^2-7x+6}$

$= \dfrac{x-4}{2-x} - \dfrac{1-6x}{(2x-3)(x-2)}$

$= \dfrac{x-4}{2-x} + \dfrac{1-6x}{(2x-3)(2-x)}$

$= \dfrac{(x-4)}{(2-x)} \cdot \dfrac{(2x-3)}{(2x-3)} + \dfrac{1-6x}{(2x-3)(2-x)}$

$= \dfrac{2x^2-11x+12+1-6x}{(2-x)(2x-3)} = \dfrac{2x^2-17x+13}{(2-x)(2x-3)}$

$= -\dfrac{2x^2-17x+13}{(x-2)(2x-3)}$

11. $x^2 + 4x - 5 \le 0$
$(x+5)(x-1) \le 0$
$\{x \mid -5 \le x \le 1\}$

12. $|2x - 5| \le 3$
$-3 \le 2x - 5 \le 3$
$-3+5 \le 2x - 5 + 5 \le 3 + 5$
$2 \le 2x \le 8$
$1 \le x \le 4$
$\{x \mid 1 \le x \le 4\}$

13. $f(x) = \left(\dfrac{1}{2}\right)^x + 1$

x	y
−1	3
0	2
2	$5/4$

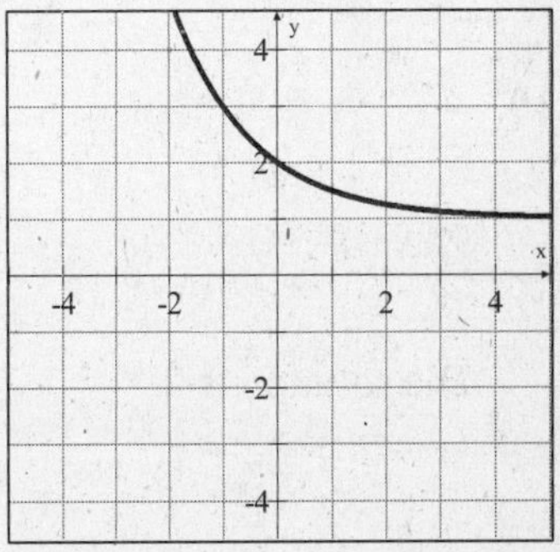

14. $f(x) = \log_2 x - 1$

$y + 1 = \log_2 x$

$x = 2^{y+1}$

x	y
1	−1
2	0
4	1

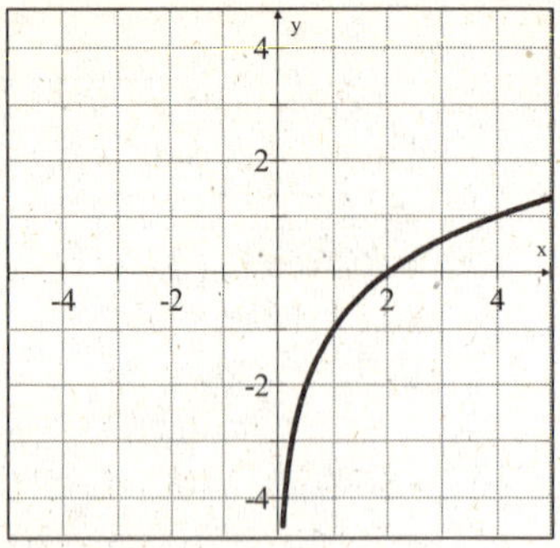

15. $f(-3) = 2^{-(-3)-1} = 2^2 = 4$

16. $\log_5 x = 3$

$x = 5^3 = 125$

17. $3\log_b x - 5\log_b y = \log_b x^3 - \log_b y^5$

$= \log_b \dfrac{x^3}{y^5}$

18. $\log_3 7 = x$

$x = \dfrac{\log 7}{\log 3} = 1.7712$

19. $4^{5x-2} = 4^{3x+2}$

$5x - 2 = 3x + 2$

$2x = 4$

$x = 2$

The solution is 2.

20. $\log x + \log(2x+3) = \log 2$

$\log(x(2x+3)) = \log 2$

$x(2x+3) = 2$

$2x^2 + 3x - 2 = 0$

$(2x-1)(x+2) = 0$

$2x - 1 = 0 \quad x + 2 = 0$

$2x = 1 \quad x = -2$

$x = \dfrac{1}{2}$

-2 does not check as a solution.

The solution is $\dfrac{1}{2}$.

21. Strategy: Let c represent the number of checks. To find the number of checks, write and solve an inequality.

Solution: $5.00 + 0.02c > 2.00 + 0.08c$

$5 > 2 + 0.06c$

$3 > 0.06c$

$50 > c$

The customer can write at most 49 checks.

22. Strategy: Let x represent the cost per pound of the mixture.

	Amount	Cost	Value
\$4.00	16	4.00	16(4.00)
\$2.50	24	2.50	24(2.50)
Mixture	40	x	$40x$

The sum of the values before mixing is equal to the value after mixing.

Solution:

$16(4.00) + 24(2.50) = 40x$

$64 + 60 = 40x$

$124 = 40x$

$x = 3.1$

The cost per pound of the mixture is \$3.10.

23. Strategy: Let x represent the rate of the wind.

	Distance	Rate	Time
With wind	1000	$225 + x$	$\frac{1000}{225+x}$
Against wind	800	$225 - x$	$\frac{800}{225-x}$

The flying time with the wind is the same as the flying time against the wind.

Solution:

$$\frac{1000}{225+x} = \frac{800}{225-x}$$

$$(225+x)(225-x)\frac{1000}{225+x} = \frac{800}{225-x}(225+x)(225-x)$$

$$(225-x)1000 = 800(225+x)$$

$$225000 - 1000x = 180000 + 800x$$

$$45000 = 1800x$$

$$25 = x$$

The rate of the wind is 25 mph.

24. Strategy: Write the basic direct variation equation replacing the variable with the given values. Solve for k.
Write the direct variation equation replacing k with its value. Substitute 34 for f and solve for d.

Solution:

$d = kf \qquad d = 0.3f$

$6 = k(20) \qquad = 0.3(34)$

$0.3 = k \qquad = 10.2$

The string will stretch 10.2 in.

25. Strategy: Let x represent the cost of redwood. The cost of fir is y.
First purchase:

	Amount	Cost	Value
Redwood	80	x	$80x$
Fir	140	y	$140y$

Second purchase:

	Amount	Cost	Value
Redwood	140	x	$140x$
Fir	100	y	$100y$

The total cost of the first purchase is \$67.
The total cost of the second purchase is \$81.

Solution:

$80x + 140y = 67$
$140x + 100y = 81$

$-5(80x + 140y) = -5(67)$
$7(140x + 100y) = 7(81)$

$-400x - 700y = -335$
$980x + 700y = 567$

$580x = 232$
$x = 0.40$

$80(0.40) + 140y = 67$
$32 + 140y = 67$
$140y = 35$
$y = 0.25$

The cost of the redwood is \$0.40 per foot.
The cost of the fir is \$0.25 per foot.

26. Strategy: To find how many years it will take for the investment to double in value.

$A = 10{,}000,\ P = 5000,\ i = \frac{7\%}{2} = \frac{0.07}{2} = 0.035$

Solution: $A = P(1+i)^n$

$10000 = 5000(1+0.035)^n$

$2 = (1.035)^n$

$\log 2 = \log(1.035)^n$

$\log 2 = n \log 1.035$

$\frac{\log 2}{\log 1.035} = n$

$n = 20$

$\frac{n}{2} = \frac{20}{2} = 10$

The investment take 10 years to double in value.

Final Exam

1. $12-8[3-(-2)]^2 \div 5-3$

$=12-8(3+2)^2 \div 5-3$

$=12-8(5)^2 \div 5-3$

$=12-8(25) \div 5-3$

$=12-200 \div 5-3$

$=12-40-3$

$=-31$

2. $\dfrac{a^2-b^2}{a-b}$

$\dfrac{3^2-(-4)^2}{3-(-4)}=\dfrac{9-16}{3+4}=\dfrac{-7}{7}$

$=-1$

3. $5-2[3x-7(2-x)-5x]$
$=5-2[3x-14+7x-5x]$
$=5-2[5x-14]$
$=5-10x+28$
$=-10x+33$

4. $\dfrac{3}{4}x-2=4$

$\dfrac{3}{4}x=6$

$\dfrac{4}{3}\cdot\dfrac{3}{4}x=6\cdot\dfrac{4}{3}$

$x=8$
The solution is 8.

5. $8-|5-3x|=1$

$-|5-3x|=-7$

$|5-3x|=7$

$5-3x=7 \quad 5-3x=-7$

$-3x=2 \qquad -3x=-12$

$x=-\dfrac{2}{3} \qquad x=4$

The solutions are $-\dfrac{2}{3}$ and 4.

6. $V=\dfrac{4}{3}\pi r^3=\dfrac{4}{3}\pi(4)^3=268.1\ ft^3$

7. $2x-3y=9$

$2x-3(0)=9$

$2x=9$

$x=\dfrac{9}{2}$

The x-intercept is $\left(\dfrac{9}{2},0\right)$.

$2(0)-3y=9$

$-3y=9$

$y=-3$
The y-intercept is (0, −3).

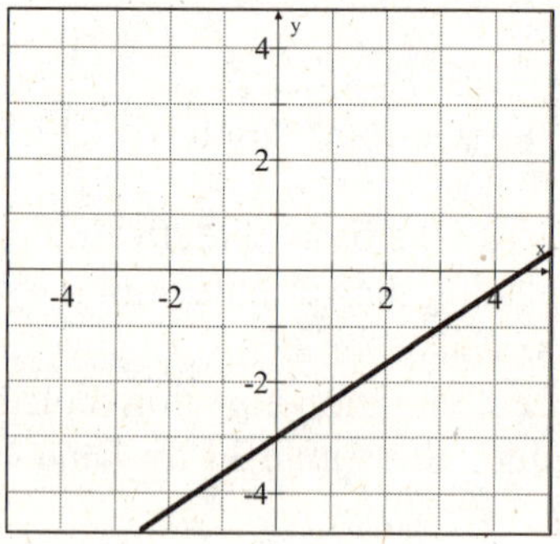

8. (3, −2) and (1, 4)

$m=\dfrac{y_2-y_1}{x_2-x_1}=\dfrac{4-(-2)}{1-3}=\dfrac{4+2}{-2}=\dfrac{6}{-2}$

$m=-3$

$y-y_1=m(x-x_1)$

$y-4=-3(x-1)$

$y-4=-3x+3$

$y=-3x+7$

9. $3x-2y=6$

$-2y=-3x+6$

$y=\frac{3}{2}x-3$

$m_1=\frac{3}{2}$

$m_1\cdot m_2=-1$

$\frac{3}{2}m_2=-1$

$m_2=-\frac{2}{3}\quad(-2,1)$

$y-y_1=m(x-x_1)$

$y-1=-\frac{2}{3}(x-(-2))$

$y-1=-\frac{2}{3}(x+2)$

$y-1=-\frac{2}{3}x-\frac{4}{3}$

$y=-\frac{2}{3}x-\frac{1}{3}$

10. $2a[5-a(2-3a)-2a]+3a^2$
$=2a[5-2a+3a^2-2a]+3a^2$
$=2a[5-4a+3a^2]+3a^2$
$=10a-8a^2+6a^3+3a^2$
$=6a^3-5a^2+10a$

11. $8-x^3y^3=2^3-(xy)^3$
$=(2-xy)(4+2xy+x^2y^2)$

12. $x-y-x^3+x^2y=x-y-x^2(x-y)$
$=1(x-y)-x^2(x-y)$
$=(x-y)(1-x^2)$
$=(x-y)(1-x)(1+x)$

13.
$$\begin{array}{r|l} & x^2-2x-3 \\ 2x-3 & 2x^3-7x^2+0x+4 \\ & \underline{2x^3-3x^2} \\ & -4x^2+0x \\ & \underline{-4x^2+6x} \\ & -6x+4 \\ & \underline{-6x+9} \\ & -5 \end{array}$$

$(2x^3-7x^2+4)\div(2x-3)=x^2-2x-3+\frac{-5}{2x-3}$

14. $\frac{x^2-3x}{2x^2-3x-5}\div\frac{4x-12}{4x^2-4}$

$=\frac{x^2-3x}{2x^2-3x-5}\cdot\frac{4x^2-4}{4x-12}$

$=\frac{x(x-3)}{(2x-5)(x+1)}\cdot\frac{4(x+1)(x-1)}{4(x-3)}$

$=\frac{x(x-3)\cdot 4(x+1)(x-1)}{(2x-5)(x+1)\cdot 4(x-3)}$

$=\frac{x(x-1)}{2x-5}$

15. The LCM is $(x-3)(x+2)$.

$\frac{x-2}{x+2}-\frac{x+3}{x-3}=\frac{x-2}{x+2}\cdot\frac{x-3}{x-3}-\frac{x+3}{x-3}\cdot\frac{x+2}{x+2}$

$=\frac{(x-2)(x-3)-(x+3)(x+2)}{(x-3)(x+2)}$

$=\frac{x^2-5x+6-x^2-5x-6}{(x-3)(x+2)}$

$=\frac{-10x}{(x-3)(x+2)}$

16. The LCM is $x(x+4)$.

$$\frac{\frac{3}{x}+\frac{1}{x+4}}{\frac{1}{x}+\frac{3}{x+4}}=\frac{\frac{3}{x}+\frac{1}{x+4}}{\frac{1}{x}+\frac{3}{x+4}}\cdot\frac{x(x+4)}{x(x+4)}$$

$$=\frac{3x+12+x}{x+4+3x}=\frac{4x+12}{4x+4}$$

$$=\frac{4(x+3)}{4(x+1)}=\frac{x+3}{x+1}$$

17. $\dfrac{5}{x-2}-\dfrac{5}{x^2-4}=\dfrac{1}{x+2}$

$$\frac{5}{x-2}-\frac{5}{(x+2)(x-2)}=\frac{1}{x+2}$$

$$(x-2)(x+2)\left(\frac{5}{x-2}-\frac{5}{(x+2)(x-2)}\right)=(x-2)(x+2)\left(\frac{1}{x+2}\right)$$

$$5(x+2)-5=x-2$$

$$5x+10-5=x-2$$

$$4x+5=-2$$

$$4x=-7$$

$$x=-\frac{7}{4}$$

The solution is $-\dfrac{7}{4}$.

18. $a_n=a_1+(n-1)d$

$$a_n-a_1=(n-1)d$$

$$d=\frac{a_n-a_1}{n-1}$$

19. $\left(\dfrac{4x^2y^{-1}}{3x^{-1}y}\right)^{-2}\left(\dfrac{2x^{-1}y^2}{9x^{-2}y^2}\right)^3$

$$\frac{4^{-2}x^{-4}y^2}{3^{-2}x^2y^{-2}}\cdot\frac{2^3x^{-3}y^6}{9^3x^{-6}y^6}$$

$$=\frac{2^3\cdot 3^2y^4x^3}{4^2\cdot 9^3x^6}$$

$$=\frac{y^4}{162x^3}$$

20. $\left(\dfrac{3x^{2/3}y^{1/2}}{6x^2y^{4/3}}\right)^6=\dfrac{3^6x^4y^3}{6^6x^{12}y^8}=\dfrac{1}{64x^8y^5}$

21. $x\sqrt{18x^2y^3}-y\sqrt{50x^4y}$

$$=x\sqrt{3^2x^2y^2(2y)}-y\sqrt{5^2x^4(2y)}$$

$$=3x^2y\sqrt{2y}-5x^2y\sqrt{2y}$$

$$=-2x^2y\sqrt{2y}$$

22. $\dfrac{\sqrt{16x^5y^4}}{\sqrt{32xy^7}} = \sqrt{\dfrac{16x^5y^4}{32xy^7}} = \sqrt{\dfrac{x^4}{2y^3}}$

$= \sqrt{\dfrac{x^4}{y^2(2y)}} = \dfrac{x^2}{y}\sqrt{\dfrac{1}{2y}} \cdot \sqrt{\dfrac{2y}{2y}}$

$= \dfrac{x^2}{y}\sqrt{\dfrac{2y}{(2y)^2}}$

$= \dfrac{x^2\sqrt{2y}}{2y^2}$

23. $\dfrac{3}{2+i} \cdot \dfrac{2-i}{2-i} = \dfrac{6-3i}{4-i^2} = \dfrac{6-3i}{4+1}$

$= \dfrac{6-3i}{5} = \dfrac{6}{5} - \dfrac{3}{5}i$

24. $(x-r_1)(x-r_2) = 0$

$\left(x - \left(-\dfrac{1}{2}\right)\right)(x-2) = 0$

$\left(x + \dfrac{1}{2}\right)(x-2) = 0$

$x^2 - \dfrac{3}{2}x - 1 = 0$

$2(x^2 - \dfrac{3}{2}x - 1) = 0(2)$

$2x^2 - 3x - 2 = 0$

25. $2x^2 - 3x - 1 = 0$

$a = 2, b = -3, c = -1$

$x = \dfrac{-b \pm \sqrt{b^2 - 4ac}}{2a}$

$x = \dfrac{-(-3) \pm \sqrt{(-3)^2 - 4(2)(-1)}}{2(2)}$

$x = \dfrac{3 \pm \sqrt{9+8}}{4}$

$x = \dfrac{3 \pm \sqrt{17}}{4}$

The solutions are $\dfrac{3+\sqrt{17}}{4}$ and $\dfrac{3-\sqrt{17}}{4}$.

26. $x^{2/3} - x^{1/3} - 6 = 0$

$\left(x^{1/3}\right)^2 - x^{1/3} - 6 = 0$

$u^2 - u - 6 = 0$

$(u-3)(u+2) = 0$

$u - 3 = 0 \quad u + 2 = 0$

$u = 3 \qquad u = -2$

Replace u with $x^{1/3}$.

$x^{1/3} = 3 \qquad x^{1/3} = -2$

$\left(x^{1/3}\right)^3 = (3)^3 \quad \left(x^{1/3}\right)^3 = (-2)^3$

$x = 27 \qquad x = -8$

The solutions are −8 and 27.

27. $f(x) = -x^2 + 4$

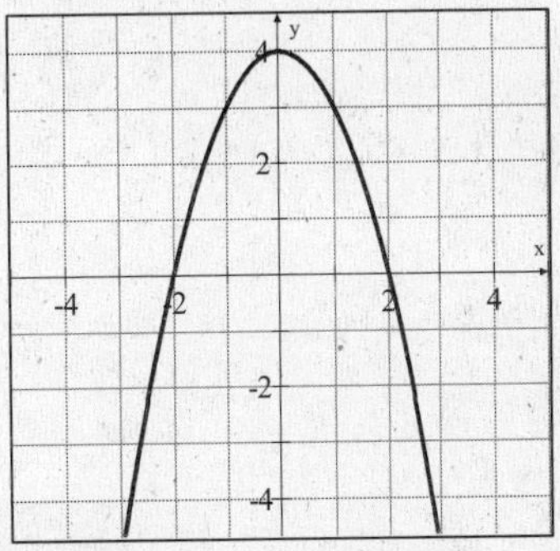

28. $f(x) = -\dfrac{1}{2}x - 3$

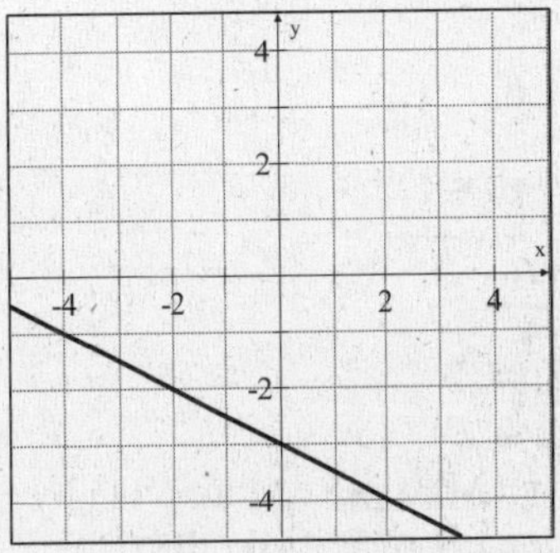

29. $\dfrac{2}{x}-\dfrac{2}{2x+3}=1$

$x(2x+3)\left(\dfrac{2}{x}-\dfrac{2}{2x+3}\right)=1x(2x+3)$

$2(2x+3)-2x=2x^2+3x$

$4x+6-2x=2x^2+3x$

$0=2x^2+x-6$

$(2x-3)(x+2)=0$

$2x-3=0 \quad x+2=0$

$2x=3 \quad x=-2$

$x=\dfrac{3}{2}$

The solutions are -2 and $\dfrac{3}{2}$.

30. $f(x)=\dfrac{2}{3}x-4$

$y=\dfrac{2}{3}x-4$

$x=\dfrac{2}{3}y-4$

$x+4=\dfrac{2}{3}y$

$\dfrac{3}{2}(x+4)=\dfrac{3}{2}\left(\dfrac{2}{3}y\right)$

$y=\dfrac{3}{2}x+6$

$f^{-1}(x)=\dfrac{3}{2}x+6$

31. (1) $3x-2y=1$
(2) $5x-3y=3$
Eliminate y. Multiply equation (1) by -3 and equation (2) by 2. Add the two equations.
$-3(3x-2y)=-3(1)$
$2(5x-3y)=2(3)$

$-9x+6y=-3$
$10x-6y=6$
$x=3$

Substitute 3 for x in equation (1).
$3(3)-2y=1$
$9-2y=1$
$-2y=-8$
$y=4$
The solution is (3, 4).

32. $\sqrt{49x^6}=\sqrt{7^2x^6}=7x^3$

33. $2-3x<6 \qquad 2x+1>4$
$-3x<4 \qquad 2x>3$
$x>-\dfrac{4}{3} \qquad x>\dfrac{3}{2}$

The solution is $\left(\dfrac{3}{2},\infty\right)$.

34. $|2x+5|<3$

$-3<2x+5<3$

$-3-5<2x+5-5<3-5$

$-8<2x<-2$

$-4<x<-1$

$\{x\,|-4<x<-1\}$

35. $3x+2y>6$

$2y>-3x+6$

$y>-\dfrac{3}{2}x+3$

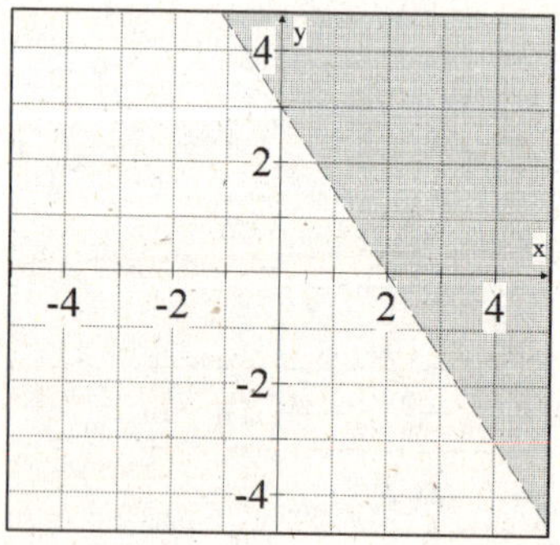

36. $f(x) = 3^{-x} - 2$

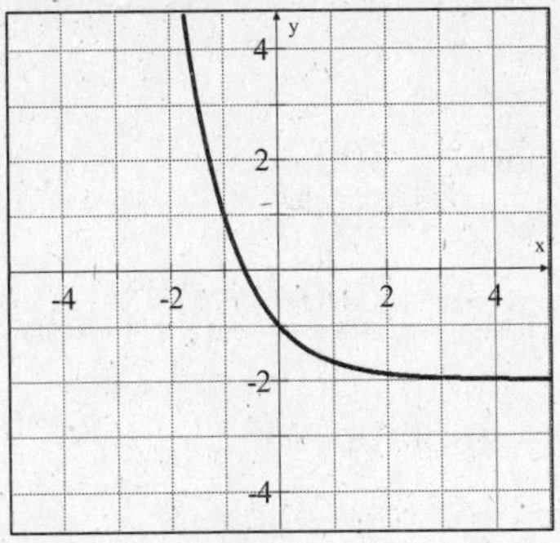

37. $f(x) = \log_2(x+1)$

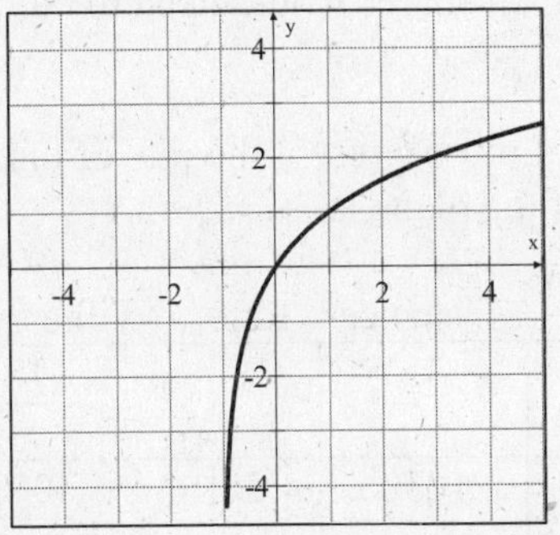

38. $2(\log_2 a - \log_2 b) = 2\log_2 \frac{a}{b} = \log_2 \frac{a^2}{b^2}$

39. $\log_3(x) - \log_3(x-3) = \log_3 2$

$\log_3 \frac{x}{x-3} = \log_3 2$

$\frac{x}{x-3} = 2$

$(x-3)\frac{x}{x-3} = 2(x-3)$

$x = 2x - 6$

$-x = -6$

$x = \frac{-6}{-1} = 6$

The solution is 6 .

40. Strategy: Let x represent the score on the last test.
To find the range of scores solve the inequality.

Solution:

$70 \le \frac{64+58+82+77+x}{5} \le 79$

$70 \le \frac{281+x}{5} \le 79$

$350 \le 281 + x \le 395$

$69 \le x \le 114$

Since 100 is the maximum score, the range of scores is $69 \le x \le 100$.

41. Strategy: Let x represent the average speed of the jogger.
The average speed of the cyclist is $2.5x$.

	Rate	Time	Distance
Jogger	x	2	$2(x)$
Cyclist	$2.5x$	2	$2(2.5x)$

The distance traveled by the cyclist is 24 miles more than the distance traveled by th2 jogger.

Solution: $2x + 24 = 2(2.5x)$

$2x + 24 = 5x$

$24 = 3x$

$x = 8$

$2(2.5x) = 5(8) = 40$

The cyclist traveled 40 mi.

42. Strategy: Let x represent the amount invested at 8.5%.
The amount invested at 6.4% is $12000 - x$.

	Principal	Rate	Interest
8.5%	x	0.085	$0.085x$
6.4%	$12000 - x$	0.064	$0.064(12000 - x)$

The sum of the interest earned from the two investments is $936.

Solution:
$0.085x + 0.064(12000 - x) = 936$
$0.085x + 768 - 0.064x = 936$
$0.021x + 768 = 936$
$0.021x = 168$
$x = 8000$
$12000 - x = 12000 - 8000 = 4000$
$8,000 is invested at 8.5% and $4,000 is invested at 6.4%.

43. Strategy: Let x represent the width of the rectangle.
The length of the rectangle is $3x - 1$.
The area of the rectangle is 140 ft^2.

Solution: $A = LW$
$140 = x(3x - 1)$
$140 = 3x^2 - x$
$0 = 3x^2 - x - 140$
$0 = (3x + 20)(x - 7)$
$3x + 20 = 0 \quad x - 7 = 0$
$3x = -20 \quad x = 7$
$x = -\frac{20}{3}$
The width cannot be a negative number.
$3x - 1 = 3(7) - 1 = 20$
The width is 7 ft.
The length is 20 ft.

44. Strategy: Let x represent the number of additional shares. Write and solve a proportion.

Solution: $\frac{300}{486} = \frac{300 + x}{810}$
$(810 \cdot 486)\frac{300}{486} = \frac{300 + x}{810}(810 \cdot 486)$
$300(810) = 486(300 + x)$
$243,000 = 145,800 + 486x$
$97200 = 486x$
$x = 200$
200 additional shares would need to be purchased.

45. Strategy: Let x represent the rate of the car.
The rate of the plane is $7x$.

	Distance	Rate	Time
Car	45	x	$\frac{45}{x}$
Plane	1050	$7x$	$\frac{1050}{7x}$

The total time traveled is $3\frac{1}{4}$ h.

Solution: $\frac{45}{x} + \frac{1050}{7x} = 3\frac{1}{4}$
$\frac{45}{x} + \frac{150}{x} = \frac{13}{4}$
$4x\left(\frac{45}{x} + \frac{150}{x}\right) = \left(\frac{13}{4}\right)4x$
$180 + 600 = 13x$
$780 = 13x$
$x = 60$
$7x = 7(60) = 420$
The rate of the plane is 420 mph.

46. Strategy: To find the distance the object has fallen, substitute 75 ft/s for v and solve for d.

Solution: $v = \sqrt{64d}$

$75 = \sqrt{64d}$

$75^2 = \left(\sqrt{64d}\right)^2$

$5625 = 64d$

$d = 87.89$

The distance traveled is 88 ft.

47. Strategy: Let x represent the rate traveled during the first 360 mi.
The rate traveled during the next 300 mi is $x + 30$.

	Distance	Rate	Time
First part of the trip	360	x	$\frac{360}{x}$
Second part of the trip	300	$x + 30$	$\frac{300}{x+30}$

The total time traveled was 5 h.

Solution: $\frac{360}{x} + \frac{300}{x+30} = 5$

$x(x+30)\left(\frac{360}{x} + \frac{300}{x+30}\right) = 5(x)(x+30)$

$360(x+30) + 300x = 5x^2 + 150x$

$360x + 10800 + 300x = 5x^2 + 150x$

$0 = 5x^2 - 510x - 10800$

$0 = 5(x^2 + 102x - 2160)$

$0 = 5(x+18)(x-120)$

$x + 18 = 0 \quad x - 120 = 0$

$x = -18 \quad\quad x = 120$

The rate cannot be a negative number.
The rate of the plane for the first 360 mi is 120 mph.

48. Strategy: Write the basic inverse variation equation replacing the variable with the given values. Solve for k.
Write the inverse variation equation replacing k with its value. Substitute 4 for d and solve for I.

Solution:

$I = \frac{k}{d^2}$

$8 = \frac{k}{20^2}$

$8 = \frac{k}{400}$

$3200 = k$

$I = \frac{3200}{d^2} = \frac{3200}{4^2} = \frac{3200}{16} = 200$

The intensity is 200 foot-candles.

49. Strategy: Let x represent the rate of the boat in calm water.
The rate of the current is y.

	Rate	Time	Distance
With current	$x + y$	2	$2(x + y)$
Against current	$x - y$	3	$3(x - y)$

The distance traveled with the current is 30 miles. The distance traveled against the current is 30 miles.
$2(x + y) = 30$
$3(x - y) = 30$

Solution:

$2(x + y) = 30$

$3(x - y) = 30$

$\frac{1}{2} \cdot 2(x + y) = \frac{1}{2} \cdot 30$

$\frac{1}{3} \cdot 3(x - y) = \frac{1}{3} \cdot 30$

$x+y=15$

$x-y=10$

$2x=25$

$x=12.5$

$x+y=15$

$12.5+y=15$

$y=2.5$

The rate of the boat in calm water is 12.5 mph. The rate of the current is 2.5 mph.

50. Strategy: To find the value of the investment after two years use the compound interest formula.

$A = 4000,\ n = 24,\ i = \frac{9\%}{12} = \frac{0.09}{12} = 0.0075$

Solution: $P = A(1+i)^n$

$P = 4000(1+0.0075)^{24}$

$P = 4000(1.0075)^{24}$

$P = 4785.65$

The value of the investment after 2 years is \$4785.65.

Algebra Review

Section R.1

Objective A Exercises

1. $a-2c$
$2-2(-4)=2+8=10$

3. $3b-3c$
$3(3)-3(-4)=9+12=21$

5. $16\div(2c)$
$16\div[(2)(-4)]=16\div(-8)=-2$

7. $3b-(a+c)^2$
$3(3)-(2+(-4))^2=9-(-2)^2=9-4=5$

9. $(b-3a)^2+bc$
$(3-3(2))^2+3(-4)=(3-6)^2-12$
$=(-3)^2-12=9-12=-3$

11. $\frac{d-b}{a}$
$=\frac{4-3}{-1}=\frac{1}{-1}=-1$

13. $\frac{b-d}{c-a}$
$\frac{3-4}{-2-(-1)}=\frac{-1}{-2+1}=\frac{-1}{-1}=1$

15. $3(b-a)-bc$
$3[3-(-1)]-3(-2)=3(3+1)+6$
$=3(4)+6=12+6=18$

17. $\frac{abc}{b-d}$
$\frac{(-1)(3)(-2)}{3-4}=\frac{-3(-2)}{-1}=\frac{6}{-1}=-6$

19. $(-b+d)^2+(-a+c)^2$
$(-3+4)^2+(-(-1)+(-2))^2=(1)^2+(1-2)^2$
$=1+(-1)^2=1+1=2$

21. $3cd-(4a)^2$
$3(-2)(4)-[4(-1)]^2=-6(4)-(-4)^2$
$=-24-16=-40$

23. $(a+b)^2-c$
$[2.7+(-1.6)]^2-(-0.8)=(1.1)^2+0.8$
$=1.21+0.8=2.01$

Objective B Exercises

25. $x+7x=8x$

27. $8b-5b=3b$

29. $-12a+17a=5a$

31. $4x+5x+2x=9x+2x=11x$

33. $6x-2y+9x=(6x+9x)-2y=15x-2y$

35. $5a+6a-2a=11a-2a=9a$

37. $12y^2+10y^2=22y^2$

39. $\frac{3}{4}x-\frac{1}{4}x=\frac{2}{4}x=\frac{1}{2}x$

41. $-4(5x)=-20x$

43. $(6a)(-4)=-24a$

45. $\frac{1}{4}(4x)=\frac{4}{4}x=x$

47. $\frac{1}{3}(21x)=7x$

49. $(36y)\left(\frac{1}{12}\right)=\frac{36y}{12}=3y$

51. $-3(a+5)=-3a-15$

53. $(-2x-6)8=-16x-48$

55. $-5(2y^2-1)=-10y^2+5$

57. $6(3x^2-2xy-y^2)=18x^2-12xy-6y^2$

59. $3-(10+8y)=3-10-8y=-8y-7$

61. $-5[2x+3(5-x]=-5[2x+15-3x]$
$=-5[-x+15]=5x-75$

63. $-5a-2[2a-4(a+7)]=-5a-2[2a-4a-28]$
$=-5a-2[-2a-28]$
$=-5a+4a+56=-a+56$

Section R.2

Objective A Exercises

1.
$$x+7=-5$$
$$x+7-7=-5-7$$
$$x=-12$$
The solution is –12.

3.
$$-9=z-8$$
$$-9+8=z-8+8$$
$$-1=z$$
The solution is –1.

5.
$$-48=6z$$
$$\frac{-48}{6}=\frac{6z}{6}$$
$$-8=z$$
The solution is –8

7.
$$-\frac{3}{4}x=15$$
$$-\frac{4}{3}\left(-\frac{3}{4}x\right)=-\frac{4}{3}(15)$$
$$x=-20$$
The solution is –20.

9.
$$-\frac{x}{4}=-2$$
$$-4\left(-\frac{x}{4}\right)=-4(-2)$$
$$x=8$$
The solution is 8.

11.
$$4-2b=-2-4b$$
$$4-2b+4b=2-4b+4b$$
$$4+2b=2$$
$$4-4+2b=2-4$$
$$2b=-2$$
$$\frac{2b}{2}=\frac{-2}{2}$$
$$b=-1$$
The solution is –1.

13.
$$5x-3=9x-7$$
$$5x-9x-3=9x-9x-7$$
$$-4x-3=-7$$
$$-4x-3+3=-7+3$$
$$-4x=-4$$
$$\frac{-4x}{-4}=\frac{-4}{-4}$$
$$x=1$$
The solution is 1.

15.
$$6a-1=2+2a$$
$$6a-2a-1=2+2a-2a$$
$$4a-1=2$$
$$4a-1+1=2+1$$
$$4a=3$$
$$\frac{4a}{4}=\frac{3}{4}$$
$$a=\frac{3}{4}$$
The solution is $\frac{3}{4}$.

17.
$$2-6y=5-7y$$
$$2-6y+7y=5-7y+7y$$
$$2+y=5$$
$$2-2+y=5-2$$
$$y=3$$
The solution is 3.

19.
$$2(x+1)+5x=23$$
$$2x+2+5x=23$$
$$7x+2=23$$
$$7x+2-2=23-2$$
$$7x=21$$
$$\frac{7x}{7}=\frac{21}{7}$$
$$x=3$$
The solution is 3.

21.
$$7a-(3a-4)=12$$
$$7a-3a+4=12$$
$$4a+4=12$$
$$4a+4-4=12-4$$
$$4a=8$$
$$\frac{4a}{4}=\frac{8}{4}$$
$$a=2$$
The solution is 2.

23.
$$9-7x=4(1-3x)$$
$$9-7x=4-12x$$
$$9-7x+12x=4-12x+12x$$
$$9+5x=4$$
$$9-9+5x=4-9$$
$$5x=-5$$
$$\frac{5x}{5}=\frac{-5}{5}$$
$$x=-1$$
The solution is –1.

25.
$$\begin{aligned} 2z-2&=5-(9-6z) \\ 2z-2&=5-9+6z \\ 2z-2&=-4+6z \\ 2z-2z-2&=-4+6z-2z \\ -2&=-4+4z \\ -2+4&=-4+4+4z \\ 2&=4z \\ \frac{2}{4}&=\frac{4z}{4} \\ \frac{1}{2}&=z \end{aligned}$$

The solution is $\frac{1}{2}$.

27.
$$\begin{aligned} 5(6-2x)&=2(5-3x) \\ 30-10x&=10-6x \\ 30-10x+10x&=10-6x+10x \\ 30&=10+4x \\ 30-10&=10-10+4x \\ 20&=4x \\ \frac{20}{4}&=\frac{4x}{4} \\ 5&=x \end{aligned}$$

The solution is 5.

29.
$$\begin{aligned} 2(3b-5)&=4(6b-2) \\ 6b-10&=24b-8 \\ 6b-6b-10&=26b-6b-8 \\ -10&=18b-8 \\ -10+8&=18b-8+8 \\ -2&=18b \\ \frac{-2}{18}&=\frac{18b}{18} \\ -\frac{1}{9}&=b \end{aligned}$$

The solution is $-\frac{1}{9}$.

Objective B Exercises

31.
$$\begin{aligned} x-5&>-2 \\ x-5+5&>-2+5 \\ x&>3 \end{aligned}$$

The solution is set is $\{x \mid x>3\}$.

33.
$$\begin{aligned} -2+n&\ge 0 \\ -2+2+n&\ge 0+2 \\ n&\ge 2 \end{aligned}$$

The solution is set is $\{n \mid n\ge 2\}$.

35.
$$\begin{aligned} 8x&\le -24 \\ \frac{8x}{8}&\le\frac{-24}{8} \\ x&\le -3 \end{aligned}$$

The solution is set is $\{x \mid x\le -3\}$.

37.
$$\begin{aligned} 3n&>0 \\ \frac{3n}{3}&>\frac{0}{3} \\ n&>0 \end{aligned}$$

The solution is set is $\{n \mid n>0\}$.

39.
$$\begin{aligned} 2x-1&>7 \\ 2x-1+1&>7+1 \\ 2x&>8 \\ \frac{2x}{2}&>\frac{8}{2} \\ x&>4 \end{aligned}$$

The solution is set is $\{x \mid x>4\}$.

41.
$$\begin{aligned} 4-3x&<10 \\ 4-4-3x&<10=4 \\ -3x&<6 \\ \frac{-3x}{-3}&>\frac{6}{-3} \\ x&>-2 \end{aligned}$$

The solution is set is $\{x \mid x>-2\}$.

43.
$$\begin{aligned} 3x-1&>2x+2 \\ 3x-2x-1&>2x-2x+2 \\ x-1&>2 \\ x-1+1&>2+1 \\ x&>3 \end{aligned}$$

The solution is set is $\{x \mid x>3\}$.

45.
$$\begin{aligned} 8x+1&\ge 2x+13 \\ 8x-2x+1&\ge 2x-2x+13 \\ 6x+1&\ge 13 \\ 6x+1-1&\ge 13-1 \\ 6x&\ge 12 \\ \frac{6x}{6}&\ge\frac{12}{6} \\ x&\ge 2 \end{aligned}$$

The solution is set is $\{x \mid x\ge 2\}$.

47.
$$\begin{aligned}-3-4x&>-11\\-3+3-4x&>-11+3\\-4x&>-8\\\frac{-4x}{-4}&<\frac{-8}{-4}\\x&<2\end{aligned}$$
The solution is set is $\{x \mid x<2\}$.

49.
$$\begin{aligned}4x-2&>3x+1\\4x-3x-2&>3x-3x+1\\x-2&>1\\x-2+2&>1+2\\x&>3\end{aligned}$$
The solution is set is $\{x \mid x>3\}$.

51.
$$\begin{aligned}9x+2&\ge 3x+14\\9x-3x+2&\ge 3x-3x+14\\6x+2&\ge 14\\6x+2-2&\ge 14-2\\6x&\ge 12\\\frac{6x}{6}&\ge\frac{12}{6}\\x&\ge 2\end{aligned}$$
The solution is set is $\{x \mid x\ge 2\}$.

53.
$$\begin{aligned}-5-2x&>-13\\-5+5-2x&>-13+5\\-2x&>-8\\\frac{-2x}{-2}&<\frac{-8}{-2}\\x&<4\end{aligned}$$
The solution is set is $\{x \mid x<4\}$.

55.
$$\begin{aligned}4(2x-1)&>3x-2(3x-5)\\8x-4&>3x-6x+10\\8x-4&>-3x+10\\8x+3x-4&>-3x+3x+10\\11x-4&>10\\11x-4+4&>10+4\\11x&>14\\\frac{11x}{11}&>\frac{14}{11}\end{aligned}$$
The solution is set is $\left\{x \mid x>\frac{14}{11}\right\}$.

57.
$$\begin{aligned}3(4x+3)&\le 7-4(x-2)\\12x+9&\le 7-4x+8\\12x+9&\le 15-4x\\12x+4x+9&\le 15-4x+4x\\16x+9&\le 15\\16x+9-9&\le 15-9\\16x&\le 6\\\frac{16x}{16}&\le\frac{6}{16}\\x&\le\frac{3}{8}\end{aligned}$$
The solution is set is $\left\{x \mid x\le\frac{3}{8}\right\}$.

59.
$$\begin{aligned}3-4(x+2)&\le 6+4(2x+1)\\3-4x-8&\le 6+8x+4\\-4x-5&\le 8x+10\\-4x-8x-5&\le 8x-8x+10\\-12x-5&\le 10\\-12x-5+5&\le 10+5\\-12x&\le 15\\\frac{-12x}{-12}&\ge\frac{15}{-12}\\x&\ge-\frac{5}{4}\end{aligned}$$
The solution is set is $\left\{x \mid x\ge-\frac{5}{4}\right\}$.

Section R.3

Objective A Exercises

1. Graphing the points (2, 3), (4, 0), (–4, 1), and (–2, –2).

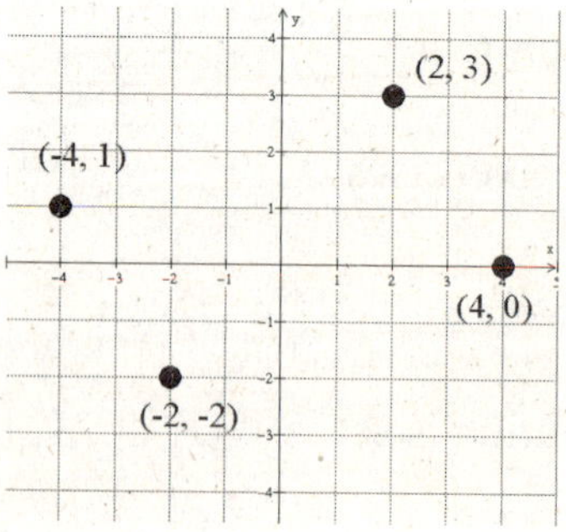

3. Graphing the points (–2, 5), (3, 4), (0, 0), and (–3, –2).

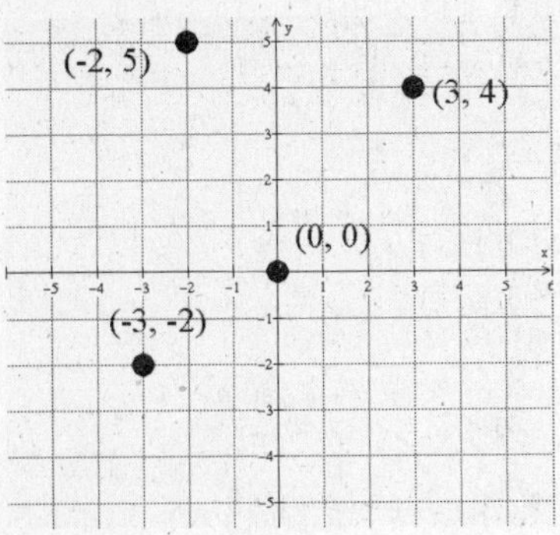

Objective B Exercises

5. $y = 2x + 1$

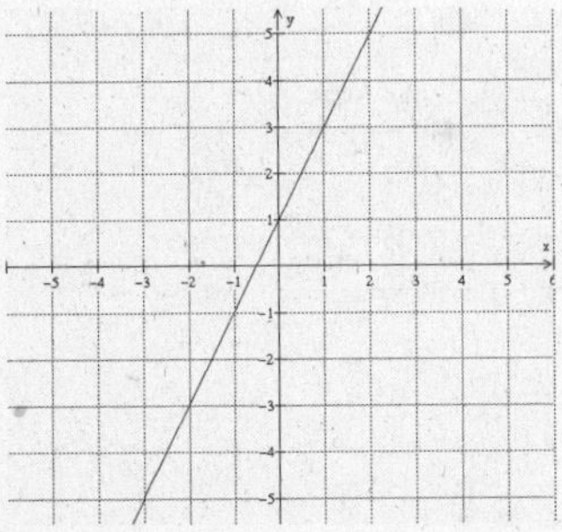

7. $y = -3x + 4$

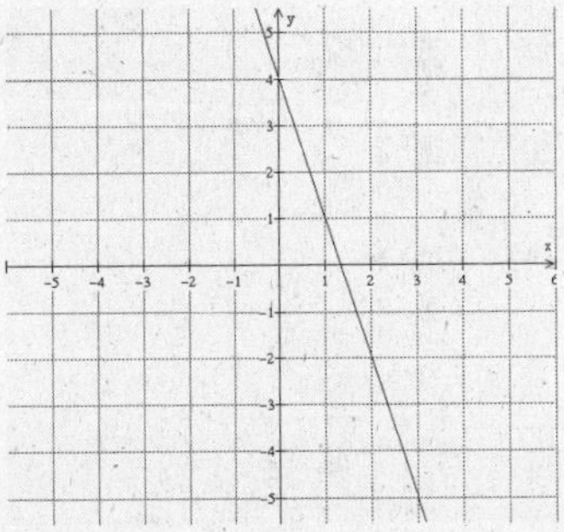

9. $y = 3x$

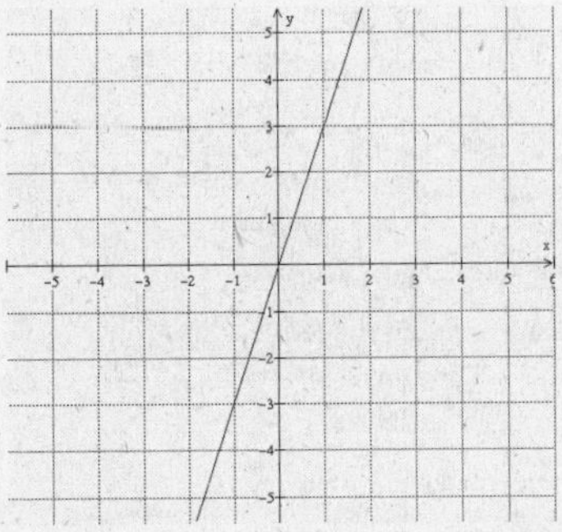

11. $y = -\frac{4}{3}x$

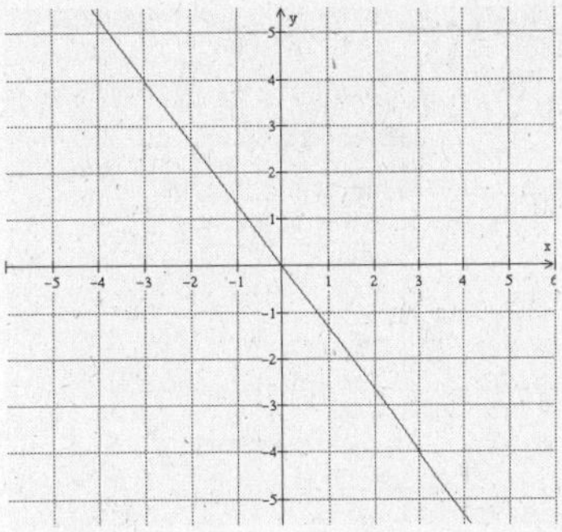

13. $y = \frac{3}{2}x - 1$

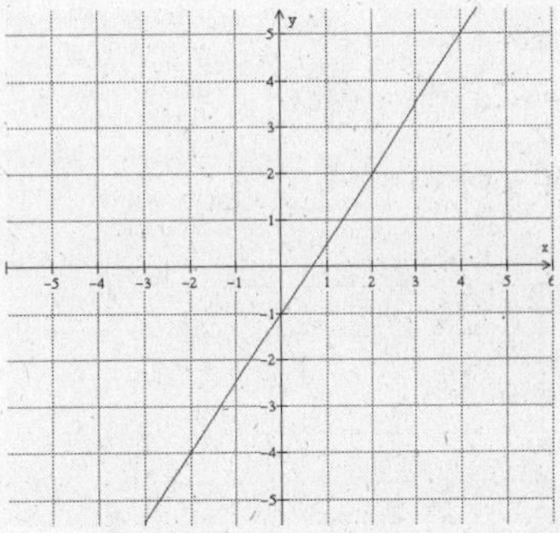

15. $y = -\frac{2}{3}x + 1$

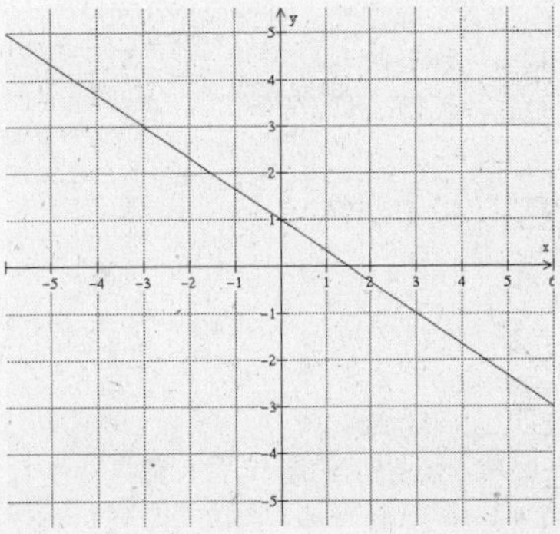

17. $2x + y = -3$

$y = -2x - 3$

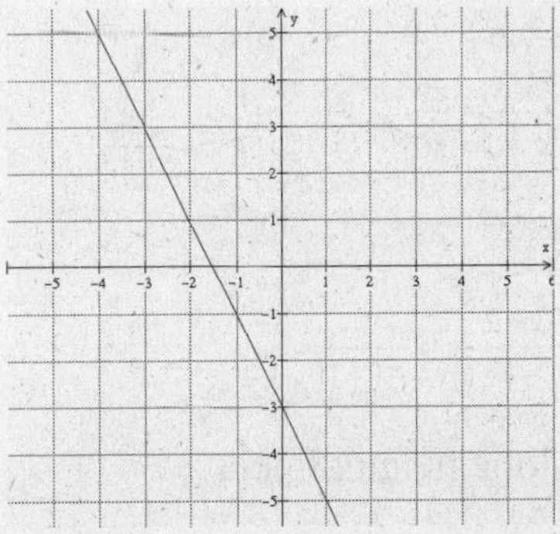

19. $x - 4y = 8$

$-4y = -x + 8$

$y = \frac{1}{4}x - 2$

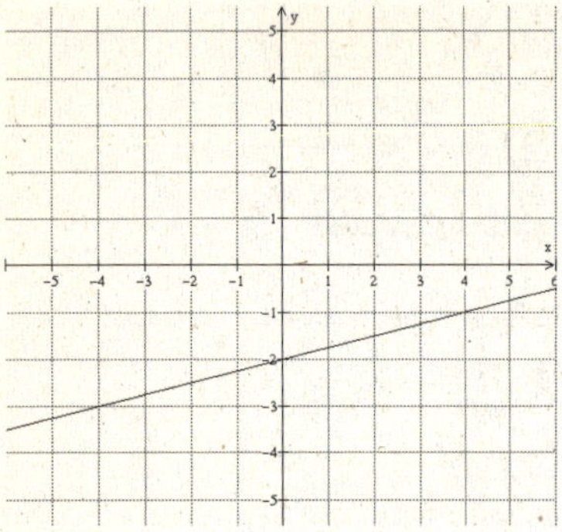

21. $3x - 2y = 8$

$-2y = -3x + 8$

$y = \frac{3}{2}x - 4$

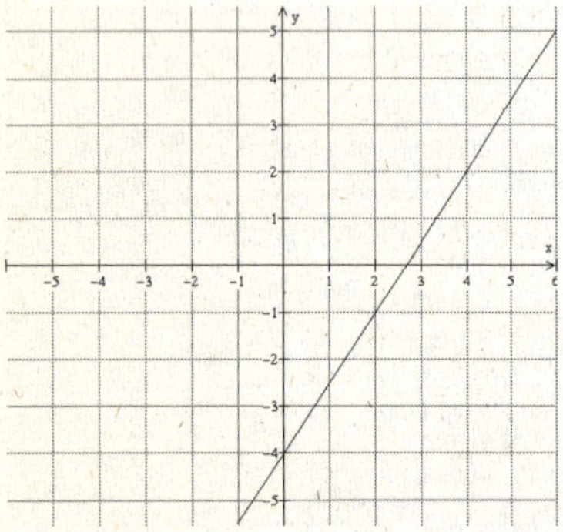

23. $m = \frac{y_2 - y_1}{x_2 - x_1}$

$m = \frac{4-1}{3-2} = \frac{3}{1} = 3$

25. $m = \frac{y_2 - y_1}{x_2 - x_1}$

$m = \frac{2-1}{2-(-2)} = \frac{1}{2+2} = \frac{1}{4}$

27. $m = \frac{y_2 - y_1}{x_2 - x_1}$

$m = \frac{-3-3}{5-1} = \frac{-6}{4} = -\frac{3}{2}$

29. $m = \frac{y_2 - y_1}{x_2 - x_1}$

$m = \frac{3-2}{-1-(-1)} = \frac{1}{-1+1} = \frac{1}{0}$

The slope is undefined.

31. $m = \frac{y_2 - y_1}{x_2 - x_1}$

$m = \frac{1-1}{-2-5} = \frac{0}{-7} = 0$

33. $m = \frac{y_2 - y_1}{x_2 - x_1}$

$m = \frac{-1-0}{2-3} = \frac{-1}{-1} = 1$

35. For $y = \frac{5}{2}x - 4$, $m = \frac{5}{2}$, and $b = (0, -4)$.

37. For $y = x$, $m = 1$, and $b = (0, 0)$.

39. $y = \frac{1}{2}x + 2$

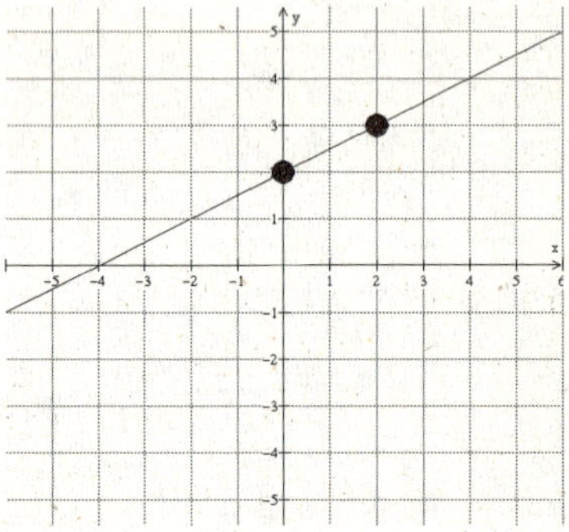

41. $y = -\frac{3}{2}x$

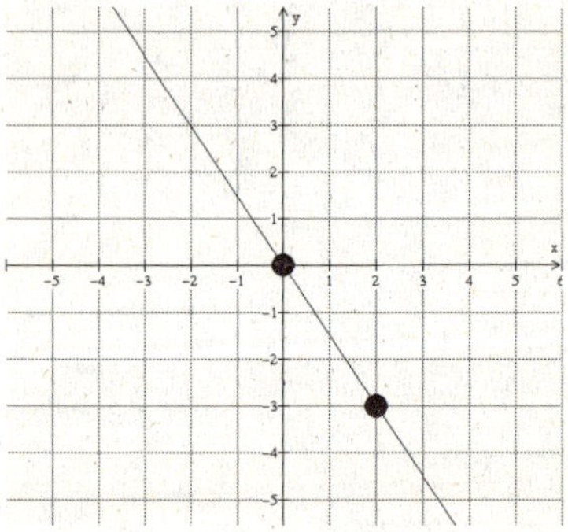

43. $y = -\frac{1}{2}x + 2$

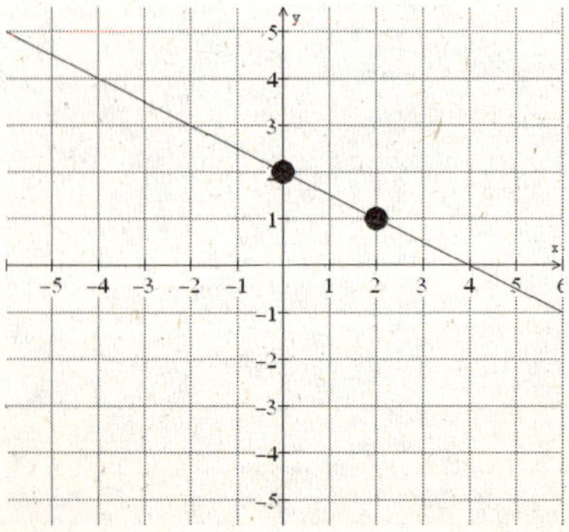

45. $x-3y=3$

$-3y=-x+3$

$y=\frac{1}{3}x-1$

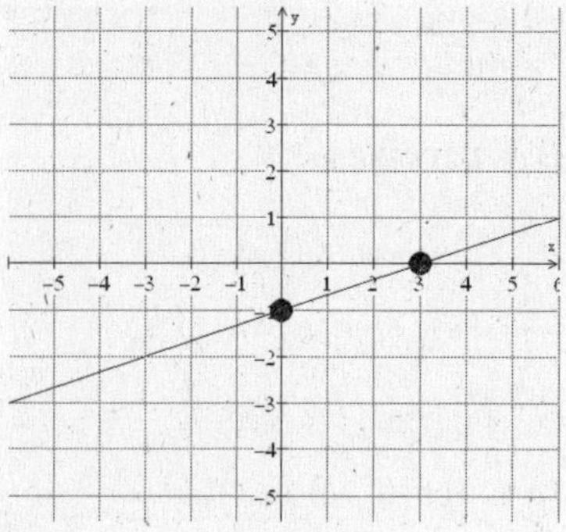

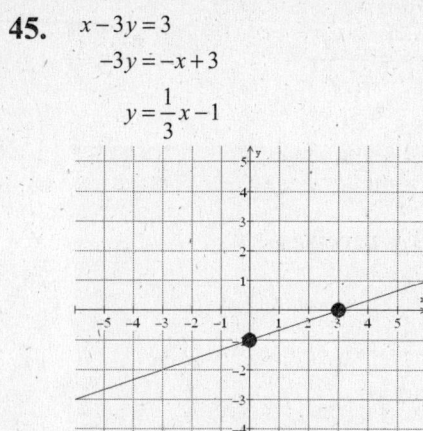

Objective C Exercises

47. $g(x)=-3x+1$

$g(-4)=-3(-4)+1$

$g(-4)=12+1$

$g(-4)=13$

$(-4, 13)$

49. $p(x)=6-8x$

$p(-1)=6-8(-1)$

$p(-1)=6+8$

$p(-1)=14$

$(-1, 14)$

51. $f(t)=t^2-t-3$

$f(2)=2^2-2-3$

$f(2)=4-2-3$

$f(2)=-1$

$(2, -1)$

53. $h(x)=-3x^2+x-1$

$h(-2)=-3(-2)^2+(-2)-1$

$h(-2)=-3(4)-2-1$

$h(-2)=-12-2-1$

$h(-2)=-15$

$(-2, -15)$

55. $g(t)=4t^3-2t$

$g(-1)=4(-1)^3-2(-1)$

$g(-1)=4(-1)+2$

$g(-1)=-4+2$

$g(-1)=-2$

$(-1, -2)$

Objective D Exercises

57. $y-y_1=m(x-x_1)$

$y-2=-3(x-(-1))$

$y-2=-3(x+1)$

$y-2=-3x-3$

$y=-3x-1$

59. $y-y_1=m(x-x_1)$

$y-(-5)=-2(x-4)$

$y+5=-2x+8$

$y=-2x+3$

61. $y-y_1=m(x-x_1)$

$y-(-3)=-\frac{3}{5}(x-5)$

$y+3=-\frac{3}{5}x+3$

$y=-\frac{3}{5}x$

63. $y-y_1=m(x-x_1)$

$y-(-2)=-\frac{2}{3}(x-(-3))$

$y+2=-\frac{2}{3}x(x+3)$

$y+2=-\frac{2}{3}x-2$

$y=-\frac{2}{3}x-4$

Section R.4

Objective A Exercises

1. $z^3 \cdot z \cdot z^4 = z^{3+1+4} = z^8$

3. $(x^3)^5 = x^{3\cdot 5} = x^{15}$

5. $(x^2y^3)^6 = x^{2\cdot 6}y^{3\cdot 6} = x^{12}y^{18}$

7. $\frac{a^8}{a^2} = a^{8-2} = a^6$

9. $(-m^3n)(m^6n^2) = -m^{3+6}n^{1+2} = -m^9n^3$

11. $(-2a^3bc^2)^3 = (-2)^3 a^{3\cdot3}b^{1\cdot3}c^{2\cdot3} = -8a^9b^3c^6$

13. $\dfrac{m^4n^7}{m^3n^5} = m^{4-3}n^{7-5} = mn^2$

15. $\dfrac{-16a^7}{24a^6} = -\dfrac{2a^{7-6}}{3} = -\dfrac{2a}{3}$

17. $(9mn^4p)(-3mp^2) = -27m^{1+1}n^4p^{1+2} = -27m^2n^4p^3$

19. $(-2n^2)(-3n^4)^3 = (-2n^2)((-3)^3n^{4\cdot3})$
$= (-2n^2)(-27n^{12})$
$= 54n^{2+12} = 54n^{14}$

21. $\dfrac{14x^4y^6z^2}{16x^3y^9z} = \dfrac{7x^{4-3}y^{6-9}z^{2-1}}{8} = \dfrac{7xy^{-3}z}{8} = \dfrac{7xz}{8y^3}$

23. $(-2x^3y^2)^3(-xy^2)^4 = ((-2)^3x^{3\cdot3}y^{2\cdot3})((-1)^4x^{1\cdot4}y^{2\cdot4})$
$= (-8x^9y^6)(x^4y^8)$
$= -8x^{9+4}y^{6+8} = -8x^{13}y^{14}$

25. $4x^{-7} = 4\cdot\dfrac{1}{x^7} = \dfrac{4}{x^7}$

27. $d^{-4}d^{-6} = d^{-10} = \dfrac{1}{d^{10}}$

29. $\dfrac{x^{-3}}{x^2} = \dfrac{1}{x^{3+2}} = \dfrac{1}{x^5}$

31. $\dfrac{1}{3x^{-2}} = \dfrac{1}{3}\cdot x^2 = \dfrac{x^2}{3}$

33. $(x^2y^{-4})^3 = x^6y^{-12} = \dfrac{x^6}{y^{12}}$

35. $(3x^{-1}y^{-2})^2 = 3^2x^{-2}y^{-4} = \dfrac{9}{x^2y^4}$

37. $(2x^{-1})(x^{-3}) = 2x^{-4} = \dfrac{2}{x^4}$

39. $\dfrac{3x^{-2}y^2}{6xy^2} = \dfrac{1}{2x^3}$

41. $\dfrac{2x^{-1}y^{-4}}{4xy^2} = \dfrac{1}{2x^2y^6}$

43. $(x^{-2}y)^2(xy)^{-2} = (x^{-4}y^2)(x^{-2}y^{-2}) = x^{-6} = \dfrac{1}{x^6}$

45. $\left(\dfrac{x^2y^{-1}}{xy}\right)^{-4} = \dfrac{x^{-8}y^4}{x^{-4}y^{-4}} = x^{-8+4}y^{4+4} = x^{-4}y^8 = \dfrac{y^8}{x^4}$

47. $\left(\dfrac{4a^{-2}b}{8a^3b^{-4}}\right)^2 = \left(\dfrac{b^5}{2a^5}\right)^2 = \dfrac{b^{10}}{4a^{10}}$

Objective B Exercises

49. $(4b^2 - 5b) + (3b^2 + 6b - 4)$
$= (4b^2 + 3b^2) + (-5b + 6b) - 4$
$= 7b^2 + b - 4$

51. $(2a^2 - 7a + 10) + (a^2 + 4a + 7)$
$= (2a^2 + a^2) + (-7a + 4a) + (10 + 7)$
$= 3a^2 - 3a + 17$

53. $(x^2 - 2x + 1) - (x^2 + 5x + 8)$
$= x^2 - 2x + 1 - x^2 - 5x - 8$
$= -7x - 7$

55. $(-2x^3 + x - 1) - (-x^2 + x - 3)$
$-2x^3 + x - 1 + x^2 - x + 3$
$= -2x^3 + x^2 + 2$

57. $(x^3 - 7x + 4) + (2x^2 + x - 10)$
$= x^3 + 2x^2 - 6x - 6$

59. $(5x^3 + 7x - 7) + (10x^2 - 8x + 3)$
$= 5x^3 + 10x^2 - x - 4$

61. $(2y^3 + 6y - 2) - (y^3 + y^2 + 4)$
$= 2y^3 + 6y - 2 - y^3 - y^2 - 4$
$= y^3 - y^2 + 6y - 6$

63. $(4y^3 - y - 1) - (2y^2 - 3y + 3)$
$= 4y^3 - y - 1 - 2y^2 + 3y - 3$
$= 4y^3 - 2y^2 + 2y - 4$

Objective C Exercises

65. $4b(3b^3 - 12b^2 - 6) = 12b^4 - 48b^3 - 24b$

67. $3b(3b^4 - 3b^2 + 8) = 9b^5 - 9b^3 + 24b$

69. $-2x^2y(x^2 - 3xy + 2y^2) = -2x^4y + 6x^3y^2 - 4x^2y^3$

71.
$$\begin{array}{r} x^2+3x+2 \\ \times \qquad x+1 \\ \hline x^2+3x+2 \\ x^3+3x^2+2x \qquad \\ \hline x^3+4x^2+5x+2 \end{array}$$

73.
$$\begin{array}{r} a^2-3a+4 \\ \times \qquad a-3 \\ \hline -3a^2+9a-12 \\ a^3-3a^2+4a \qquad \\ \hline a^3-6a^2+13a-12 \end{array}$$

75.
$$\begin{array}{r} -2b^2-3b+4 \\ \times \qquad b-5 \\ \hline 10b^2+15b-20 \\ -2b^3-3b^2+4b \qquad \\ \hline -2b^3+7b^2+19b-20 \end{array}$$

77.
$$\begin{array}{r} x^3 \qquad -3x+2 \\ \times x-4 \\ \hline -4x^3 \qquad +12x-8 \\ x^4 \qquad -3x^2+2x \qquad \\ \hline x^4-4x^3-3x^2+14x-8 \end{array}$$

79.
$$\begin{array}{r} y^3+2y^2-3y+1 \\ \times \qquad y+2 \\ \hline 2y^3+4y^2-6y+2 \\ y^4+2y^3-3y^2 \quad +y \qquad \\ \hline y^4+4y^3+y^2-5y+2 \end{array}$$

81. $(a-3)(a+4)=a^2+4a-3a-12$
$=a^2+a-12$

83. $(y-7)(y-3)=y^2-3y-7y+21$
$=y^2-10y+21$

85. $(2x+1)(x+7)=2x^2+14x+x+7$
$=2x^2+15x+7$

87. $(3x-1)(x+4)=3x^2+12x-x-4$
$=3x^2+11x-4$

89. $(4x-3)(x-7)=4x^2-28x-3x+21$
$=4x^2-31x+21$

91. $(3y-8)(y+2)=3y^2+6y-8y-16$
$=3y^2-2y-16$

93. $(7a-16)(3a-5)=21a^2-35a-48a+80$
$=21a^2-83a+80$

95. $(x+y)(2x+y)=2x^2+xy+2xy+y^2$
$=2x^2+3xy+y^2$

97. $(3x-4y)(x-2y)=3x^2-6xy-4xy+8y^2$
$=3x^2-10xy+8y^2$

99. $(5a-3b)(2a+4b)=10a^2+20ab-6ab-12b^2$
$=10a^2+14ab-12b^2$

101. $(4x-7)(4x+7)=16x^2-49$

Objective D Exercises

103.
$$\begin{array}{r} x+2 \\ x-3\overline{)x^2-x-6} \\ \underline{x^2-3x} \qquad \\ 2x-6 \\ \underline{2x-6} \\ 0 \end{array}$$

$(x^2-x-6)\div(x-3)=x+2$

105.
$$\begin{array}{r} 2y-7 \\ y-3\overline{)2y^2-13y+21} \\ \underline{2y^2-6y} \qquad \\ -7y+21 \\ \underline{-7y+21} \\ 0 \end{array}$$

$(2y^2-13y+21)\div(y-3)=2y-7$

107.
$$\begin{array}{r} x-2 \\ x+2\overline{)x^2+0+4} \\ \underline{x^2+2x} \qquad \\ -2x+4 \\ \underline{-2x-4} \\ 8 \end{array}$$

$(x^2+4)\div(x+2)=x-2+\frac{8}{x+2}$

109.

$$\begin{array}{r} 3y-5 \\ 2y+4\overline{)6y^2+2y+0} \\ \underline{2y^2+12y} \\ -10y+0 \\ \underline{-10y-20} \\ 20 \end{array}$$

$$\left(6y^2+2y\right)\div\left(2y+4\right)=3y-5+\frac{20}{2y+4}$$

111.

$$\begin{array}{r} b-5 \\ b-3\overline{)b^2-8b-9} \\ \underline{b^2-3b} \\ -5b-9 \\ \underline{-5b+15} \\ -24 \end{array}$$

$$\left(b^2-8b-9\right)\div\left(b-3\right)=b-5-\frac{24}{b-3}$$

113.

$$\begin{array}{r} 3x+17 \\ x-4\overline{)3x^2+5x-4} \\ \underline{3x^2-12x} \\ 17x-4 \\ \underline{17x-68} \\ 64 \end{array}$$

$$\left(3x^2+5x-4\right)\div\left(x-4\right)=3x+17+\frac{64}{x-4}$$

115.

$$\begin{array}{r} 5y+3 \\ 2y+3\overline{)10y^2+21y+10} \\ \underline{10y^2+15y} \\ 6y+10 \\ \underline{6y+9} \\ 1 \end{array}$$

$$\left(10y^2+21y+10\right)\div\left(2y+3\right)=5y+3+\frac{1}{2y+3}$$

117.

$$\begin{array}{r} x^2-5x+2 \\ x-1\overline{)x^3-6x^2+7x-2} \\ \underline{x^3-x^2} \\ -5x^2+7x \\ \underline{-5x^2+5x} \\ 2x-2 \\ \underline{2x-2} \\ 0 \end{array}$$

$$\left(x^3-6x^2+7x-2\right)\div\left(x-1\right)=x^2-5x+2$$

119.

$$\begin{array}{r} x^2+5 \\ x^2+0-2\overline{)x^4+0+3x^2+0-10} \\ \underline{x^4+0-2x^2} \\ 5x^2+0-10 \\ \underline{5x^2+0-10} \\ 0 \end{array}$$

$$\left(x^2+3x^2-10\right)\div\left(x^2-2\right)=x^2+5$$

Objective E Exercises

121. $12y^2=2\cdot2\cdot3y^2$

$5y=5y$

The GCF is y.

$$\begin{aligned} 12y^2-5y &= y(12y)-y(5) \\ &= y(12y-5) \end{aligned}$$

123. $10x^2yz^2=2\cdot5x^2yz^2$

$15xy^3z=3\cdot5xy^3z$

The GCF is $5xyz$.

$$\begin{aligned} 10x^2yz^2+15xy^3z &= 5xyz(2xz)+5xyz\left(3y^2\right) \\ &= 5xyz\left(2xz+3y^2\right) \end{aligned}$$

125. $5x^2=5x^2$

$15x=3\cdot5x$

$35=5\cdot7$

The GCF is 5.

$$\begin{aligned} 5x^2-15x+35 &= 5\left(x^2\right)+5(-3x)+5(7) \\ &= 5\left(x^2-3x+7\right) \end{aligned}$$

127. The GCF is $3y^2$.

$$\begin{aligned} 3y^4-9y^3-6y^2 &= 3y^2\left(y^2\right)+3y^2(-3y)+3y^2(-2) \\ &= 3y^2\left(y^2-3y-2\right) \end{aligned}$$

129. The GCF is xyz.

$$\begin{aligned} & x^4y^4-3x^3y^3+6x^2y^2 \\ &= x^2y^2\left(x^2y^2\right)+x^2y^2(-3xy)+x^2y^2(6) \\ &= x^2y^2\left(x^2y^2-3xy+6\right) \end{aligned}$$

131. $16x^2y = 2\cdot 2\cdot 2\cdot 2x^2y$

$8x^3y^4 = 2\cdot 2\cdot 2x^3y^4$

$48x^2y^2 = 2\cdot 2\cdot 2\cdot 2\cdot 3x^2y^2$

The GCF is $8x^2y$.

$16x^2y - 8x^3y^4 - 48x^2y^2$

$= 8x^2y(2) + 8x^2y(-xy^3) + 8x^2y(-6y)$

$= 8x^2(2 - xy^3 - 6y)$

133.

Factors	Sum
+1, –2	–1
–1, +2	1

$x^2 + x - 2 = (x+2)(x-1)$

135.

Factors	Sum
–1,+12	11
+1, –12	–11
–2, +6	4
+2, –6	–4
–3, +4	1
+3, –4	–1

$a^2 + a - 12 = (a+4)(a-3)$

137.

Factors	Sum
–1, –2	–3

$a^2 - 3a + 2 = (a-1)(a-2)$

139.

Factors	Sum
–1, +8	7
+1, –8	–7
–2, +4	2
+2, –4	–2

$b^2 + 7b - 8 = (b+8)(b-1)$

141.

Factors	Sum
–1, +45	44
+1, –45	–44
–3, +15	12
+3, –15	–12
–5, +9	4
+5, –9	–4

$z^2 - 4z - 45 = (z+5)(z-9)$

143.

Factors	Sum
–1, –45	–46
–3, –15	–18
–5, –9	–14

$z^2 - 14z + 45 = (z-5)(z-9)$

145.

Factors	Sum
+1, +20	21
+2, +10	12
+4, +5	9

$b^2 + 9b + 20 = (b+4)(b+5)$

147.

Factors	Sum
–1, –81	7
–3, –27	–30
–9, –9	–18

$y^2 - 9y + 81$ is nonfactorable over the integers.

149.

Factors	Sum
–1, –56	–57
–2, –28	–30
–4, –14	–16
–7, –8	–15

$x^2 - 15x + 56 = (x-7)(x-8)$

151. Factors of 2: 1, 2 Factors of 3: 1, 3

Trial Factors	Middle Term
$(1y+1)(2y+3)$	$3y + 2y = 5y$
$(1y+3)(2y+1)$	$y + 6y = 7y$

$2y^2 + 7y + 3 = (y+3)(2y+1)$

153. Factors of 3: 1, 3 Factors of 1: –1, –1

Trial Factors	Middle Term
$(1a-1)(3a-1)$	$-1 - 3a = -4a$

$3a^2 - 4a + 1 = (a-1)(3a-1)$

155. Factors of 2: 1, 2

Factors of –3: –1, +3 or +1, –3

Trial Factors	Middle Term
$(1x-1)(2x+3)$	$3x - 2x = x$
$(1x+3)(2x-1)$	$-x + 6x = 5x$
$(1x+1)(2x-3)$	$-3x + 2x = -x$
$(1x-3)(2x+1)$	$x - 6x = -5x$

$2x^2 - 5x - 3 = (x-3)(2x+1)$

157. Factors of 10: 1, 10 or 2, 5
Factors of 3: +1, +3

Trial Factors	Middle Term
$(1t+1)(10t+3)$	$3t+10t=13t$
$(1t+3)(10t+1)$	$t+30t=31t$
$(2t+1)(5t+3)$	$6t+5t=11t$
$(2t+3)(5t+1)$	$2t+15t=17t$

$10t^2+11t+3=(2t+1)(5t+3)$

159. Factors of 10: 1, 10 or 2, 5
Factors of –4:–1,+4 or +1,–4 or +2,–2

Trial Factors	Middle Term
$(1z-1)(10z+4)$	Common factor
$(1x+4)(10z-1)$	$-z+40z=39z$
$(1z+1)(10z-4)$	Common factor
$(1z-4)(10z+1)$	$z-40z=-39z$
$(1z-2)(10z+2)$	Common factor
$(1z+2)(10z-2)$	Common factor
$(2z-1)(5x+4)$	$8z-5z=3z$
$(2z+4)(5z-1)$	Common factor
$(2x+1)(5z-4)$	$-8z+5z=-3z$
$(2z-2)(5z+2)$	Common factor
$(2z-2)(5z+2)$	Common factor
$(2z+2)(5z-2)$	Common factor

$10z^2+3z-4=(2z-1)(5z+4)$

161. Factors of 3: 1, 3
Factors of 10: +1, +10 or +2, +5

Trial Factors	Middle Term
$(1z+1)(3z+10)$	$10z+3z=13z$
$(1z+10)(3z+1)$	$z+30z=31z$
$(1z+2)(3z+5)$	$5z+6z=11z$
$(1z+5)(3z+2)$	$2z+15z=17z$

$3z^2+95z+10$ is nonfactorable over the integers.

163. $2t^2-t-10$

$2(-10)=-20$

The factors of –20 whose sum if –1: 4 and –5

$$\begin{aligned}2t^2-t-10&=2t^2+4t-5t-10\\&=(2t^2+4t)+(-5t-10)\\&=2t(t+2)-5(t+2)\\&=(t+2)(2t-5)\end{aligned}$$

165. $12y^2+19y+5$

$12(5)=60$

The factors of 60 whose sum is 19: 4 and 15

$$\begin{aligned}12y^2+19y+5&=12y^2+4y+15y+5\\&=(12y^2+4y)+(15y+5)\\&=4y(3y+1)+5(3y+1)\\&=(3y+1)(4y+5)\end{aligned}$$

167. $11a^2-54a-5$

$11(-5)=-55$

The factors –55 whose sum is –54: –55 and 1

$$\begin{aligned}11a^2-54a-5&=11a^2-55a+a-5\\&=(11a^2-55a)+(a-5)\\&=11a(a-5)+1(a-5)\\&=(a-5)(11a+1)\end{aligned}$$

169. $6b^2-13b+6$

$6(6)=36$

The factors of 36 whose sum is –13: –9 and –4

$$\begin{aligned}6b^2-13b+6&=6b^2-9b-4b+6\\&=(6b^2-9b)+(-4b+6)\\&=3b(2b-3)-2(2b-3)\\&=(2b-3)(3b-2)\end{aligned}$$

171. The GCF is 3.

$3x^2+15x+18=3(x^2+5x+6)$

Factor the trinomial.

Factors	Sum
+1, +6	7
+2, +3	5

$3x^2+15x+18=3(x+2)(x+3)$

173. The GCF is a.

$ab^2+7ab-8a=a(b^2+7b-8)$

Factor the trinomial.

Factors	Sum
–1, +8	7
+1, –8	–7
–2, +4	2
+2, –4	–2

$ab^2+7ab-8a=a(b+8)(b-1)$

175. The GCF is $2y^2$.

$2y^4-26y^3-96y^2=2y^2(y^2-13y-48)$

Factor the trinomial.

Factors	Sum
–1, +48	47
+1, –48	–47
–2, +24	22
+2, –24	–22
–3, +16	13
+3, –16	–13
–4, +12	8
+4, –12	–8
–6, +8	2
+6, –8	–2

$2y^4-26y^3-96y^2=2y^2(y+3)(y-16)$

177. $2x^3-11x^2+5x$

The GCF is x.

$2x^3-11x^2+5x=x(2x^2-11x+5)$

$2(5)=10$

The factors of 10 whose sum is –11: –1 and –10

$$\begin{aligned}2x^3-11x^2+5x&=x(2x^2-11x+5)\\&=x[2x^2-x-10x+5]\\&=x[(2x^2-x)+(-10x+5)]\\&=x[x(2x-1)-5(2x-1)]\\&=x(2x-1)(x-5)\end{aligned}$$

179. $10t^2-5t-50$

The GCF is 5.

$10t^2-5t-50=5(2t^2-t-10)$

$2(-10)=-20$

The factors of –20 whose sum is –1: –5 and 4

$$\begin{aligned}10t^2-5t-50&=5(2t^2-t-10)\\&=5[2t^2+4t-5t-10]\\&=5[(2t^2+4t)+(-5t-10)]\\&=5[2t(t+2)-5(t+2)]\\&=5(t+2)(2t-5)\end{aligned}$$

181. $6p^3+5p^2+p$

The GCF is p.

$6p^3+5p^2+p=p(6p^2+5p+1)$

$6(1)=6$

The factors of 6 whose sum is 5: 3 and 2

$$\begin{aligned}6p^3+5p^2+p&=p(6p^2+5p+1)\\&=p(6p^2+2p+3p+1)\\&=p[(6p^2+2p)+(3p+1)]\\&=p[2p(3p+1)+1(3p+1)]\\&=p(3p+1)(2p+1)\end{aligned}$$